AF658419

Lecture Notes in Electrical Engineering

Volume 335

Board of Series editors

Leopoldo Angrisani, Napoli, Italy
Marco Arteaga, Coyoacán, México
Samarjit Chakraborty, München, Germany
Jiming Chen, Hangzhou, P.R. China
Tan Kay Chen, Singapore, Singapore
Rüdiger Dillmann, Karlsruhe, Germany
Haibin Duan, Beijing, China
Gianluigi Ferrari, Parma, Italy
Manuel Ferre, Madrid, Spain
Sandra Hirche, München, Germany
Faryar Jabbari, Irvine, USA
Janusz Kacprzyk, Warsaw, Poland
Alaa Khamis, New Cairo City, Egypt
Torsten Kroeger, Stanford, USA
Tan Cher Ming, Singapore, Singapore
Wolfgang Minker, Ulm, Germany
Pradeep Misra, Dayton, USA
Sebastian Möller, Berlin, Germany
Subhas Mukhopadyay, Palmerston, New Zealand
Cun-Zheng Ning, Tempe, USA
Toyoaki Nishida, Sakyo-ku, Japan
Bijaya Ketan Panigrahi, New Delhi, India
Federica Pascucci, Roma, Italy
Tariq Samad, Minneapolis, USA
Gan Woon Seng, Nanyang Avenue, Singapore
Germano Veiga, Porto, Portugal
Haitao Wu, Beijing, China
Junjie James Zhang, Charlotte, USA

About this Series

"Lecture Notes in Electrical Engineering (LNEE)" is a book series which reports the latest research and developments in Electrical Engineering, namely:

- Communication, Networks, and Information Theory
- Computer Engineering
- Signal, Image, Speech and Information Processing
- Circuits and Systems
- Bioengineering

LNEE publishes authored monographs and contributed volumes which present cutting edge research information as well as new perspectives on classical fields, while maintaining Springer's high standards of academic excellence. Also considered for publication are lecture materials, proceedings, and other related materials of exceptionally high quality and interest. The subject matter should be original and timely, reporting the latest research and developments in all areas of electrical engineering.

The audience for the books in LNEE consists of advanced level students, researchers, and industry professionals working at the forefront of their fields. Much like Springer's other Lecture Notes series, LNEE will be distributed through Springer's print and electronic publishing channels.

More information about this series at http://www.springer.com/series/7818

Koushik Maharatna · Goutam Kumar Dalapati
P.K. Banerjee · Amiya Kumar Mallick
Moumita Mukherjee
Editors

Computational Advancement in Communication Circuits and Systems

Proceedings of ICCACCS 2014

Springer

Editors
Koushik Maharatna
School of Electronics and Computer Science
University of Southampton
Southampton
UK

Goutam Kumar Dalapati
Department of Design and Growth
Institute of Materials Research and Engineering
Singapore
Singapore

P.K. Banerjee
Electronics and Telecommunication Engineering
Jadavpur University
Kolkata, West Bengal
India

Amiya Kumar Mallick
Electronics and Electrical Communication
Indian Institute of Technology Kharagpur
Kharagpur, West Bengal
India

Moumita Mukherjee
Centre for Millimeter Wave Semiconductor Devices and Systems
Defence Research and Development Organisation
Kolkata, West Bengal
India

ISSN 1876-1100 ISSN 1876-1119 (electronic)
Lecture Notes in Electrical Engineering
ISBN 978-81-322-2273-6 ISBN 978-81-322-2274-3 (eBook)
DOI 10.1007/978-81-322-2274-3

Library of Congress Control Number: 2015931531

Springer New Delhi Heidelberg New York Dordrecht London
© Springer India 2015
This work is subject to copyright. All rights are reserved by the Publisher, whether the whole or part of the material is concerned, specifically the rights of translation, reprinting, reuse of illustrations, recitation, broadcasting, reproduction on microfilms or in any other physical way, and transmission or information storage and retrieval, electronic adaptation, computer software, or by similar or dissimilar methodology now known or hereafter developed.
The use of general descriptive names, registered names, trademarks, service marks, etc. in this publication does not imply, even in the absence of a specific statement, that such names are exempt from the relevant protective laws and regulations and therefore free for general use.
The publisher, the authors and the editors are safe to assume that the advice and information in this book are believed to be true and accurate at the date of publication. Neither the publisher nor the authors or the editors give a warranty, express or implied, with respect to the material contained herein or for any errors or omissions that may have been made.

Printed on acid-free paper

Springer (India) Pvt. Ltd. is part of Springer Science+Business Media (www.springer.com)

Message from Convener

It is our pleasure to introduce the proceedings of the First International Conference on Computational Advancement in Communication Circuits and Systems (ICCA-CCS 2014) in relation to complex issues of communication circuit and system with the analysis of computational techniques. The conference aims to create a forum for further discussion on integrated information in the significant field incorporating a series of issues.

The relevance of the conference theme, to a wide variety of disciplines, is reflected in the diverse range of papers submitted. The link between Electronics and Communication Engineering and Soft Computing strengthens the area of research to be presented and provides the precise information required for assessment. The international delegates like Profs. Christophe Fumeaux, Australia; Arokiaswami Alphones, Singapore; I-Fang Chung, Taiwan; Chia-Feng Juang, Taiwan; and Sajjan G. Shiva, Memphis were highly impressed and have shown a high level of international interest in the subject.

The level of interest in the subject matter of the conference was maintained by submitting 122 suitable papers at the conference. Every submitted paper went through a precise review process. Each paper received at least three reviews; where issues remained, additional reviews were commissioned. Finally, 62 papers were selected by 40 reviewers for presentation in four different tracks like Microwave and Devices, Communication and Networking, Signal and Image processing, and Computations, Mathematics and Control.

Finally, we would like to record our appreciation to the Organizing Committee members for their work in securing a substantial input of papers to make the conference successful. We are also indebted to those who served as reviewers and chairmen; without their support, the conference could not have been the success that it was. We also acknowledge the authors themselves, without their expert input there would have been no conference.

November 2014

Prof. Dr. M.R. Kanjilal
Convener ICCACCS 2014

Conference Organizing Committee

Chief Patron

Sardar Jodh Singh, Chairman, JIS Group, India

Patron

Mr. Taranjit Singh, Managing Director, JIS Group, India
Prof. Dr. S.M. Chatterjee, Chairman—BOG, JIS Group
Prof. Dr. Asit Guha, Director (Academic), JIS Group, India
Dr. Sajal Dasgupta, Director Technical Education
Mr. U.S. Mukherjee, Deputy Director, JIS Group, India
Manpreet Kaur, CEO, JIS Group, India
Jaspreet Kaur, Trusty Member, JIS Group

Conference Chair

Prof. Dr. Ajoy Roy, Director, IIEST Shibpur, Howrah, India

Advisory Committee

Prof. Christophe Fumeaux, School of Electrical and Electronic Engineering, University of Adelaide, Australia
Prof. Arokiaswami Alphones, School of Electrical and Electronic Engineering, Nanyang Technology University, Singapore
Prof. I-Fang Chung, Institute of Biomedical Informatics, National Yang-Ming University, Taiwan
Prof. Chia-Feng Juang, Department of Electrical Engineering, National Chung-Hsing University, Taiwan
Prof. Sajjan Shiva, Department of Computer Science, University of Memphis, Memphis, TN, USA
Prof. Dr. Koushik Maharatna, University of Southampton, Southampton, UK
Prof. Dr. Goutam Kumar Dalapati, IMRE, Singapore
Prof. Prasad Shastry, Bradley University, Illinois, USA
Prof. Dr. Tapan Kumar Sarkar, Syracuse University, USA
Prof. Dr. N.R. Pal, Indian Stastical Institute, Kolkata, India
Prof. Dr. Debatosh Guha, Chairman, IEEE Kolkata Section, India
Dr. M.H. Kori, Technology Advisor of Validus Technologies, USA
Prof. Dr. Iti Saha Misra, Jadavpur University, Kolkata, India
Prof. D. Saha, IIM Calcutta, Kolkata, India
Dr. P. Dhak, Seoul National University, South Korea
Dr. Bharat Panigrahi, IIT Hyderabad, India
Prof. Dr. S.K. Lahiri, Ex. IIT Kharagpur, India
Prof. Dr. B.K. Sarkar, Consultant KCSTC, ISRO Chair, Professor (Rtd.) IIT Kharagpur, India
Prof. Dr. Swapna Banerjee, IIT Kharagpur, India
Prof. Dr. Mrityunjay Chakraborty, IIT Kharagpur, India
Prof. Dr. Ajit Pal, IIT Kharagpur, India
Prof. Dr. H. Saha, IIEST Shibpur, Howrah, India
Prof. Dr. S. Bhadra Chowdhuri, IIEST Shibpur, Howrah, India
Prof. Dr. Subhananda Chakraborty, IIT Bombay, Mumbai, India
Prof. Dr. Bhaskar Gupta, Jadavpur University, Kolkata, India
Prof. Dr. Amit Konar, Jadavpur University, Kolkata, India
Prof. Dr. S.K. Sanyal, Jadavpur University, Kolkata, India
Prof. Dr. Sanatan Chattopadhyay, Calcutta University, Kolkata, India
Dr. Soma Das, Guru Ghasidas Central University, Bilaspur, India
Dr. Swapna. S. Nair, Central University, Kerala, India

Local Advisory Committee

Prof. Dr. D.K. Bhattacharya, Ex. Calcutta University, India
Dr. M. Mukherjee, Scientist, DRDO, India
Prof. Dr. P.K. Banerjee, Ex. Jadavpur University, Kolkata, India
Prof. Dr. J.K. Das, Advisor R&D, Narula Institute of Technology, Kolkata, India
Prof. Dr. Amiya Kumar Mallick, Ex. IIT Kharagpur, India
Prof. Dr. D. Ghosh Dastidar, Ex. Jadavpur University, Kolkata, India
Prof. Dr. D.K. Bhattacharya, Ex. Calcutta University, Kolkata, India
Prof. Dr. B.L. Eshpuniyani, Dean (R&D)
Prof. Dr. S. Hati, Professor-CSE
Prof. S.C. Bera, HOD-ME
Prof. J.C. Guha, HOD-CE
Prof. A. Chakraborti, HOD-EE
Dr. I. Sarkar, HOD-Physics
Dr. P. Paul, HOD-MCA
Dr. P. Basak, HOD-Math

Convener

Prof. Dr. M. Ray Kanjilal, Principal, NIT

Jt. Convener

Prof. Dr. B.K. Medya, HOD-IT

Secretary

Mr. Kaushik Sarkar, Assistant Professor, ECE
Mr. Sohan Ghorai, Assistant Professor, ECE

Organizing Committee

Prof. C.S. Majumdar, Professor, CE
Prof. Dr. M. Ray Kanjilal, Principal, NIT
Prof. Dr. B.K. Medya, Head-IT
Dr. Sumit Nandi, HOD, Chemistry
Mr. Jayanta Paul, TIC-CSE
Mr. Saradindu Panda, Assistant Professor, ECE
Mr. Surajit Bari, Assistant Professor, ECE
Mr. Anilesh Dey, Assistant Professor, ECE and TIC-EIE
Mr. Kaushik Sarkar, Assistant Professor, ECE
Mr. Pranab Hazra, Assistant Professor, ECE
Mr. Subhram Das, Assistant Professor, CSE
Mr. Soumen Ghosh, Assistant Professor, IT
Mrs. Sandhya Pattanayak, Assistant Professor, ECE
Mr. Soumen Pal, Assistant Professor, ECE
Mrs. Swati Banerjee, Assistant Professor, ECE
Mrs. Sangita Roy, Assistant Professor, ECE
Mrs. Dola B. Gupta, Assistant Professor, ECE
Mrs. Arpita Barman Santra, Assistant Professor, ECE
Mrs. Payel Biswas, Assistant Professor, ECE
Mrs. Arnima Das, Assistant Professor, ECE
Ms. Piyu Sarcar, Assistant Professor, ECE
Mr. Abhijit Ghosh, Assistant Professor, ECE
Mr. Sohan Ghorai, Assistant Professor, ECE
Mr. Puspak Pain, Assistant Professor, ECE
Mrs. Moupali Roy, Assistant Professor, ECE
Ms. Rimpi Datta, Assistant Professor, ECE
Mrs. Tamasree Biswas, Assistant Professor, IT
Mr. Sudhansu Sarkar, Assistant Professor, EE

Contents

Part VIII Advances in Image Processing

About the Editors

Dr. Koushik Maharatna received his B.Sc. in Physics and M.Sc. in Electronic Science from Calcutta University, Calcutta, India in 1993 and 1995 respectively. He received his Ph.D. degree from Jadavpur University, Calcutta, India, in 2002 for his thesis, "CORDIC-based signal processors for biomedical applications".

In February 2000 he joined the Institute for High-Performance Microelectronics (IHP), Frankfurt (Oder), Germany, as a Research Scientist where he was involved in the BMBF-funded projects "Wireless Broadband Networks" (WBN) and IBMS2, both of these projects targeted low-cost low-power implementations of the IEEE 802.11a and Hyperlan/2 standards. His work in IHP resulted into four patents and several research publications in prestigious IEEE journals and conferences. In August 2003, Dr. Maharatna was appointed as a Lecturer in the Dept. of Electrical and Electronics Engineering (EE), University of Bristol, UK and in October 2006, he joined the School of Electronics and Computer Science (ECS), University of Southampton, UK, as a Senior Lecturer where he is a Reader (Associate Prof.) at present. He is currently pursuing his research vision for next-generation mobile healthcare system development applying his several years of experience in VLSI Circuits & Systems design and signal processing. As part of that he took part in several high-profile ARTEMIS and FP7 funded projects.

Dr. Maharatna has been a member of a number of several prestigious conference programme committees and acted as a session chair in conferences such as IEEE ISCAS 2005, 2007, 2008, 2012, VLSI design conference 2006 etc. He is a member of IEEE VLSI System Application (VSA) Technical Committee and IEEE Plagiarism committee. He has published over 90 research papers in internationally reputed Journals and Conferences. He has also edited the book "Systems design for remote healthcare" published by Springer in November 2013.

Dr. Goutam Kumar Dalapati is working as Scientist-II in the Department of Design and Growth at Institute of Materials Research and Engineering (IMRE, A*STAR), Singapore. He has completed his Ph.D. from Jadavpur University, Kolkata, India in 2005. His research interests include Next generation solar cells; Earth abundant materials for Photovoltaic application ($FeSi_2$, CuO, and CuS);

Inorganic solar cells (Si- and III–V-based); Heterogeneous integration (III–V on Si platform) for electronic and optical applications; Advanced CMOS front-end technology; Semiconductor process and technology; ALD High-K dielectrics for photovoltaic and electronic applications; High mobility channel materials (GaAs, SiGe, strained-Si); and GaN power transistor. Dr. Dalapati has published several papers in international journals. Dr. Dalapati has filed a patent for "Photoelectric Transducer Using Iron Silicide and Aluminium", US Provisional Patent Application No. 61/316,696.

Prof. P.K. Banerjee retired Professor of Electronics and Telecommunication Engineering, Jadavpur University, Kolkata, completed his undergraduate, post graduate and Ph.D. programme in the same University in the years 1965, 1967 and 1972 respectively. After completion of his master's degree he joined as senior research fellow to undertake research activities leading to the Ph.D. degree. He joined the department as lecturer in the year 1971 and was subsequently promoted to Reader and Professor. His academic interest is in the field of Communication Engineering with special activities in Digital Communication Systems, RF engineering and allied fields. Currently he is working in the field of mobile Ad hoc network (MANET), MIMO system and its use in different environments. Professor Banerjee has guided a large number of students for their Master's project and also four students for their Ph.D. (ongoing) work. He is also actively involved in guiding both M.Tech and Ph.D. students. Professor Banerjee along with his fellow associates and co-workers published more than 100 technical papers in national and international journals and has one patent to his credit.

Prof. Amiya Kumar Mallick joined All India Radio, Calcutta as Assistant Engineer through the UPSC examination after graduation in Electrical Engineering from Jadavpur University and thereafter postgraduation in Microwave Engineering from Indian Institute of Technology, Kharagpur. On completion of exhaustive training under the Technical Teacher Training Scheme, Government of India, Professor Mallick joined the Department of Electronics and Electrical Communication Engineering, Indian Institute of Technology, Kharagpur as Lecturer in 1967. Finally, Prof. Mallick became Professor of the Department and carried out all his responsibilities, with dignity, as the Head of the Centre for Research and Training in Radar and Communication Engineering. He was also a member of the Senate—the highest academic body of the Institute. Apart from teaching, Prof. Mallick provided leadership and took active initiatives in research and development of the Institute. He was Chairman of various Committees connected with a variety of R&D activities—departmental as well as interdisciplinary in nature. Professor Mallick retired in 2000 from IIT, Kharagpur. Professor Mallick obtained his Ph.D. (Engineering) degree from the Indian Institute of Technology, Kharagpur and produced a number of PhDs under his supervision. He is a Fellow and life member of number of Professional Societies like IE (India), IETE, SEMCE and Associate member of IEEE. Professor Mallick published a number of high quality technical research papers in national and international journals.

Dr. Moumita Mukherjee is working as Senior Scientist at Centre for Millimeter Semiconductor Devices and Systems, Defence Research and Development Organisation, Kolkata, India. Besides her R&D job, she likes teaching and out of that interest, she is attached with the Applied Physics Department, Calcutta University as 'guest faculty'. She completed her Bachelor's and Master's (Physics) from Presidency College, Kolkata, University of Calcutta. Dr. Mukherjee completed Ph.D. (Technology) in Radio Physics and Electronics from Calcutta University. She has authored more than 120 research papers in international journals and also in a number of international research papers/book-chapters/books. Dr. Mukherjee received the Visiting-Scientist' offer from Newcastle University, UK and selected as PDF from Germany, and also obtained "National Merit Scholarship" Award from GOI.

Part I
Advances in RF, Microwave and Antenna

Chapter 1
Design and Development of Low-Level RF Digital Feedback Loop

Synthesized Signal Generator

Arnab Das, Bipa Datta and Moumita Mukherjee

Abstract A controlled synchrotron light source is a specialized particle accelerator, typically accelerating electrons required for scientific and technical purposes. To energize charged particles to the final energy and to compensate the synchrotron radiation loss, RF power is used. To do so, RF cavities are used and power to RF cavities is fed using high power amplifiers (like Klystron for Indus-2 and Tetrode tube for Indus-1, at RRCAT, Indore, MP, India). With the advancement in the field of programmable logical devices and the Hardware description language, digital RF feedback control system using FPGAs is adopted for providing better flexibility, reliability and stability. In phase (I) and quadrature phase, (Q) scheme is used here for extracting the amplitude and phase information about RF signal. By processing information, the EM field inside the RF cavity has stable amplitude control loop (ACL) and phase control Loop (PCL). A new, FPGA-based, digital, low-level RF system, based on an analog I/Q modulator and demodulator, is proposed here for development.

Keywords FPGA · In-phase · LLRF system · Quadrature-phase · Stability

A. Das (✉) · B. Datta
ECE Department, Brainware Group of Institutions, Barasat, West Bengal University of Technology, Kolkata, West Bengal, India
e-mail: u_call_arnab@yahoo.co.in

B. Datta
e-mail: bipa.datta@gmail.com

M. Mukherjee
Centre for Millimeter Wave Semiconductor Devices and Systems, University of Calcutta, Kolkata, West Bengal, India
e-mail: mm_drdo@yahoo.com

© Springer India 2015

K. Maharatna et al. (eds.), *Computational Advancement in Communication Circuits and Systems*, Lecture Notes in Electrical Engineering 335,
DOI 10.1007/978-81-322-2274-3_1

1.1 Introduction

Low-level RF (LLRF) control systems consist of synthesized signal source, 0–360° phase shifter, feedback loops for amplitude, phase and frequency control, and coaxial RF switch to put RF on and off and limiter. Indus-1 synthesized signal generator giving outputs at 31.613 MHz is developed, which will also be used to get synchronized RF drive signal for Indus-2. Amplitude and phase control feedback loops are incorporated to maintain the amplitude and phase of the cavity gap voltage within ±1 % and ±2°, respectively, for proper operation of the machine [1, 2]. Translation of the amplitude and phase information to I/Q is advantageous because of the symmetry of the I/Q signal paths. This analog I/Q RF system also provides a real function structure to verify the working principle, block functions and performance evaluation for the developing digital low-level RF system [2].

This paper, based on analog I/Q and digital FPGA LLRF systems, presents the designed function diagrams, measured results of the characteristics of the main RF vector components and the integration test of the low-level RF digital feedback loop, while maintaining amplitude in suitable range.

1.2 Design Consideration for Digital LLRF Feedback Loop

1.2.1 Scheme of the Digital LLRF Feedback Control System

Figure 1.1 shows a schematic of the digital LLRF control system, where cavity field is directly down-converted to baseband signals, for which I/Q detection is performed using Spartan 3 DSP protoboard. The resulting I/Q baseband signals that describe the cavity field are also processed using VHDL program for getting controlled output, one for the I signal and another for the Q signal.

In this project we only control amplitude, i.e. work as an amplitude control feedback loop. FPGA-based VHDL program helps DSP protoboard work as a broadband quadrature demodulator with an integrated intermediate frequency (IF) after baseband amplifier and controlled clock signal. It is responsible for converting the low-level RF signal into baseband differential in-phase and quadrature components.

The FPGA-based amplitude controller designs with the help of the differences between the two singles (I/Q) with the set point, because for all measurements founded on AD8345, the input level on each baseband input pin is 0.7 V ± 0.3 V peak [3].

The vector modulator module modifies the I and Q components to produce the desired RF drive signal for the klystron according to the PI controller signal. The AD8345 is used to perform I/Q up-conversion [3]. The component provides excellent specifications of amplitude and phase balance and sideband suppression.

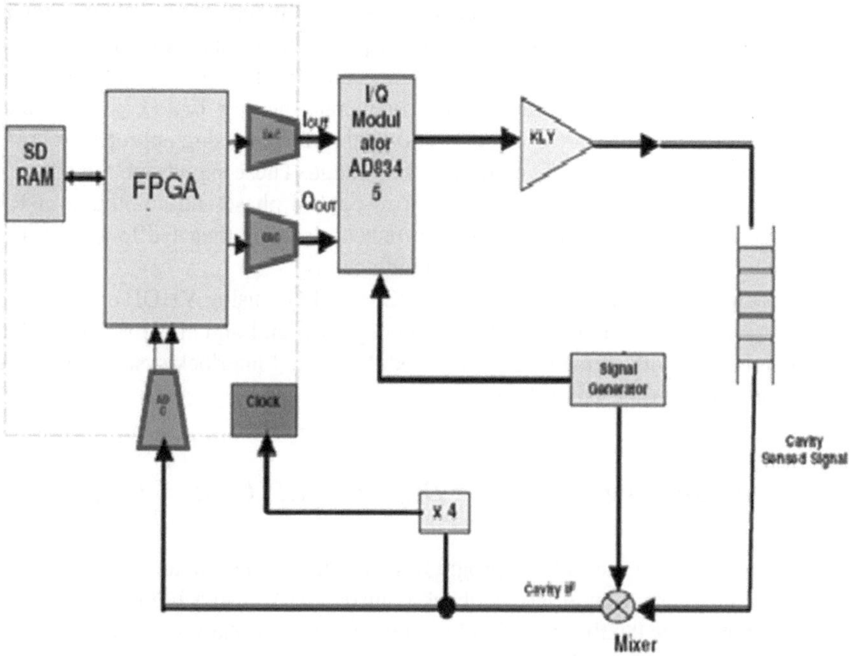

Fig. 1.1 Block diagram of the digital LLRF feedback control system

1.2.2 I/Q Feedback-Control Module

The digital I/Q feedback-control module is the central controller of the feedback system, providing the following controls.

1. Two input variables or predefined data adjust the operational set levels of I and Q for the cavity accelerating voltage; for visualizing a dual scope oscillator, the I set and Q set values can be used.
2. Maintaining the cavity frequency-tuning loop can tune the cavity frequency to a resonant frequency and the cavity-accelerating voltage is controlled by the set values of I and Q.

1.3 Integration Testing

Recent technology in System on Chip (SoC) has enabled to develop high-density FPGA devices that are suited to the needs of high performance real-time signal processing. With the addition of embedded processor cores and powerful IO interfaces they provide a valuable combination of high performance and configurability. At this point in time we process analog and digital boards individually.

Here, down-conversion of the data RF signal from the sensed cavity port is performed using a mixer to the comfortable frequency range and extracting the I, Q, I_n and Q_n information from cavity IF. After comparing the set and sensed value of amplitude an algorithm should be run to generate the new I_{out}, Q_{out}, I_n_{out} and Q_n_{out}. Generated signals are converted into analog format using appropriate DAC and they must be put within I/Q modulator safe range. These signals are fed to the I, Q, I_n and Q_n ports of the I/Q modulator to control phase and amplitude of RF generators signal. Phase and amplitude corrected signal is then fed to amplifying system which in turn corrects the field in side cavity.

The control logic is implemented in the Xilinx FPGA using VHDL coding. The hardware components can be divided into analog parts and digital parts. The analog parts mainly deal with signal mixing, IQ modulation and interlock system while the digital parts contain the control algorithm.

1.3.1 Programming Layout for Digital LLRF Control Loop

With advancement in the field of programmable logical devices and the Hardware description language, digital RF feedback control system using FPGAs is adopted for providing better flexibility, reliability and stability. In phase (I) and quadrature phase, (Q) scheme is used for extracting the amplitude and phase information about RF signal. A block diagram for FPGA-based VHDL programming layout is shown in Fig. 1.2. The development environment for FPGA coding is as follows:

- coding language: VHDL
- synthesis tool: XST in ISE 8.2i from Xilinx
- implementation: ISE 8.2i from Xilinx
- mapping and routing: ISE 8.2i from Xilinx

1.3.2 Working Steps from Programming Point of View

FPGA-based control system in VHDL code will perform the following tasks:

- Spartan-3 DSP protoboard (XC3S-PQ208) by the VHDL code converts input baseband signal (cavity IF signal) into digital format (ADC).
- By adjusting sampling frequency four times of analog input signal, collect I, Q, I_n and Q_n from different output lines with checking phase difference between every neighbour's output channel at 90°, are stored.
- To avoid synchronizing problem, one channel of AFG3102 is used for clock generation for protoboard and another is used for synchronizing the signal generator.
- Stored digital I, Q, I_n and Q_n data in different registers are compared with predefined set values and the operation is performed as desired, i.e. process according to the compared value, if instantaneous value (I, Q, I_n and Q_n) is

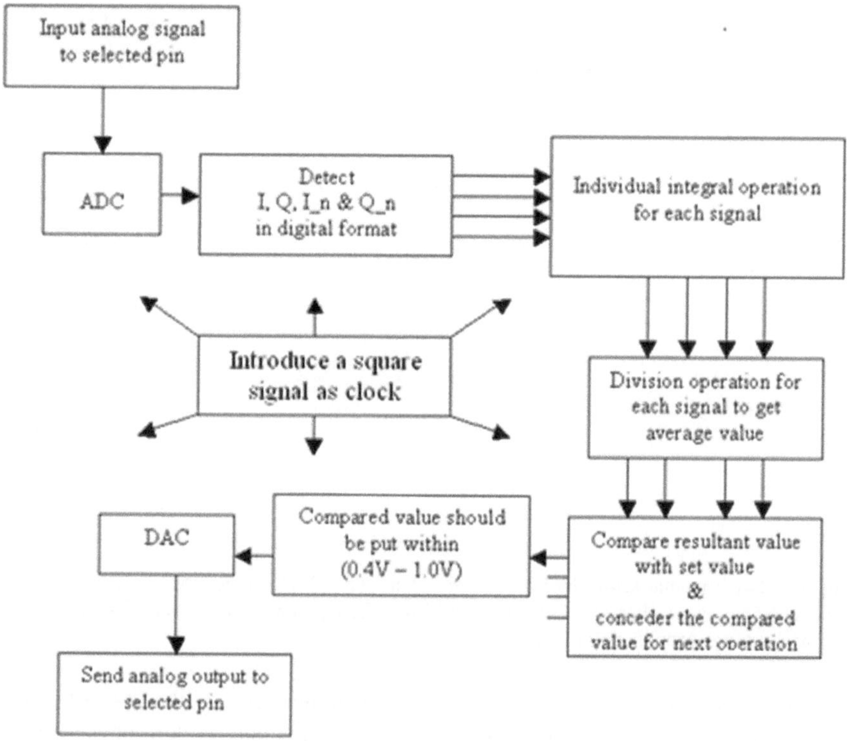

Fig. 1.2 Block diagram for FPGA-based programming layout

(a) = set value then output will be assign position voltage
(b) > set value then output will be in decrement order
(c) > set value then output will be in incremental order

Since the above processed output signals are differential I and Q input to the I/Q modulator (AD8345), their values are limited differentially between 0.4 and 1.0 V with 0.7 V in the middle position [3]. To overcome noise in instantaneous I, Q, I_n and Q_n value, get average result for each, which will compare with set value to get I_{out}, Q_{out}, I_n_{out} and Q_n_{out}.

1.4 Closed Loop Operation with Test Set-up

The main hardware components of the digital RF feedback system are ADC for sampling of the RF signal, FPGA for signal processing and DAC for driving the IQ modulator. An XC3S-PQ208 commercial Spartan-3 DSP protoboard is adopted for the ADC/DAC and FPGA board. The experimental set-up block diagram used is shown in Fig. 1.3.

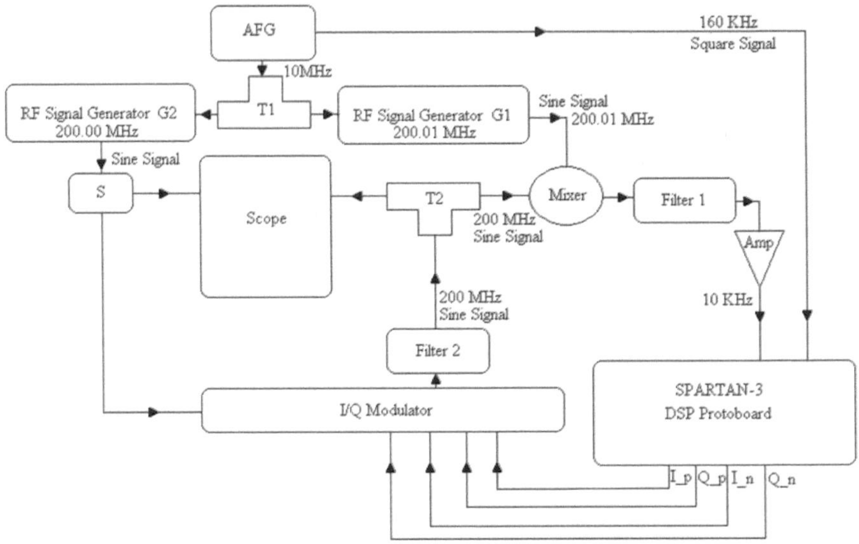

Fig. 1.3 Experimental set-up block diagram (where *T1*, *T2* TEE type BNC connector, *S* 6 dB splitter, *F1* low pass filter (dc—40 MHz): used for filtering sum freq., *F2* harmonic filter (dc—580 MHz), *Amp* Amplifier (for amplifying the IF signal))

The feedback logic based on the PI control is implemented in the FPGA by using VHDL. The I and Q components of the cavity field signal are fed into the FPGA using the ADC, which samples the RF signal four times during one period. The sampled I and Q components of the cavity signals are compared with the set value, which generates the error signal.

1.4.1 Observed Figures

Visualized and stored various waveform patterns [by Scope (TPS2024 (Tektronix) and 54832B DSO (Agilent))] for making the conclusion are shown in Figs. 1.4 and 1.5).

1.4.2 Important Observations

During the testing of complete loop many important observations are listed as:

1. Proposed VHDL program can generate I/Q signal if operating frequency of the protoboard is 16 times that of input signal.
2. After four steps (reset, write, conversion, read) analog to digital conversion occurs, i.e. sampling frequency is four times less than operating frequency.

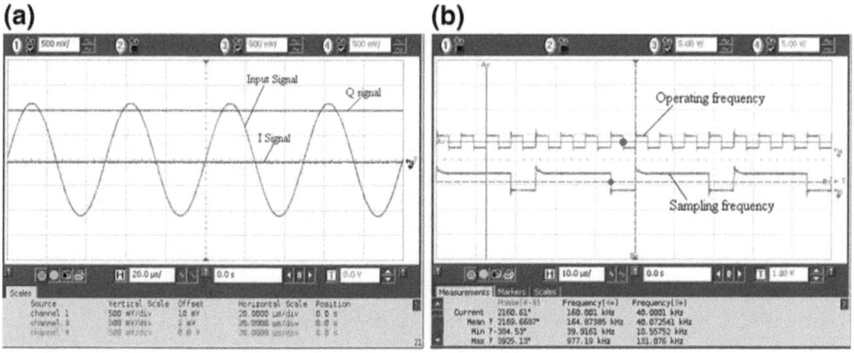

Fig. 1.4 **a** Input signal with I/Q waveform and **b** operating freq. versus sampling freq.

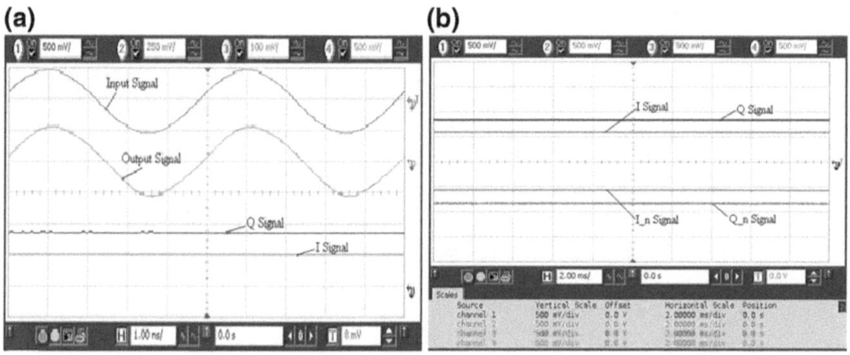

Fig. 1.5 **a** Various signals at a time and **b** I, Q, I_n and Q_n signal (*within limiting value*)

3. We may also follow the ratio between operating frequency and input signal frequency (cavity IF signal) as 16:(1 + 4*n*), where *n* is any counting number.

1.5 Conclusion

FPGA-based LLRF digital feedback loop was realized using Spartan-3 protoboard, I/Q modulator (AD8345), Mixer (SRA1WH), Arbitrary function generator (AFG) where RF cavity signal was simulated using RF signal generator in laboratory. During testing for closed-loop operation, dynamic range of 13 dB for amplitude control and 360° phase control was observed. Overall, it can be concluded that the proper working of digital feedback low level RF control system using Spartan-3 DSP protoboard can be demonstrated.

References

1. Lawrence Doolittle, Low-level RF control system design and architecture, in *APAC, RRCAT*, Indore, India (2007)
2. M.S. Yeh, L.H. Chang, F.T. Chung, Y.H. Lin, C.H. Wang, The design, fabrication and performance testing of the analog I/Q RF control system at NSRRC, in *WEPMA071, APAC 2007*, Raja Ramanna Centre for Advanced Technology (RRCAT), Indore, India (2007)
3. http://www.analog.com/static/imported-files/data_sheets/AD8345.pdf

Chapter 2
Estimation of Slot Position for a Slotted Antenna

Arnab Das, Chayan Banerjee, Bipa Datta and Moumita Mukherjee

Abstract Compact microstrip patch antennas have become quite popular nowadays. With lesser form factor requirements and for multiband applications, slotted antennas have proved useful. This paper evaluates the merits of slot positioning with respect to a fixed feed point. The evaluation is based on antenna parameters like resonating frequency, return loss and bandwidth. The simulations are run for a triangular (Δ) and a V slot. Results show that output characteristic follows almost similar in nature to slot positions with respect to fixed feeding point, irrespective of the slot shape. It is found that a slot produces maximum signal bandwidth and gain when put near the feeding point.

Keywords Microstrip antenna · Slot antenna · Slot positioning · Triangular slot · V-shaped slot

2.1 Introduction

In applications where size, weight, cost, performance, ease of installation and aerodynamic profile are constrains, low profile antennas like microstrip and printed slot antennas are required. Slot antennas exhibit wider bandwidth, lower dispersion

A. Das (✉) · C. Banerjee · B. Datta
Department of Electronics and Communication, Brainware Group of Institutions, Barasat, West Bengal, India
e-mail: u_call_arnab@yahoo.co.in

C. Banerjee
e-mail: cbanerjee929@gmail.com

B. Datta
e-mail: bipa.datta@gmail.com

M. Mukherjee
Centre for Millimeter Wave Semiconductor Devices and Systems, University of Calcutta, Calcutta, West Bengal, India
e-mail: mm_drdo@yahoo.com

© Springer India 2015

K. Maharatna et al. (eds.), *Computational Advancement in Communication Circuits and Systems*, Lecture Notes in Electrical Engineering 335,
DOI 10.1007/978-81-322-2274-3_2

and lower radiation loss than microstrip antennas [1–5]. It is better to have a single multiband antenna than to have different antennas to work at different frequencies. The most effective technique to design a multiband slot antenna is to cut a slot on the microstrip patch at a proper position and with the right dimensions. The dimensions, i.e. the length (*L*) of the conventional antenna determine the resonant frequency and the width (*W*) of the antenna has a predominant effect on input matching condition [1, 6].

The available literature clearly defines the slot antenna parameters like antenna geometry, feed line types, ground plane and substrate, but they do not provide any clear information about the positioning of the slot on the conductor plate of a slot antenna. In this paper we tried to evaluate the effect of changing of position (of a slot of arbitrary shape) on the antenna characteristics. The parameters include bandwidth, reflected impedance and resonant frequencies. For the sake of simplicity and analysis the antenna feed line is kept fixed at a corner of the substrate, throughout the entire evaluation.

2.2 Parameter with Antenna Geometry Selection

Microstrip lines feed and coaxial probe feed are popularly used in slot antenna design. The purpose of the feed is to carry energy from a connector to the actual antenna, so their proper placement is very crucial. Coaxial probe-feed (radius = 0.5 mm) is located at *W*/2 and *L*/3 to get faithful operation for a rectangular microstrip antenna [1]. From our experimental point of view, we choose feed point fixed at *P*(−3.25, −1.7), with varying slot section positions in horizontal direction.

We have used two popular slot shapes for the evaluation of antenna characteristics, a triangular slot (Δ) and a V type (non-tapered) slot. Triangular slots and its variations (e.g. Vivaldi, Sierpinski triangle) are popular and are used in a lot of applications requiring multiband application. The slots are simulated for their output characteristics like return loss, bandwidth and resonant frequencies at different locations on the conductor. There are 14 discrete locations, considered from edge AB of the microstrip top conductor, while moving towards the feeding point.

Here, the used conventional antenna dimensions are $L = 6$ mm, $W = 10$ mm, substrate (PTFE) thickness $h = 1.5847$ mm, dielectric constant $\varepsilon_r = 4.4$ with coaxial probe-feed (radius = 0.5 mm) located at *P*(−3.25, −1.7) (Fig. 2.1). The proposed V-shaped slot antenna (Fig. 2.2) and Δ-shaped slot antenna (Fig. 2.3), for both feed point is located at *P* (−3.25, −1.7).

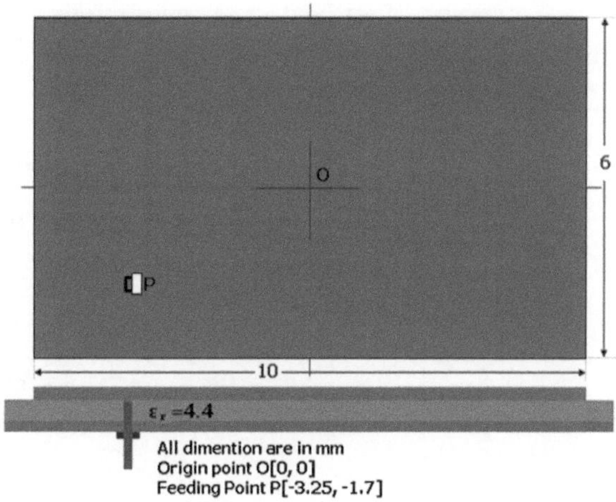

Fig. 2.1 Conventional microstrip antenna

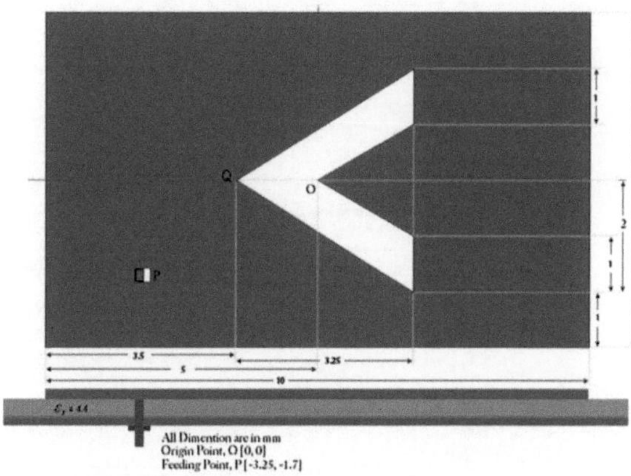

Fig. 2.2 V-slotted patch antenna

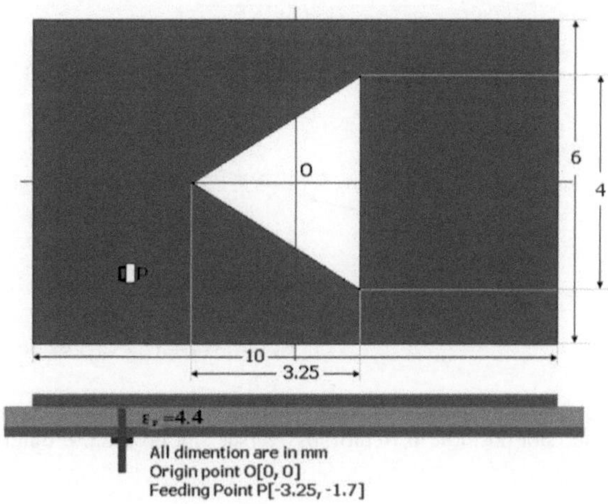

Fig. 2.3 Triangular slotted patch antenna

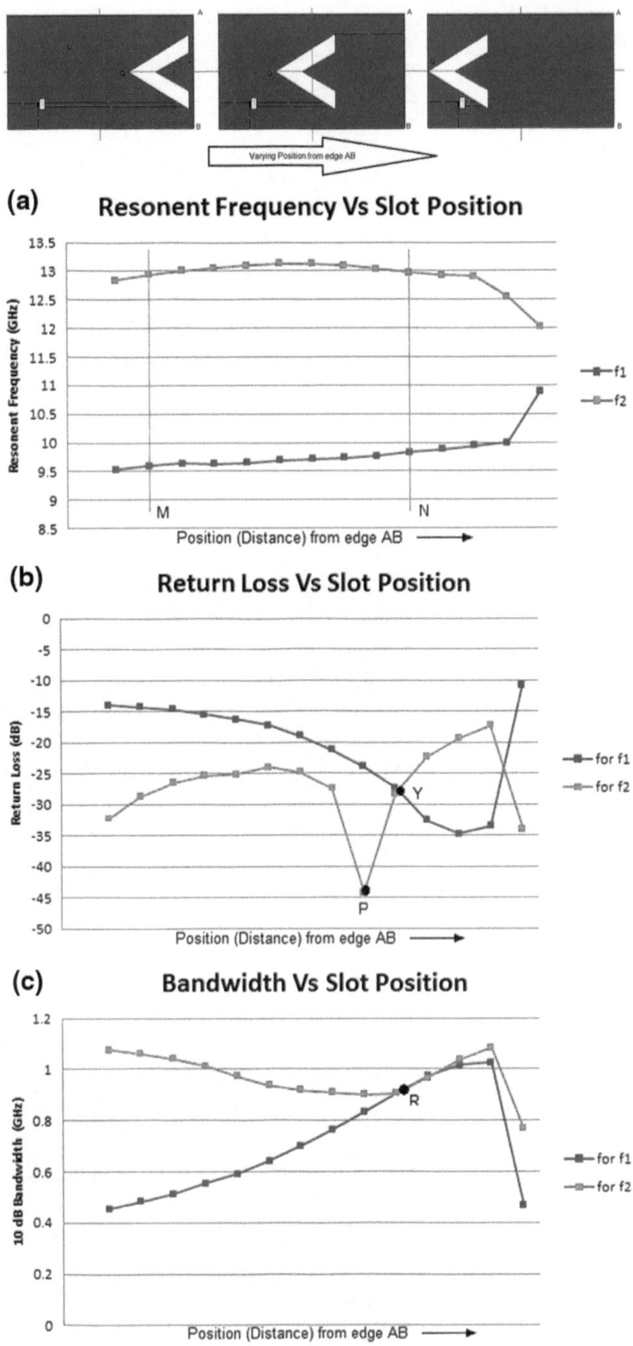

Fig. 2.4 Comparison plots of V slot antenna radiation characteristics. **a** Resonant frequency versus slot position. **b** Return loss versus slot position. **c** Bandwidth versus slot position

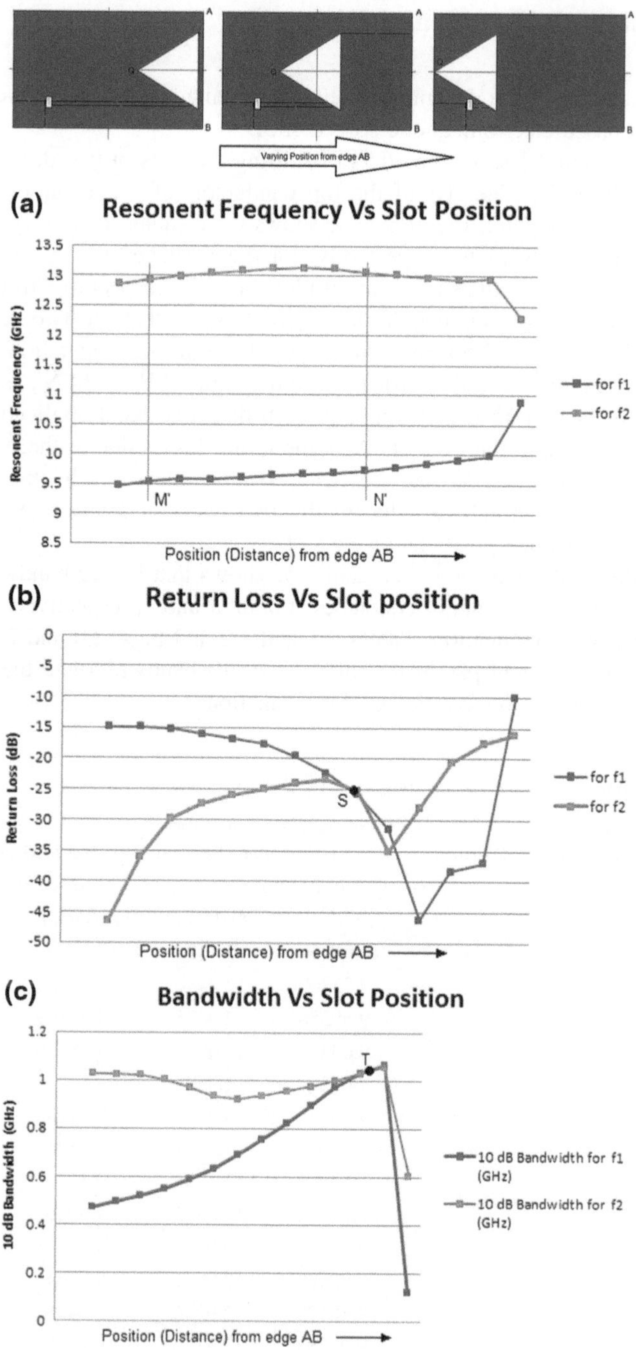

Fig. 2.5 Comparison plots of Δ slot antenna radiation characteristics. **a** Resonant frequency versus slot position. **b** Return loss versus slot position. **c** Bandwidth versus slot position

2.3 Simulation, Results and Discussion

The characteristics of the designed structures presented in this paper are simulated using MoM-based electromagnetic solver, IE3D.

Figures 2.4 and 2.5 show the effect of shifting the slots of two different shapes, away from the right edge AB of the top conductor (of a conventional antenna) towards the feeding point. Figure 2.4a–c shows the resonant frequency, return loss and 10 dB bandwidth plots, respectively, for the V-shaped slot at different positions. Figure 2.5 shows the same kind of plots but with respect to a triangular slot.

Considering a slot antenna (V or Δ), we can see that the two resonant frequencies (Figs. 2.4a or 2.5b) make it suitable for multiband operation. Dual band applications can be efficiently carried out in the region MN or M′N′, as labelled in the plot. It is clear from the plot in Fig. 2.4b that dual band applications can be properly carried out at the point *Y*. At *Y* the return losses due to the two resonant frequencies are around −30 dB and hence comparable to each other. It is also observed that the return losses due to the two resonant frequencies reach their minima at two extreme positions of the slot.

Bandwidth analysis in Figs. 2.4c and 2.5c shows that higher bandwidth can be obtained for both the resonant frequencies at point *R* and *T*, respectively, paving the way for dual band application. Tables 2.1 and 2.2 and Figs. 2.6 and 2.7 show the comparison of important parameters like the 10 dB bandwidth and the return loss for the two types of slots considered for evaluation.

Table 2.1 Comparison of resonant frequency (f_1 and f_2)

Various antennas (Varying reference point position)	Resonant frequency for: (f_1—1st resonant freq. and f_2—2nd resonant freq.)			
	V slot antenna		Δ slot antenna	
	f_1 (GHz)	f_2 (GHz)	f_1 (GHz)	f_2 (GHz)
Conventional antenna	10.0696	13.1724	10.0696	13.1724
Q (1.5, 0)	9.53809	12.8487	9.46265	12.8488
Q (1, 0)	9.59556	12.9425	9.53849	12.9249
Q (0.5, 0)	9.6333	13.0007	9.58026	12.9819
Q (0, 0)	9.63261	13.0595	9.57677	13.0388
Q (−0.5, 0)	9.65341	13.0952	9.61185	13.077
Q (−1, 0)	9.69033	13.1313	9.65266	13.1153
Q (−1.5, 0)	9.70882	13.1326	9.6732	13.132
Q (−2, 0)	9.72973	13.0962	9.69168	13.1153
Q (−2.5, 0)	9.76746	13.039	9.72994	13.0621
Q (−3, 0)	9.82499	12.9731	9.78616	13.0201
Q (−3.5, 0)	9.88172	12.9266	9.84385	12.9627
Q (−4, 0)	9.93883	12.9146	9.90101	12.9405
Q (−4.5, 0)	9.99595	12.5511	9.97696	12.9412
Q (−5, 0)	10.8884	12.0311	10.8694	12.2787

Table 2.2 Comparison of important parameters

	Q position	10 dB BW		Return loss	
		V	Triangle	V	Triangle
Position (distance) from edge AB ↓	*Reading at resonant frequency f_1*				
	(1.5, 0)	0.45489	0.47215	−13.9808	−14.9224
	(1, 0)	0.48469	0.49773	−14.2307	−14.9037
	(0.5, 0)	0.51355	0.52207	−14.6223	−15.178
	(0, 0)	0.55468	0.55139	−15.5391	−15.9629
	(−0.5, 0)	0.593	0.58911	−16.2733	−16.7812
	(−1, 0)	0.64135	0.63294	−17.2323	−17.6743
	(−1.5, 0)	0.70045	0.69045	−19	−19.5895
	(−2, 0)	0.76585	0.75495	−21.2624	−22.2803
(max)	(−2.5, 0)	0.83395	0.82379	−23.8892	−25.7986
	(−3, 0)	0.90472	0.89856	−27.3633	−31.3659
	(−3.5, 0)	*0.97357*	*0.97572*	*−32.518*	*−46.1489*
	(−4, 0)	1.01592	1.03495	−34.7746	−38.3037
	(−4.5, 0)	1.0258	1.06903	−33.4651	−36.9497
	(−5, 0)	0.4663	0.122	−10.7416	−10.0672
	Reading at resonant frequency f_2				
	(1.5, 0)	1.0759	1.0259	−32.1745	−46.4336
	(1, 0)	1.0596	1.0234	−28.7547	−35.9967
	(0.5, 0)	1.0407	1.0206	−26.509	−29.7632
	(0, 0)	1.0112	1.0015	−25.3984	−27.3375
	(−0.5, 0)	0.974	0.9706	−25.1344	−25.9813
	(−1, 0)	0.9368	0.9335	−23.9716	−24.93
	(−1.5, 0)	0.9178	0.922	−24.7998	−23.8736
	(−2, 0)	0.9097	0.9362	−27.4495	−23.2671
	(−2.5, 0)	0.9009	0.9568	−44.1331	−25.0709
	(−3, 0)	0.9052	0.9759	−28.3544	−34.9831
	(−3.5, 0)	*0.9656*	*0.9989*	*−22.3528*	*−28.0403*
	(−4, 0)	1.0361	1.02982	−19.39	−20.4924
	(−4.5, 0)	1.0856	1.0584	−17.2511	−17.4633
	(−5, 0)	0.7676	0.6049	−34.0197	−16.0221

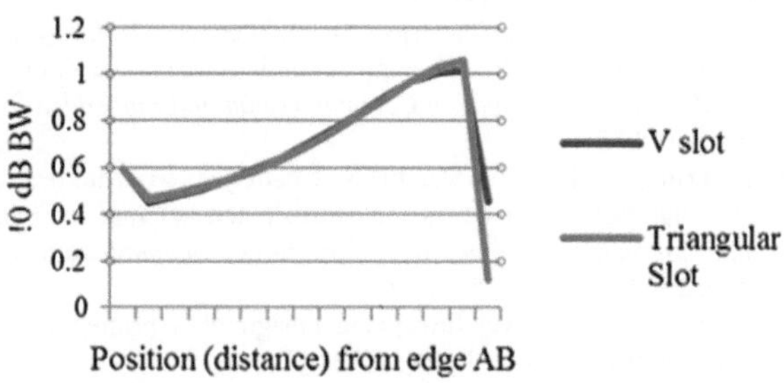

Fig. 2.6 Bandwidth versus position plots for Δ and V slots, for f_1

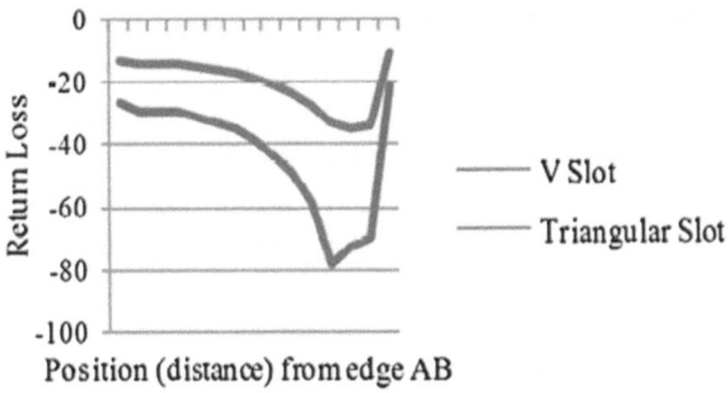

Fig. 2.7 Return loss versus position plots for Δ and V slots, for f_1

2.4 Conclusion

Positioning of the slot is crucial for the efficiency of a microstrip antenna. This paper evaluated the output characteristics of a slot antenna. The simulations are made with a fixed feeding point and two types of slot shapes. The slot positions were varied from one edge of the radiating conductor (AB), while moving towards the feeding point. The gain is maximum when the vertex (Q) of the slot is almost above the feeding point, but gain value falls drastically as the vertex crosses the feeding point. A similar nature is shown by return loss and bandwidth. It may be concluded that if the slot (irrespective of shape) is moved towards a fixed feeding point along the x axis (without moving in Y axis), then the antenna gives a maximum bandwidth and gain with slot is placed near the feed point. Return loss values though may differ in the location of their maxima and minima from slot to slot and depending on their resonant frequencies.

References

1. K.-L. Wong, *Compact and Broadband Microstrip Antennas* (Wiley, New York, 2002)
2. S.K. Padhi, N.C. Karmakar, C.L. Law, *CPW-Fed MMIC Slot Dipole for MM-Wave Applications IEEE 0-7803-7330-8/02*, pp. 414–417
3. B. Datta, A. Das, M. Mukherjee, A. Kundu, S.K. Chowdhury, Triple band slotted patch antenna for microwave communication, in *Advance Computing Conference (IACC), 2013 IEEE 3rd International*, 22–23 Feb 2013, pp. 202, 206
4. A. Das, B. Datta, S. Chatterjee, M. Mukherjee, S.K. Chowdhury, Dual-band slotted microstrip patch antenna design for application in microwave communication, in *2013 International Conference on Information Communication and Embedded Systems (ICICES)*, 21–22 Feb 2013, pp. 850, 855
5. B. Datta, A. Das, A. Kundu, S. Chatterjee, M. Mukherjee, S.K. Chowdhury, Twice-band irregular rectangular cut-in microstrip patch antenna for microwave communication, in *2013 International Conference on Information Communication and Embedded Systems (ICICES)*, 21–22 Feb 2013, pp. 598, 602
6. W.-S. Chen, A novel broadband design of a printed rectangular slot antenna for wireless applications. Microw. J. **49**(1), 122 (2006)

Chapter 3
Wide-Banding of Half-Mode Substrate Integrated Waveguide (HMSIW) Filters Using L-Slots

Sourav Moitra, Basudeb Mondal, Asish Kumar Mukhopadhyay and Partha Sarathee Bhowmik

Abstract The technique of wide-banding of Half-Mode Substrate Integrated Waveguide (HMSIW) Bandpass Filters has been presented in this paper. Single and simultaneous arrays of L-shaped slots have been used to achieve the purpose successfully. Effects of variation of several slots parameters have been studied in detail and are presented with their outcomes. The filter comes with minimal insertion loss over the entire passband. Entire experiments have been carried out over a material of dielectric constant of 3.2 and thickness of 30 mils. The filter finds applications in Ku-band operations with advantages like low insertion loss, convenient integration, compact size, low cost and ease of fabrication.

Keywords HMSIW · BPF · Ku-band · Insertion loss · L-slot array · EBG · System on Substrate (SoS)

Sourav Moitra has several publications in International Journals and Conferences.

S. Moitra (✉)
Department of Electronics and Communication Engineering,
Dr. B.C. Roy Engineering College, Durgapur 713206, India
e-mail: souravmoitra25@yahoo.in

B. Mondal
Department of Electrical Engineering, Dr. B.C. Roy Engineering College,
Durgapur, India

A.K. Mukhopadhyay
SR Group of Institutions, Jhansi, Uttar Pradesh, India

P.S. Bhowmik
Department of Electrical Engineering, National Institute of Technology,
Durgapur, India

© Springer India 2015

K. Maharatna et al. (eds.), *Computational Advancement in Communication Circuits and Systems*, Lecture Notes in Electrical Engineering 335,
DOI 10.1007/978-81-322-2274-3_3

3.1 Introduction

Waveguide slotted filters have several advantages like compact configuration, stable mechanical characteristic, low loss and enhanced efficiency. Thus they are widely used in the communication and radar systems. However, modern applications demand much more compactness and lesser weight which enables them to be integrated with complex systems. Substrate integrated waveguide (SIW) filters have been found to be a promising technology in this respect [1–4]. SIW structures are synthesized in dielectric substrate with linear arrays of metallic via which can also be treated as a periodic structure [5, 6]. The field distribution in an SIW is similar to that in a conventional rectangular waveguide and thus SIW components inherit all the advantages of the rectangular waveguide such as high Q factor, low insertion loss and high-power capability [7, 8]. Characteristics of SIW filters show a band-stop behaviour and thus wideband filters can be obtained by integrating the SIW with a periodic array structure. This study shows the effect of introducing periodic Electro-Band-Gap (EBG) structures (L-slot arrays) which enables the filter to work as a bandpass filter. The passband is obtained by selecting the dimensions of the L-slots corresponding to the higher cut-off of the BPF. The resulting filters thus have a much smaller size than the conventional SIW filter. In this paper basic SIW filter has been designed over a substrate of dielectric constant 3.2 with thickness of 30 mils. The design is then modified by insertion of single and multiple arrays of periodic L-slots which enable the filter to operate in microwave Ku-band. Several other parameters of the L-slots have been studied which offers a freedom of selecting desired passband as per the required application. This technique provides an attractive alternative to conventional metallic waveguides for low-cost, low-loss and high-density integration of microwave and millimetre wave components and subsystems.

3.2 HMSIW Filter Design

The basic design started with HMSIW structure for obtaining high-pass characteristics. Problems like the operation bandwidth, radiation leakage, dielectric and conductor losses have been considered for designing the basic structure. The basic structure with its transmission parameter has been shown in Figs. 3.1 and 3.2, respectively.

HMSIW consists of nearly half the total dimension of the SIW without any compromise in filter characteristics and thus carry the advantage of more compact layouts. HMSIW can only support $TE_{(m+0.5),0}$ ($m = 0, 1, 2, \ldots$) mode and restrains the $TE_{(m,0)}$ ($m = 1, 2, \ldots$) mode which is equivalent to the $TE_{(2m,0)}$ ($m = 1, 2, \ldots$) mode in SIW, so the first spurious passband becomes more far away from the passband of a bandpass filter. The HMSIW can be considered equivalent to a high-pass filter due to its inherent sharp cut-off in lower frequency [9].

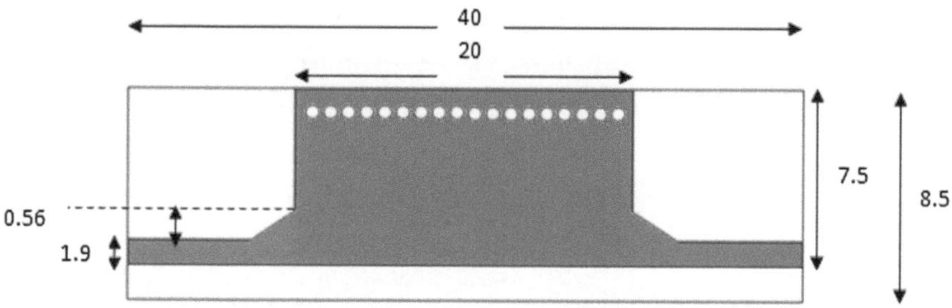

Fig. 3.1 Basic HMSIW structure

Fig. 3.2 S-parameter of basic HMSIW structure

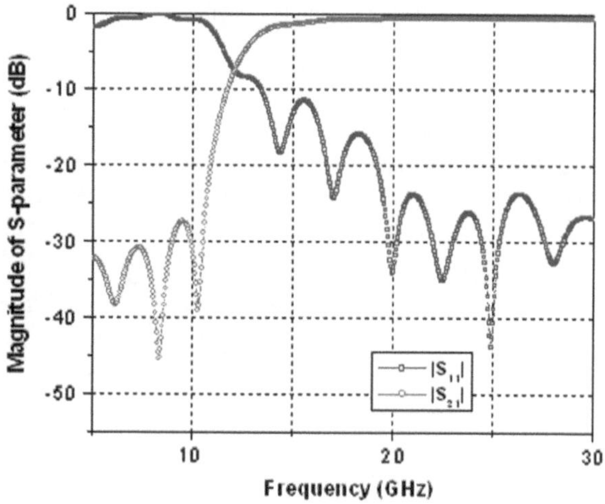

This equivalent width is the effective width of the SIW and can be obtained by the following relation:

$$W_{\text{eff}} = W - d - \frac{d^2}{(0.95)} \tag{3.1}$$

This property can be used to analyse and design various components just by knowing W_{eff} of the SIW. The cut-off frequency for the SIW may be defined as

$$f_c = \frac{c}{2\varepsilon_r \cdot W_{\text{eff}}} \tag{3.2}$$

where, c is the velocity of light in vacuum. This property proves to be an important technique for design of rectangular waveguides in a straightforward way to analyse and design various components just knowing W_{eff} of the SIW.

Over the high-pass characteristics, a stopband starting at 17.5 GHz has been created using L-shaped slots. Several other slots are also being studied. It is shorted

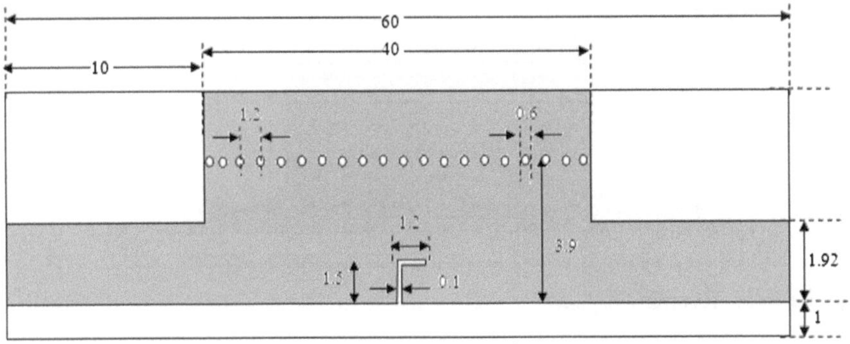

Fig. 3.3 HMSIW BPF with single L-slot

Fig. 3.4 S-parameter of HMSIW BPF with single L-slot

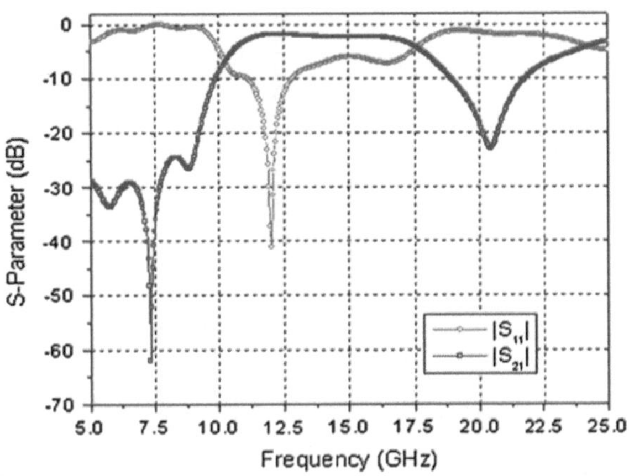

at one end and open circuited at the other end. The electrical length of the resonators is quarter wavelength long of the centre frequency of the stopband to be created. The insertion loss and the ripple properties are controlled by the external Q_e which is given by the following relation:

$$Q_e = 2Q_l = \left(\frac{2f_0}{\Delta f_{-3\ \mathrm{dB}}}\right) \tag{3.3}$$

Design of a single L-slot filter with its required S-parameters has been shown in Figs. 3.3 and 3.4 respectively.

Single L-slot has been found effectively to create a stopband from 17.5 GHz, while the passband is from 12 to 17.5 GHz with insertion loss of 1.5 dB. For better transmission characteristics of the BPF another L-slot has been inserted and the design with its characteristics has been shown in Figs. 3.5 and 3.6.

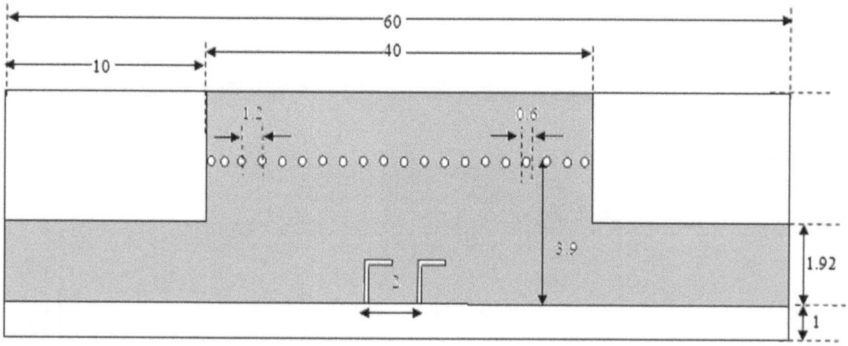

Fig. 3.5 HMSIW BPF with double L-slots

Fig. 3.6 S-parameter of HMSIW BPF with double L-slots

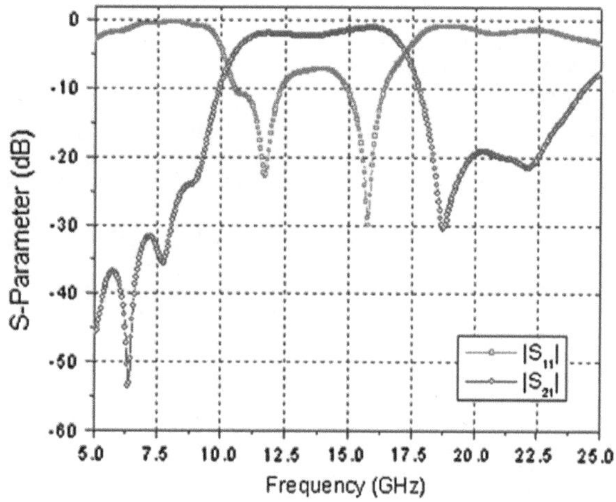

Introduction of double L-slots has been found to have better effect on the stopband characteristic at around 17.4 GHz. Further the insertion loss and broad-banding of the BPF can be obtained by variation of the height of the L-slot S_h and spacing or gap between the slots S_g. The effect of variation of these gaps has been studied further and is provided later in this article. The effects of triple L-slots over the stopband as well as passband have also been studied and are given in Figs. 3.7 and 3.8.

The results confirm a wideband characteristics covering full microwave Ku-band with better IL. However, larger IL is found near to higher cut-off which may be adjusted by varying L-slot height S_h and width of the slots S_w.

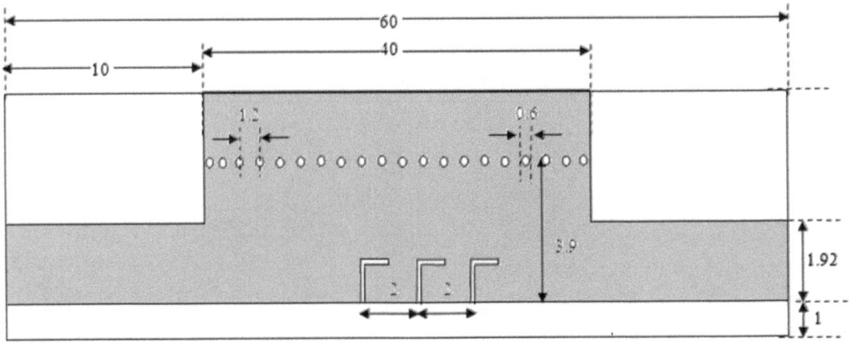

Fig. 3.7 HMSIW BPF with triple L-slots

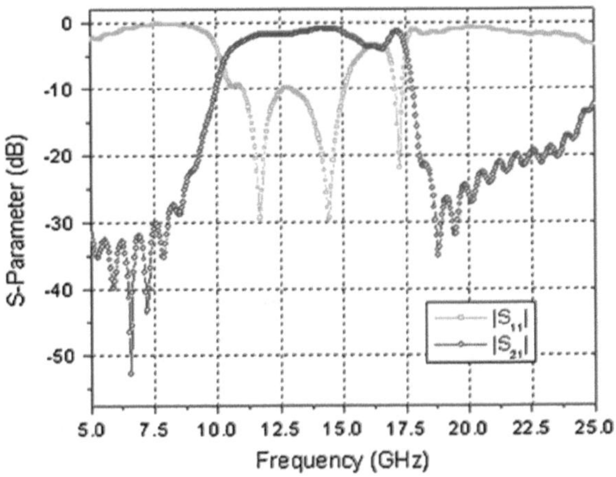

Fig. 3.8 S-parameter of HMSIW BPF with triple L-slots

3.2.1 Parametric Analysis

Parametric analyses of these parameters are shown in Figs. 3.9, 3.10 and 3.11. It has been observed that these parameters have important effects over the control of entire passband as well as the stopband characteristics. These analyses will serve to design and development of application-oriented designing by the high-frequency communication engineers.

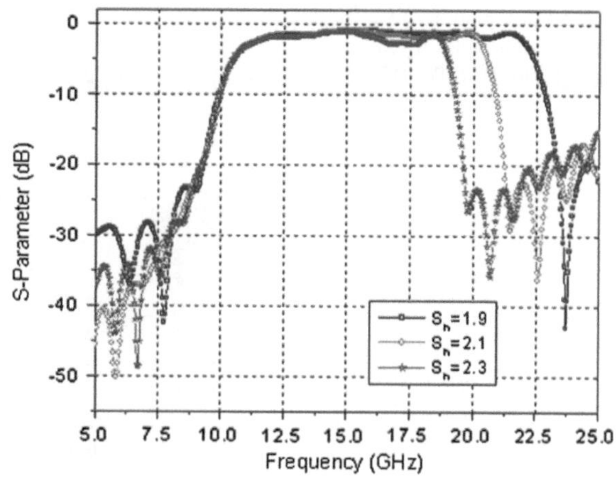

Fig. 3.9 Variation of S_{21} for different slot height S_h in triple L-slot HMSIW BPF

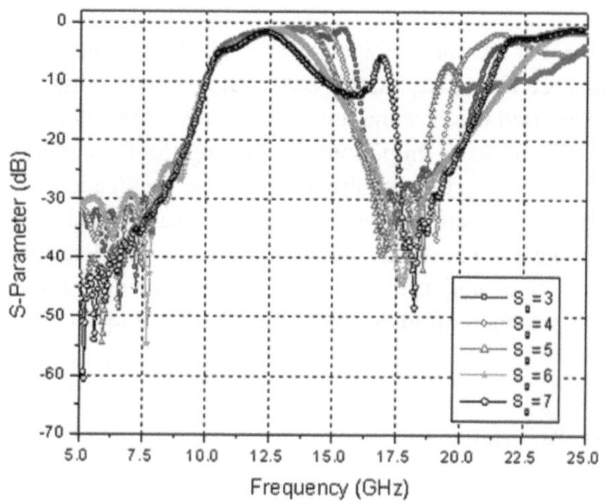

Fig. 3.10 Variation of S_{21} for different slot gaps S_g in triple L-slot HMSIW BPF

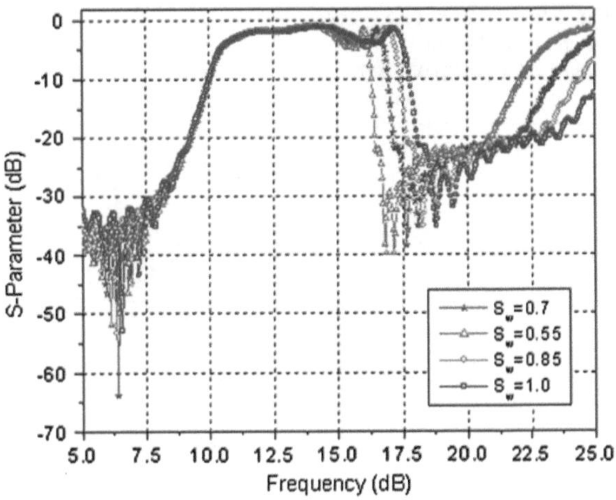

Fig. 3.11 Variation of S_{21} for different slot width S_w in triple L-slot HMSIW BPF

3.3 Conclusion

A novel technique for widebanding of band-pass characteristics in HMSIW filters has been discussed in this paper. Single and multiple periodic arrays of L-slots have been used to serve the purpose. Several parameters of the L-slots are studied which are found to be important for making the filter work as broadband or narrowband filter as per the requirement of the application. Careful alteration of the slot parameters will provide the desired passband as may be required by the design engineers. The filter can find wide-band application in radar and remote sensing operations under microwave Ku-band. The design is simple and easy to fabricate in the presence of advanced fabrication techniques.

References

1. F. Shigeki, Waveguide line. Japan Patent 06- 053 711, 25 Feb 1994 (in Japanese)
2. D. Deslandes, K. Wu, Single-substrate integration technique of planar circuits and waveguide filters. IEEE Trans. Microw. Theory Tech. **MTT-51**(2), 593–596 (2003)
3. J.-X. Chen, W. Hong, Z.-C. Hao, H. Li, K. Wu, Development of a low cost microwave mixer using a broadband substrate integrated waveguide (SIW) coupler. IEEE Microw. Wirel. Comp. Lett. **16**(2) (2006)
4. Y. Cassivi, K. Wu, Low cost microwave oscillator using substrate integrated waveguide cavity. IEEE Microw. Wirel. Comp. Lett. **13**(2) 48–50 (2003)
5. H.J. Hsu, M.J. Hill, R.W. Ziolkowski, J. Papapolymerou, A duroid based planar EBG cavity resonator filter with improved quality factor. IEEE Antennas Wirel. Propag. Lett. **1**, 67–70 (2002)
6. J.J. Simpson, A. Taflove, J.A. Mix, H. Heck, Computational and experimental study of a microwave electromagnetic bandgap structure with wave guiding defect for potential use as a bandpass wireless interconnect. IEEE Microw. Wirel. Comp. Lett. **14**(7), 343–345 (2004)

7. D. Deslandes, K. Wu, Integrated microstrip and rectangular waveguide in planar form. IEEE Microw. Wirel. Comp. Lett. **11**(2), 68–70 (2001)
8. K. Wu, D. Deslandes, Y. Cassivi, The substrate integrated circuits—a new concept for high-frequency electronics and optoeletronics. in *Proceedings of 6th Telecommunications in Modern Satellite, Cable and Broadcasting Service*, vol. 1 (2003), pp. P-III–P-X
9. L. Qiang, Y.-J. Zhao, Q. Sun, W. Zhao, B. Liu, A compact UWB HMSIW bandpass filter based on complementary split-ring resonators. Progr. Electromagnet. Res. C **11**, 237–243 (2009)

Chapter 4
Broadband Rectangular Microstrip Patch Antennas for K and EHF Bands

Piyu Sarcar, Sukla Basu and Abhijit Ghosh

Abstract In spite of many advantages of patch antenna, narrow bandwidth is one of the major drawbacks. In this paper it is demonstrated that significant bandwidth enhancement of a rectangular patch antenna can be achieved by simple modification of the rectangular patch. Two modified structures are presented here; both of them are capable of giving wider bandwidth and higher gain than conventional rectangular patch antennas. The antennas are simulated using High Frequency Structure Simulator (HFSS v12) for different feed positions and for two different substrate materials. A comparative study of these structures is made from simulation results. Proposed antenna structures can be operated in K bands and EHF region, whose frequency range is very important in microwave communication systems including radar application.

Keywords Rectangular patch antenna · Slotted rectangular patch antenna · Broadband patch antenna · K band patch antenna

4.1 Introduction

Microstrip antennas have become one of the most important topics of research interest in the fields of antenna techniques since the 1970s in conjunction with the advancement of electronic circuit miniaturization technology, due to their

P. Sarcar (✉) · A. Ghosh
Department of ECE, Narula Institute of Technology, 81, Nilgunz Road, Kolkata 700109, India
e-mail: piyusarcar@gmail.com

A. Ghosh
e-mail: jit.ghosh18@gmail.com

S. Basu
Department of ECE, Kalyani Government Engineering College, Nadia 741235, West Bengal, India
e-mail: sbasu1996@gmail.com

© Springer India 2015

K. Maharatna et al. (eds.), *Computational Advancement in Communication Circuits and Systems*, Lecture Notes in Electrical Engineering 335,
DOI 10.1007/978-81-322-2274-3_4

compactness and compatibility with MMIC design [1]. They have been widely employed for civilian and military applications such as television, broadcast, radio, mobile systems, global positioning system (GPS), radio-frequency identification (RFID), multiple-input multiple-output (MIMO) systems, vehicle collision avoidance system, satellite communications, surveillance systems, direction founding, radar systems, remote sensing, biological imaging, missile guidance and so on. However, narrow bandwidth and low gain are the main drawbacks of these types of antennas [2]. Demand for small antennas with wide bandwidth is increasing at a rapid pace. Conventional rectangular patch antenna is easy to fabricate but provide narrow bandwidth. A straightforward method of improving the bandwidth is increasing the substrate thickness. However, surface wave power increases and radiation power decreases with the increasing substrate thickness [3], which leads to poor radiation efficiency. Increase in dielectric constant of substrate material leads to decrease in bandwidth [4]. A number of methods are found in the literature for increasing bandwidth of patch antenna [5–13]. Almost all of these methods increase complexity of the antenna structure. Towards the traditional millimeter-wave, microstrip antenna having narrowband and high side-lobe causes severe influence to radiation of antenna [14]. In this paper simple modified rectangular patch antenna structures are proposed, which are capable of giving reasonable gain and a wide range of frequency coverage in Ku, K, Ka bands and EHF region.

4.2 Antenna Design

4.2.1 Conventional Rectangular Patch Antenna

A conventional rectangular patch with length $L = 5$ mm, width $W = 4$ mm is shown in Fig. 4.1. Coaxial probe-feed is located at the corner of the patch for the best impedance matching. Substrate thickness is taken as 1 mm.

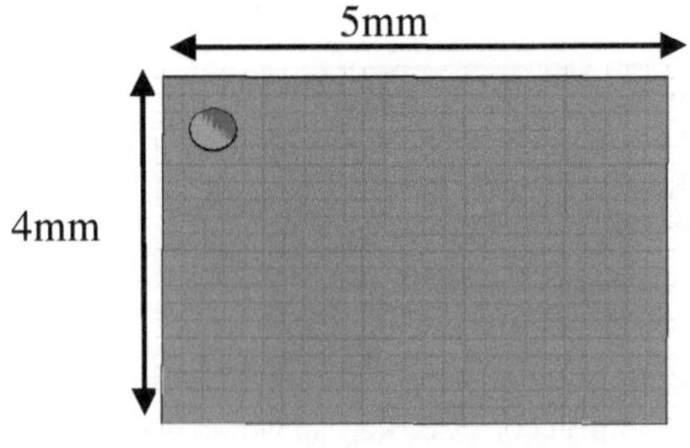

Fig. 4.1 A conventional rectangular patch antenna with coaxial feed

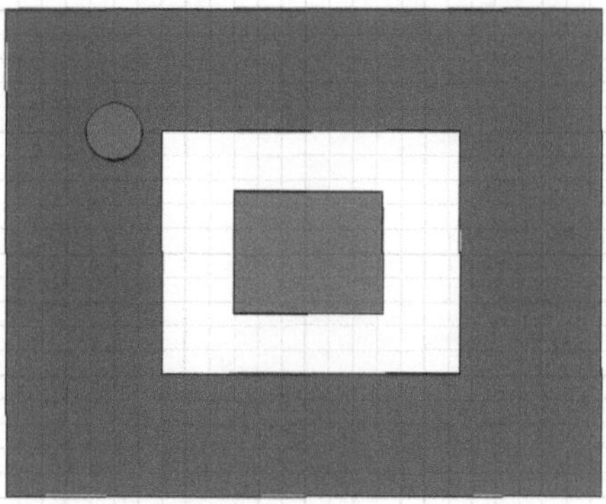

Fig. 4.2 Modified rectangular patch antenna with slot

4.2.2 Modified Rectangular Patch Antenna with Slot

Structure of a modified rectangular patch antenna with slot, derived from the conventional rectangular patch of Fig. 4.1 is shown in Fig. 4.2. Thus area covered by the patch is reduced while outer perimeter remains the same as for the rectangular patch. The length of the inner patch is 1.25 mm and width is 1 mm.

4.2.3 Modified Rectangular Patch Antenna, Without Central Patch (Rectangular Ring)

Area covered by the patch is further reduced by removing the central patch of the slotted patch antenna of Fig. 4.2, However, the outer perimeter of the patch remains the same as before.

Each of these three antenna structures is analysed for substrate material Rogers RT/duroid 5,880 of dielectric constant 2.2 and Taconic RF-60 of dielectric constant 6.15.

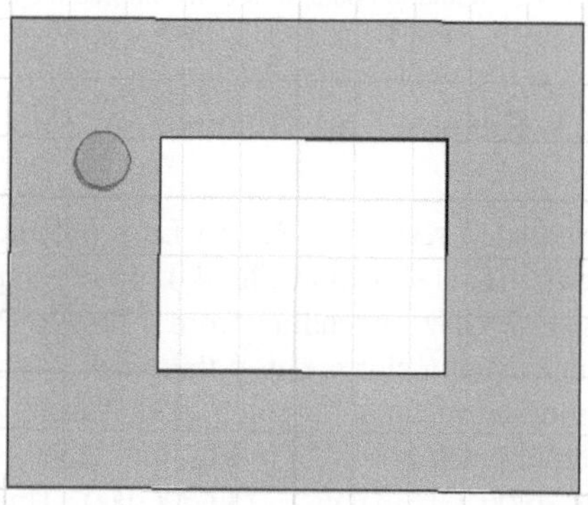

Fig. 4.3 Modified rectangular patch antenna without central patch (*rectangular ring*)

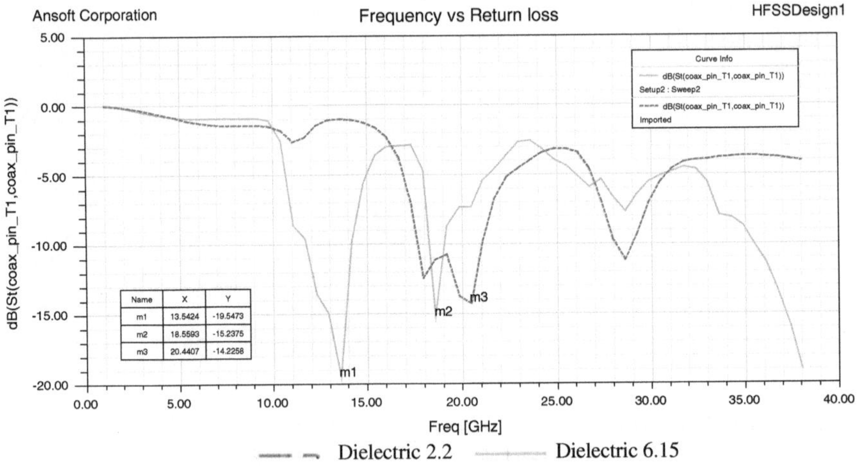

Fig. 4.4 Frequency versus return loss of conventional patch of Fig. 4.1 with substrate material dielectric constants 2.2 and 6.15

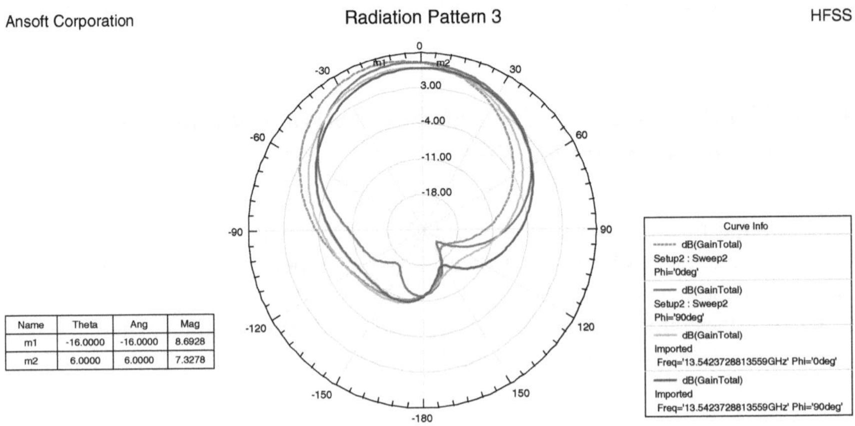

Fig. 4.5 Radiation pattern of conventional rectangular patch of Fig. 4.1 with feed at the corner of the patch. Substrate material dielectric constants 2.2 and 6.15

4.3 Results and Discussion

Simulated results of Return Loss obtained for the conventional rectangular patch (Fig. 4.1) is shown in Fig. 4.4. 10 dB impedance bandwidth (BW) for this antenna is 3.37 GHz, extending from 17.62 to 20.99 GHz. Keeping all other parameters unchanged a slot is cut at the patch as shown in Fig. 4.2. Results obtained for the modified rectangular patch of dielectric 2.2 with coaxial feed at the middle of the centre patch is shown in Fig. 4.6. Here 10 dB impedance band obtained from 62.0 to 76.06 GHz (BW = 14.02 GHz). Thus with the introduction of slot operating

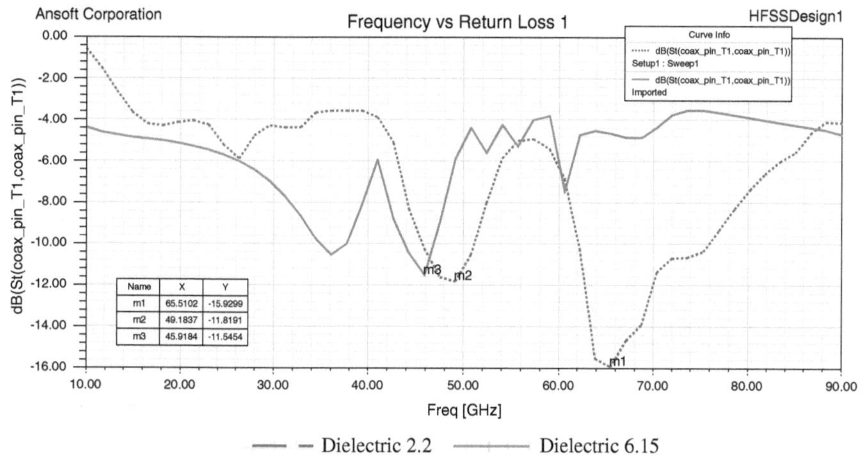

Fig. 4.6 Frequency versus return loss of centre feed slotted patch of Fig. 4.2 with substrate material dielectric constants 2.2 and 6.15

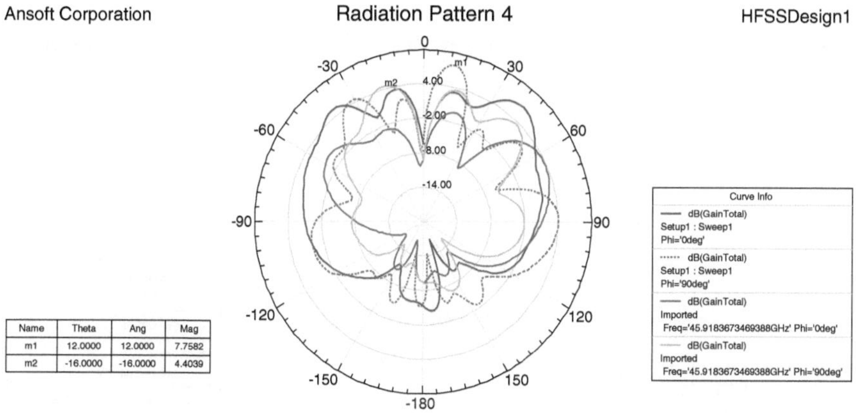

Fig. 4.7 Radiation pattern of centre feed slotted patch of Fig. 4.2. Substrate material dielectric constants 2.2 and 6.15

frequency range is shifted to the higher side of frequency spectrum as well as a significant increase in bandwidth is obtained. Results obtained for modified rectangular patch with coaxial feed at the corner of the patch (Fig. 4.2) are shown in Fig. 4.8. Here 10 dB impedance band is obtained from 55.92 to 70.01 GHz (BW = 14.09 GHz). Thus with this corner feed a continuous wideband operating region is obtained in Ku, K, Ka and near mm wave bands. When central patch is removed (Fig. 4.3) then operating frequency range varies from 54.19 to 70.92 GHz (Fig. 4.10). However, BW becomes 16.73 GHz which is significantly greater than the previously mentioned cases. In the figures dotted lines are used as dielectric 2.2 and solid lines are used as dielectric 6.15 (Figs. 4.5, 4.7, 4.9 and 4.11).

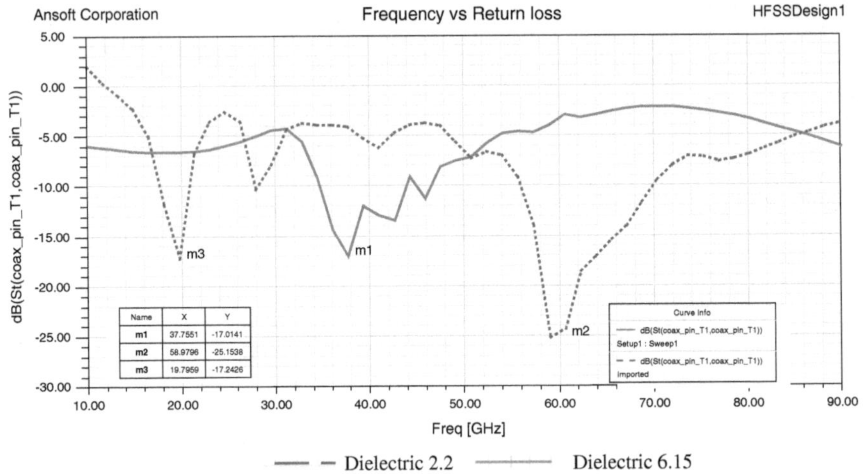

Fig. 4.8 Frequency versus return loss of corner feed slotted rectangular patch of Fig. 4.2 with substrate material dielectric constants 2.2 and 6.15

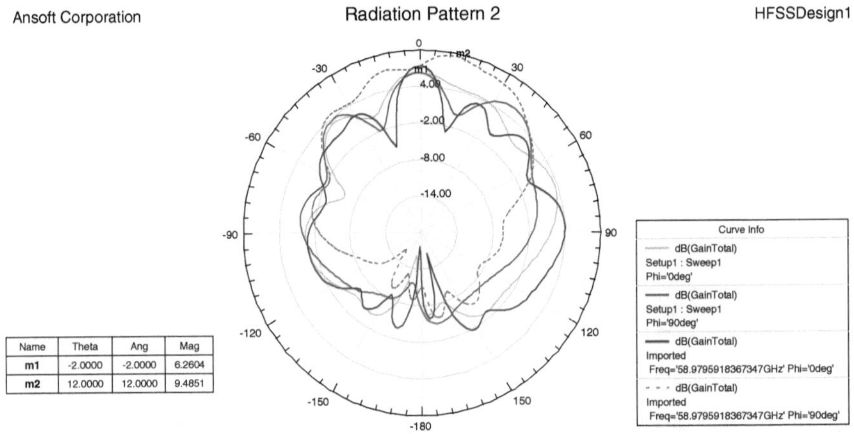

Fig. 4.9 Radiation pattern of corner feed slotted rectangular patch of Fig. 4.2. Substrate material dielectric constants 2.2 and 6.15

In order to operate at lower frequency range with these antenna structures, substrate material dielectric constant is changed to 6.15, keeping other parameters unchanged. Figure 4.4 shows the conventional rectangular patch that gives operating frequency range from 11.69 to 14.13 GHz (BW = 2.43 GHz). Results obtained for the centre feed slotted patch of Fig. 4.2 with dielectric constant 6.15 are shown in Fig. 4.6. Inclusion of slot shifts the operating frequency range to Ka band and EHF. The operating frequency range is 43.69–46.86 GHz (BW = 3.17 GHz), Fig. 4.8

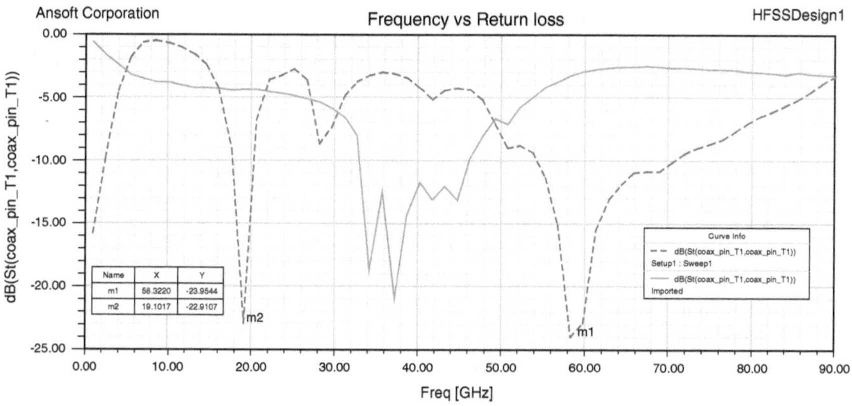

Name	X	Y
m1	58.3220	-23.9544
m2	19.1017	-22.9107

Fig. 4.10 Frequency versus return loss of rectangular ring of Fig. 4.3 with feed at the corner of the outer patch. Substrate material dielectric constants 2.2 and 6.15

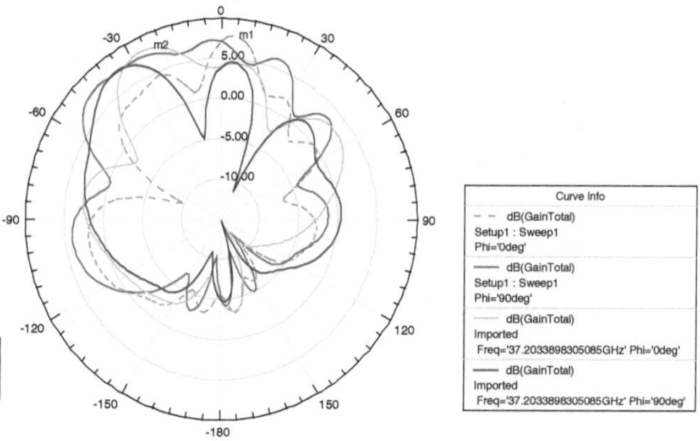

Name	Theta	Ang	Mag
m1	6.0000	6.0000	7.6966
m2	-22.0000	-22.0000	7.9546

Fig. 4.11 Radiation pattern of rectangular ring of Fig. 4.3. Substrate material dielectric constants 2.2 and 6.15

shows the results for the slotted antenna with corner feed of Fig. 4.2. Here the operating frequency range becomes 34.63–43.84 GHz (BW = 9.21 GHz). Figure 4.10 shows the results for the antenna of Fig. 4.3. Here the operating frequency range becomes 32.85–47.18 GHz (BW = 14.33 GHz). Thus better frequency coverage in Ka and EHF band is obtained with this structure. Table 4.1 shows the comparison of simulation results for different structures.

From the simulation results of different patch structures shown in Table 4.1, it is evident that the structures are capable of operating at different ranges of Ku, K, Ka bands and in Extremely High Frequency (EHF) region towards mm wave band. Modified structures can operate at broader frequency range than conventional ones.

Table 4.1 Simulation results for different structures

Sl no.	Antenna structure	Substrate material dielectric constant	Frequency band	Operating frequency range in GHz	% BW w.r.t centre band frequency	Gain in dB at minimum return loss frequency
1	Conventional rectangular patch	2.2	Ku, K	17.62–20.99	17.45	8.69
2	Slotted rectangular patch with centre feed	2.2	EHF	62.04–76.06	20.30	7.76
3	Slotted rectangular patch with corner feed	2.2	EHF	55.92–70.02	22.38	9.48
4	Rectangular ring	2.2	EHF	54.19–70.92	26.74	7.95
5	Conventional rectangular patch	6.15	X, Ku	11.69–14.13	18.82	7.33
6	Slotted rectangular patch with centre feed	6.15	EHF	43.69–46.86	7.01	4.40
7	Slotted rectangular patch with corner feed	6.15	Ka, EHF	34.63–43.84	23.48	6.26
8	Rectangular ring	6.15	Ka, EHF	32.85–47.18	35.82	7.69

The structure of Fig. 4.3 shows 10 dB impedance bandwidth of 26.74 % at centre frequency of 39.79 GHz with substrate of dielectric constants 2.2 and 35.82 % at centre frequency of 14.85 GHz with substrate of dielectric constant 6.15.

4.4 Conclusion

Narrow bandwidth is a disadvantage of conventional microstrip patch antennas. The proposed modified rectangular patch antenna structures are simple and capable of overcoming this disadvantage providing wide bandwidths in Ku, K and Ka bands and EHF region with reasonable gain values. Rectangular ring structure gives much greater bandwidth and significantly more gain than the slotted structure.

References

1. M. Bugaj, R. Przesmycki, L. Nowosielski, K. Piwowarczyk, Analysis of different methods of microstrip antennas feeding for their electrical parameters, in *PIERS Proceedings*, Kuala Lumpur, Malaysia, (2012), pp. 27–30
2. Y. Liu, L.-M. Si, M. Wei, P. Yan, P. Yang, H. Lu, C. Zheng, Y. Yuan, J. Mou, X. Lv, H. Sun, Some recent developments of microstrip antenna. Int. J. Antennas Propag. **2012** (Article ID 428284), 10 (2012)
3. D.M. Pozar, Microstrip antennas. Proc. IEEE **80**(1), 79–91 (1992)
4. R. Garg, P. Bhartia, I. Bahl, A. Ittipiboon, *Microstrip Antenna Design Handbook* (Artech House inc., Norwood, 2001)
5. C.A. Balanis, *Antenna Theory Analysis and Design* (Wiley, New York, 2005)
6. G. Kumar, K.P. Ray, *Broadband Microstrip Antennas*. Artech House antennas and propagation library (2003)
7. D.M. Pozar, B. Kaufman, Increasing the bandwidth of a microstrip antenna by proximity coupling. Electron. Lett. **23**(8), 368–369 (1987)
8. H.F. Pues, A.R. Van de Capelle, Impedance-matching technique for increasing the bandwidth of microstrip antennas. IEEE Trans. Antennas Propag. **37**(11), 1345–1354 (1989)
9. A. Kaya, E.Y. Yüksel, Investigation of a compensated rectangular microstrip antenna with negative capacitor and negative inductor for bandwidth enhancement. IEEE Trans. Antennas Propag. **55**(5), 1275–1282 (2007)
10. K.L. Wong, N.H. Hsu, A broad-band rectangular patch antenna with a pair of wide slits. IEEE Trans. Antennas Propag. **49**(9), 1345–1347 (2001)
11. S.I. Latif, L. Shafai, S.K. Sharma, Bandwidth enhancement and size reduction of microstrip slot antennas. IEEE Trans. Antennas Propag. **53**(3), 994–1003 (2005)
12. P.M.T. Ikonen, S.I. Maslovski, C.R. Simovski, S.A. Tretyakov, On artificial magnetodielectric loading for improving the impedance bandwidth properties of microstrip antennas. IEEE Trans. Antennas Propag. **54**(6), 1654–1662 (2006)
13. A.S. Jadon, J. Sharma, A. Prajapat, A. Bhadauria, Coplanar rectangular patch antenna for X band applications using inset fed technique. Int. J. Electron. Commun. Eng. Technol. 95–102 (2013)
14. J.-H. Fu, Q. Wu, G.-H. Yang, The research of Ka band broadband low side-lobe microstrip antenna array, in *Microwave Conference, 2008. APMC 2008*. Asia-Pacific, 16–20 Dec 2008

Chapter 5
Design of Dielectric Resonator Antenna with Different Dielectric Constants

Abhijit Ghosh, Madhumita Pal, Pratyusha Bhadra and Piyu Sarcar

Abstract Probe-fed rectangular dielectric resonator antennas (DRAs) are investigated by varying their relative permittivity to study different parameters of antennas. With an optimum combination of antenna geometry and permittivity, the variation of different parameters are observed and plotted with dielectric constants.

Keywords Dielectric constant · Dielectric resonator antennas · Percentage bandwidth · Resonant frequency · VSWR · Input impedance

5.1 Introduction

Today, there is a deep interest in antenna systems which operate at frequencies in the millimetre wave region (100–300 GHz) [1]. Conventional metallic antennas suffer problems with regard to power losses, radiated power capabilities and fabrication difficulties when reduced to the sizes necessary to operate in this frequency band. These obstacles can be overcome if a simply shaped antenna with few conducting surfaces is designed. The dielectric resonator antenna (DRA) meets these requirements and has been shown to be a good choice for use in this band [2, 3].

The rectangular-based DRA is evaluated with two degrees of freedom, the length-to-height ratio and the length-to-width ratio. This additional degree of

A. Ghosh (✉) · M. Pal · P. Bhadra · P. Sarcar
Department of ECE, Narula Institute of Technology, Kolkata 700109, India
e-mail: jit.ghosh18@gmail.com

M. Pal
e-mail: madhuece.pal@gmail.com

P. Bhadra
e-mail: pratyushabhadra@gmail.com

P. Sarcar
e-mail: piyusarcar@gmail.com

© Springer India 2015

K. Maharatna et al. (eds.), *Computational Advancement in Communication Circuits and Systems*, Lecture Notes in Electrical Engineering 335,
DOI 10.1007/978-81-322-2274-3_5

freedom gives more flexibility to the experimenter in manipulation of the resonant frequency of the excited modes. Taking these results a step further, antennas with the following specifications are numerically evaluated [4].

5.2 Antenna Geometry

In this paper, we shall observe the changes in the parameters by changing dielectric constant of the dielectric material in a small amount [5, 6]. Also, we are going to observe the changes when the dielectric material is changed. The rectangular DRA under investigation has length, width and height a = 1.8 mm, b = 2.29 mm and h = 1.15 mm, respectively (shown in Fig. 5.1).

The radius and width of the ground plane is 2.5 and 0.2 mm, respectively. The air gap between the dielectric material and the ground plane is 0.5 mm.

The radius and height of the probe is 0.16 and 1.85 mm, respectively. The radius and height of the coax pin is 0.16 and 0.5 mm, respectively. The radius and height of the coax is 0.34 and 0.5 mm, respectively.

5.3 Simulated Results

All the measurements of the antenna are kept constant and the relative permittivity of the antenna is varied from 4 to 51. The dielectrics used are sapphire, taconic CER-10 (tm), rogers R03210 (tm), gallium arsenide, lead monoxide, maleic anhydride, etc. These are only some of them while many other dielectrics are used. The parameters which are being observed with the dielectric constant are centre frequency and 3 dB bandwidth from return loss plot, centre frequency and bandwidth from VSWR plot where VSWR is less than 2 [2] and the resonant frequency at which impedance (imaginary) is zero. Here are some figures that show the differences in the parameters when the dielectric material is changed (Fig. 5.2).

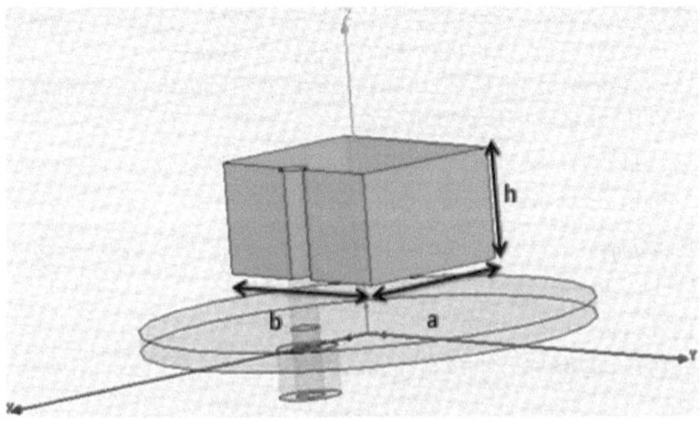

Fig. 5.1 Rectangular DRA

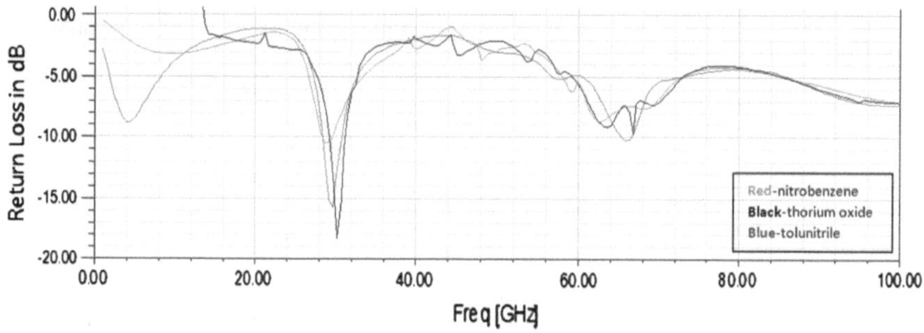

Fig. 5.2 Plot for frequency versus return loss (in dB) for three different dielectric constants

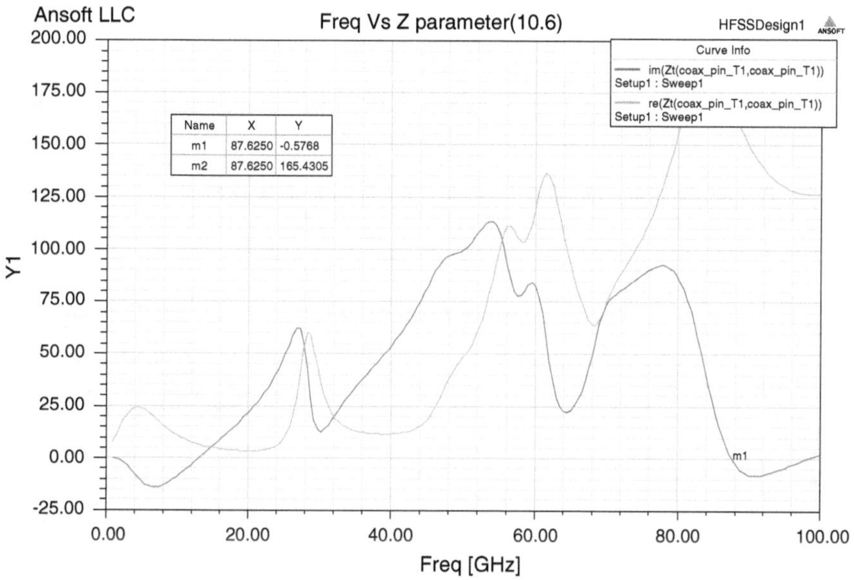

Fig. 5.3 Plot for frequency versus Z parameter for thorium oxide ($\varepsilon_r = 10.6$) (colour figure online)

The black line shows that 4.1 GHz bandwidth with centre frequency 29.71 GHz and one notch occurs at 64 GHz frequency for thorium oxide (ε_r = 10.6). The blue line shows that 3.75 GHz bandwidth with one centre frequency 29.215 GHz for tolunitrile (ε_r = 18.8). The red line shows that 2 GHz bandwidth with centre frequency 29.215 GHz as notch for nitrobenzene (ε_r = 35.7) (Figs. 5.3, 5.4, 5.5 and Table 5.1).

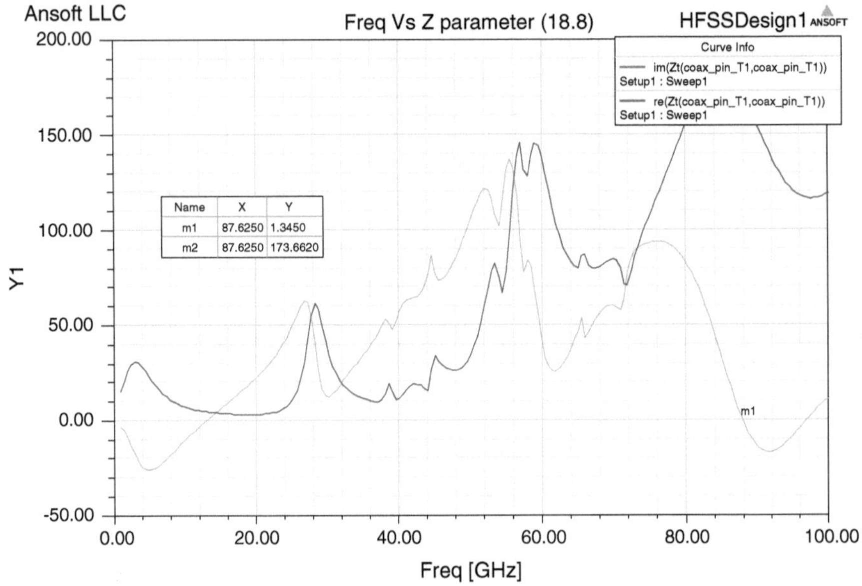

Fig. 5.4 Plot for frequency versus Z parameter for tolunitrile (ε_r = 18.8)

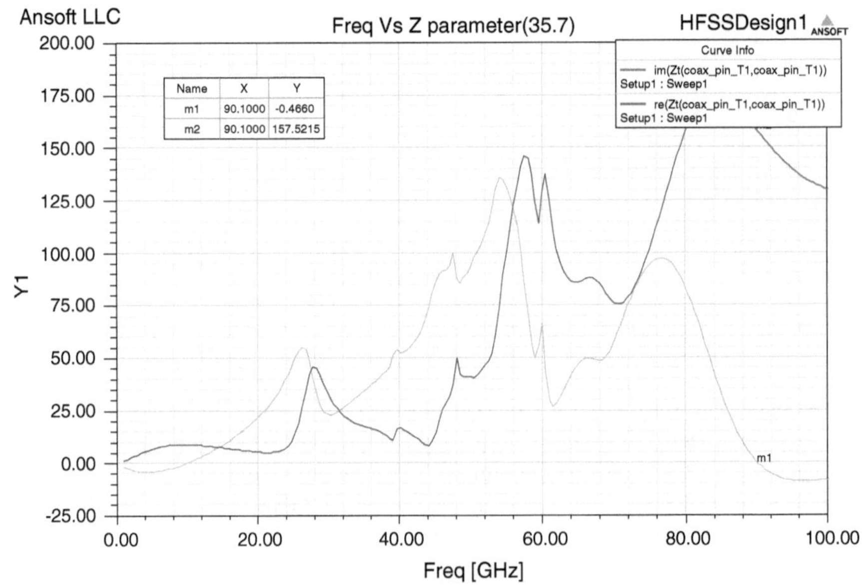

Fig. 5.5 Plot for frequency versus Z parameter for nitrobenzene (ε_r = 35.7)

Table 5.1 Table showing observation on the parameters of RDRA by using different dielectric materials

Name of dielectric	Dielectric constant (relative permittivity)	Return loss		% BW (GHz) w.r.t centre frequency	VSWR	BW for VSWR <2 (GHz)
		Resonant frequency (GHz)	S_{11} (min) in dB			
Silicon dioxide	4	29.8	−15.04	6.71	1.43	2.75
Sapphire	10	29.8	−10.89	5.03	1.79	1.79
Taconic CER-10 (tm)	10	29.8	−15.5	6.71	1.38	2.46
Rogers R03210 (tm)	10.2	29.8	−15.92	8.38	1.38	2.47
Roger RT/duroid 6010/6010LM (tm)	10.2	29.8	−15.8	8.38	1.38	2.46
Thorium oxide	10.6	29.71	−15.77	6.73	1.38	2.97
Silicon	11.9	29.8	−15.50	5.03	1.38	2.44
Zirconium oxide	12.5	29.8	−14.93	6.71	1.43	2.51
Gallium arsenide	12.9	29.8	−15.96	8.38	1.37	2.51
Titanium oxide	40	66.34	−8.5	–	2.20	–
Thallium chloride	46.9	65.35	−9.73	–	1.96	1.48
Maleic anhydride	51	31.19	−19.24	6.41	1.24	0.99

5.4 Analysis

The results from the above study are summarized and plotted against dielectric constant to observe their characteristics.

From the graph in Fig. 5.6 we can infer that as the relative permittivity of the dielectric material increases, the % bandwidth (from the return loss plot) increases up to a certain level, take a slight dip and then again increases. But after 40, it decreases gradually. We obtain the % bandwidth from the formula (*f*1 − *f*2)/ fc × 100 where (*f*1 − *f*2) is the bandwidth and fc is the centre frequency.

From the graph in Fig 5.7 it is understood that the % bandwidth changes in a discontinuous manner with the increasing relative dielectric constant. It also consists of irregular crests and troughs.

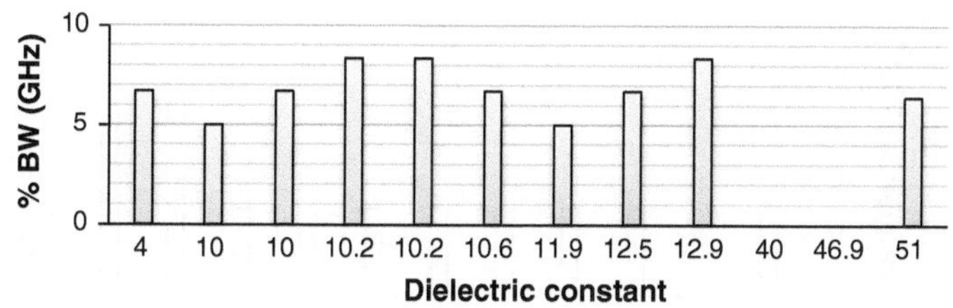

Fig. 5.6 Plot for relative dielectric constant versus 3 dB % BW from return loss plot

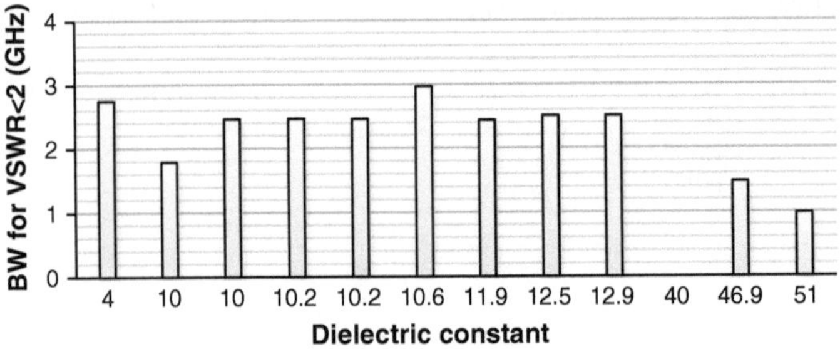

Fig. 5.7 Plot for relative dielectric constant versus % BW from VSWR plot

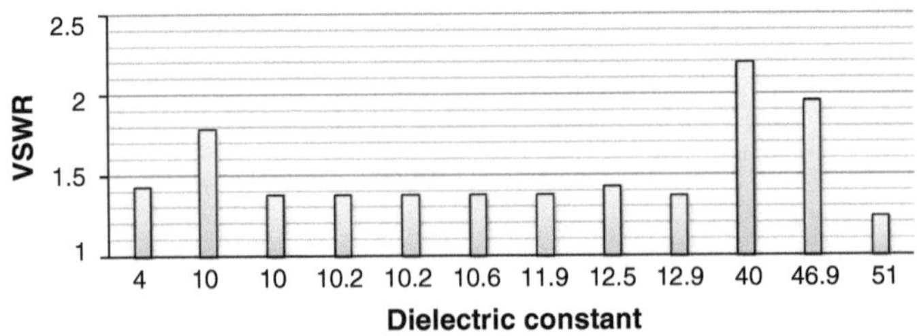

Fig. 5.8 Plot for relative dielectric constant versus VSWR

From the graph in Fig. 5.8 it can be concluded that the VSWR varies in an irregular manner with the increasing relative dielectric constant but it is always greater than 1, as we know that VSWR always ranges between 1 and infinity. It consists of irregular crests and troughs.

The graph in Fig. 5.9 shows the relationship between the relative permittivity of the dielectric material and the resonant frequency from the return loss plot. It can be inferred that the frequency remains almost constant up to dielectric constant of 20 and then it starts oscillating.

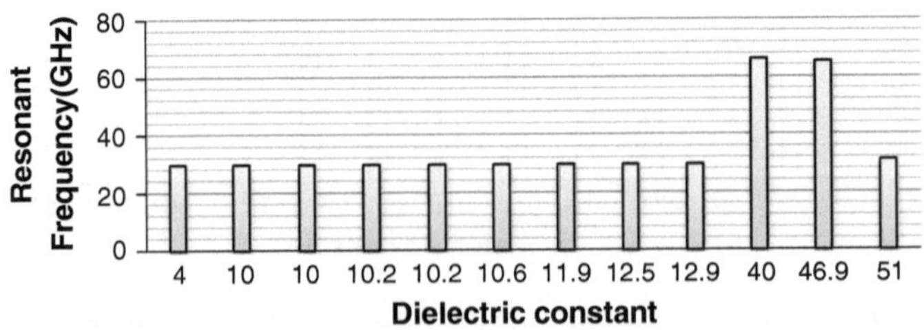

Fig. 5.9 Plot for relative dielectric constant versus resonant frequency from s parameter plot

5.5 Conclusion

From this table and analyses we have learnt that the resonant frequency from the return loss plot remains almost unchanged for the lower values of dielectric constants, but in the higher values it fluctuates. The VSWR remains almost in the range of 1–2 which is desirable, but for lead monoxide and titanium oxide the minimum VSWR is 2.02 and 2.2 respectively. The impedance (real) ranges between 130 and 200 Ω. The bandwidth, when VSWR is less than 2, ranges from approximately 1–3 GHz. The maximum bandwidth is for thorium oxide. The % bandwidth from the return loss plot ranges from 1 to 13 GHz and the maximum is for titanium oxide.

References

1. K.M. Luk, K.W. Leung, *Dielectric Resonator Antenna* (Research Studies Press Ltd. Copyright ©, Baldock, 2003)
2. C.S. De Young, S.A. Long, Wideband cylindrical and rectangular dielectric resonator antennas. IEEE Antennas Wirel. Propag. Lett. **5**, 426−429 (2006)
3. A. Petosa, A. Ittipiboon, Dielectric resonator antennas: a historical review and the current state of the art. IEEE Antennas Propag. Mag. **52**(5), 91−116 (2010)
4. M.H. Neshati, Z. Wu, Numerical modeling and experimental study of probe-fed rectangular dielectric resonator antenna (RDRA) supported by finite circular ground plane. IJE Trans. A: Basics **17**(3), 271−280 (2004)
5. X.S. Fang, K.W. Leung, Designs of single-, dual-, wide-band rectangular dielectric resonator antennas. IEEE Trans. Antennas Propag. **59**(6), 2409−2414 (2011)
6. K.A. O'Connor, R.D. Curr, A dielectric resonator antenna based on high dielectric constant composites for high power UHF antenna applications. 978-1-4673-1225-7/12/©2012 IEEE

Chapter 6
Overview of Various Bandwidth Enhancement Techniques for Ultra-Wide Band Antennas

Arpita Barman Santra and Amiya Kumar Mallick

Abstract This article deals with the Ultra-Wide Band (UWB) antennas that are appraised in respect of different geometries, design parameters and their experimental results. Several types of UWB antennas including special horn, micro-strip patch radiators and array antennas in recent works are elucidated, while comparing their measured return loss, gain and radiation patterns. Projections to the future developments of UWB antenna technology are also specified.

Keywords Ultra-wide band · Bandwidth · Gain

6.1 Introduction

Ultra-wide Band (UWB) systems [1] concern with the systems which use electromagnetic signals having greater than 20 % bandwidth [2] around its centre frequency. The systems provide high data rates for wireless communications, and accurate radar and geolocation systems. Since the allocation of unlicensed various frequency bands, such as 0–960 MHz, 3.1–10.6 GHz, 22–29 GHz, etc., to UWB systems [3] by Federal Communications Committee (FCC) of USA, research as on UWB has gained much attention of research workers in the academia and industry.

The UWB system covers the frequency range from 3.1 to 10.6 GHz, which is based on narrow pulses to transmit data at extremely low power and looks like random noise to most conventional radio systems. The UWB technology offers several advantages over conventional communications systems. For instance, there is no carrier frequency. Instead, UWB emits timed "pulses" (a train of RF pulses of

A.B. Santra (✉)
ECE, Narula Institute of Technology, Kolkata 700109, India
e-mail: arpitasantra@yahoo.com

A.K. Mallick
Indian Institute of Technology, Kharagpur 721302, India
e-mail: akmallick@aol.com

© Springer India 2015
K. Maharatna et al. (eds.), *Computational Advancement in Communication Circuits and Systems*, Lecture Notes in Electrical Engineering 335,
DOI 10.1007/978-81-322-2274-3_6

EM energy) of electromagnetic energy. Therefore transmitter and receiver hardware can be made very simple, which is necessary for portable devices. There is a wide range of applications for UWB technology, which includes wireless communication systems, position and tracking, sensing and imaging and radar. The antenna plays an essential role in the UWB system, which is different from narrowband system. UWB systems transmit extremely narrow pulses of the order of 1 ns or less resulting in bandwidths in excess of 1 GHz or more, i.e. giving the rise of bandwidth or making the system broadband [4] type. However, the design and fabrication of high-performance transmitting/receiving antennas often present significant challenges in the implementation of these systems. The challenge lies in the development of an antenna, capable of handling these high-speed pulse trains. The design of a UWB antenna is difficult, because the fractional bandwidth [5] is actually large and antenna must cover multiple-octave bandwidths in order to transmit pulses that are of the order of nanoseconds in duration. Since data may be contained in the shape of the UWB pulse, antenna pulse distortion must be kept to a minimum. From a system design perspective, the impulse response of the antenna is of particular interest, because it has the ability to alter or shape the transmitted or received pulses. In practice, attempt must be made to limit the amplitude and group delay distortion below a certain threshold which will ensure reliable system performance.

6.2 UWB Antenna: In a Proper Perspective

There are several types of wireless antenna, divided into two main groups: directional and omni-directional, which might be used in narrowband or wideband [6] systems. Directional antenna is suitable for long distance communication as they have focused beam with high gain, while omni-directional antenna covers a wide area with reasonable gain. Hence, omni-directional antenna is suitable for short distances and in indoor environments, such as in offices or rooms.

Wireless antennas may be classified into two classes, narrowband and wideband. The narrowband class demonstrates tremendous smallness for a given operating bandwidth. The wideband [7, 8] class possesses extreme bandwidth capability, capable of covering multiple octaves. Both classes achieve performance very close to the theoretical Chu-Harrington limit [9, 10], indicating that they are as small as possible for the exhibited bandwidth. The Chu-Harrington graph is a theoretical limit concerning the volumetric size of an antenna element to its quality factor or bandwidth of operation. This relationship gives the antenna designer an approximation of a switch between the size and the desired bandwidth. There are many issues involved in designing of UWB systems, such as antenna design, channel model and interference. UWB antennas must cover an extremely wideband of 3.1–10.6 GHz (lower band 3.1–5.1 GHz, upper band 5.85–10.6 GHz) for indoor and handheld applications, have electrically small size and high efficiency. In addition, they are required to have a non-dispersive characteristic in time and frequency domain, providing narrow pulse duration to enhance a high data throughput. Antennas in the frequency domain are typically characterized by radiation pattern, directivity, impedance matching and bandwidth. However, there are certain requirements for antennas in the wireless system, irrespective of ultra-wideband or narrowband, the same as regulatory issues, antenna gain, antenna efficiency and group delay of antenna.

There has been continuing interest in designing a UWB antenna that operates over a wide frequency range and that can be used for multiple channels or systems. Regulatory restrictions on bandwidth and the greater technical challenges and cost involved in developing a UWB antenna, as well as the corresponding operating systems, have largely limited the development of UWB antennas to special research organizations and agencies in military applications. A recent FCC ruling allowing the operation of UWB signals in various sensing, imaging, and communication applications has once again revived interest in designing UWB antennas and in their potential applications. Technology advances in high-speed analog-to-digital and digital-to-analog devices, digital frequency synthesis, and digital receivers also allow for rapid and low-cost implementation of UWB systems below 3 GHz. Therefore, more attention is now being given to designing compact and low-cost UWB antennas.

The term UWB has often been used loosely. The original FCC definition of UWB specifies 25 % or more fractional bandwidth. That is, the ratio of a signal's −10 dB bandwidth to centre frequency. In this paper, UWB mainly refers to a bandwidth of more than 2:1. That is, the ratio of the highest frequency to the lowest frequency of the −10 dB bandwidth. Some people determine the antenna's bandwidth based on the frequency range where the return loss (or reflection coefficient) level is less than −10 dB. However, note that a low return loss does not necessarily imply a strong radiation. Nor does it say anything about the radiation pattern. The return loss value has even less meaning if the antenna contains absorptive mechanisms. Since the main function of an antenna is to radiate sufficient energy in the desired directions, it makes sense to use realized gain and radiation pattern to define the bandwidth.

In February 2002, the Federal Communications Commission (FCC) of the United States of America conditionally liberalized unlicensed operation of personal UWB devices in the private sector. In the meantime, various antennas have been developed to be used for UWB systems, to name a few, double-ridged waveguide horn, log periodic, biconical and monopole antennas. None of these antennas, however, simultaneously meet omni-directional and low voltage-standing-wave-ratio (VSWR) requirements, essential for some measurement applications such as UWB channel sounding. In 2002, J. Farserotu, A. Hutter et al. used two potentially important technologies for realization of nomadic PANs: Ultra Wideband (UWB) and MIMO antennas. In this approach low cost UWB technology is considered for the link between simple personal devices, such as sensors, to a mobile bridge, operating at 2.4 GHz or higher frequencies, while small multiple antenna systems are considered for the link between the mobile bridge, router or gateway to wireless local area networks (WLANs) and ultimately wide area networks (WANs) in order to facilitate efficient use of the available spectrum, improve coexistence with other systems operating in the same band and support robust, high data rate connectivity between the users PAN and external networks.

In 2007, A.H.M. Zahirul Alam, Rafiqul Islam and Sheroz Khan proposed a tuning fork type structure of UWB antenna which offers excellent performance for UWB system, ranging from 3.7 to 13.8 GHz, exhibiting a 10 dB return loss bandwidth over the entire frequency band. The rectangular patch antenna is designed on FR4 substrate and fed with 50 Ω micro strip line by optimizing the width of partial ground, the width of position of feed line to operate in UWB.

6.3 Conclusion

The aim of this study is to understand the significant aspects of UWB antenna design and how they are related to the system performance. UWB antennas should be designed with specification of flat amplitude and linear phase response over the desired bandwidth, as UWB system antenna is the significant part of the system. Its characteristics have an effect on the overall system performance.

References

1. http://apps.fcc.gov/edocs_public/attachmatch/FCC--02-48A1.pdf. p. 12
2. G. Vetharatnam, B.K. Chung, H.T. Chuah, Design of a microstrip patch antenna array for airborne SAR applications. J. Electromagnet. Waves Appl. **19**(12), 1687–1701 (2005)
3. F. Yang, X.X. Zhang, X. Ye, Y. Rahmat-Samii, Wide-band E-shaped patch antennas for wireless communications. IEEE Trans. Antennas Propag. **49**(7), 1094–1100 (2001)
4. K.L. Wong, W.H. Hsu, A broad-band rectangular patch antenna with a pair of wide slits. IEEE Trans. Antennas Propagat. **49**(9), 1345–1347 (2001)
5. A. Yu, X.X. Zhang, A method to enhance the bandwidth of microstrip antennas using a modified E-shaped patch. in *Proceedings of Radio and Wireless Conference*, pp. 261–264, 10–13 Aug 2003
6. G. Rafi, L. Shafai, Broadband microstrip patch antenna with V-slot. IEE Proc. Microw. Antenna Propag. **151**(5), 435–440 (2004)
7. K.F. Tong, K.M. Luk, K.F. Lee, S.M. Shum, Analysis of a broadband U-slot microstrip antenna. in *10th International Conference on Antenna and Propagation. Conference Publication no. 436, IEEE* (1997), pp 14–17
8. A.M. Hadian, Wideband rectangular microstrip patch antenna with U-slot. in *The Second European Conference on Antennas and Propagation, EuCAP 2007* (2007), pp. 1–5
9. L.J. Chu, Physical limitations on omni-directional antennas. J. Appl. Phys. **19**, 1163–1175 (1948)
10. R.F. Harrington, Effect of antenna size on gain, bandwidth and efficiency. J. Res. Nut. Bur. Stund. **64-D**, 1–12 (1960)

Chapter 7
Investigation on Microwave MEMS Capacitive Shunt Switch by Using Coventor and CST Software

Balaka Biswas and Amiya Kumar Mallick

Abstract In this paper, a detailed analysis of electrostatic and electromechanical aspects of RF capacitive shunt switch is presented. The effects of the membrane materials on different parameters of the switch are also studied for aluminium and gold membranes. The analysis is carried out for charges, electric fields, power handling capacity and hysteresis using MATLAB and further full wave analysis using COVENTOR.

Keywords Shunt switch · CPW · MEMS switch

7.1 Introduction

Switches are the integral and important parts of RF systems. Conventional MICs and MMIC-based switches have limitations for broadband operations. Broadband RF design poses challenges in the realization of RF systems for multiband communication system, phased array antenna, T/R module, phase shifter or spatial diversity antenna. The recent trend is to design a compact and reliable system, which is more affordable, integrable than ever before and permits a wider degree of RF functionality and have a broader bandwidth. Microelectromechanical systems (MEMS) are becoming increasingly popular in RF application due to their attractive advantages such as low insertion loss, high isolation, wide bandwidth operation, low power consumption, simple biasing network and no inter-modulation products compared to FET's and PIN diodes.

B. Biswas (✉)
Jadavpur University, Kolkata, India
e-mail: balaka.biswas@gmail.com

A.K. Mallick
Narula Institute of Technology, Kolkata, India
e-mail: akmallick@aim.com

© Springer India 2015

K. Maharatna et al. (eds.), *Computational Advancement in Communication Circuits and Systems*, Lecture Notes in Electrical Engineering 335,
DOI 10.1007/978-81-322-2274-3_7

Basically two types of switches, namely capacitive and ohmic contact, have been demonstrated [1]. Stiction limits the operational life of the ohmic contact switches [2]. Capacitive switches can be broadly classified into series and shunt switches. The series switch has limitations in regard to operation in 1–10 GHz range and power handling capability. This paper analyses the shunt capacitive switch actuated by a voltage for electrostatic and electromechanical properties. The process steps are made compatible to CMOS process for cost effectiveness. The effect of bridge material on different parameters is investigated for the commonly used gold and aluminium bridges.

7.1.1 Electrostatic Analysis of the Shunt Switch

7.1.1.1 Coplanar Waveguide (CPW)

The switch studied here is based upon CPW structure, as shown in Fig. 7.1. The configuration of the Tx-line is denoted by G/W/G. Commonly used microstrip structure gives higher losses and dispersive behaviour at high frequencies. So, CPW configuration is preferred for shunt switch configurations.

The study of different gap width and line width has been done and optimum value of 90/120/90 (μm) has been chosen considering pull-down voltage and testing limitations.

7.1.1.2 Switch Design

The physical structure of the switch is shown in Fig. 7.2a, b. A stack of oxide-nitride-oxide (ONO) is implemented to reduce the losses and dispersion in the substrate, above which CPW structure is realized. The top metal membrane of thickness '*t*', supported by two corner metal posts, is suspended at a distance '*g*' above the bottom metal surface. A small portion of the lower electrode is covered with a thin layer of SiN. When a DC bias is applied between the two electrodes, charges are induced on the metal which pulls down the top metal membrane by electrostatic force.

Fig. 7.1 CPW Configuration

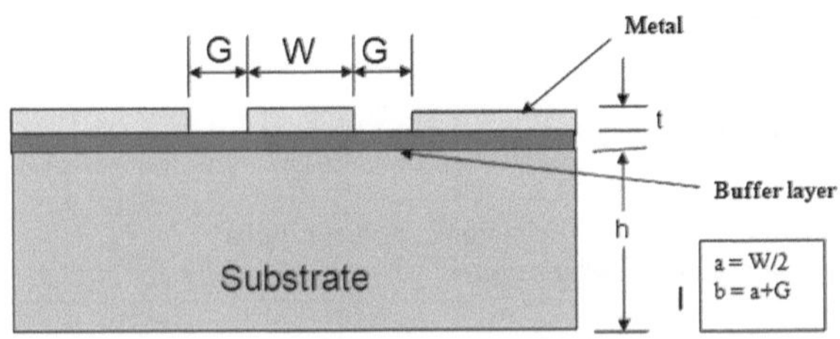

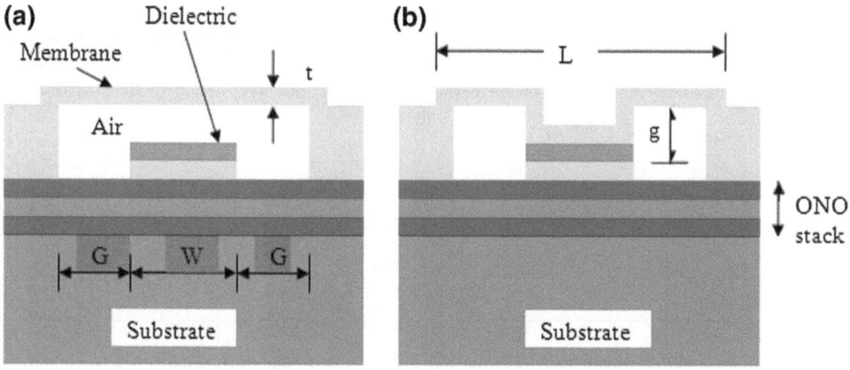

Fig. 7.2 Cross-section of a shunt capacitive switch in **a** up-state and **b** down-state

Fig. 7.3 MEMS shunt capacitive switch in CPW implementation

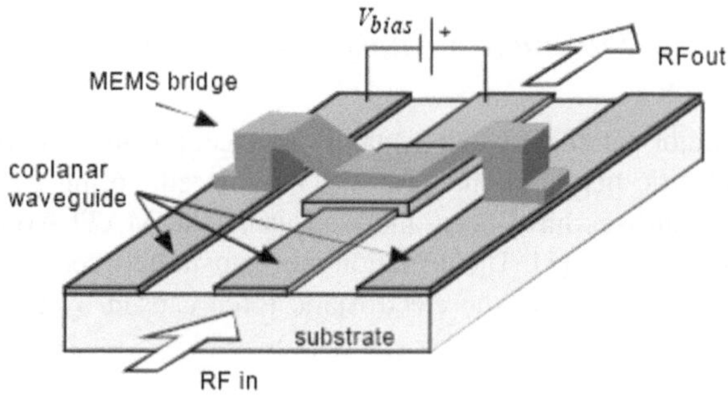

Above a certain threshold voltage (pull-down voltage; *Vp*), the membrane snaps down to the "closed" position when the force of attraction overcomes mechanical stresses in the material as shown on Fig. 7.2b.

The switch operates as a digitally tunable capacitor with two states: smaller capacitance of the order of fF when the membrane is in up-state and larger value in the range of pF in down state. A perspective view of the switch in closed or pulled down state is shown in Fig. 7.3.

Parameters related to switch designing are—length, width and thickness of the membrane, air gap height, CPW configuration (G/W/G), dielectric thickness and beam material. These parameters determine the electrical as well as mechanical characteristics of the beam. The typical parameters of the switch analysed are given in Table 7.1.

Table 7.1 Typical parameters for RF shunt switch analysis

Parameters	Value (μm)
Length of the membrane '*l*'	280
Width of the membrane '*w*'	100
Gap b/w electrode and membrane	2.5
Thickness of the lower electrode '*t*'	0.5
CPW configuration	90/120/90

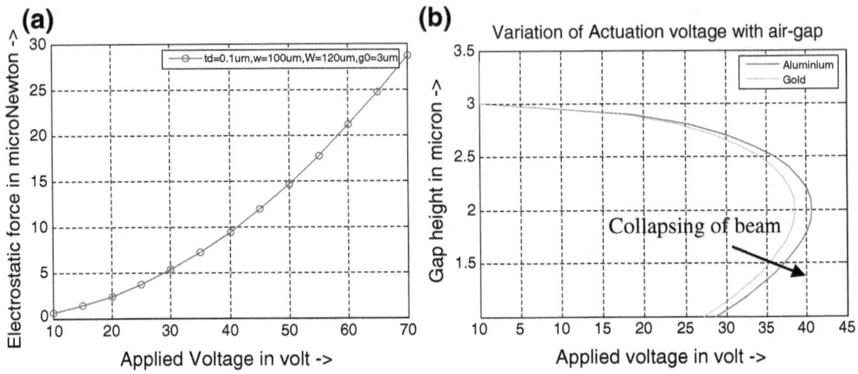

Fig. 7.4 **a** Variation of the electrostatic force with voltage, **b** change in voltage due to gap

7.1.2 Parametric Analysis of the Switch

Major parameters like actuation voltage, pull-in and hold-down voltages, generated electric field and charges, switching speed, spring constant variation and its effect on stresses have been analysed using the MATLAB for both the aluminium and gold beams [1]. The actuation mechanism between the top and bottom electrodes is achieved due to the electrostatic force caused by the actuation voltage V_a and is given by (Fig. 7.4)

$$F = \frac{\varepsilon_0 w W V_a^2}{2\left[g_0 + \left(\frac{t_d}{\varepsilon_r}\right)\right]^2} \quad V_a = \sqrt{\frac{2Kg^2(g_0 - g)}{\varepsilon_0 w W}} \tag{7.1}$$

where g_0 and g are the pre- and post-actuation gap heights of the beam. The variation in the electrostatic force with applied voltage is shown in Fig. 7.5.

Lowering the gap height lowers the actuation voltage as well as electrostatic force but at the expense of smaller capacitance ratio and higher probability of stiction. At 2/3 g_0, the increase in the electrostatic force is greater than the restoring force. This results in collapsing of the beam and corresponding pull-down voltage depending on length, width, gap, height and type of the membrane material. As evident from the figures, there is not much difference in force and actuation voltage for gold and aluminium membrane as these parameters have very little dependency on the mechanical properties of the beam. For a switch to stay in the down state, the electrostatic force must be larger than the mechanical restoring force. The hold-on voltage depends on the spring constant and the gap height represented as

$$V_h = \left[\frac{2K(g_0 - g)}{\varepsilon_0 w W}\left(g + \left(\frac{t_d}{\varepsilon_r}\right)\right)^2\right]^{\frac{1}{2}} \tag{7.2}$$

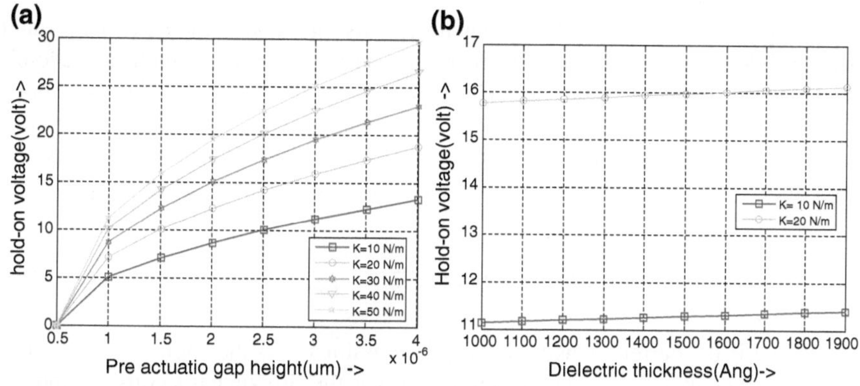

Fig. 7.5 Variation of hold-on voltage **a** with g_0 and **b** with t_d

This dependency is illustrated in Fig. 7.5a, b for different spring constants.

One can choose the dielectric thickness according to process limitations for a particular switch configuration regarding the hold-on voltage.

In capacitive membrane switches, the limiting mechanism is dielectric charging. The dielectric layer of SiN in this study is selected as 1200 Å to get higher up-state capacitance. This quantum of charge accumulated on the membrane due to electrostatic force is obtained using the relations:

$$Q = \sqrt{2gKwW\varepsilon_0} \quad E = \frac{\varepsilon_0 V_p}{C_u\left[g + \left(\frac{t_d}{\varepsilon_r}\right)\right]^2} \tag{7.3}$$

where

$$C_u = C_{pp} + 0.35\frac{\varepsilon_0 A}{\left[g + \left(\frac{t_d}{\varepsilon_r}\right)\right]}; \quad \mathrm{A} = \mathrm{w} \times \mathrm{W} \tag{7.4}$$

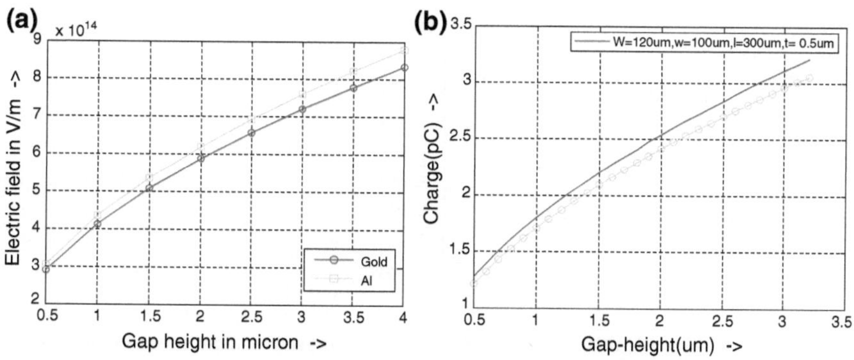

Fig. 7.6 **a** Variation of the electric field with gap, **b** variation in charge with gap height

The operational lifetime of the switch depends on the accumulated charges and electric field. The failure mechanism in capacitive membrane switch is predominantly due to dielectric charging. The field penetrated into the device is trapped within the dielectric, as there is no conduction path. As the recombination time for these charges is very long, the operational life of the switch is affected by the magnitude of the applied field. The Frenkel–Poole emission is given by [3]

$$\mathrm{J} = \mathrm{V}\exp\left(2a\frac{\sqrt{V}}{T} - \frac{q\phi_B}{kT}\right) \tag{7.5}$$

where J is the current density due to FP emissions, V is the applied voltage, T is absolute temperature, ϕ_B is the barrier height, k is the Boltzmann's constant and 'a' is a constant composed of electron charge, insulator dynamic permittivity and film thickness.

This analysis quantifies the field intensity as well as generated charges in the dielectric layer for both aluminium and gold membranes (Fig. 7.6).

The increase in the electrostatic force results in the decrease of the beam height and consecutive increase in induced charges in the dielectric. So the gap and dielectric has to be chosen accordingly. Figure 7.7 shows the variation in filed and charges due to different gap heights.

The spring constant of the beam has a greater role in electrostatic force, hold-on voltage and charge storage. The spring constant of the beam depends on mechanical properties as given [1]

$$K = 32E'w\left(\frac{t}{l}\right)^3 + 8\sigma_0(1 - v)w\left(\frac{t}{l}\right) \tag{7.6}$$

$$\text{and} \quad E' = E\left(1 - v^2\right) \tag{7.7}$$

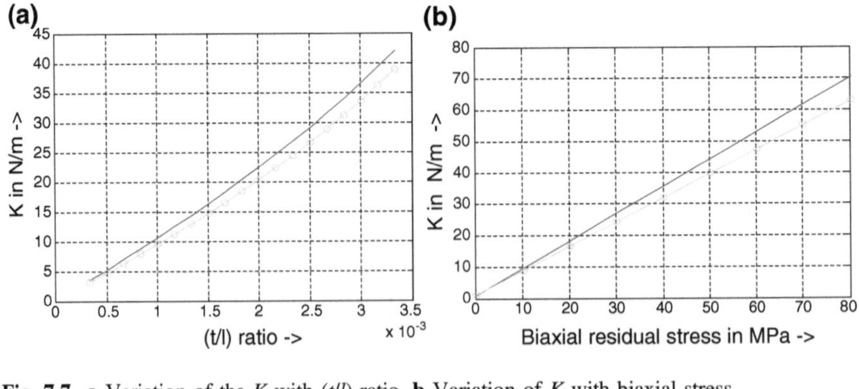

Fig. 7.7 **a** Variation of the K with (t/l) ratio. **b** Variation of K with biaxial stress

Where, E, ν, σ_0 are Young's modulus of the beam material, Poisson's ratio of the material and the biaxial residual stress of the structure. The variation in spring constant with thickness-to-length ratio and residual stress are shown in Fig. 7.7a, b, respectively, for both gold and aluminium beams.

Due to pulling-down of the membrane, a compressive stress is developed within the structure, which may lead to buckling of the membrane. Before buckling, the critical stress is estimated as (Figs. 7.8 and 7.9).

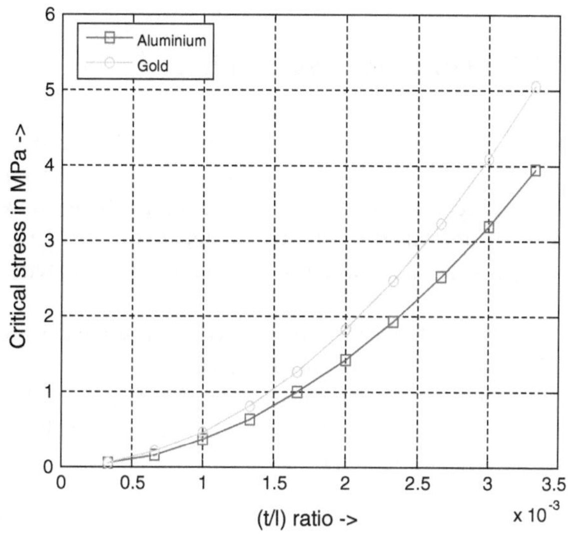

Fig. 7.8 Variation of the critical stress with (t/l) ratio

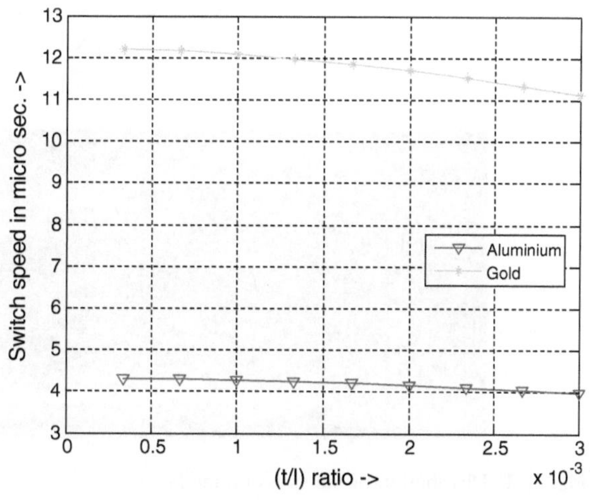

Fig. 7.9 Variation of the switching speed

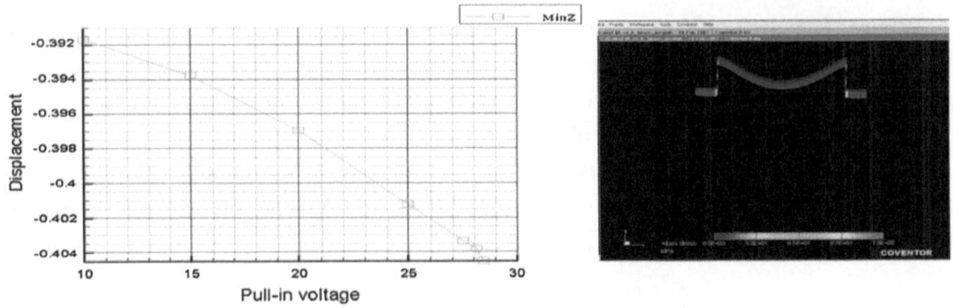

Fig. 7.10 Graphical and pictorial representation of the pull-in phenomena for the beam

7.1.3 Electromagnetic Analysis

The FEM-based software, Coventorware [4], has been used to analyse the hysteresis and pull-down voltage of the switch. The schematic is constructed with the foundry details using the typical parameters as shown in Table 7.1. The pull-in phenomenon has been shown graphically in Fig. 7.10. It reveals that the pull-down voltage is 28.75 V, which is very close to its analytical MATLAB predicted value (30 V).

Table 7.2 shows the close relevance of some design parameters derived from the analytical method and simulated value.

The majority of the stresses are on the anchors. Distribution of the Mises stress on the beam structure after applying different levels of uniformly distributed load is highlighted in Fig. 7.11.

Table 7.2 Analytical and simulated results

Parameters	Analytical method	Simulated result
Actuation voltage (V)	30	28.75
Electrostatic force (μN)	6	5.05

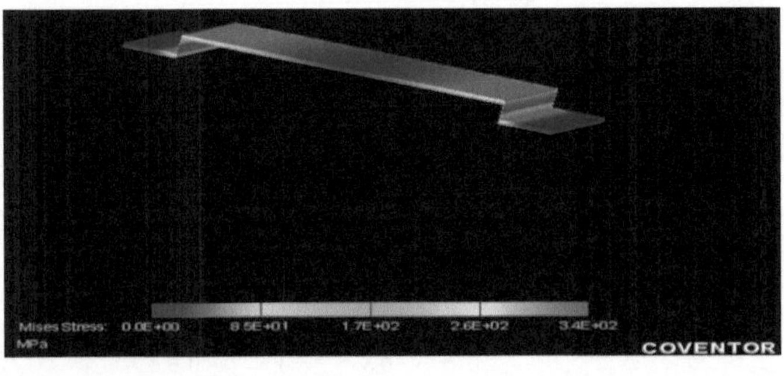

Fig. 7.11 Distribution of the stress on the beam

The hysteresis characteristic of the switch is also analysed by observing the effects of an increasing and decreasing voltage ramp on the switch response.

7.2 Conclusion

This paper analysed the variation of pull-down voltage, hysteresis phenomena, power handling etc. for different switch dimensions. The lifetime phenomena associated with the dielectric charging is also detailed. The parameters have to be chosen very carefully as large capacitance ratio and large pullout and pull-in voltages are less sensitive to a given dielectric charging condition. Both MATLAB and full wave simulations have been used and shown to have nearby values. The membranes of gold and aluminium have been used for the analysis and it was found out that major difference in selection comes through switching speed criteria. The main feature of the analysed switch is its compatibility with the CMOS foundry. The authors feel that this analysis will provide useful database for the designers and engineers involved in this area. Further electromagnetic analysis has also been done for the above-mentioned switch.

Acknowledgments The authors are thankful for the suggestions and help of Mr. A. Karmakar, ISRO.

References

1. G.M. Rebeiz, *RF MEMS Theory, Design, and Technology* (Wiley, New York, 2003)
2. K.C. Gupta, R. Garg, I. Bahl, *Microstrip Lines and Slot Lines*, 2nd edn. (Artech House, Norwood, 1996)
3. C. Goldsmith, J. Ehmke et al., Lifetime characterization of capacitive RF-MEMS switches. IEEE MTT-S Digest **2001**, 227–230 (2001)
4. (www.coventor.com/rf)

Chapter 8
Design of Microstrip Lowpass Filter in Combination with Defected Ground and Defected Microstrip Structures

P. Mondal, H. Dey and S.K. Parui

Abstract An asymmetric defected ground structure (DGS) in combination with back to back connected two spiral Defected microstrip structure (DMS) cells is proposed here to design of a lowpass filter with elliptic roll-off. The unit cell of DGS unit consists of one rectangular slot and two square shaped slots connected by two thin slots underneath the microstrip line on the ground plane. The passband insertion loss has been improved to 0.2 dB using High–Low impedance line. Both IE3D simulated and measured S-parameter responses are found a very good agreement. Finally, a lowpass filter at 3 dB cutoff frequency at 2.5 GHz with a 20 dB rejection bandwidth of 40 % and sharpness factor of 60 dB/GHz has been achieved.

Keywords Lowpass filter · DGS · DMS · Hi–lo line · Elliptic response

8.1 Introduction

A defected structure etched on the ground metallic plane of a microstrip line is an attractive way for having finite pass band, rejection band and also slow-wave characteristics. The defected structure on the ground effectively disturbs the shield current distribution on the ground plane and thus, introduces high line inductance and capacitance of the microstrip line. Thus, the structure produces wide stopband with reduce size. Dumb-bell shaped Defected structure is traversed first time by Ahn and applied to design a lowpass filter [1, 2]. Unit cell was delineated as a

P. Mondal (✉) · H. Dey · S.K. Parui
Department of Electronics and Telecommunication Engineering,
Indian Institute of Engineering Science and Technology, Shibpur,
Howrah 711103, India
e-mail: pratik665@gmail.com

S.K. Parui
e-mail: susanta_p@telecom.becs.ac.in

© Springer India 2015

K. Maharatna et al. (eds.), *Computational Advancement in Communication Circuits and Systems*, Lecture Notes in Electrical Engineering 335,
DOI 10.1007/978-81-322-2274-3_8

one-pole Butterworth type filter, where the capacitances are created from the gap and so as the inductances from the loop. As compared to all pole filters, the obvious knowledge of a filter with attenuation poles and zeros are at finite frequencies, thus showing higher selectivity. Such defected ground structures (DGS) with quasi-elliptical response and asymmetric nature was proposed in recent time [3–5]. The radiation in distinction to the ground plane is the major compel to design DGS based circuit. A high selective filter would be favourable owing to the requirement for recently expanding communication systems in reach of finite spectrum resources. A DGS filter with quasi-elliptical response having high selectivity was proposed first by Chen et al. [3].

The microstrip line is always a good transmission line. The microstrip line performance may also be improved by etching signal plane geometry and such structures are widely known as Defected Microstrip Structure (DMS). The impedance of microstrip line is increased due to the discontinuities introduced in the path of EM-wave which has been raised by the slow wave factor of a defected microstrip line. This phenomenon may be used to bring down the size of planar passive circuits like coupled lines, microstrip-line length and microstrip antennas. Thus filtering characteristics improves, which helps to meet emerging applications and challenges.

In this paper, an asymmetric DGS underneath the microstrip transmission line is proposed. Its unit cell consists of one rectangular slot and two square slots connected by two thin slots underneath microstrip line on the ground plane. It was observed that the asymmetric DGS produces 3rd order elliptical response and thus, yields very sharp filtering characteristics. A spiral DMS shows one pole band rejection characteristics with higher attenuation around −30 dB. Two spiral DMS cells cascaded back to back on the signal plane of the microstrip line yield two pole response with close pole frequencies. Finally a lowpass filter has been realized using proposed DGS unit in ground plane in combination with pair of proposed DMS units on the signal plane which provide a wide and deep stopband. The microstrip line has been replaced by hi–lo impedance line to reduce insertion loss in the passband.

8.2 Frequency Characteristics of Proposed Asymmetrical DGS

Defected Ground Structure (DGS) is an etched lattice shape, which locates on the ground plane. DGS are of increasing importance in filter applications. Figure 8.1a shows the schematic diagram of an investigated asymmetric DGS unit pattern consists of one rectangular slot and two square slots connected by two thin slots underneath microstrip line on the ground plane. Thin transverse slots underneath microstrip line increases effective capacitance value, whereas rectangular/square slots provide effective inductive loading.

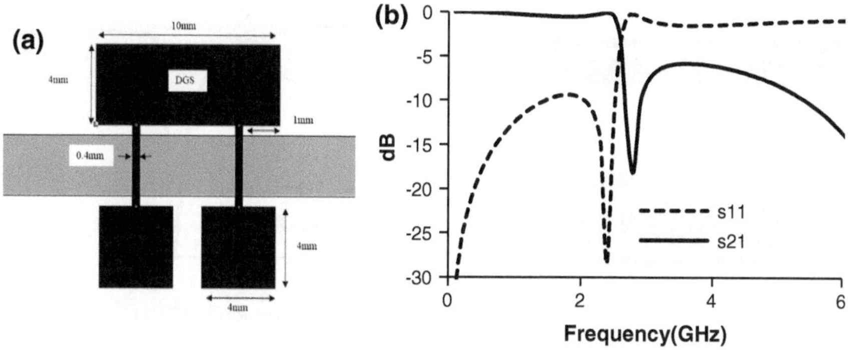

Fig. 8.1 **a** Proposed asymmetric DGS structure, **b** simulated S-parameters

In order to investigate the frequency characteristics of the proposed DGS unit, different dimensions are taking as: upper rectangle is of length 10 mm and width of 4 mm, the lower square is of 4 mm × 4 mm, and for the connecting slots of 4 mm × 0.4 mm as shown in Fig. 8.1a. The FR4 substrate with dielectric constant of 4.4, height of 1.59 mm and loss tangent of 0.02 is considered here for the design. Width of the microstrip line is found to be 3 mm corresponding to 50 Ω characteristics impedance. The structure is simulated by MOM based IE3D E-M Simulator. The simulated S-parameter results show a 3rd order lowpass filter with elliptic response and having pole frequency at 2.8 GHz and 3 dB cutoff frequency at 2.5 GHz.

8.3 Frequency Characteristics of Proposed Both Sided Symmetrical Spiral DMS

The line inductance and capacitance are added due to the disturbances created in the shield current distribution which has been made by the defected structure on the signal plane of a microstrip line. These new brand of slow-wave structures are called DMS. The schematic diagram of proposed spiral DMS cell and back to back connected DMS pair of cells are shown in Fig. 8.2a. The back to back connected DMS on the microstrip line shows a stop band with pole frequency at 3.6 GHz with a maximum attenuation of −40 dB as shown in Fig. 8.2b.

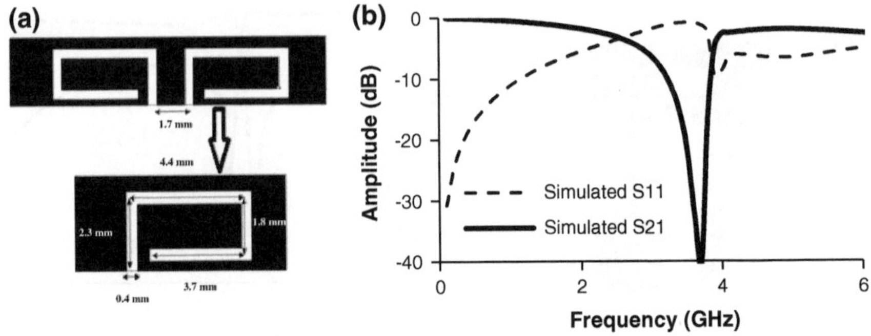

Fig. 8.2 **a** Proposed double sided symmetric spiral DMS structure, **b** simulated S-parameters

8.4 Lowpass Filter in Combination with DGS and DMS on a Microstrip Hi–Lo Line

One remarkable fact of using both the DGS and DMS under or on the microstrip line is that it can increase the overall characteristic impedance incorporating the additional effective inductance produced by the proposed structure. The inclusion of this ancillary inductance causes the characteristic impedance to be above from that of a stereotypical microstrip line having same line width. So, the microstrip line impedance go high in case of DGS and DMS use. As the matching ports are of 50 Ω line only, thus it can't match properly with the high impedance line. So, the

Fig. 8.3 **a** Schematic of the proposed lowpass filter, **b** simulated S-parameters for 50 Ω line and Hi–Lo line

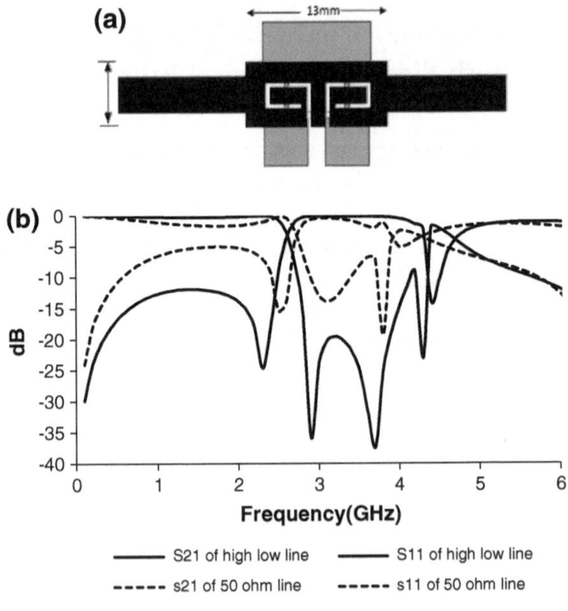

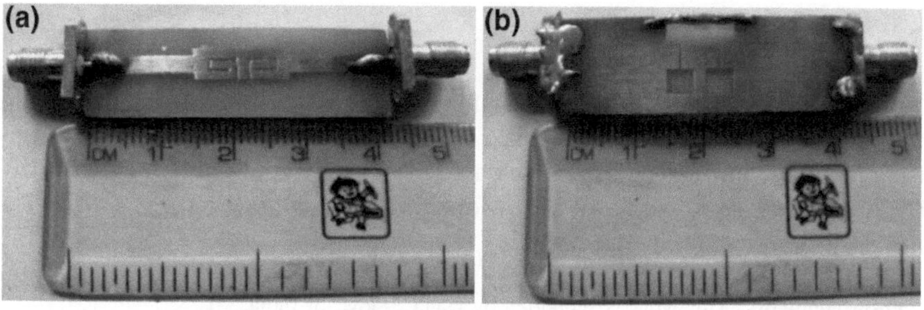

Fig. 8.4 Photographic view of the fabricated prototype of lowpass filter **a** front and **b** back

structure becomes lossy throughout the pass band. So, as to remove this problems we have used here hi–lo line as the signal plane instead of simple microstrip as shown in Fig. 8.3a. The shunt capacitances of high–low impedance line due to fringing effect improve the insertion losses in the passband.

The high–low line reduces the losses by 1.2 dB that was presented in the case of normal 50 Ω microstrip line as shown in Fig. 8.3b. The combination of DGS and DMS cells shows a 5 pole lowpass filter having 20 dB rejection bandwidth of 40 % and 3 dB cutoff at 2.5 GHz.

The prototype of the proposed lowpass filter has been fabricated on the FR4 substrate with dielectric constant of 4.4, height 1.59 mm and loss tangent of 0.02 as shown in Fig. 8.4. The fabricated unit has been measured using Agilent make vector network analyser (model N5230A). The measured result shows the 3 dB cutoff frequency at 2.4 GHz in comparison with the simulated result of 2.5 GHz. Also, the measured selectivity or roll-off of 60 dB/GHz is achieved in comparison with simulated result of 68 dB/GHz. The measured 20 dB rejection bandwidth is 45 % with respect to simulated values of 40 % (Fig. 8.5).

Fig. 8.5 Simulated and measured S-parameters of the proposed filter

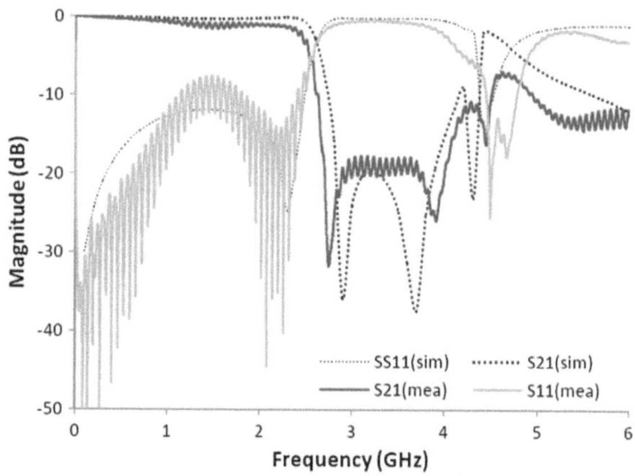

8.5 Conclusion

In this paper a new method of designing lowpass filter is proposed with the combination of DGS and DMS cells. The passband insertion loss is improved by using high–low impedance line and obtained the passband insertion loss of 0.6 dB. The filter offers 40 % fractional bandwidth (rejection) with cutoff at 2.4 GHz. The filter is very miniature in size and easy to fabricate.

Acknowledgments This work is supported by U.G.C, Govt. of India.

References

1. C.S. Kim, J.S. Park, D. Ahn, J.B. Lim, A novel one dimensional periodic defected ground structure for planar circuits. IEEE Microw. Guid. Wave Lett. **10**(4), 131–133 (2000)
2. D. Ahn, J.S. Park, C.S. Kim, J. Kim, Y. Qian, T. Itoh, A design of the lowpass filter using the novel microstrip defected ground structure. IEEE Trans. Microw. Theor. Tech. **49**(1), 86–93 (2001)
3. J.-X. Chen, J.-L. Li, K.-C. Wan, Q. Xue, Compact quasi-elliptic function filter based on defected ground structure, in *IEEE Proceedings-Microwave Antennas propagation*, vol. 153, no. 4 (2006) pp. 320–324
4. S.K. Parui, S. Das, An asymmetric defected ground structures with elliptical response and its application as a lowpass filter. Int. J. Electron. Commun. (AEÜ) **63**, 483–490 (2009)
5. T. Moyra, S.K. Parui, S. Das, Application of a defected ground structure and alternative transmission Line for designing a quasi-elliptic lowpass filter and reduction of insertion loss. Int. J. RF Microw. Comput.-Aided Eng. **20**(6), 682–688 (2010)

Part II
Advances in Communication System

Chapter 9
Adaptive Power Control Scheme for the Cognitive Radio System Based on Receiver Sensitivity

Indu Bala, Manjit Singh Bhamrah, Vanita Rana, Neelu Jain and Ghanshyam Singh

Abstract Cognitive radio has been proposed as the most promising technology to overcome the spectrum scarcity problem by exploiting the vacant frequency band of assigned licensed spectra; though some tradeoffs exist to protect the licensed primary user communication and to enhance the throughput of unlicensed secondary cognitive user. Through spectrum sensing and power control scheme, this issue can be addressed well as it allows the cognitive user to access the licensed band when primary user is not using it. In most of the previous works, it has been assumed that for transmission power control, perfect channel state information is available beforehand to the cognitive transmitter. This assumption does not hold good in practical scenarios and moreover it is difficult to implement spectrum sensing in device. To address this issue, a sensing free power control scheme based on CR receiver sensitivity is proposed in this paper. It allows cognitive user to controls its transmission power adaptively by approximating the distance from primary through cognitive receiver sensitivity. Simulation results are presented to verify the theoretical analysis.

Keywords Cognitive radio · Signal to interference plus noise ratio · Transmission power · Receiver sensitivity

---

I. Bala (✉) · M.S. Bhamrah
ECE Department, Punjabi University, Patiala, India
e-mail: i.rana80@gmail.com

M.S. Bhamrah
e-mail: manjitsingh_manjit@rediffmail.com

V. Rana · N. Jain
ECE Department, PEC University, Chandigarh, India
e-mail: vanita.rana@juit.ac.in

N. Jain
e-mail: neelujain@pec.ac.in

G. Singh
ECE Department, JUIT, Solan, H.P, India
e-mail: ghanshyam.singh@juit.ac.in

© Springer India 2015

K. Maharatna et al. (eds.), *Computational Advancement in Communication Circuits and Systems*, Lecture Notes in Electrical Engineering 335,
DOI 10.1007/978-81-322-2274-3_9

9.1 Introduction

Cognitive Radio is an emerging wireless technology that has grabbed the attention of researchers, academician and telecommunication industries worldwide. Recent survey reports show that unlicensed frequency bands have become overcrowded from past few years with widespread acceptance of diverse wireless technology applications like smartphones, tablets, etc., whereas, licensed frequency bands are being used intermittently. To deal with the conflicts between unlicensed spectrum overuse and underutilization of licensed frequency bands, Cognitive Radio technology has been proposed as a viable solution to this scarcity problem [1]. The term Cognitive Radio was first introduced by J. Mitola and it refers to the radio that uses real-time interaction with its environment to determine transmitter parameters such as frequency, power and modulation scheme, etc. It allows unlicensed subscribers to access licensed frequency band with some power constraints such that minimum or no interference may be experienced by the primary user and its quality of service (QoS) may not get affected while sharing its spectrum with unlicensed secondary subscriber. Few performance criteria to evaluate QoS of secondary user have been enlisted in [2]. To minimize the interference from SU to PU and in order to have guaranteed QoS to the PU, transmitted power of SU must be controlled intelligently based on its distance from the licensed primary user [3].

The rest of the paper is organized as follows: In Sect. 9.2, literature review is given. System model is discussed in Sects. 9.3 and 9.4 which gives mathematical modelling for the proposed model followed by system parameters and simulation results in Sect. 9.5. Conclusion is given in Sect. 9.7.

9.2 Literature Survey

In [4], the effective transmission area for CR user has been approximated from geometrical point of view. Due to its highly irregular shape, it is difficult to apply power control scheme over it. In [5], a technique is described to approximate the distance between primary and secondary user by using sensing side information and to control its transmission power. The effectiveness of this approach has dependence upon the sensing data, i.e. how fast the data sensed and how accurate is the sensing data. Several literature studies have dealt with an issue of reducing sensing time [6] and sensing techniques such as energy detection, matched filter detection, cyclostationary feature detection, etc. [7]. Due to the ease in implementation, energy detection scheme is most popular. The performance of any spectrum sensing scheme is evaluated in terms of probability of false alarm, probability of detection and probability of missed detection, where the probability of missed detection may be defined as the probability that secondary or unlicensed user may not be able to detect the presence of primary user when in actual it is present. In such situations, secondary user transmission may cause severe interference to primary legacy user.

On contrary, probability of false alarm may be defined as the probability of detecting primary user by secondary when it is not present. This may lead to the spectrum underutilization and precious natural resource may be wasted. Generally, there exist a tradeoff between probability of the false alarm and probability of the missed detection based on sensing threshold.

Many researchers have proposed power control techniques based on spectrum sensing schemes and assumed that perfect channel state information (CSI) is available. All these assumptions do not hold good in practical scenarios due to time varying channel, feedback delay, quantization error and estimation errors [8, 9]. Moreover, due to the non-zero probability of the false alarm and missed detection and implementation complexity of the spectrum sensing [3], in this paper sensing free power control scheme has been proposed based on CR receiver sensitivity.

9.3 System Model

Figure 9.1 shows the proposed system model of cognitive ad hoc network. Red, Blue and Green triangles are representing cognitive radio terminal, primary transmitter and primary receiver, respectively. R represents maximum coverage area of primary transmitter. R_p is representing protected radius within which primary receiver will have guaranteed quality of services, and therefore no secondary user is allowed to communicate inside this region. For an area $\pi\left(R^2 - R_p^2\right)$, both primary and secondary users may coexist with some power constraints on secondary users.

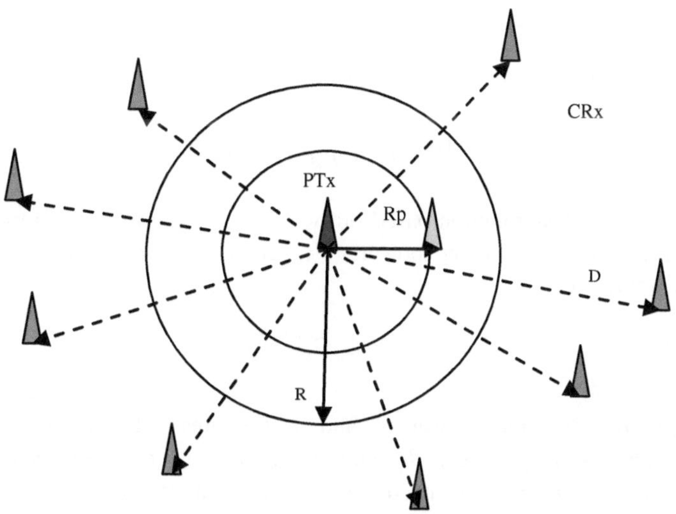

Fig. 9.1 Proposed system model

Since primary and secondary users are accessing same channel, legacy primary user will receive the interference from secondary cognitive user. Therefore, it has been assumed that if signal plus interference noise ratio (SINR) received at primary receiver is more than the minimum SINR required to interpret signal from primary transmitter, the secondary interference is acceptable. A worst-case scenario is considered when primary receiver is at the boundary of protected radius, i.e. R_p away from the primary transmitter where SINR for primary receiver is lowest. In this paper, overlay transmission scheme is considered for which cognitive user could present anywhere at a distance D away from the primary transmitter. In Fig. 9.1, all red triangles are representing the plausible positions of single CR user.

9.4 Mathematical Modelling for Proposed Power Control Scheme

For the spectrum sharing between legacy primary user and secondary user in same frequency band, SINR of primary receiver δ_p must remain all the time higher than $\delta_{\min}$ [5]. If P_p is transmitted power of primary user, P_s is secondary transmitted power, P_n is noise power signal to noise plus interference power δ_p may be defined as:

$$\delta_p = {P_p f(R_p)}/{P_n + P_s f(D)} \tag{9.1}$$

For both types of users to coexist

$$\delta_p \geq \delta_{\min} \tag{9.2}$$

Using (9.2), (9.1) could be written as

$$P_p f(R_p)/P_n + P_s f(D) \geq 10^{\frac{\delta_{\min}}{10}} \tag{9.3}$$

where $f(\cdot)$ is path-loss function based on two ray ground propagation model [10]. According to this, the received power P_r from transmitter situated at a distance r may be written as

$$P_r = P_p G_p G_s h_p^2 h_s^2 \Big/ r^n \tag{9.4}$$

where P_p is the transmitted power, G_p and G_s are gains of PU transmitter and CR receiver antennas, respectively; h_p and h_s are heights of antennas of the PU transmitter and CR receiver, respectively; r is the distance between transmitter and receiver and n is path loss factor. Without loss of generality, it has been assumed that gains of transmitting and receiving antennas are unity, heights of both antennas are equal and both types of users are experiencing same path loss. The value of path-loss exponent is assumed 3 until stated otherwise. Based on these assumptions, path-loss function in (9.3) may be defined as

$$f(x) == x^{-n} \text{ or } - 10n\log_{10}(x)[\text{dB}], \tag{9.5}$$

where x represents distance and n represents path loss exponent. Generally, it ranges between 2 and 4 for dense urban environment. Using (9.4), (9.3) become

$$\left[\left\{\left(P_p R_p^{-n}/10^{\frac{\delta_{\min}}{10}}\right) - P_n\right\}/P_s\right]^{\frac{-1}{n}} = D_{\min} \tag{9.6}$$

$$\left\{\left(P_p R_p^{-n}\Big/10^{\frac{\delta_{\min}}{10}}\right) - P_n\right\}D_{\min}^n = P_s \tag{9.7}$$

Minimum value of power received by SU receiver from PU transmitter when SU is positioned nearest to it (i.e. $D_{\min}$) will be given by

$$P_s = \min[P_s, P_{\max}], \tag{9.8}$$

where P_s calculated power is level from (9.7) and $P_{\max}$ is maximum allowed transmitted power for secondary user. Maximum transmission power limit has been imposed on secondary transmitter otherwise it could be health endangering situation. If we define a function $g(R_p, D_{\min})$ to decide the current power level, such that

$$g(R_p, D_{\min}) = P_p - 10n\log(R_p + D_{\min}) \tag{9.9}$$

Depending upon physical separation between primary receiver and secondary transmitter, two cases arise. *Case I:* If $g(R_p, D) > g(R_p, D_{\min})$ means secondary user is operating near to the primary transmitter (as shown in Fig. 9.2a), and therefore spectrum sharing is not possible. *Case II:* If $g(R_p, D) \leq g(R_{p,} D_{\min})$ means secondary user is operating away from primary transmitter (as shown in Fig. 9.2b), and therefore can transmit at power level calculated from (9.9).

Here, protected radius R_p is fixed and $D_{\min}$ is minimum distance calculated from the PU receiver position at the boundary of the protected radius when PU and SU receiver are positioned in straight line and SINR is just sufficient to provide guaranteed quality of service to the primary user. At a distance $D < D_{\min}$, the communication from secondary user must be strictly prohibited so that it may not deteriorate the service quality of primary. In other words, the existence of the optimal power control relies on R_p and $D_{\min}$ and therefore for optimal power control *Case I* must be avoided. Thus, for optimum power control, condition is (Fig. 9.3)

$$g(R_p, D) \leq g(R_{p,} D_{\min}) \tag{9.10}$$

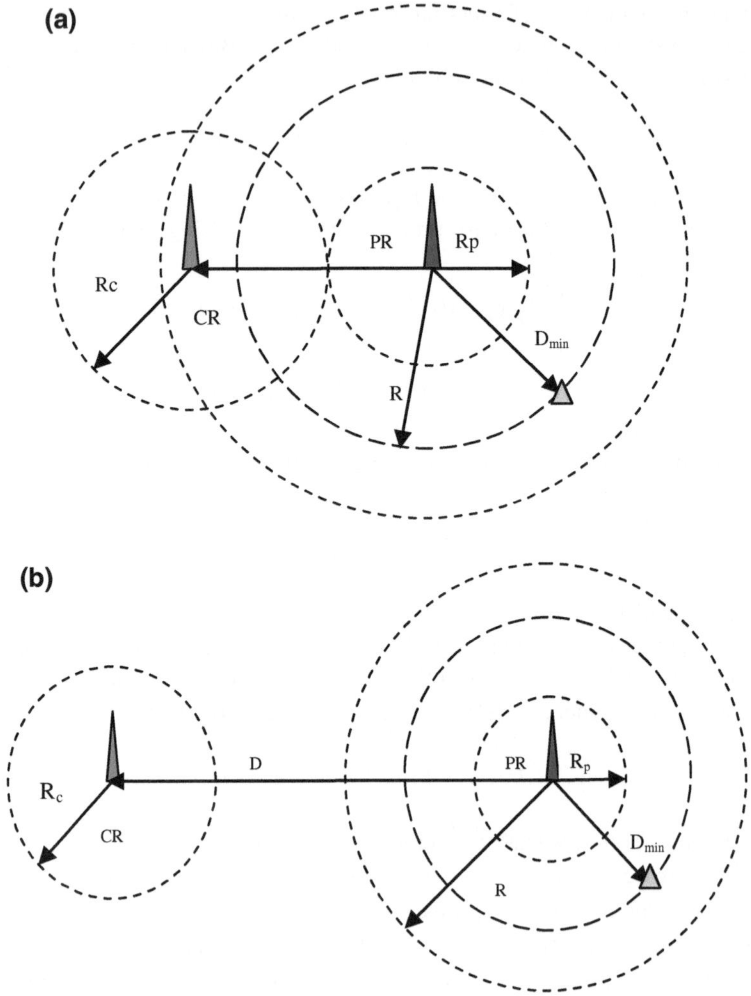

Fig. 9.2 **a** $g(R_p, D) > g(R_p, D_{\min})$ (Case I). **b** $g(R_p, D) \leq g(R_p, D_{\min})$ (Case II)

The maximum decodable radius of primary transmitter is R as shown in Fig. 9.1. Therefore, the distance between primary and secondary receivers can be calculated as

$$D = \sqrt{R_p^2 + R_1^2 + 2R_pR_1 \cos\theta}, \tag{9.11}$$

where θ is relative angle between PU and SU receivers.

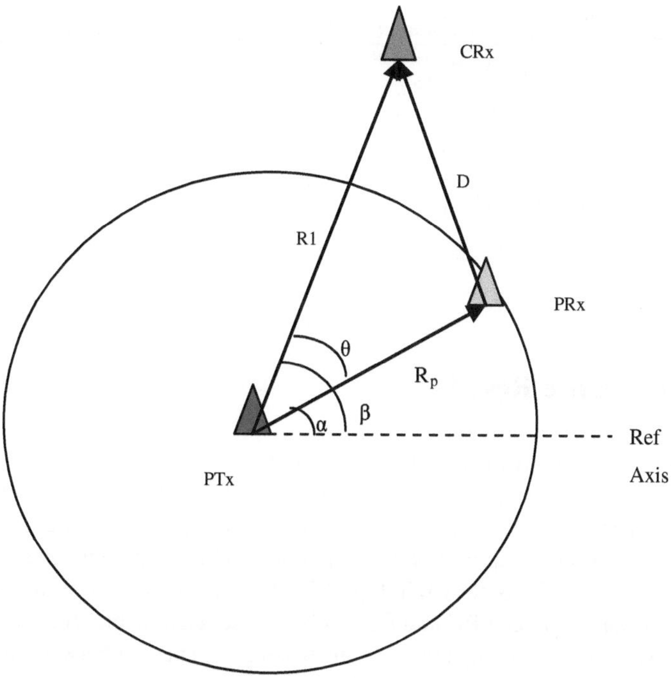

Fig. 9.3 Unknown distance calculation between PU and SU receiver

9.5 Shadowing and Fading Effect

In this section, the shadowing effect is considered which mainly occurs due to the shadowing by building and other objects. Under shadowing, the path loss can be written as [10]:

$$PL(d)[dB] = 10\log[P_0(d_0)] + 10\alpha\log\left(\frac{d}{d_0}\right) + X \quad (9.12)$$

Here, $PL(d)$ is the path loss between primary transmitter and secondary transceiver separated by distance d. d_0 is the reference distance; P_0 is the signal power at d_0 and X is the shadowing parameter described by a zero-mean Gaussian random variable with standard deviation σ which is location and distance dependent. Therefore, the received power at SU receiver is

$$g\left(R_p, D_{\min}\right)[dB] = P_p[dB] - PL(d)[dB] \quad (9.13)$$

Algorithm1:Transmission Power Control Algorithm of Proposed Scheme

```
Initialize R_p, P_p, P_s, SINR_p, R_1, α, β and P_max;
Calculate D_min and g(R_p, D_min) using (5) and (9);
Calculate D using (11);
If g(R_p, D) ≤ g(R_p, D_min)
  Calculate P_s using (7); // SU Transmission is possible
    P_Current = min(P_s, P_max);
Else
  Stop Transmission; // SU Transmission not possible
Endif
```

9.6 Performance Results

In this section, the performance of the proposed power control scheme is evaluated through simulation. The parameters used in simulation are listed in Table 9.1.

Figure 9.4 shows the variation in SU transmission power level corresponds to the minimum distance required between PU and SU for different threshold values without power control. As shown in Fig. 9.5, SU transmission power and minimum separable distance between PU and SU holds inverse relationship. It is clear that SU transmission power can be increased by increasing the spacing between PU and SU only. Figure 9.5 shows power received by secondary user from primary user (i.e. receiver sensitivity) and corresponding transmission power of secondary user. As expected, received primary power at SU location decreases as secondary user transmission power increases. In other words, SU receiver sensitivity must be increased as the SU transmission power increased. This is because, with an increase in secondary user transmission power, interference to the primary user also increases. Therefore, secondary receiver sensitivity must be increased to make detection of the weakest primary signal possible under noisy environment. In addition to this, it has also been observed that, receiver sensitivity must be increased as the minimum decodable SINR for the primary user increases.

Table 9.1 Simulation parameters

S. No.	Parameter	Symbol	Value
1.	Primary transmission power	P_p	30 dBm (1 W)
2.	Secondary transmission power	P_s	0–30 dBm (0.1–1 W)
3.	Shadow parameter	X	0 or 2 dBm
4.	Path loss factor	n	3
5.	Primary threshold	$\gamma_{\min}$	5 dBm
6.	Maximum transmission power for SU	$P_{\max}$	30 dBm (1 W)

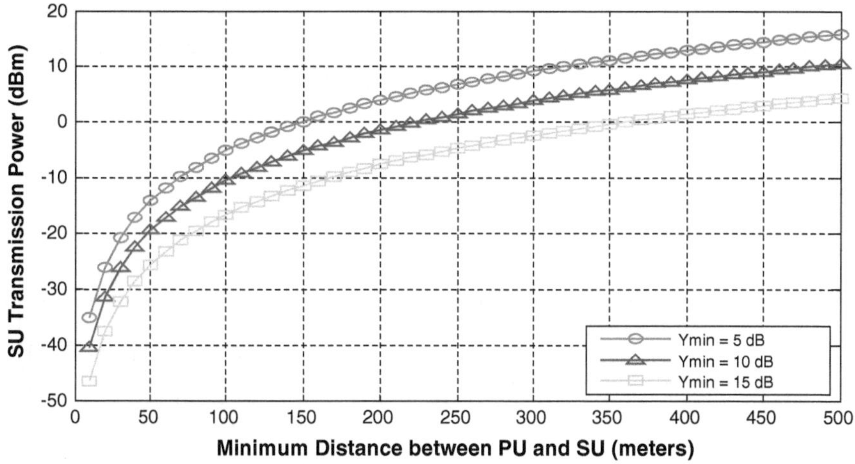

Fig. 9.4 SU transmission power versus $D_{\min}$ without power control

Figure 9.6 shows secondary user transmission power with respect to the distance between incumbent primary user and secondary user. The impact of minimum detectable SINR has also been observed. It is clear from the graph that CR transmitter power increases as it is moving away from the primary user location. Figure 9.7 shows the plane of optimum power of cognitive transmitter for the proposed power control algorithm. It can be observed that optimum transmission power for SU remains constant, i.e. maximum 30 dB for certain range of θ when the separation between primary transmitter and cognitive receiver is far greater than from R_p. Therefore, to avoid health endangering situation, highest transmit power level is restricted to the maximum value of 30 dB. Similarly, an optimum power for the

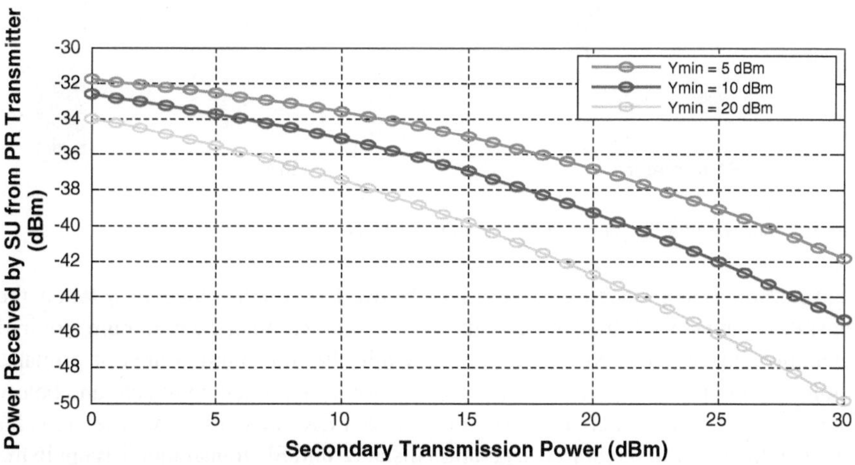

Fig. 9.5 Receiver sensitivity versus SU transmission power

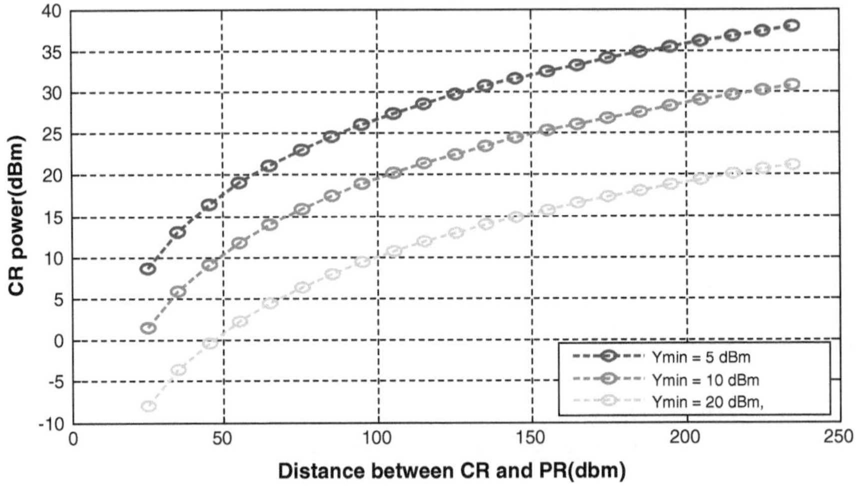

Fig. 9.6 SU transmission power versus distance between CR and PR

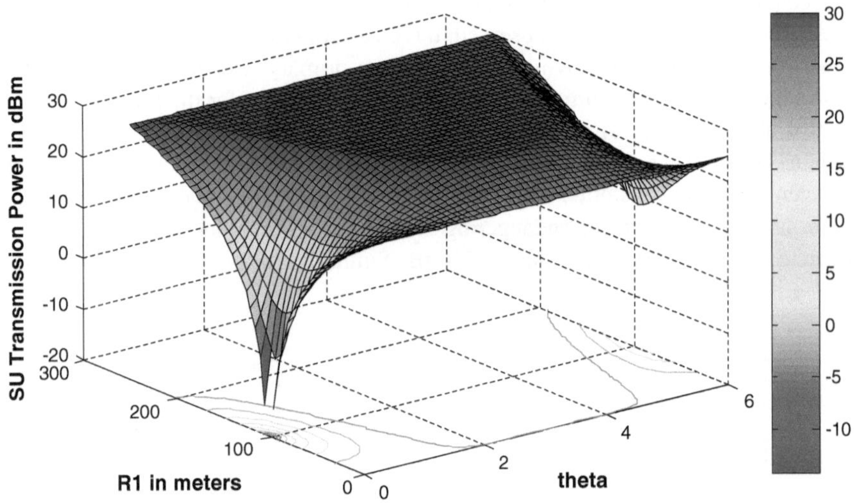

Fig. 9.7 Surface plot for proposed power control scheme

transmission is approaching to zero and below for certain range of θ when the separation between CR transmitter and receiver is far less than the minimum required separation distance and SU transmitter is within the transmission area of primary. Similarly, an optimum power for the transmission is approaching to zero and below for certain range of θ when the separation between CR transmitter and receiver is far less than the minimum required separation distance and SU transmitter is lying in the transmission area of primary.

9.7 Conclusion

In this paper, sensing free, adaptive power control scheme for cognitive radio networks is proposed for SUs in which the SU makes transmission power level decision based on the receiver sensitivity. The distance between PU and SU is approximated through SU receiver sensitivity. Simulation results are verifying theoretical results that the optimum power control for cognitive user can only be achieved if and only if the separation between PU and SU is significantly large and the same can be approximated through SU receiver sensitivity, i.e. the minimum detectable power by SU from PU. The idea proposed in this paper can be extended further and generalized for multiple secondary users. A distributed power control scheme and multiple access protocol could be designed jointly to make cognitive radio network much more robust and efficient.

References

1. I.F. Akyildiz, W.Y. Lee, M.C. Vuran, S. Mohanty, Next generation/dynamic spectrum access/cognitive radio wireless networks: a survey. Comput. Netw. **50**, 2127–2159 (2006)
2. V. Rana, I. Bala, N. Jain, Resource allocation models for cognitive radio networks: a study. Int. J. Comput. Appl. **91**(12), 51–55 (2014)
3. I. Bala, M.S. Bhamrah, G. Singh, Analytical modeling of ad hoc cognitive radio environment for optimum power control. Int. J. Comput. Appl. **92**(7), 19–22 (2014)
4. L.C. Wang, A. Chen, Effect of location awareness on concurrent transmission for cognitive ad-hoc networks overlaying infrastructure based systems. IEEE Trans. Mobile Comput. **8**(5), 577–589 (2009)
5. K. Hamadi, W. Zhang, K. B. Letaief, Power control in cognitive radio systems based on spectrum sensing side information, in *Proceedings of IEEE Conference on Communication* (2007), pp. 5161–5165
6. Y.C. Liang, Y. Zeng, E.C.Y. Peh, A.T. Hoang, Sensing-throughput tradeoff for cognitive radio networks. IEEE Trans. Wirel. Commun. **7**(4), 1326–1337 (2008)
7. F.F. Digham, M.S. Alouini, M.K. Simon, On the energy detection of unknown signals over fading channels. IEEE Trans. Commun. **55**(1), 21–24 (2007)
8. A. Bagayoko, I. Fijalkow, P. Tortelier, Power control of spectrum-sharing in fading environment with partial channel state information. IEEE Trans. Signal Process. **59**(5), 2244–2256 (2011)
9. R. Zhang, Optimal power control over fading cognitive radio channel by exploiting primary user CSI, in *Proceedings of IEEE GLOBECOM* 2008, pp.1–5
10. T.S. Rappaport, *Wireless Communications: Principles and Practice*, 2nd edn. (Prentice Hall, 2002)

Chapter 10
A Random Bit Generator Using Rössler Chaotic System

Chayan Banerjee, Debanjana Datta and Debarshi Datta

Abstract In this paper, a random bit generator is proposed which utilizes the chaotic nature of a Rossler System. The generator is modelled and the generated random sequence is tested using statistical tests like autocorrelation test, runs test and monobit test. The sequence shows a high degree of randomness, besides being simple for real-world implementation.

Keywords Chaotic system · Random bit generator · Rossler system

10.1 Introduction

Nowadays with the increase in dependence on the Internet and telecommunication systems, the security of the transmitted information has become a major issue of concern. Cryptographic techniques are thereby used to enhance the security of these systems. Random numbers play a very essential role in enhancing cryptographic systems. A random generator may be defined as an algorithm which generates a stream of unbiased binary digits or shows randomness. Random bit generators are also used in fields like computer simulation, numerical analysis, transfer of data via channel coding and watermarking [1, 2]. The classical method of generating random numbers is by employing Linear Feedback Shift Registers. But alternatively, chaotic systems can also be used for designing random bit generators. Chaotic systems possess certain interesting features like sensitivity to initial conditions, ergodicity and mixing (stretching and folding) [3], which are quite desirable

C. Banerjee (✉) · D. Datta · D. Datta
Department of ECE, Brainware Group of Institutions, Barasat, West Bengal, India
e-mail: cbanerjee929@gmail.com

D. Datta
e-mail: debanjana_08@yahoo.com

D. Datta
e-mail: debarshidatta7@gmail.com

© Springer India 2015

K. Maharatna et al. (eds.), *Computational Advancement in Communication Circuits and Systems*, Lecture Notes in Electrical Engineering 335,
DOI 10.1007/978-81-322-2274-3_10

characteristics for developing RNGs and encryption systems. There are certain specifications of a chaotic dynamic system [4–6].

1. Chaotic movement is neither periodic nor convergent.
2. The output of the chaotic system is very erratic and resembles random noise
3. Chaotic systems are very sensitive to their initial state or conditions. A partial change to their initial state causes a big difference in the outputs of the chaotic system.
4. Outputs of a chaotic system are unpredictable.

There are a lot of literatures [3, 7–9] that proposes the application of chaotic systems or chaotic maps in generation of pseudo random bitstream. Most of these RNGs employ two simple chaotic maps and XOR the generated sequence from the two different chaotic functions. This XOR process increases the complexity of the sequences and makes it hard for the attacker to extract sensitive information from the sequences [10]. In dynamical systems, a map denotes an evolution function that creates the system. Maps may be parameterized by a discrete-time or a continuous-time parameter. Discrete maps usually take the form of iterated functions. Chaotic maps may also be classified according to their space dimensions. [10] uses two, 1D chaotic maps whereas [3, 11] use a single 1D chaotic map (tent map) for random binary string generation. 3D chaotic maps are also used [12, 13] (Lorentz attractor and Chua's circuit) for cryptographic applications and for generation of random bit sequences.

10.2 Rossler System

Rossler System is a chaotic system comprising of three nonlinear ordinary differential equations. The equations define a continuous-time dynamical system that shows chaotic dynamics. Rossler attractor is a chaotic attractor solution to the following system

$$\dot{x} = -y - z \tag{10.1}$$

$$\dot{y} = x + ay \tag{10.2}$$

$$\dot{z} = b + z(x - c) \tag{10.3}$$

This system is minimal for continuous chaos for at least three reasons: its phase space has the minimal dimension 3, its nonlinearity is minimal because there is a single quadratic term and lastly it generates a chaotic attractor with a single lobe, in contrast to the Lorenz attractor which has two lobes. In Eq. (10.1), (x, y, z) are the three variables that evolve in the continuous time t and (a, b, c) are the three parameters. The linear terms of the two first equations create oscillations in the variables x and y. These oscillations can be amplified if $a > 0$, which results into a

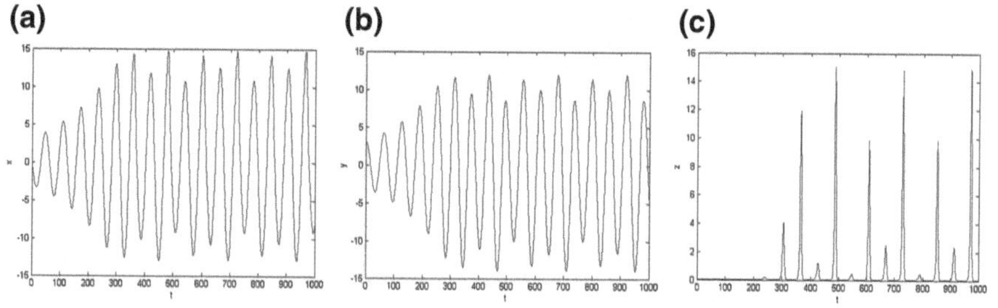

Fig. 10.1 Plots showing the solutions to the system of differential equations, x, y and z plotted against time

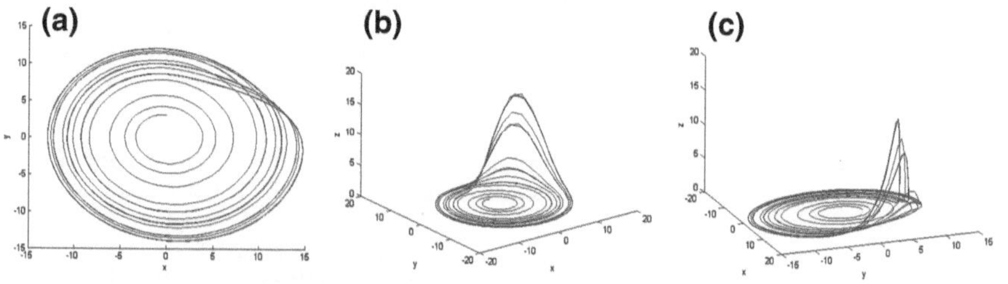

Fig. 10.2 Plot showing the chaotic nature of the system. **a** Plots x and y, **b**, **c** shows the 3D plots of the system with x, y, and z

spiraling-out motion. The motion in x and y is then coupled to the z variable ruled by the third equation, which contains the nonlinear term and which induces the reinjection back to the beginning of the spiraling-out motion [14], see Figs. 10.1 and 10.2.

10.3 Rossler-Based Pseudo Random Number Generator (RPRNG)

The random number generator is based on the solutions to the three differential equations, namely x, y and z. Each of these continuous outputs is sampled and the sequences of integers are generated. From the sampled values only positive integers are retained A $\times$ 10^3 multiplier, works to convert the samples into large integer values. In the next step, the integers are rounded off to the nearest zero and then these series of the integers are converted to binary bits and appended together to form the sequences of required size.

So now we have three binary sequences generated from the actual solutions x, y and z. These solutions directly depend on the initial values of the Rossler System, as is the general characteristic of a chaotic system. So at start, a and b are fed to the system as constant (both kept at 0.1) while c is kept constant only for a cycle and it may change in the next cycle so that the sequences generated over different cycles

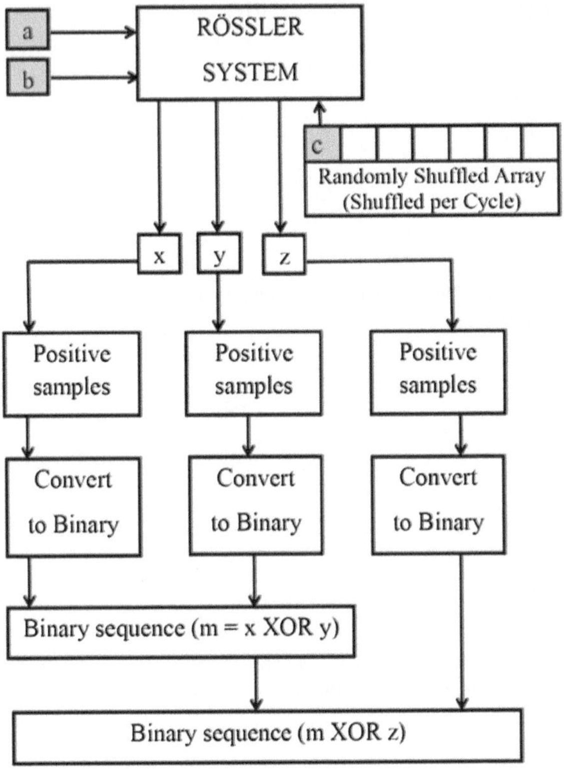

Fig. 10.3 The complete random sequence generation process

preserve a particular amount of variety (see Fig. 10.3). We have identified certain states or values of *c* (keeping *a*, *b* constant) where the Rossler behaves truly chaotic, e.g. at $c = 9$, 10, 13, 18, etc. So for deciding on the value of *c* for a particular cycle, we have kept a 'Randomly shuffled array' [15] which is shuffled every cycle and the first value after each shuffle is forwarded to the Rossler system as the c value for that cycle. As the next step, the two binary sequences belonging to *x* and *y* are XORed together and the result '*m*' is saved. Next the *z* sequence is also XORed with the previous result *m*. The final sequence thus generated is the pseudorandom sequence generated with aid of the Rossler System. Figure 10.3, represents the process as described above.

Figure 10.4 shows the SIMULINK model of the Rossler System that has been used for generation of the random sequence. The model can be explained in relation to the Rossler System equations, mentioned in the previous section. Figure 10.5 is the random sequence (140 bits) generated after going through the complete process as described in Fig. 10.3. The sequence is generated by taking the initial values to the Rossler system as $a = b = 0.1$ and $c = 9$.

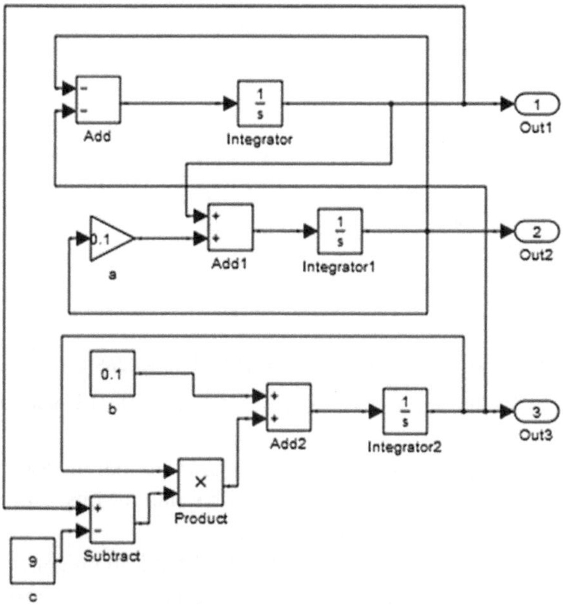

Fig. 10.4 Simulink model of the Rossler system

0 0 1 0 0 0 0 0 0 0 1 1 0 1 0 1 0 0 1 0 0 0 0 0 0

0 1 1 0 0 0 0 0 0 1 0 0 1 0 0 0 0 1 1 0 0 0 1 0 0

0 1 0 1 0 0 0 0 0 0 0 1 0 1 0 1 1 1 0 1 0 0 0 1 1

1 0 0 0 0 0 1 0 0 0 0 0 0 0 1 0 0 0 0 0 0 0 0 0 0

0 1 0 1 0 1 1 0 0 0 0 0 0 0 1 1 0 0 0 0 0 1 0 0 0

0 0 0 0 0 1 1 0 0 1 0 1 1 1 1

Fig. 10.5 Random sequence generated, $a = b = 0.1$, $c = 9$, length = 140 bits

10.4 Analysis of Randomness and Conclusion

Statistical tests are performed to check the randomness of the sequence. The results of the tests are mentioned as follows.

Runs Test: Success, with a p-value = 0.4932

Autocorrelation and partial Autocorrelation: Pass, see Fig. 10.6

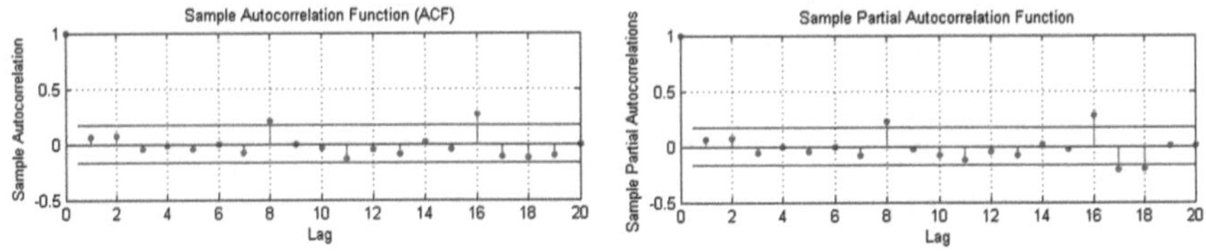

Fig. 10.6 ACF and PACF plots

Monobit Test:

$$\begin{aligned}\text{Bits}(n) &= 10,\ s_{10} = 0,\ s_{\text{obs}} = 0,\ p = 1,\ \text{success}\\ &= 40,\ s_{40} = -16,\ s_{\text{obs}} = 2.52,\ p = 0.012,\ \text{success.}\end{aligned}$$

From the basic tests performed, it has been observed that in case of Monobit test the sequence fails as the bit length is increased beyond 40 bits, while the sequence shows perfect randomness for bit lengths of around 10.

Thus the sequence generated from the proposed random bit generator is found to have good randomness. The advantages of the generator are sensitivity to initial inputs, different sequences for different cycles (using randomly shuffled array) and simplicity for implementation and the portability architecture (only positive integers used).

References

1. M.E. Yalcin, J.A.K. Suykens, J. Vandewalle, A double scroll based true random bit generator, in *Proceedings of the 2004 International Symposium on Circuits and Systems, 2004*. ISCAS'04, vol. 4. IEEE (2004)
2. X. Wang, C. Yu, Cryptanalysis and improvement on a cryptosystem based on a chaotic map. Comput. Math. Appl. **57**(3), 476–482 (2009)
3. N.K. Pareek, V. Patidar, K.K. Sud, A random bit generator using chaotic maps. Int. J. Netw. Secur. **10**(1), 32–38 (2010)
4. A.J. Menezes, P.C. Van Oorschot, S.A. Vanstone, *Handbook of Applied Cryptography* (CRC press, New York, 2010)
5. T. Stojanovski, L. Kocarev, Chaos-based random number generators-part I: analysis [cryptography]. IEEE Trans. Circuits Syst. I: Fundam. Theor. Appl. **48**(3):281–288 (2001)
6. G. Alvarez, S. Li, Some basic cryptographic requirements for chaos-based cryptosystems. Int. J. Bifurcat. Chaos **16**(08), 2129–2151 (2006)
7. H.P. Hu, L.F. Liu, N.D. Ding, Pseudorandom sequence generator based on the Chen chaotic system. Comput. Phys. Commun. **184**(3), 765–768 (2013)
8. A.-C. Dsscslescu, R.E. Boriga, A.-V. Diaconu, Study of a new chaotic dynamical system and its usage in a novel pseudorandom bit generator. Math Probl Eng (2013)
9. A.B.O. López et al., *Trident, a New Pseudo Random Number Generator Based on Coupled Chaotic Maps*. Computational Intelligence in Security for Information Systems 2010 (Springer, Berlin, 2010) pp. 183–190

10. M. Francois, T. Grosges, D. Barchiesi, R. Erra, A new pseudo-random number generator based on two chaotic maps. Informatica **24**(2), 181–197 (2013)
11. X. Wang, X. Qin, L. Teng, A novel true random number generator based on mouse movement and a one-dimensional chaotic map. Math. Probl. Eng. (2012)
12. W. Yu, G. Bai. A novel random number generator based on continuous-time chaos, in *2010 2nd IEEE International Conference on Network Infrastructure and Digital Content*. IEEE (2010)
13. L. Merah, A. Ali-Pacha, N.H. Said, M. Mamat, A pseudo random number generator based on the chaotic system of Chua's circuit, and its real time FPGA implementation. Appl. Math. Sci. **7**(55), 2719–2734 (2013)
14. P. Gaspard, Rössler systems. *Encyclopedia of Nonlinear Science*, pp. 808–811 (2005)
15. D.E. Knuth, *The Art of Computer Programming*, vol. 2 (Addison-Wesley, Reading, Mass, 1969)

Chapter 11
Optimized WiMAX Network Development in India: Specification and Implementation

Angana Chakraborty, Sajal Saha, Indrajit Banerjee and Arnab Gupta

Abstract WiMAX became popular due to simplicity of installation and cost reduction compared with traditional DSL cable. It is very important to provide Quality of Service (QoS) to end user to guarantee diversifying requirements of different applications in WiMAX. The system requirement of next-generation mobile WiMAX is supposed to be based on IEEE 802.16 m which is still in letter ballot stage specifically considering Indian environment. This paper presents an optimized architecture and scenario based on the recently standardized IEEE 802.16 m framework, integrating both mobility management and QoS in Indian scenario. Moreover, we analyse the QoS issues like throughput, uplink and downlink packet drop. OPNET-based simulation platform has been introduced for the verification of analytical model and results. The numerical results found in simulation can be used to select proper network configuration.

Keywords QoS service classes · Jitter · Throughput

A. Chakraborty (✉)
Indian Institute of Engineering Science and Technology (IIEST), Howrah 711103, India
e-mail: angana.chakraborty9@gmail.com

S. Saha
Narula Institute of Technology, 81, Nilgunj Road, Kolkata 700109, India
e-mail: sajalkrsaha@ieee.org

I. Banerjee
Department of IT, Indian Institute of Engineering Science and Technology (IIEST), Howrah 711103, India
e-mail: askforindra@gmail.com

A. Gupta
Department of Mathematics, Narula Institute of Technology, Kolkata 700109, India
e-mail: arnab_math@yahoo.co.in

© Springer India 2015

K. Maharatna et al. (eds.), *Computational Advancement in Communication Circuits and Systems*, Lecture Notes in Electrical Engineering 335,
DOI 10.1007/978-81-322-2274-3_11

11.1 Introduction

Ever-increasing demand for higher bandwidth cannot be met with the existing Digital Subscriber Line (DSL), cable Internet and other non-optical wired solutions requiring installation of transmission cables. Wireless technologies eliminate need for such installations. WiMAX is a solution for wireless broadband communication. WiMAX is flexible, robust, affordable and economically feasible and therefore assures to provide high bandwidth communication and networking solutions up to 48 (fixed downlink) and 7 Mbps (fixed uplink) even in the remote areas with a coverage area of 8 km [1]. Mobile WiMAX provides 9.4 Mbps for downlink and 3.3 Mbps for uplink across the coverage area of 3 km. Its high-speed data enables various multimedia applications along with the conventional telephony service.

The original WiMAX standard, IEEE 802.16 (2001), specified in the 10–66 GHz range for fixed and nomadic services has gradually [2] evolved to IEEE 802.16 m (2011) as shown in [3]. WiMAX operates in both licensed and unlicensed bands. Unlicensed band operates on 2.4 and 5.8 GHz bands. Licensed band operates on 700 MHz, 2.5 GHz. Bharat Sanchar Nigam Limited (BSNL) gave a trial run [4] of WiMAX in some metropolitan cities of India in 2011 but did not move further. The primary focus of this paper is to analyse the QoS issues of WiMAX network in Indian scenario and check technical viability.

11.2 Background Study

A group was formed by the industry and standardization forums and agencies in telecommunication such the Internet Engineering Task Force (IETF), 3rd Generation Partnership Project (3GPP2), DSL forum, Open Mobile Alliance (OMA) and Intel (more than 470 members till date). Aldmour [5] proposed some features of WIMAX and its future perspectives. Singh [6] illustrated benefits, drawbacks and applications of WiMAX. Etemad [7] and Ahmedi [8] illustrate the brief overview, technology, architecture, interface and network specifications of WiMAX and its evolution through the WiMAX forum. Chen et al. 9] proposed design and implementation of WiMAX module for ns-2 simulator.The main features of WiMAX are its high throughput and large coverage area [10, 11]. The QoS classes of WiMAX are, unsolicited grant service (UGS), real-time polling service (rtPS), non-real-time polling service (nrtPS) and best effort (BE). A different type of data uses a different type of QoS class to achieve the maximum throughput using the minimum bandwidth [12]. Scheduling of the four QoS classes are defined in IEEE 802.16 m for various applications. Khalil and Ksentini [13] and Jain and Verma [14] compared different scheduling services like Proportional Fair Scheduling algorithm, Cross-layer Scheduling Algorithm and TCP aware uplink Scheduling algorithm. Liu et al. [15] tries to optimize the resources in Cross-layer scheduling algorithm by providing fixed slots k to the UGS out of the n slots ($n > k$) of radio resources. ($n - k$) slots are allocated to the rtPS, nrtPS and are dynamically based on an algorithm defined in [10, 13, 15, 16]. Author in [4, 12] investigated the behaviour of different scheduling algorithms for OFDMA/MIMO-based WiMAX (IEEE 802.16 m) architecture and analyse the quality of service (QoS) issues like throughput, end-to-end delay, interference, signal-to-noise ratio, call blocking probability and call dropping probability using OPNET modeller for only 2 nodes.

11.3 WiMAX Architecture

Each WiMAX network architecture consists of two prime entities: ASN (Access Service Network) and CSN (Connectivity Service Network) as shown in Fig. 11.1. ASN provides the radio interfaces that connect MNs with the network. ASN handles the radio interface and contains the Base Station (BS). ASN gateway which has a one-to-many relationship with BS. ASN gateway handles the mobility between BS.

WiMAX architecture comes up with three ASN profile (profile A, B, C). Profile A controls overall handover procedures. The CSN is a set of network functions that provide IP connectivity to WiMAX MNs. The CSN contains gateways for internet

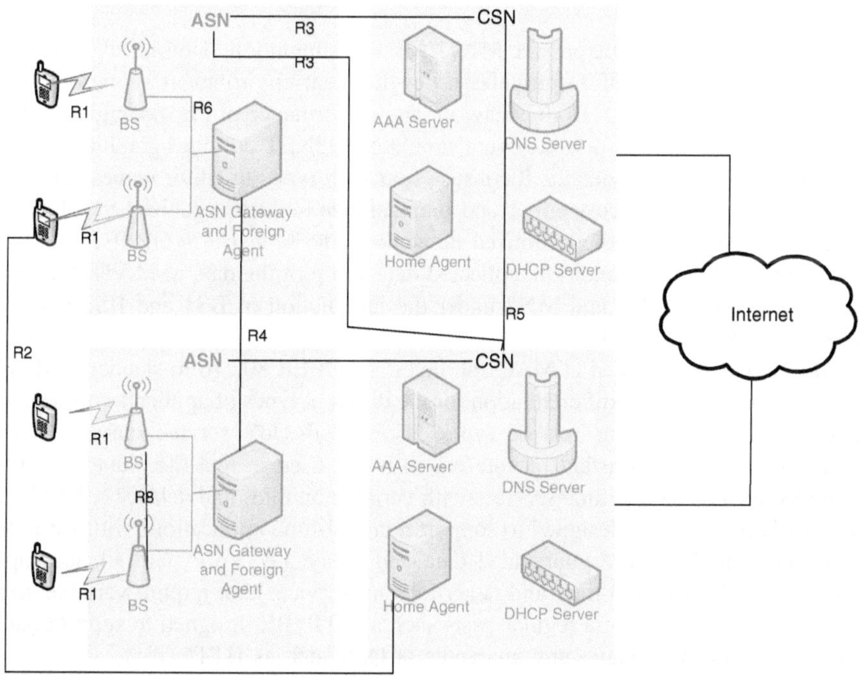

Fig. 11.1 IEEE 802.16 m network architecture developed by the WiMAX forum

Table 11.1 802.16 m interfaces

R1 interface	MN to BS connection, radio link
R2 interface	MN to HA connection, supports roaming
R3 interface	ASN to CSN connection, supports authentication, billing and MIP messages
R4 interface	ASN to ASN connection, defines mobility procedures when a MN crosses from one ASN to another ASN
R5 interface	CSN to CSN connection, supports roaming
R6 interface	MN to ASN connection, supports mobility messages
R7 interface	Internal to ASN gateway, not shown in Fig. 11.1
R8 interface	BS to BS connection, supports handoff

access, routers, servers or proxies for AAA, IP-allocation, user databases and internetworking devices. HA and AAA together provide authentication. AAA also provides billing records. CSN also handles admission and policy control, mobility between ASN and specific WiMAX services such as location-based services.

Interfaces are used to communicate between different entities. Specifications of the different interfaces are shown in Table 11.1.

11.4 Network Model

We used OPNET simulator and MATLAB for simulation. Optimized Network Engineering Tool (OPNET) provides a development environment supporting the communication networks. Both behaviour and performance of the modelled system can be analysed through discrete event simulator. OPNET provides graphical editor to enter network model details. It consists four such types of editor names network editor, node editor, process editor and parameterized editor organized in a hierarchical way. We design the optimized network model scenario in OPNET, run the model under different condition, collected data and plot the data in MATLAB. We set up two nodes MN1 and MN2 under the jurisdiction of BS1 and BS2 respectively as shown in Fig. 11.2.

QoS issues are handled at MAC sub layer. The IEEE 802.16 m slandered MAC sub layer provides QoS differentiation for the different types of applications through five different scheduling service types also called QoS service classes. UGS designed to support constant bit rate which is not used in real life situation. rtPS designed to support real time services with variable bit rates and delay, i.e. VoIP on a periodic basis. nrtPS designed to support non-real-time applications with variable data rates which require guaranteed data and delay, i.e. VoIP with silence suppression. nrtPS non-real-time and delay-tolerant services that require variable size data grant burst types on a regular basis such as FTP. BE designed to support data streams that do not require any guarantee in QoS such as HTTP.

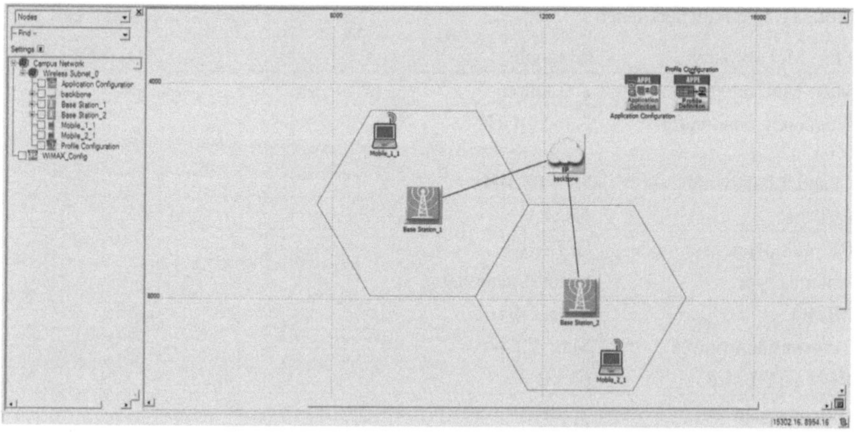

Fig. 11.2 WiMAX architecture

rtPS is taken as initial QoS service class to support real-time services with variable bit rates. We consider the following parameters as shown in Table 11.2.

11.5 Simulation Results and Analyses

Two mobile nodes Mobile_1_1 and Mobile_2_1 are located under Base stations BS_1 and BS_2, respectively.

Mobile_1_1 making a call under WiMAX environment with Mobile_2_1 using 4 Mbps as maximum available bandwidth and 0.5 Mbps as minimum available bandwidth.

Figures 11.3 and 11.4 show the throughput and voice packet end-to-end delay under static condition. Figure 11.3 shows the throughput to be 380 Kbps compared to GSM 9.6 Kbps voice data transfer.

Mobile_1_1 and Mobile_2_1 are moving following the node movement path as shown in Fig. 11.5.

Test cases has been taken into consideration and each of the cases we try to figure out the fluctuation in throughput and packet dropped at 60, 90, 100 kmph speed to define the complete agility of the system.

Urban area in India maximum vehicular speed limit is 60 kmph. We keep MN's speed at 60 kmph and run the simulation as shown in Fig. 11.6 and drop in throughput is found to be 26 % during handoff after 5 min which is below the breakeven point.

This speed is achieved by the express train in India. We observed that throughput drop is 19.67 % during the handoff as shown in Fig. 11.7.

Table 11.2 Parameters taken

Min. MN velocity	60 kmph	
Max. MN velocity	100 kmph	
Frequency bandwidth used	2.3–2.4 GHz	
Channel bandwidth	8.75/10 MHz	
FFT size	1,024	
RF multiple access mode	OFDMA	
Antenna type	Omni directional	
MIMO	Matrix 4/4	
Transmission power	Max. 20 Wt	
RCV buffer size	64 kB	
Antenna gain	23 dBi	
Base station parameters:		
	Maximum number of SS nodes	15
	Service class name	Gold (rtPS)
	Mac address	Auto assigned
	Maximum power transmission	0.5 W
	PHY profile	wireless OFDMA 20 MHz
	PHY profile type	OFDM
Mobile station parameters:		
Handover parameters:		
	MS handover retransmission timer	30 ms
	Maximum handover request retransmit	6 ms
	Maximum sustained traffic rate	4 Mbps
	Minimum reserved traffic rate	0.5 Mbps

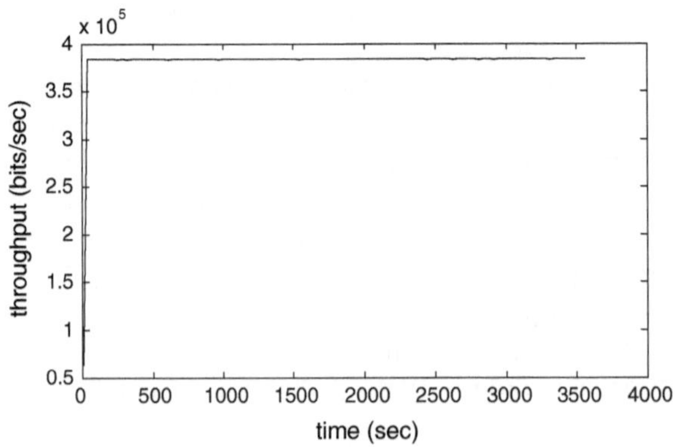

Fig. 11.3 Throughput at static condition. *Scenerio1* MNs are in static condition

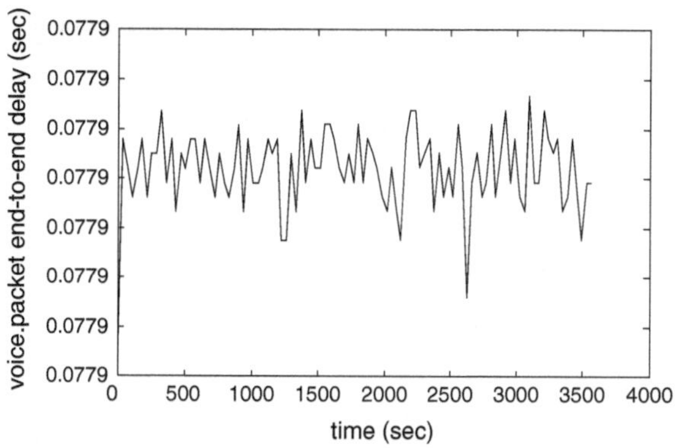

Fig. 11.4 Discrete packets sending from mobile_1_1 to mobile_2_1

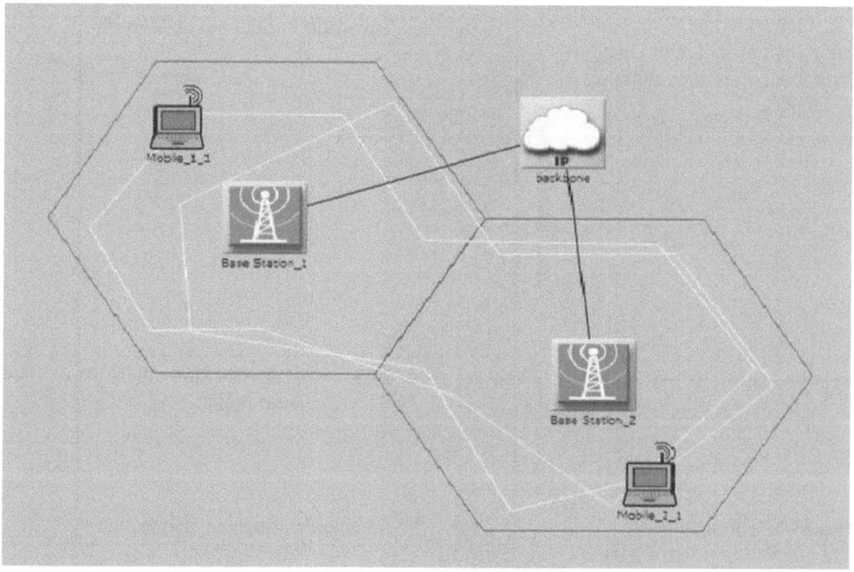

Fig. 11.5 Node movement path. *Scenerio2* MNs are moving

As the speed further increased, throughput dropped drastically as shown in Fig. 11.8 to 73.68 % which is much beyond the break-even point and call drop and call block occurs. Therefore, from Table 11.3 it is evident that 90 kmph is the maximum achievable speed for a MN under vehicular condition.

A comparative analysis of throughput has been made with other four available scheduling algorithms like UGS, rtPS, nrtPS, BE as shown in Fig. 11.9. Simulation is performed keeping the speed at 90 kmph.

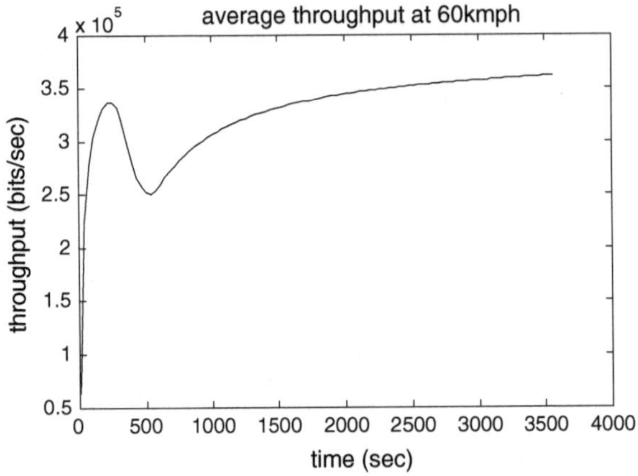

Fig. 11.6 Average throughput at 60 kmph. *Test Case 1* MNs are moving at 60 kmph

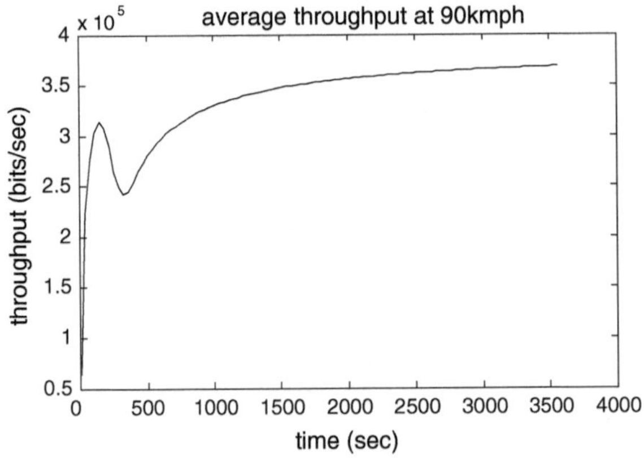

Fig. 11.7 Average throughput at 90 kmph. *Test Case 2* MNs are moving at 90 kmph

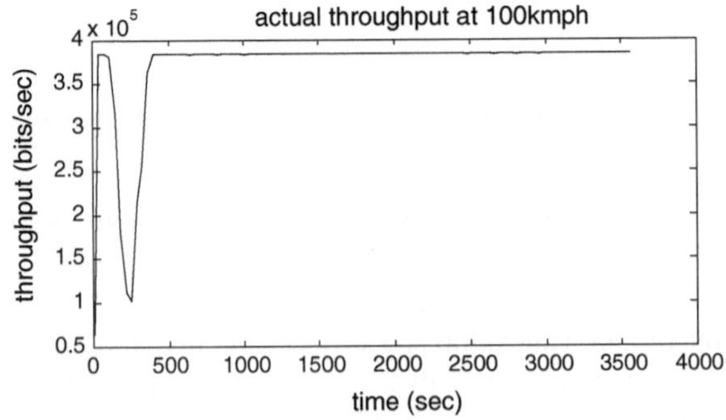

Fig. 11.8 Throughput at 100 kmph. *Test Case 3* MNs are moving at 100 kmph

Table 11.3 Drop in throughput under various moving conditions

	Drop in throughput	
	bits/s	Percentage
60	90,000	26
90	61,000	19.67
100	280,000	73.68

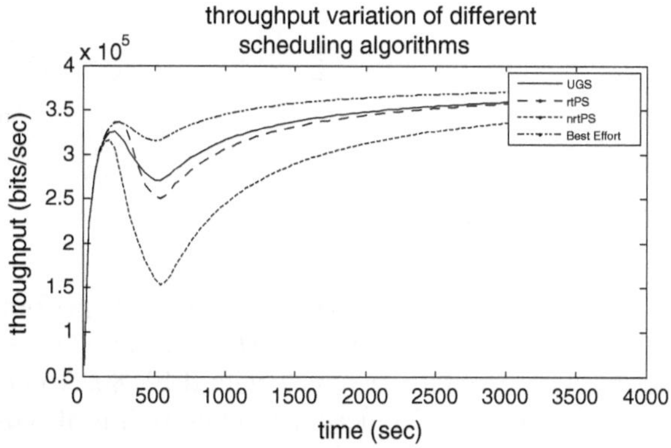

Fig. 11.9 Throughput variation of different scheduling algorithms

Figure 11.9 shows the comparative throughput analysis of different QoS service classes. It is evident from Fig. 11.9 that throughput for service class BE is maximum and nrtPS is minimum. Figures 11.10 and 11.11 depict that uplink packet drop and downlink packet drop variation of different scheduling algorithm. We can conclude from Figs. 11.10 and 11.11 that we can use BE service class except some specific conditions.

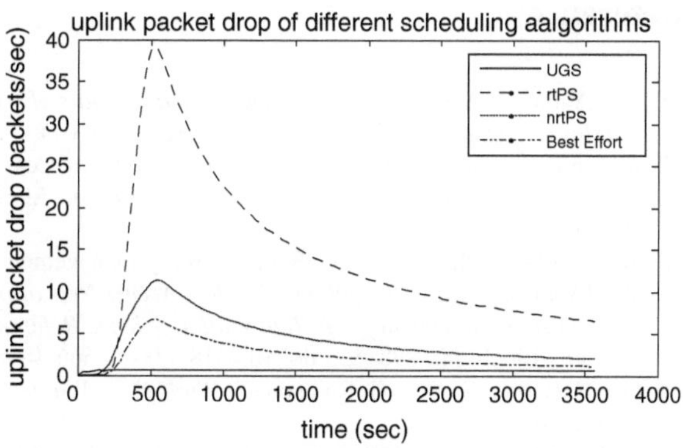

Fig. 11.10 Uplink packet drop of different scheduling algorithms

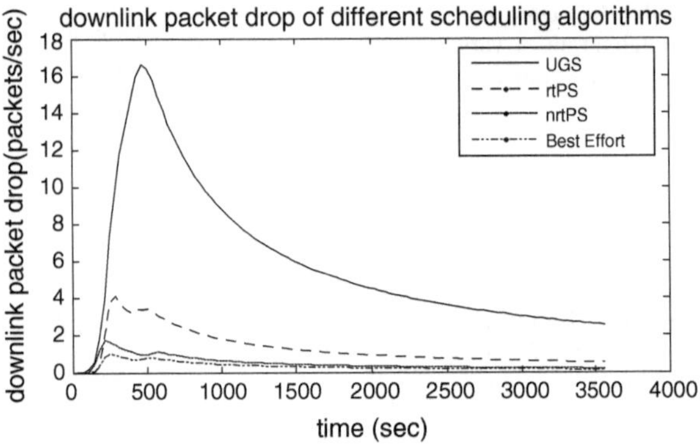

Fig. 11.11 Downlink packet drop of different scheduling algorithms

11.6 Conclusions

In this paper we made an in-depth study of the performance of the mobile WiMAX network with respect to different MAC service classes under vehicular condition. It has been observed that a vehicle with IEEE 802.16 m connectivity can move up to 90 kmph avoiding significant packet drop. It is also observed that BE QoS service class can be used in general condition to get maximum throughput, low latency and low BER.

WiMAX may be a viable solution for India as current government policy targets to have broadband connection in all secondary schools, healthcare centres. WiMAX will provide a sustainable solution for the Indian service providers as some Indian operators having with 3G licenses cannot adequately develop wireless broadband service due to the limited 2X5 MHz bandwidth scarcity. Therefore utilizing 2X20 MHz spectrum in the 2.3 GHz band for the operation of a WiMAX network is a promising candidate solution for realizing next generation all-IP networks.

References

1. J.G. Andrews, A. Ghosh, R. Muhamed, *Fundamentals of WiMAX: Understanding Broadband Wireless Networking* (Prentice Hall, New Jersey, 2007)
2. A. Yarali, S. Rahman, WiMAX broadband wireless access technology: services, architecture and deployment models. in *Canadian Conference on Electrical and Computer Engineering (CCECE)* (2008), pp.77–82
3. A.K. Mukhopadhyay et al., QoS and mobility management issues on next generation mobile WiMAX networks. in *Broadband Wireless Access Networks for 4G*: *Theory, Application, and Experimentation, Hershey, PA*: *Information Science Reference*, ed. by R. Santos, V. Licea, A. Edwards-Block. pp. 298–323 doi:10.4018/978-1-4666-4888-3.ch016
4. G.S.V. Radha, K. Rao, G. Radhamani, *WiMAX : a Wireless Technology Revolution* (Aurbach Publication)
5. I. Aldmour, LTE and WiMAX: comparison and future perspective. Commun. Netw. **5**, 360–368 (2013)

6. R.K. Singh, WiMAX networks: service and their quality matrics. Int. J. Emerg. Technol. Adv. Eng. **2**(6), 329–333 (2012). ISSN 2250-2459
7. K. Etemad, Overview of mobile WiMAX technology and evolution. IEEE Commun. Mag. **46** (10), 31–40 (2008)
8. S. Ahmadi, An overview of next-generation mobile WiMAX technology. IEEE Commun. Mag. **47**, 84–98 (2009)
9. J. Chen, C.-C. Wang, F.C.-D. Tsai, C.-W. Chang, S.-S. Liu, J. Guo, W.-J. Lien, J.-H. Sum, C.-H. Hung, *The Design and Implementation of WiMAX Module for ns-2 Simulator*. WNS2 (Pisa, Italy, 2006)
10. A.H. Rashwan, H.M.E. Badawy, H.H. Ali, Comparative assessments for different WiMAX scheduling algorithms, in *Proceedings of the World Congress on Engineering and Computer Science* (*WCECS*), vol. I (San Francisco, USA, 2009)
11. A. Belghith, L. Nuaymi, Comparison of WiMAX scheduling algorithms and proposals for the rtPS QoS class, in *14th European Wireless Conference* (*EW*) (2008), pp. 1–6
12. A.K. Mukhopadhyay, et al., Performance analysis of service classes for IEEE 802.16 m QoS optimization, in *Proceedings of the 1st International Conference on Emerging Trends and Applications in Computer Science (ICETACS)* (2013), pp. 149–155
13. A. Khalil, A. Ksentini, Classification of the uplink scheduling algorithms in IEEE 802.16, in *IRISIA* (2007)
14. A. Jain, A.K. Verma, Comparative study of scheduling algorithms for WiMAX, in *Proceedings of the National Conference on Mobile and Pervasive Computing (CoMPC)* (2008), pp. 7–8
15. Q. Liu, X. Wang, G.B. Giannakis, A cross-layer scheduling algorithm with QoS support in wireless networks. IEEE Trans Veh. Technol. **55**(3), 839–847
16. M. Aguado, E. Jacob, M. Berbineau, I.L. Samper, in *QoS challenges in handover process*, ed. by S.-Y. Tang, P. Muller, H. Sharif. WiMAX Security and Quality of Service: An End-to-End Perspective. (Wiley, 2010) pp. 213–238

Chapter 12
Performance Analysis of RTS/CTS Protocol in Accessing Control Channel in Distributed Cognitive Radio Networks

Subhasree Bhattacharjee and Swarup Mandal

Abstract In this paper, we analyse the delay performance of RTS/CTS protocol in DCF mode to find out whether it is suitable for sensing inter packet white space. For simultaneous sharing of channels by primary users (PUs) and secondary users (SUs), the white space has duration of approximately 50 ms. In distributed cooperative sensing, the SUs perform RTS/CTS protocol with distributed coordination function (DCF) for accessing common control channel to share sensing information among each other. Total communication delay of this sensing approach is given as the sum of delays of broadcasting RTS, CTS and sensing results by SUs to other SUs. We find that RTS/CTS protocol causes much more delay than basic CSMA/CA in distributed cooperative sensing approach.

Keywords RTS/CTS · DCF · CRN

12.1 Introduction

As the growth in wireless communications increases, demand for deployment of new wireless devices in both licensed and unlicensed frequency band increases. Reports from FCC reveal that static spectrum assignment policy reduces spectrum utilization. Cognitive radio network (CRN) is proposed as one of the emerging techniques to enhance the spectrum utilization. There are two types of users: licensed or primary users (PU) and unlicensed users or secondary users (SU) in CRN. The SUs should not cause interference to the PUs. The SUs periodically sense a list of channels to find out which of the licensed channels are empty and then opportunistically use the available

S. Bhattacharjee (✉)
Narula Institute of Technology, Kolkata, India
e-mail: bhattacharjeesubhasree@gmail.com

S. Mandal
Wipro Technologies, Kolkata, India
e-mail: swarup.mandal@wipro.com

© Springer India 2015

K. Maharatna et al. (eds.), *Computational Advancement in Communication Circuits and Systems*, Lecture Notes in Electrical Engineering 335,
DOI 10.1007/978-81-322-2274-3_12

bands. Thus, one objective of sensing is to reduce harmful interference to the communications of PUs while using the white space of a channel and at the same time improve the spectrum utilization by making effective use of available white spaces.

In distributed cooperative sensing, each SU shares its own sensing results with other SUs. Then each SU combines its own sensing data with the sensing results of other peers to decide on the presence or absence of a PU in a channel. The sharing of sensing information takes place through a common control channel (CCC). It is to be mentioned that SUs use Distributed coordination function (DCF) mode of carrier sense multiple access with collision avoidance (CSMA/CA) protocol to access the CCC. In this paper, we consider that SUs use Request-To-send/Clear-To-send (RTS/CTS)-based access technique to share their sensing results.

In the recent past, a number of research works have been carried out to analyse different performance metrics of DCF. Bianchi proposed a markov chain model [1] to calculate the throughput of IEEE 802.11 DCF. Wu et al. [2] modified the model of [2] by considering packet retransmission limit. Non-adaptive characteristics of DCF are discussed in many literatures [3, 4]. Adaptivity is the method of improving DCF in MANET. However, to the best of our knowledge, no research has been carried out in the recent past to analyse the effectiveness of RTS/CTS protocol during sharing of sensing results.

Our objective is to propose analytical models to measure the performance of RTS/CTS protocol in DCF mode. In particular, we calculate the amount of time required by the SUs to calculate CCC access time. This measurement will guide us for determining the suitability of RTS/CTS based CSMA/CA access strategy.

The rest of the paper is organized as follows. In Sect. 12.2, system model is discussed. Results and graphs are discussed in Sect. 12.3. Finally, we conclude the paper in Sect. 12.4.

12.2 System Model

In this paper we consider a single band CRN which consists of a single channel and one dedicated CCC. We consider that CCC lies in unlicensed band. In the distributed system, SUs perform local sensing and share their sensing results with each other using the dedicated CCC. When N number of cooperating SUs try to access the CCC, they use RTS/CTS protocol in DCF mode. At any given time, the CCC may be in ON or OFF state. When the CCC is used by one SU then it is in ON state and when it is idle (i.e. not used by any SU) then it is in OFF state. We make the following assumptions to model the access to CCC using RTS/CTS in DCF mode:

We consider that each SU broadcasts single bit control messages to all other cooperative SUs. CCC is assumed as transmission error-free. In case of collision, colliding SUs double their contention window for the retransmissions. All SUs are time synchronized. So, all SUs sense the CCC at the same epoch.

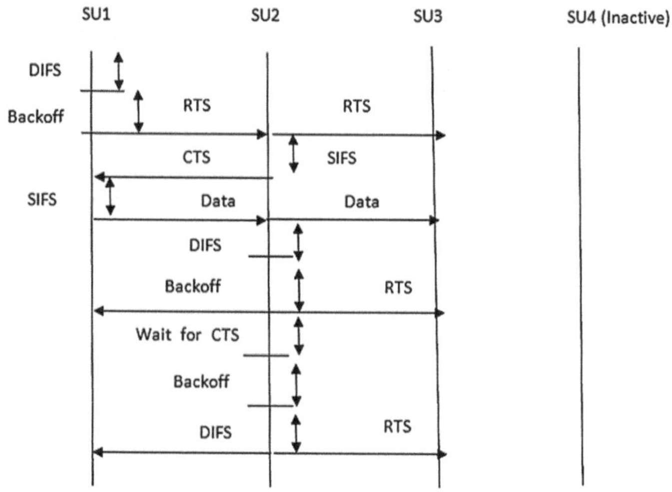

Fig. 12.1 Timing diagram of RTS/CTS protocol

Figure 12.1 shows the timing diagram of RTS/CTS protocol. Average RTS or CTS or data transmission delay can be expressed from [5] as

$$E[D] = \sum_{j=0}^{R} (E[D_j] \times Q_j) \tag{12.1}$$

$E[D_j]$ and Q_j can be derived from [5]. For RTS transmission, time for sending a control packet is denoted by Ts and duration of collision, i.e. average time spent in collision is Tc. Ts = TRTS + €. TRTS is the time for sending RTS and € is the propagation delay. Tc = TRTS + DIFS + €.

Similarly, for CTS transmission Ts = TCTS + € and Tc = Ts + DIFS. TCTS is the time for sending CTS. For data packet transmission Ts = H + L + € and Tc = Ts + DIFS. H and L is the time needed for transmit packet header and payload. For a single round total delay is the sum of the three above-mentioned delays. For N SUs number of rounds for getting consensus is 0.6 × N [5]. Total communication delay can be obtained by multiplying the total delay for a particular round with 0.6 × N × N, where N is the number of SUs.

12.3 Result and Discussion

In this section, we first describe system parameters which have been used for numerical analysis. Table 12.1 depicts the various system parameters.

Figure 12.2 shows the variation in total delay (communication delay and queuing delay) of distributed cooperative sensing approach with number of SUs

Table 12.1 System parameters

Parameters	Values
DIFS	50 μs
Slot time	20 μs
SIFS	10 μs
Collision probability	0.55
Backoff window	16, 32 slots
Retransmission limit	6
Phy header	192 bits
RTS	160 bits
CTS	112 bits
Data rate	11 Mbps
Propagation delay	1 μs
MAC header	224 bits

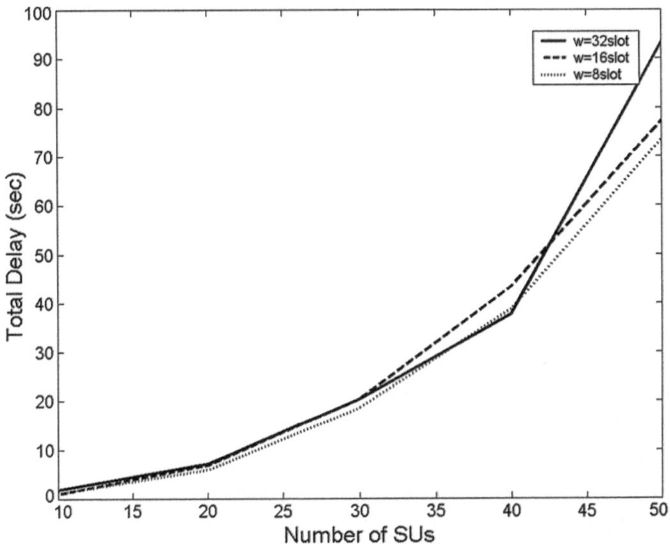

Fig. 12.2 Variation in total delay in distributed sensing approach using basic CSMA/CA

using basic CSMA/CA [5]. It is revealed that total delay varies from 1.898 to 93.299 s when the number of SUs varies from 10 to 50. Three different contention window sizes have been used for comparison.

Figure 12.3 shows the comparison of basic CSMA/CA and RTS/CTS protocols. It is revealed from the graph that RTS/CTS protocol incurs much more communication delay than basic CSMA/CA in distributed cooperative sensing environment. For contention window size 16 slots, RTS/CTS gives 9.257 s communication delay for 15 SUs, whereas basic CSMA/CA causes only 2.782 s communication delay for the same contention window size and the same number of SUs.

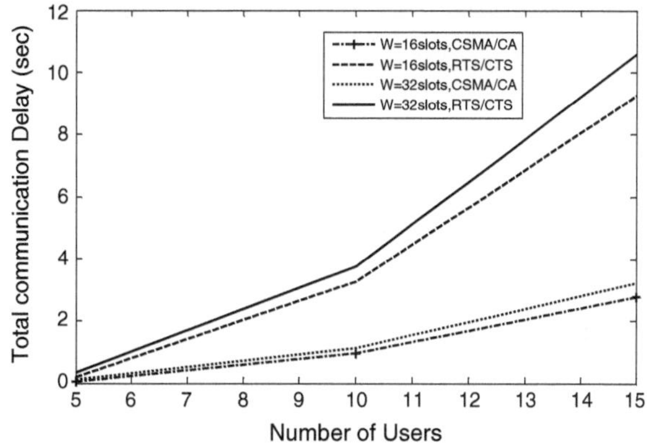

Fig. 12.3 Comparison of total communication delay with the number of users

12.4 Conclusion

In this paper we analyse delay performance analysis of RTS/CTS protocol in distributed cooperative sensing environment. We find that RTS/CTS protocol incurs much more delay than basic CSMA/CA protocol. From the literature it is found that for simultaneous sharing of channels by PUs and SUs, white space duration is approximately 50 ms. However, RTS/CTS protocol causes 195 ms (contention window = 16 slots) delay for 5 SUs in distributed cooperative sensing environment. So, RTS/CTS protocol is not a suitable choice in the distributed environment.

References

1. G. Bianchi, Performance analysis of the IEEE 802.11 distributed coordination function. IEEE J. Selected Area Commun. **18**(3), 535–547 (2000)
2. H. Wu, Y. Peng, K. Long, S. Cheng, J. Ma, Performance of reliable transport protocol over IEEE 802.11 wireless LAN: analysis and enhancement. in *Proceedings of IEEE INFOCOM '02* (2002) pp. 599–607
3. S. Khurana, A. Kahol, S.K.S. Gupta, P.K. Srimani, Performance evaluation of distributed coordination function for IEEE 802.11 wireless LAN protocol in presence of mobile and hidden terminals. in *MASCOT'99* (1999) pp. 40–47
4. V. Bharghavan, A. Demers, S. Shenker, L. Zhang, MACAW: a media access protocol for wireless LANs. in *ACM SigComm* (1994) pp. 212–225
5. S. Bhattacharjee, S. Mandal, B. Sardar, Performance analysis of CSMA/CA protocol during white space identification in cognitive radio networks. in *Proceedings of Applications and Innovations in Mobile Computing (AIMoC)* (2014) pp. 91–96

Chapter 13
Winner Determination Algorithm in Auction Framework of Cognitive Radio Network

Subhasree Bhattacharjee and Arunava Bhattacharya

Abstract In this paper we consider the auction framework of cognitive radio network consisting of a set of primary users and a set of secondary users (SUs). The spectrum has been divided into channels with the help of frequency division multiple access (FDMA). The primary users (PUs) are licensed users who lease the vacant channels to secondary user (SUs) temporarily by auction mechanism. In this paper we try to form an algorithm to determine the secondary user who wins the auction. We also discuss the effect of change of number of users on bandwidth utilization, channel utility and revenue generation by using the algorithm.

Keywords Auction · Cognitive radio · Channel utility

13.1 Introduction

Increasing demand in wireless devices leads to scarcity of spectrum resources. But still there are number of available spectrum bandwidths that are not efficiently used [1]. Dynamic spectrum access (DSA) solves the problem by allowing unlicensed or secondary users (SU) to use the idle (i.e. not used by primary users (PUs)) band opportunistically. Cognitive radio (CR) technology which provides DSA has become the effective approach of usage of spectrum [2]. Thus, CR is an autonomous unit which senses the environment and changes its operating parameters to obtain efficient spectrum utilization. Many researches are going on regarding DSA [2–4]. Spectrum auction is another novel aspect of spectrum sharing [5, 6]. In the spectrum trading approach, PUs lease the vacant spectrum to SUs for certain

S. Bhattacharjee (✉)
Narula Institute of Technology, Kolkata, India
e-mail: bhattacharjeesubhasree@gmail.com

A. Bhattacharya
Kovair Software Pvt Ltd, Kolkata, India
e-mail: arunavabhattacharya89@gmail.com

© Springer India 2015

K. Maharatna et al. (eds.), *Computational Advancement in Communication Circuits and Systems*, Lecture Notes in Electrical Engineering 335,
DOI 10.1007/978-81-322-2274-3_13

amount of time to earn some revenue. SUs bid for getting access to the spectrum. The SU who provides the highest bid value will be the winner. Game theory and pricing mechanism are efficiently been used in the design of efficient spectrum sharing skills. In [7–9] spectrum auction is proposed for SUs. Authors in [10] discuss a multi-winner spectrum auction algorithm. References [11–14] discuss various researches of spectrum sharing on CR. Double auction framework was discussed in [13]. Wang et al. [13] does not consider the interaction between the primary users and the secondary users in the auction framework. Authors in [14] allow the multiple winners to achieve the spectrum item at the same time.

The rest of the paper is organized as follows. In Sect. 13.2, system model is discussed. In Sect. 13.3, parameters for graph are discussed. Section 13.4 casts light on experiments and results. Finally, we conclude the paper in Sect. 13.5.

13.2 System Model

In this paper we try to solve the problem of determining the winner when an auction for specific channel/channels takes place with multiple secondary users bidding for them. We determine the winner in the auction strategy. We consider the following matrices for our purpose. We first consider that a channel can be broken down into some fixed length blocks. The secondary users have to specify the number of blocks they demand as well as the price they are willing to pay for the same.

We keep track of the demand of the user by using a demand matrix $d(m \times n)$ where d_{ij} = number of blocks of channel j that is demanded by user i. The minimum price that any secondary user should pay while bidding is got by the minimum bid matrix bn where b_i = minimum bidding price for a unit block of channel i. The total price paid by the secondary user for the number of blocks he demands is stored in the price matrix $p(m \times n)$ where p_{ij} is the total price user i pays for the channel j. Once we have the price, minimum bid matrix and the demand matrix, we can solve the problem to get which user gets which channel. This is achieved by the following formula:

$$z = \text{Max}\left\{ \left(\frac{p_{ij}}{d_{ij}} \right) - b_i \right\} \tag{13.1}$$

(p_{ij}/d_{ij}) that is price paid per unit demand by secondary user i gives us the price user i is paying for each block of the channel j. b_j gives the minimum bid value of the channel for each block of the channel j. From Z we can get the identity of the secondary user i who is willing to pay the maximum price for the channel j. Given, the minimum bid matrix bn where n is the number of channels. The proposed winner determination algorithm is as follows:

```
BEGIN
Step 1: [INPUT] The bandwidth the secondary users are
demanding are taken as input and stored in the matrix d
(mxn) where m is the number of users and n is the number of
channels.
[INPUT] The price that the secondary users are paying is
taken. The values are stored in the price matrix p(mxn)
where m is the number of users and n is the number of
channel.
Step 2: Loop (For each user i)
Step 3: Find the price paid per unit demand for user i
bidding for any channel j. Find, (pij/dij)
Step 4: Calculate the difference of the price per unit
demand for user I and minimum bid value of user i for
channel j.
Step 5: End Loop
Step 6: For any user i find the maximum deviation calculated
in step 3.
Step 7: END
```

13.3 Parameters for Simulation

We are taking into consideration of 10 SUs and 10 channels. There are no constant values assumed for construction of this graph. The following will be used as characteristics for drawing the graph. The parameters are

1. Channel Utility
2. Revenue Generated
3. Bandwidth Utilization

Bandwidth Utilization percentage is found by the amount of bandwidth allocated to the secondary users after the auction versus the total bandwidth associated with the channels.

For a single user the amount of bandwidth consumed by the user is calculated and then the total available bandwidth is taken. The percentage of the used bandwidth by the secondary user against the total available bandwidth gives us the total utilization of the channels.

Channel utility as perceived by user is the ratio of utility of a channel to a particular user and the total utility of that channel.

Revenue generated is the amount of revenue collected by PUs by auctioning the channel to the secondary users who are determined as winner by the algorithm.

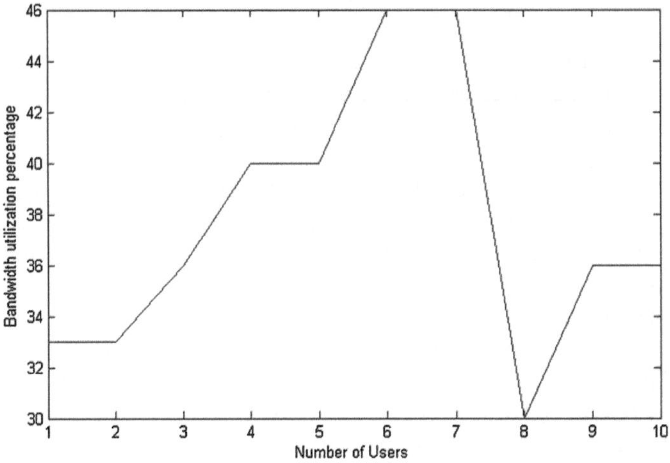

Fig. 13.1 Number of users versus bandwidth utilization (%)

13.4 Experiments and Results

As shown in Fig. 13.1, the bandwidth utilization has increased as the number of users increased. The maximum utilization was during the time when we had the number of users of 6 and 7. The utilization was 46 %, however, as the number of the users was increased from 7 to 8 the utilization percent has decreased to 30 %. According to the Fig. 13.2, we see revenue is increasing with the number of users

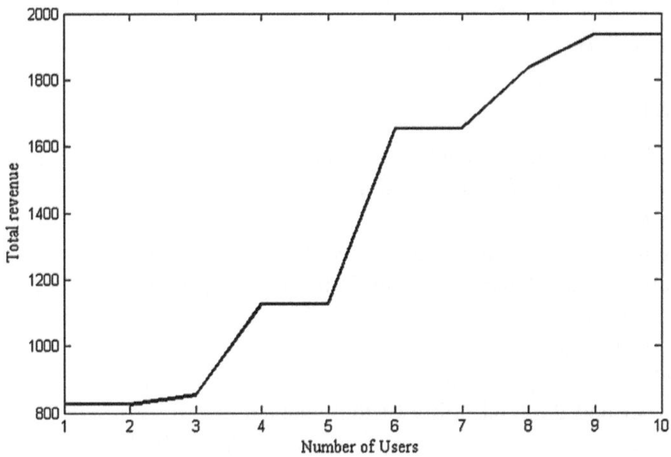

Fig. 13.2 Number of users versus total revenue

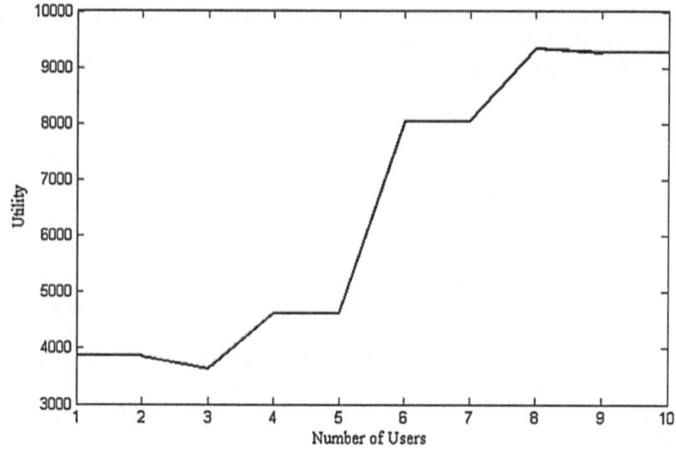

Fig. 13.3 Number of users versus utility

following our algorithm for winner determination. Figure 13.3 depicts the variation of utility with number of SUs. Utility is the rating of usability as perceived by the users who are participating in the auction for channels.

13.5 Conclusion

In this paper we consider a cognitive radio network with a set of primary users (PUs) and a set of secondary users (SUs). When the channels are not used by the PUs then PUs lease the vacant spectrum for monetary gain. SUs bid for the channels and the primary users (PUs) select the purchaser offering the highest bid values. The main objective is to maximize the benefit of the PUs. In this paper we propose an algorithm to determine the secondary user who wins the auction. Results show the effect of change of number of users on bandwidth utilization, channel utility and revenue generation by using the algorithm.

References

1. Federal communications commission, spectrum policy task force, in *Report ET Docket* no. 02–135 (2002)
2. N. Chang, M. Liu, Competitive analysis of opportunistic spectrum access strategies, in *Proceedings of IEEE Infocom* (2008), pp. 1535–1542
3. C. Santivanez, Transport capacity of opportunistic spectrum access (OSA) MANETs, in *Proceedings of Crowncom* (2007), pp. 9–18
4. S. Huang, X. Liu, Z. Ding, Opportunistic spectrum access in cognitive radio networks, in *Proceedings of IEEE Infocom* (2008), pp. 1427–1435

5. D. Niyato, E. Hossain, Spectrum trading in cognitive radio networks: a market-equilibrium-based approach. IEEE Wirel. Commun. **15**(6), 71–80 (2008)
6. W. Beibei, W. Yongle, Game theoretical mechanism design methods. IEEE Signal Process. Mag. **25**(6), 74–84 (2008)
7. J. Huang, R.A. Berry, M.L. Honig, Auction-based spectrum sharing. Mob. Netw. Appl. **11**(3), 418 (2006)
8. S. Gandhi, C. Buragchain, L. Cao, H. Zheng, S. Suri, Towards real time dynamic spectrum auctions. Comput. Netw. **52**(4), 879–897 (2008)
9. H. Xu, J. Jin, D. Li, A secondary market for spectrum, in *Proceedings of IEEE Infocom* (2010), pp. 1–5
10. Y. Wu, B. Wang, K. Liu, T.C. Clancy, A multi-winner cognitive spectrum auction framework with collusion-resistant mechanisms, in *Proceedings of 3rd IEEE Symposium on New Frontiers in Dynamic Spectrum Access Networks* (2008), pp. 1–9
11. H.-B. Chang, K.-W. Chen, Auction-based spectrum management of cognitive radio networks. in IEEE Trans. Veh. Technol. **59**(4), 1923–1935 (2010)
12. X. Zhou, H. Zheng, TRUST: a general framework for truthful double spectrum auctions, in *IEEE INFOCOM'09* (2009), pp. 999–1007
13. S. Wang, P. Xu, X. Xu, S. Tang, X. Li, X. Liu, TODA: truthful online double auction for spectrum allocation in wireless networks, in *IEEE Symposium on New Frontiers in Dynamic Spectrum* (2010), pp. 1–10
14. M. Li, X. Li, H. Ji, Virtual bidder group auction mechanism for dynamic spectrum access, in *IEEE PIMRC'09* (2009), pp. 2797–2801

Chapter 14
An Improved Energy Detector for Spectrum Sensing in Cognitive Radio System with Adaptive Noise Cancellation and Adaptive Threshold

Aritra Das, Bhaswati Chatterjee, Sandhya Pattanayak and Manisha Ojha

Abstract Over the years the usage of wireless communication systems has increased rapidly leading to scarcity of bandwidth. Hence the concept of utilizing the existing system to its fullest has become very important. Cognitive radio is a technique based on this concept which identifies the unutilized white spaces in the spectrum, and are allotted to the secondary user in a non-interfering manner. The energy detection technique does not work at low SNR. In this paper we attempted to improve the efficiency by performing adaptive noise cancellation and adaptive threshold in the energy detector.

Keywords Cognitive radio · Spectrum sensing energy detector · Adaptive threshold · Adaptive noise cancellation

14.1 Introduction

Underutilization of spectra exclusively allocated to conventional wireless communication created a sphere of research in efficient spectrum usage. The cognitive radio identifies the unused licensed bands and allots them to the secondary user in a non-interfering manner [1]. But the goal of spectrum sensing is not only to find out

A. Das (✉) · B. Chatterjee · S. Pattanayak
ECE Department, Narula Institute of Technology, Kolkata, India
e-mail: aritra.ece5@gmail.com

B. Chatterjee
e-mail: bhaswationline@gmail.com

S. Pattanayak
e-mail: sandhya.pattanayak@yahoo.com

M. Ojha
ETCE Department, Jadavpur University, Kolkata, India
e-mail: manisha.jolly@gmail.com

© Springer India 2015

K. Maharatna et al. (eds.), *Computational Advancement in Communication Circuits and Systems*, Lecture Notes in Electrical Engineering 335,
DOI 10.1007/978-81-322-2274-3_14

the white space of licensed spectra but also to protect the primary licensed users from intrusion caused by cognitive radio communications. So a dynamic, intelligent and adaptive access and utilization of spectrum is necessary. The cognitive radio (CR) will be the most auspicious technology due to its innate spectrum sensing competency and frequency-lithe radio utilities. The CR is an adaptive, smart and spectrally flexible radio and network technology that can automatically perceive available channels in a wireless spectrum and change transmission parameters enabling efficient bandwidth utilization. Cognitive radio is used to identify the primary user (PU) in the channel with a high probability of detection (Pd), and a low probability of false alarm (Pfa).

In energy detector-based CR, meeting both these requirements involves a trade-off. Probability of false alarm (Pfa) depends on number of samples (N), noise variance (Varn) and threshold (Th) selected for making the decision about the presence or absence of signal. To estimate (Pd), in addition to the parameters required for the estimation of Pfa, the signal-to-noise ratio (SNR) of the channel is also required. From various studies made it is found that the energy detector does not perform well at low SNR [2].

In this paper we have tried to enhance the performance of energy detector by using adaptive noise cancellation and adaptive threshold along with the energy detector. This enhances the detection of signal at low SNR, hence reducing the probability of false alarm.

In Sect. 14.2 a brief literature survey of energy detector and spectrum sensing techniques is given. In Sect. 14.3 we discuss the application of adaptive noise cancellation and adaptive threshold to enhance the performance of energy detector. In Sect. 14.4 we present a comprehensive conclusion indicating some future research directions.

14.2 Literature Review

Spectrum sensing is a process of identification of white spaces in the licensed band. The detection of the primary signal in the licensed band is based on the well-known hypothesis [2].

H_0 $\eta = 0$, primary signal absent
H_1 $\eta = 1$, primary signal present

where η is 0 or 1 which determines the presence or absence of the primary signal, respectively.

$$\begin{aligned} H_0 &: y(t) = n(t) \\ H_1 &: y(t) = hx(t) + n(t) \end{aligned} \tag{14.1}$$

where $x(t)$ is the primary user's signal to be detected
$n(t)$ is the additive white Gaussian noise,
h is the channel gain.

H_0 is a null hypothesis indicating white space and H_1 indicates the presence of primary signal in the channel. There are various spectrum sensing techniques like energy detection technique [2], matched filter detection technique [2] and cyclostationary detection techniques [2]. The various spectrum sensing methods and implementation issues are discussed in [3]. Each technique has its pros and cons, like the energy detector technique does not require prior knowledge of the signal but the disadvantage is that the technique cannot detect the presence of low SNR signals.

In order to enhance the performance of energy detector we are introducing the adaptive noise cancellation and adaptive threshold technique into it. The adaptive threshold for energy detection is discussed in [4], however, they have detected the threshold directly from the signal received from the channel. In order to improve the performance we have introduced adaptive noise cancellation prior to the adaptive threshold.

So we make adaptive filter as an aid to the energy detector. Unlike conventional filter design techniques, adaptive filters do not have constant filter coefficients and have no priori information of the input signal. Such a filter with adjustable parameters is called an adaptive filter. Adaptive filters adjust their coefficients to minimize an error signal. The most common form of adaptive filter is the transversal filter using least mean square (LMS) algorithm and NLMS algorithm.

14.3 Problem Formulation

In this paper we identify the presence of primary signal in a channel with low SNR. Spectrum sensing is the process to identify the white space in a licensed frequency band. The major problem in spectrum sensing is to identify the signals which have low SNR and have faded. This is required to reduce the probability of false alarm and increase the probability of detection. In this paper we have tried to enhance the energy detector performance with adaptive noise cancellation algorithm and adaptive threshold. This process does not require prior knowledge of the signal and noise characteristics.

In a conventional energy detector the input signal is the signal corrupted with noise. The mean square value of the signal is determined and the presence and absence of the signal are determined based on the threshold. But as the performance of the energy detector is low in case of low SNR we use adaptive noise canceller (Fig. 14.1).

The Adaptive noise cancellation (ANC) efficiently attenuates low frequency noise for which passive methods are ineffective. After ANC determination of threshold level of energy detector is important. If the energy of the scanned channel is more than the threshold we assume that the channel is occupied with a primary

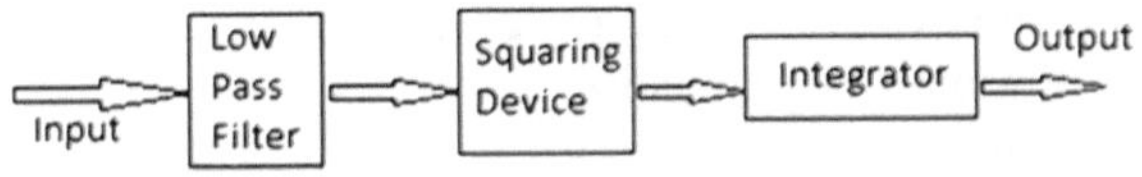

Fig. 14.1 Conventional energy detector

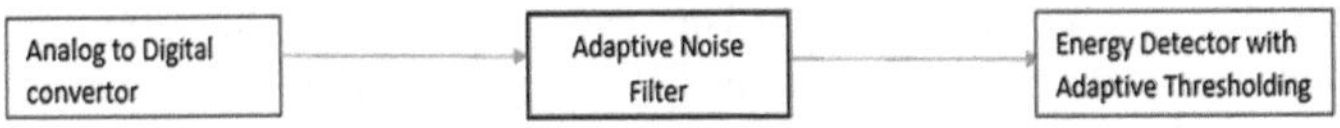

Fig. 14.2 Proposed technique

signal else we assume it as a white space. There are two widely used principles to find the threshold required to meet a set of target performance metrics (Pd, Pfa). Conventionally, the threshold of energy detector is fixed by a target Pfa, also known as constant false alarm rate (CFAR) principle [4]. Moreover, the threshold is calculated by setting a target Pd, also known as constant detection rate (CDR) principle [4]. The threshold is determined adaptively, hence it increases the performance of energy detector at low SNR and suggests strategies to maintain the operational relevance of the CR. Hence a combination of AN and adaptive threshold will enhance the performance of energy detector (Fig. 14.2).

14.4 Proposed Technique

1. The ANC receives two signals as primary inputs: an FM signal transmitted through a channel associated with the sensor and a noise $d(n) + r(n)$. A second input to the ANC is the noise $r'(n)$ uncorrelated with the signal but correlated with the noise as shown in Fig. 14.3. The second input acts as a reference canceller. This noise is filtered to produce an output $e(n)$. The system output is produced by detracting $y(n)$ from the primary input.
2. As in Fig. 14.1 the reference input is administered by an adaptive filter. An adaptive filter is that which instinctively adjusts its own impulse response.

Fig. 14.3 Adaptive noise cancellation

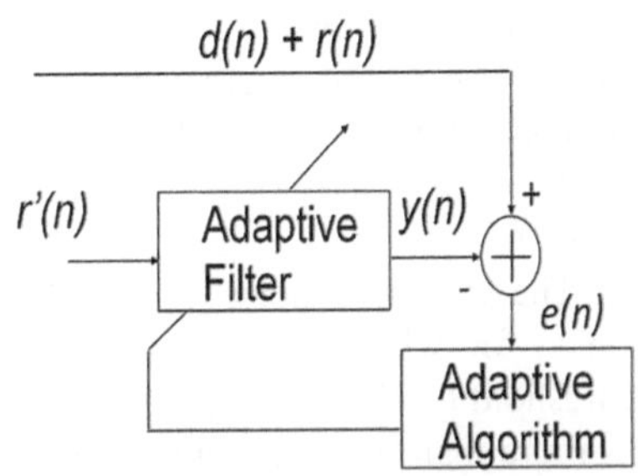

Amendment is accomplished through an algorithm. The filter can operate under changing conditions and can continually readjust its weight to minimize the error signal. The practical objective is to produce a system output

$$e(n) = d(n) + r(t) - y(n) \tag{14.2}$$

that is a best fit in the least squares sense to the signal *d*(*n*). This is accomplished by feeding the system output back to the adaptive filter and adjusting the filter through an LMS adaptive algorithm.
The signal *s(t)* is uncorrelated with *r*(*n*) and *r*′(*n*), and *r*(*n*) is allied with *r*′(*n*). The output *z* will be

$$e(n) = d(n) + r(n) - y(n) \tag{14.3}$$

The output of the equalizer minimizes the noise and the output we get is the signal with minimum noise power. Hence the output of the ANC is a signal with a better SNR than the SNR of the incoming signal.

$$d(n) = e(n) \quad \text{when} \quad (r(n) - r'(n)) = 0 \tag{14.4}$$

At *y*(*n*) = *r*(*n*) the noise is minimized and hence an output SNR is present. The output of the ANC is applied to the energy detector where the threshold is set dynamically. *e*(*n*) will have signal and some amount of noise and this is the input to the energy detector with adaptive threshold. So we can get a figure of merit of the performance of the adaptive noise chancellor as the ratio of output SNR and input SNR.

3. The adaptive threshold λd is determined by the channel SNR and the performance metrics, Pd and Pfa of the system [4]. SNR is given as the ratio of mean (Ωs) and variance of the signal (Ωn).

$$\text{SNR } \Upsilon = \Omega s2/\Omega n2 \tag{14.5}$$

Fix the target Pfa and Pd and use (14.4 and 14.5) to compute λf the adaptive threshold.

$$\text{Pfa} = \text{P}(\text{M}(y) > \lambda f|H_0) = \text{Q}\left(\frac{\lambda f|-\mu_0}{\sigma_0}\right) \tag{14.6}$$

$$\text{Pd} = \text{P}(\text{M}(y) > Zd\,|H_1) = \text{Q}\left(\frac{\lambda d - \mu_1}{\sigma_1}\right) \tag{14.7}$$

N, Pfa and Pdet to calculate critical threshold γ_c by using (14.8). Where *N* is the number of samples.

$$\gamma_c = \frac{[\mathrm{Q}^{-1}(\mathrm{Pfa}) - \mathrm{Q}^{-1}(\mathrm{Pd})]}{\mathrm{Q}^{-1}(\mathrm{Pd}) + \sqrt{N}} \tag{14.8}$$

where Q(x) is defined by the complementary distribution function of the standard Gaussian and is given as:

$$\mathrm{Q}(x) = \frac{1}{2\pi} \int_x^\infty e^{-u^2/2} du \tag{14.9}$$

The adaptive threshold λ new is calculated by [4]

$$\lambda\,\mathrm{new} = \lambda f + \beta * (\lambda d - \lambda f) \quad 0 < \beta < 1 \tag{14.10}$$

The Adaptive Noise Canceller and Adaptive Thresholding system cannot work together, because the noise on the basis of which the threshold is being determined is the channel noise, but we are actually feeding the filtered output to the adaptive threshold, i.e. input with less noise than the actual channel noise. This will introduce error in the system and Adaptive Thresholding will not work at all. To overcome this difficulty we are proposing a new technique to make both adaptive noise canceller and adaptive threshold together.

In this technique the input to the energy detector being the output of the adaptive noise chancellor the noise content will be less than the original noise n_0. So the relation between ω/n_0 will be proportional to input SNR/output SNR of the ANC (noise figure). We can determine the amount of noise in the output of the ANC ω using the noise figure. This will assist the determination of threshold adaptively which is dependent on the SNR.

14.5 Result Analysis

As the main problem with the conventional energy detector remains in it's inability not to work in low SNR situation we have introduced quite a few measures to increase it's efficiency in low SNR situation. Complete noise removal is not possible but The Adaptive Noise filter working on NLMS algorithm removes noise and the output is applied to adaptive threshold. So the energy detector works efficiently at low SNR.

Hence this reduces the probability of false alarm by great means. Whether the noise is low or high the threshold is being set accordingly (Fig. 14.4).

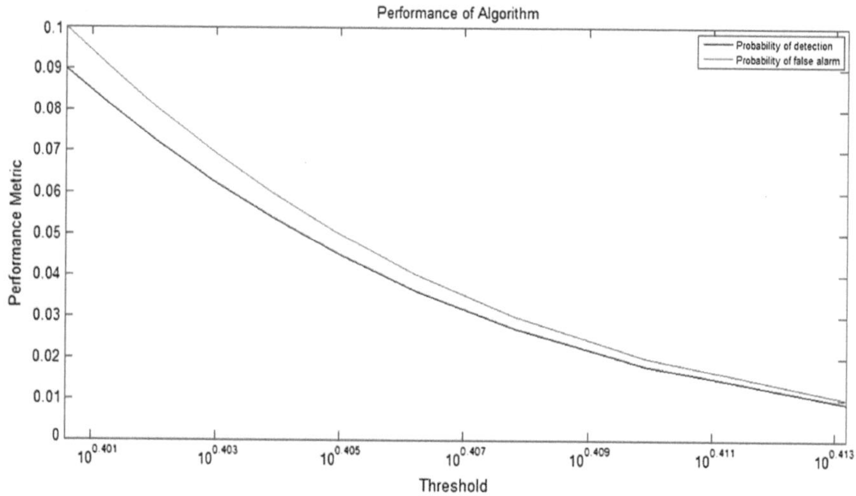

Fig. 14.4 Performance of proposed energy detector

14.6 Conclusion

In this paper we have tried to enhance the performance of energy detector by adaptive threshold and adaptive noise cancellation. It is observed that at low SNR the energy detector performs well. Further analysis on this technique will be done in future.

Acknowledgments We express our gratitude to Prof. R. Nandi and Prof. P. Venkateswaran of ETCE Dept. Jadavpur university for their guidance and support.

References

1. Federal Communications Commission, Notice of proposed rule making and order: facilitating opportunities for flexible, efficient, and reliable spectrum use employing cognitive radio technologies, ET Docket No 03-108, Feb 2005-73 (2005)
2. T. Yucek, H. Arslan, A survey of spectrum sensing algorithms for cognitive radio applications. IEEE Commun. Surv. Tutor. **11**, 116–130 (2009)
3. D. Cabric, S. Mishra, R. Brodersen, Implementation issues in spectrum sensing for cognitive radios, in *Proceedings Asilomar Conference on Signals, Systems and Computers*, vol. 1 (Pacific Grove, California, USA), 2004 pp. 772–776
4. P.R. Nair, A.P. Vinod, A.K. Krishna, An energy detector for cognitive radios in channels at low SNR using adaptive threshold, in *8th International Conference on Information, Communications and Signal Processing (ICICS) 2011*, 13–16 Dec 2011 pp. 1, 5. doi: 10.1109/ICICS.2011.6173586

Chapter 15
A Comparative Analysis of PAPR in OFDM System for Different Parameters

Rimpi Datta

Abstract The Peak to Average Power Ratio (PAPR) plays an important role in Orthogonal Frequency Division Multiplexing (OFDM) in communication systems, especially in wireless cellular systems. This article illustrates the effectiveness of the CCDF curve to study the signal performance and effect of PAPR in orthogonal frequency division multiplexing. I performed some analysis of signal on power level basis and found that PAPR is an effective analysis tool to detect noise in signal.

Keywords PAPR · CCDF · OFDM

15.1 Introduction

Orthogonal Frequency Division Multiplexing (OFDM) scheme is a popular standard for next generation mobile radio communication system [1]. In OFDM system, serial data stream is multiplexed into a large number of orthogonal sub-channels producing a bandwidth efficient signal spectrum. As it is robust against multipath fading, the orthogonal frequency division multiplexing system can transmit high-speed data using a number of orthogonal sub-carriers [2]. Various studies show that when OFDM sub-carriers are added with the same phase [2], it produces high peak to average power ratio (PAPR). PAPR is the ratio between the maximum power of a sample OFDM signal and the average power of that OFDM symbol. High PAPR introduces signal distortion in the nonlinear region of power amplifier and the signal distortion introduces the degradation of bit error rate [1]. A major study has been carried out on effect of PAPR in OFDM, which shows that PAPR depends on the SNR for channel with signal and noise. The effect has been tested with introducing a Gaussian noise in OFDM system.

R. Datta (✉)
Narula Institute of Technology, Kolkata, India
e-mail: rimpidatta@yahoo.com

© Springer India 2015

K. Maharatna et al. (eds.), *Computational Advancement in Communication Circuits and Systems*, Lecture Notes in Electrical Engineering 335,
DOI 10.1007/978-81-322-2274-3_15

15.2 Background

15.2.1 OFDM System

OFDM has been developed as the modulation method in the new wireless technologies. It mitigates the serious problem of multipath propagation, which causes massive data errors, and loss of signal in the microwave and UHF spectrum. OFDM is a broadband multicarrier modulation technique that gives very good performance and is beneficial over single-carrier modulation methods as it fits better with today's high-speed data requirements and operation in the UHF and microwave spectrum [3]. The main features of OFDM systems are processing on the source data, symbols are modulated onto orthogonal sub-carriers, orthogonality is maintained during channel transmission, synchronization, demodulation of the received signal using FFT, channel equalization, decoding and de-interleaving. In the OFDM system, Inverse Fast Fourier Transform/Fast Fourier Transform (IFFT/FFT) algorithms are used in the modulation and demodulation of the signal [4, 5]. The length of the IFFT/FFT vector determines the resistance of the system to errors caused by the multipath channel [4]. OFDM is generated by first choosing the spectrum required, based on the input data and modulation scheme used. Each carrier to be produced is assigned some data to transmit. The required amplitude and phase of the carrier is then calculated based on the three modulation schemes (typically differential BPSK, QPSK, or QAM) [3, 4]. Then, the IFFT converts this spectrum into a time domain signal. The FFT transforms a cyclic time domain signal into its equivalent frequency spectrum. This is done by finding the equivalent waveform, generated by a sum of orthogonal sinusoidal components. The amplitude and phase of the sinusoidal components represent the frequency spectrum of the time domain signal. OFDM signal has high peak-to-average-power ratio (PAPR) value (Fig. 15.1).

15.2.2 Peak to Average Power Ratio (PAPR)

Let us consider a data block of length N as a vector $X = [x_0, x_1, x_2, \ldots\ldots, x_{n-1}]^T$ where N denotes the number of sub-carriers. The duration of a symbol x_n in X is T.

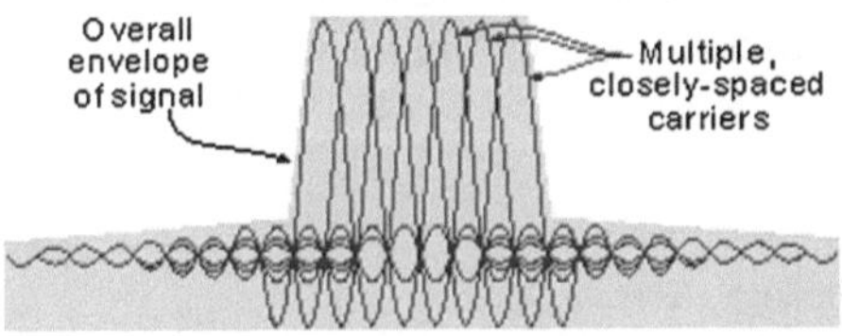

Fig. 15.1 OFDM spectrum

Each symbol in X modulates one of a set of sub-carriers f_n where $\{n = 0, 1, 2, \ldots\ldots, N-1\}$. The modulation scheme takes either quadrature amplitude modulation (QAM) or quadrature phase shift keying (QPSK) by definition. The sub-carriers are orthogonal to each other, that is, $f_n = n\nabla f$, where $\nabla f = 1/NT$ and NT is the duration of OFDM [6] data block X. The complex envelop of the OFDM signal is expressed as

$$x(t) = \frac{1}{\sqrt{N}} \sum_{n=0}^{N-1} x_n e^{-j2\pi f_n t}, \quad 0 \leq t < NT, \tag{15.1}$$

PAPR is defined as the ratio of the maximum divided by the average power of the signal [1]. Now the PAPR of the signal expressed in (15.1) can be defined as

$$\text{PAPR} = \frac{\max_{0 \leq t < NT}(|x(t)|)^2}{\frac{1}{NT}\int_0^{NT}(|x(t)|)^2 \mathrm{d}t} \tag{15.2}$$

The effects of PAPR need to be studied for a continuous time OFDM signals, as the cost and power dissipation of the analog components dominated by this factor. Reducing the value of PAPR dramatically increases the performance of OFDM system and reduces the cost of system. However, most existing PAPR tuning methods can only be implemented on discrete time OFDM signals [7]. For better calculation of true PAPR, the OFDM signal samples are obtained by oversampling Eq. (15.1) by a factor of L to approximate the PAPR of the true OFDM signals better [7]. The L-times oversampled time domain samples can be obtained by an LN-point inverse discrete fourier transform (IDFT) of the data block with $(L-1)\ N$ zero padding [6]. It was shown in [8] that $L \geq 4$ is sufficient to get $\text{PAPR}(x) \approx \text{PAPR}(x(t))$. Therefore, I have considered $L = 4$ for my numerical calculation in this paper [8].

15.2.3 Complimentary Cumulative Distribution Function (CCDF)

Complimentary Cumulative Distribution Function is a parameter that represents how much time the signal spends at or above a given power level [9]. It gives an idea of probability that the PAPR exceeds a given threshold. It is already defined in Eq. (15.1) that x_n is quadrature amplitude modulation data symbol transmitted through the nth sub-carrier. Input information symbol x_n from Eq. (15.1) is assumed to be stastistically independent and identically distributed [3]. Let us assume that $x_n \in \{1, -1, j, -j\}$. When N is large ($N \gg 64$), the real ($\text{Re}\{x_n\}$) and imaginary ($\text{Im}\{x_n\}$) part of x_n are independent of each other and identically distributed Gaussian random variables with zero mean and a common variance $\sigma^2 = E\left[|x_n|^2\right]/2$, according to the central limit theorem [3]. Thus the amplitude of x_n, which can be denoted as $x_n = \sqrt{\mathrm{Re}^2\{x_n\} + \mathrm{Im}^2\{x_n\}}$, has a Rayleigh distribution with cumulative distribution function [3] as follows:

$$F_{|s_n|}(x) = \mathrm{prob}\{|x_n| \leq x\} = 1 - e^{-\frac{x^2}{\sigma^2}} \tag{15.3}$$

From Eq. (15.3), it can be said that power distribution becomes a central chi-square distribution with two degrees of freedom [2, 3]. Finally, assuming that $|x_n|_{n=0}^{N-1}$ are mutually uncorrelated from Eq. (15.3), the CCDF of the PAPR can be obtained as

$$\mathrm{CCDF}_{\mathrm{PAPR(t)}} = 1 - (1 - e^{-\mathrm{PAPR}_0})^N \tag{15.4}$$

where PAPR_0 is a given threshold. Since PAPR is a random value, a certain threshold value is required to calculate the probability of getting a higher value than threshold by eliminating noise.

15.3 Simulation Result

Simulation has been performed using MATLAB taking number of carrier $N = 64$, 128, 256, 512, 1,024. The PAPR values for different number of carriers are obtained for practical as well as theoretical values (Figs. 15.2, 15.3, 15.4, 15.5 and 15.6).

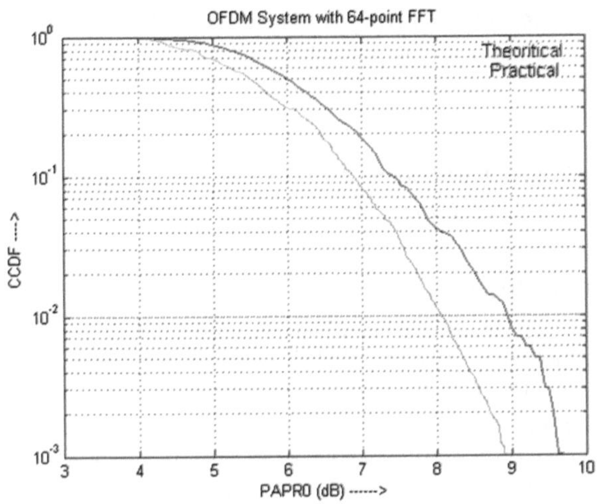

Fig. 15.2 CCDF of PAPR for 64-point FFT

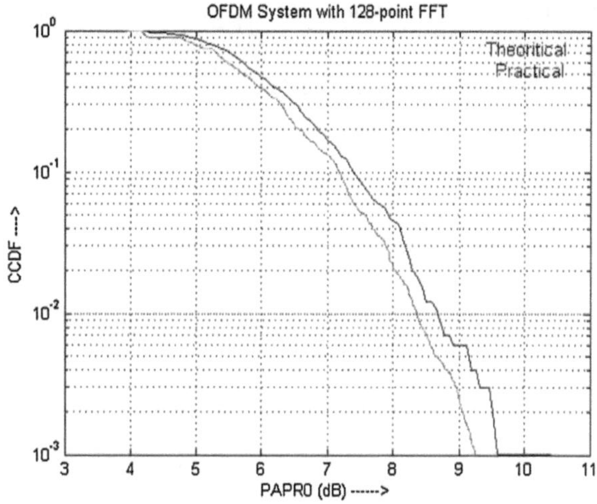

Fig. 15.3 CCDF of PAPR for 128-point FFT

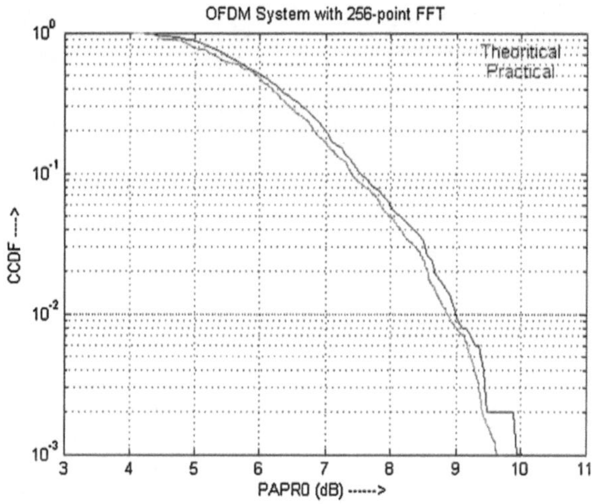

Fig. 15.4 CCDF of PAPR for 256-point FFT

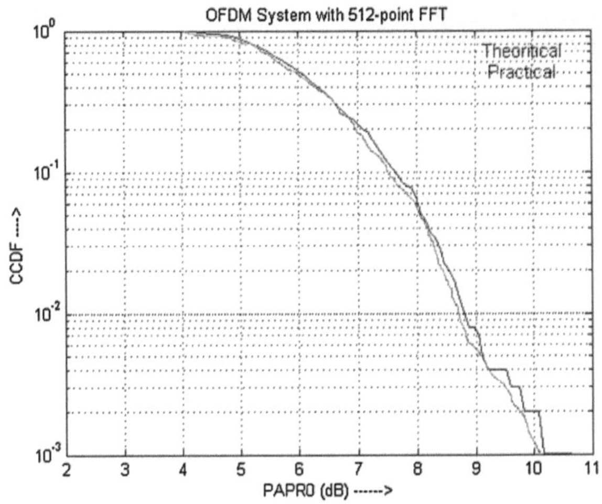

Fig. 15.5 CCDF of PAPR for 512-point FFT

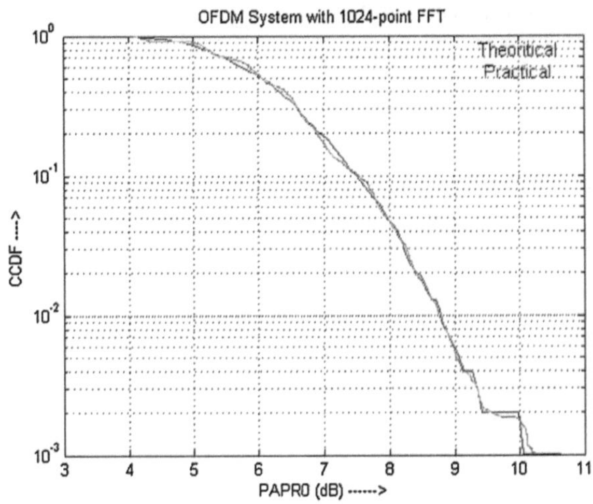

Fig. 15.6 CCDF of PAPR for 1024-point FFT

15.4 Conclusion

The simulation result shows that when the number of carriers becomes small, the practical values of CCDFs deviate from the theoretical value. The CCDF improves with increasing value of *N*. From the result, it can be concluded that high carrier number improves signal quality of OFDM output. CCDF plots provide an efficient method for the analysis of the PAPR of the OFDM signal and it allows us to detect noise in channel. The equation for the CDF of the signals is accurate only when the number of carriers is sufficiently large.

References

1. D.-W. Lim, S.-J. Heo, J.-S. No, H. Chung, A new PTS OFDM scheme with low complexity for PAPR reduction. IEEE Trans. Broadcast. **52**(1) (2006)
2. Y. Seungsoo, Y. Seokho, S.Y. Kim, I. Song, A novel PAPR reduction scheme for OFDM systems selective mapping of partial tones (SMOPT). IEEE Trans. Consum. Electron. **52**(1), 40–43 (2006)
3. T. Jiang, Y. Yang, Y.-H. ong, Exponential companding technique for PAPR reduction in OFDM systems. IEEE Trans. Broadcast. **51**(2), 244–248 (2005)
4. D. Wulich, Definition of efficient PAPR in OFDM. IEEE Commun. Lett. **9**(9), 832–834 (2005)
5. D.-W. Lim, J.-S. No, C.-W. Lim, A new SLM OFDM scheme with low complexity for PAPR reduction. IEEE Signal Process. Lett. **12**(2), 93–96 (2005)
6. S.H. Han, J.H. Lee, PAPR reduction of OFDM signals using a reduced complexity PTS technique. IEEE Signal Process. Lett. **11**(11), 887–890 (2004)
7. T. Jiang, W. Xiang, P.C. Richardson, J. Guo, G. Zhu, PAPR reduction of OFDM signals using partial transmit sequences with low computational complexity. IEEE Trans. Broadcast. **53**(3), 719–724 (2007)
8. J. Tellado, Peak to average ratio reduction for multicarrier modulation. Ph.D. dissertation, University of Stanford, (1999)
9. K. Parneet, S. Ravinder, Complementary cumulative distribution function for performance analysis of OFDM signals. IOSR J. Electron. Commun. Eng. **2**(5), 851–854 (2012)

Part III
Advances in Computation and Mathematics

Chapter 16
Identification of Protein Coding Region of DNA Sequence Using Multirate Filter

S. Singha Roy and S. Barman

Abstract Gene predication is an important part of Genomic Signal Processing (GSP) to identify the protein coding region of a DNA sequence. Discrete Fourier Transform (DFT) technique is most popular method for period-3 peaks identification for long time which generates various harmonics, generally known as 1/f noise along with sharp peaks, which may leads to false predication of coding region. Previous researchers used different types of filter to suppress this noise. The author in this article used a multirate filter along with anti-notch filter to suppress the noise and performance of the filter is measured by Signal-to-Noise Ratio (SNR).

Keywords GSP · DNA · Multirate filter · Lattice filter

16.1 Introduction

Genomic signal processing (GSP) is gaining popularity in the field of bio-informatics. One of the crucial topics for GSP is to detect Gene. A DNA is a long molecule comprising of gene and intergenic regions. Genes are the only segments that content genetic information. The main issue of DNA sequence analysis is to predict location of protein-coding region. Genetic information is stored in DNA in a particular order consisting of purine bases Adenine (A), Guanine (G) and pyrimidine bases Thymine (T), Cytosine (C). Different types of gene predication algorithms have been proposed [1–4] which are based on period-3 property [5] of fourier power spectrum of the DNA chain. In order to identify the period-3 peaks, it is essential to suppress the background 1/f noise which is always

S. Singha Roy (✉) · S. Barman
Institute of Radio Physics and Electronics, University of Calcutta,
92, APC Road, Kolkata 700009, India
e-mail: saikat.sroy@yahoo.in

S. Barman
e-mail: barmanmandal@gmail.com

© Springer India 2015

K. Maharatna et al. (eds.), *Computational Advancement in Communication Circuits and Systems*, Lecture Notes in Electrical Engineering 335,
DOI 10.1007/978-81-322-2274-3_16

present in the Power Spectral Density (PSD) spectrum of DNA sequence. Filtering of PSD is necessary to detect exact location of coding region. A number of researchers have addressed this problem in different ways. Vaidyanathan and Yoon [6] used multi stage IIR anti-notch filter whereas Tuqan and Rushdi [7] used a poly-phase FIR filter with different windows to suppress background noise. Epps et al. [8] investigated the effect of window shape and length on a DFT—based method whereas Kakumani and Devabhaktuni [9] proposed a model based technique and Roy et al. [10] introduced PFDN algorithm for exon detection. Barman (Mandal) et al. [11] introduce cascaded lattice filter to suppress the back ground noise. In this paper authors have presented a multirate filtering approach to suppress the noise and compared its performance with cascaded lattice filter. Matlab 7.9 with simulink toolbox has been used for filter realization. This paper is organized as introduction, methodology, result discussions and conclusion.

16.2 Methodology

16.2.1 Representation of DNA Sequences Using Different Numerical Mapping Techniques

The simplest form of DNA sequence representation is FASTA representation, uses the alphabet *A*, *C*, *T* and *G*. Detection of period-3 property of DNA sequences using DSP need to convert the character string into numerical sequences. Many rules have been proposed for DNA sequence numerical conversion. One of the simplest and oldest methods is the binary or Voss [12] representation. The Voss representation maps the nucleotides *A*, *C*, *T* and *G* into four binary indicator sequences $x_A(n)$, $x_C(n)$, $x_T(n)$ and $x_G(n)$.

In this article several mapping techniques have been used by the authors to analyse the Period-3 property in the protein coding region. The gene C-Elegan Cosmid F56F11.4a [13], Accession No. AF099922.1, sequence length 7990 bp comprising of 1st coding segment: 929–1039 bp, 2nd coding segment: 2528–2857 bp, 3rd coding segment: 4114–4377 bp, 4th coding segment: 5465–5644 bp and 5th coding segment: 7255–7605 bp is used as sample for comparative analysis throughout the article. Table 16.1 shows a comparative study of average deviation from period-3 position for different mapping technique.

16.2.2 DFT Analysis of DNA Sequence

In the present paper, the authors have concentrated on Pyrimidine Purine Complex mapping rule for DFT analysis of DNA sequence because this mapping rule ($a = j$, $c = 1$, $t = -1$ and $g = -j$) provides less average deviation from period-3 position compare to other mapping rules (Table 16.1).

Table 16.1 Average deviation from period-3 position for various mapping method

Sl. no.	Mapping technique	Numerical value	Average deviation from period-3 position
1	Binary indicator	$x_a = 1$; $x_c = 1$; $x_t = 1$; $x_g = 1$;	170.2
2	Integer number	$a = 2$; $t = 0$; $c = 1$; $g = 3$;	189.8
3	EIIP code	$a = 0.1260$; $t = 0.1335$; $c = 0.1340$; $g = 0.0806$	179.8
4	Molecular mass	$a = 134$; $t = 125$; $c = 110$; $g = 150$	316.6
5	Atomic number	$a = 70$; $t = 66$; $c = 58$; $g = 78$	151.8
6	Paired nucleotide atomic number	$a = 42$; $t = 42$; $c = 62$; $g = 62$	224.6
7	Complex number	$a = 1 + j$; $t = 1 - j$; $c = -1 - j$; $g = -1 + j$;	264.4
8	Pyrimidine purine complex	$a = j$; $c = 1$; $t = -1$; $g = -j$	139.6

For example given DNA sequence is $x(n) = a\ g\ a\ a\ t\ c\ a\ a\ g\ a\ t\ g\ t$. Using Pyrimidine Purine Complex mapping rule:

$$x(n) = j - jjj - 1\ 1jj - jj - 1 - j - 1$$

Let $X_S(k)$ be the FFT of mapped sequence given by

$$X_s(k) = \sum x_s[n] e^{-2\pi nk/N}, \quad n = 0, 1, 2, \ldots, N - 1 \tag{16.1}$$

The power spectral density of the sequence is

$$P_s(k) = \sum |X_s(k)|^2 \tag{16.2}$$

$P_S(k)$ plot may be used as preliminary indicator to detect probable coding region in DNA sequence. It has been observed that a protein coding region in DNA sequence typically has a peak at the frequency $k = N/3$, where N is the length of the sequence. As PSD plot is very much noisy in nature, therefore filtering of PSD is necessary to detect exact location of coding region.

16.2.3 Filtering Method

The authors have attempted two types of filter to suppress the noise in the PSD plots of DNA sequences and compare the performance of filtering in terms SNR. The overall method is represented in Fig. 16.1.

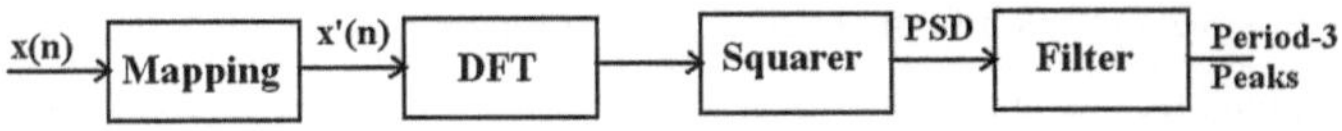

Fig. 16.1 Block representation of filtering technique

16.2.3.1 Cascaded Lattice Filter

The previous researchers [11] used IIR anti-notch filter with cascaded lattice structure to suppress PSD spectrum noise. IIR anti-notch filter cascaded with FIR Low Pass Filter (LPF) with Kaiser window functions having specification: Direct Form Structure, order = 100, W_C = 0.004 (Normalized) has been used for smoothing the filter response. For implementation of different types of filters Matlab Simulink environment has been used. Figure 16.3 depicted the output response of cascaded lattice filter which shows noise component crossed the amplitude level of 4 V. The author attempted to reduce the noise level less than 4 V using multirate filter along with IIR anti-notch filter.

16.2.3.2 Multirate Filter

The two principal operations in multirate filter banks are down-sampling and up-sampling by a factor of N. Table 16.2 shows the basic operations in multirate filter banks with their input output relations are given below [14].

Here for designing a filter, we used Up-sampler followed by Down-sampler by N with cascade of LPF and BPF are used in between N = 2 for both the down and up-sampling. In order to smooth the filter response, FIR low pass filter with Blackman window function having specifications: Direct form structure, order = 400, F_S = 8000 Hz and F_C = 0.003 has been used and FIR band pass filter (BPF) with Blackman function having specifications: Direct form FIR, order = 35, $WC1$ = 0.6666 and $WC2$ = 0.6667 (Normalized) is used to identify the period-3

Table 16.2 Basic operations in multirate filter and their input output relations

Operation	Input-output relation	Schematics
Down-sampling by N	$Y(z) = 1/N \sum_{K=0}^{N-1} X(W^k z^{1/N})$	X(z) — (N↓) — Y(Z)
Up-sampling by N	$Y(z) = X(z^N)$	X(z) — (N↑) — Y(Z)
Down sample followed by up-sampling by N with filter in between	$Y(z) = \frac{1}{N} H(z^N) \sum_{K=0}^{N-1} X(W^k z)$	X(z) — (N↓) — H(z) — (N↑) — Y(Z)
Up-followed by down-sampling by N with filter in between	$Y(z) = \frac{1}{N} X(z) \sum_{K=0}^{N-1} X(W^k z^{1/N})$	X(z) — (N↑) — H(z) — (N↓) — Y(Z)

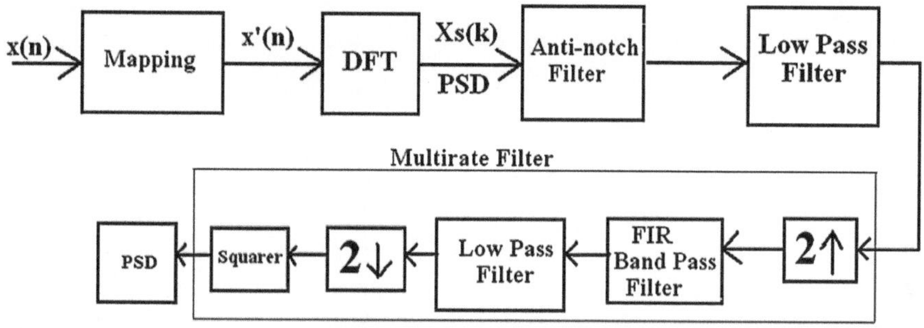

Fig. 16.2 Block diagram of proposed multirate filter

property. An IIR anti-notch filter having specification: Direct form with numerator length 151 has been used. Block diagram of proposed multirate-based anti-notch filter for identification of exon region is given in Fig. 16.2.

16.3 Result and Discussion

In this article authors have compared the performance between cascaded lattice filter and multirate filter with respect to noise suppression. From Table 16.1 it can be easily seen that the performance of Pyrimidine Purine Complex mapping is better because it gives lesser deviation from original period-3 position. The simulated PSD plot of two different types of filter i.e. IIR anti-notch filter with cascaded lattice and IIR anti-notch filter with multirate filter are shown in Figs. 16.3 and 16.4 respectively. From Plots, it is clear that spectrum resolution is higher and lower noise level in case of multirate filter compared to lattice filter (Table 16.3).

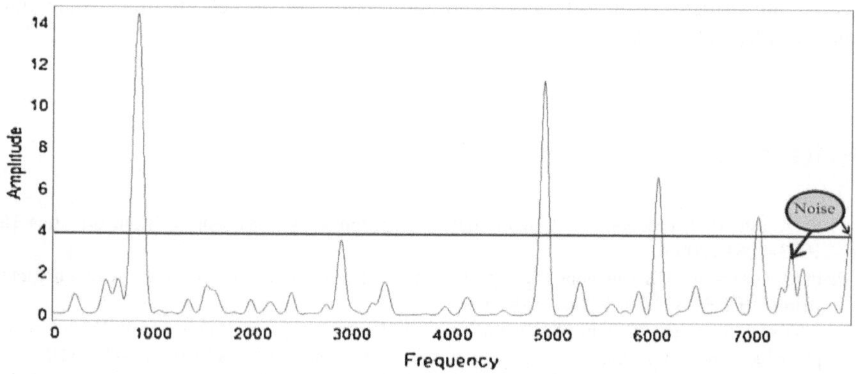

Fig. 16.3 Output of IIR anti-notch filter with cascaded lattice

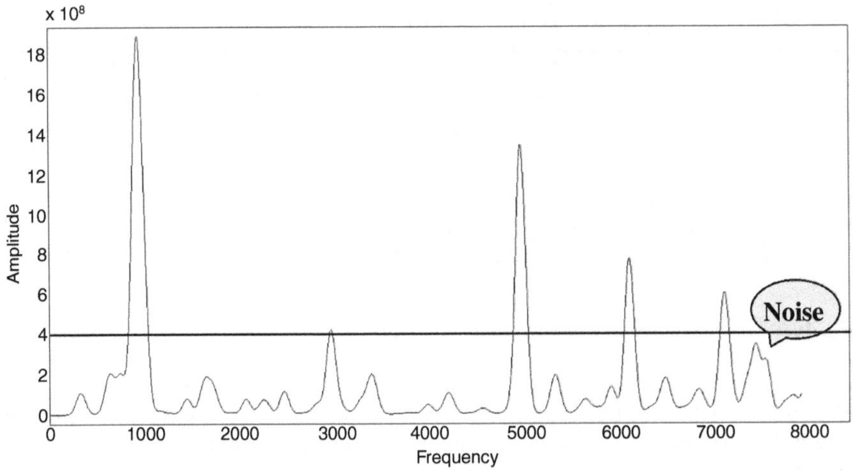

Fig. 16.4 Output of IIR anti-notch with multirate filter

Table 16.3 Performance comparison of different types of filters

Filter type	SNR
IIR anti-notch filter with cascaded lattice	12.9
IIR anti-notch with multirate filter	34.5

16.4 Conclusion

A multirate filter along with IIR anti notch filter has been explored to identify protein coding region of a DNA sequence. The spectrum plots showed Pyrimidine Purine Complex mapping provides less deviation of coding regions locations. Performance of multirate filter is better compared with IIR anti-notch filter with cascaded lattice by measuring SNR. Location accuracy and noise suppression which are critical issues in gene prediction both considered in this article and successfully achieved.

References

1. D. Anastassiou, Frequency—domain analysis of biomolecular sequences. Bioinformatics **16**, 1073–1081 (2000)
2. D. Anastassiou, DSP in genomics: processing and frequency domain analysis of character strings. IEEE-7803-7041-2001 (2001)
3. S Tiwari, S. Ramachandran, A. Bhattacharya, S. Bhattacharya, R. Ramaswamy, Prediction of probable genes by fourier analysis of genomic sequences. CABIOS **3**(3), 263–270 (1997)
4. C. Yin, S. Stephen, T. Yau, Prediction of protein coding regions by the 3 base periodicity analysis of a DNA sequence. J. Theoret. Biol. **247**, 687–694 (2007)

5. E.N. Trifonov, J.L. Sussman, The pitch of chromatin DNA is reflected in its nucleotide sequence. Proc. Natl. Acad. Sci. USA **77**, 3816–3820 (1980)
6. P.P. Vaidyanathan, B.J. Yoon, The role of signal-processing concepts in genomics and proteomics. J. Franklin Inst. (Special Issue on Genomics) (2004)
7. J. Tuqan, A. Rushdi, A DSP perspective to the period—3 detection problem, in *Proceedings of the IEEE International Workshop on Genomic Signal Processing and Statistics, GENSIPS* (2006), pp. 53–54
8. J. Epps, E. Ambikairajah, M. Akhtar, An integer period DFT for biological sequence processing, in *Proceedings of the IEEE International Workshop on Genomic Signal Processing and Statistics GENSIPS* (2008), pp. 1–4
9. R. Kakumani, V. Devabhaktuni, Prediction of protein coding regions in DNA sequence using a model based approach, in *IEEE Explore* (2008) pp. 1918–1920. doi:978-1-4244-1684-4/08
10. M. Roy, S. Biswas, S. Barman(Mandal), Identification and analysis of coding and non-coding regions of a DNA sequence by positional frequency distribution of nucleotides (PFDN) algorithm, in *International Conference on Computers and Devices for Communication (CODEC)* (2009)
11. S. Barman(Mandal), S. Biswas, S. Das, M. Roy, Performance analysis and simulation of IIR anti-notch filter with various structures for gene predication application, in *5th International Conference on Computer and Devices for Communication (CODEC)* (2012)
12. R.F. Voss, Evolution of long-range fractal correlation and 1/f noise in DNA base sequences. Phys. Rev. Lett. **68**(25), 3805–3808 (1992)
13. National Centre for Biotechnology Information (NCBI). http://www.ncbi.nlm.nih.gov
14. M. Vetterli, A theory of multirate filter banks. IEEE Trans. Acoust. Speech Signal Process. **35**(3), 356–372 (1987)

Chapter 17
E-Waste Recycling as Criteria for Green Computing Approach: Analysis by QFD Tool

Biswajit Debnath, Rahul Baidya, Namrata T. Biswas, Rayan Kundu and Sadhan Kumar Ghosh

Abstract Green computing is an environmentally responsible approach to reduce electronic waste and power consumption that helps in use of computing resources efficiently. With the increase in use of computer and other electronic devices the energy consumption and carbon footprint are also increasing. E-waste recycling is one of the important approaches towards green computing. This paper focuses on the approaches of green computing and how it minimizes the environmental impacts of computers and other electronic devices effectively by e-waste recycling. Quality Function Deployment (QFD) analytical tool is used to find different parameters from primary research data those affect the e-waste recycling practice as green computing approach. The result will help the stakeholders in implementing green computing approach.

Keywords Green computing · E-waste recycling · QFD

B. Debnath (✉) · N.T. Biswas
Department of Chemical Engineering, Jadavpur University, Kolkata, India
e-mail: bisuworld@gmail.com

N.T. Biswas
e-mail: namratab92@gmail.com

R. Baidya · S.K. Ghosh
Department of Mechanical Engineering, Jadavpur University, Kolkata, India
e-mail: rahulbaidya.ju@gmail.com

S.K. Ghosh
e-mail: sadhankghosh9@gmail.com

R. Kundu
Department of Computer Science and Engineering Department,
Heritage Institute of Technology, Kolkata, India
e-mail: me.rayankundu@gmail.com

© Springer India 2015

K. Maharatna et al. (eds.), *Computational Advancement in Communication Circuits and Systems*, Lecture Notes in Electrical Engineering 335,
DOI 10.1007/978-81-322-2274-3_17

17.1 Introduction

Green computing is the study and practice of using computing resources efficiently. The main goals of green computing are to maximize energy efficiency during the product lifetime, reduce the use of hazardous materials and promote recyclability or biodegradability of defunct products and factory waste [1]. The approach includes the implementation of energy-conserving instruments and peripherals as well as reduced resource consumption and proper disposal of electronic waste (e-waste) [2]. A model is proposed by Dougherty et al. [3] driven by engineering approach to optimize the configuration, energy consumption, and operating cost of cloud auto-scaling infrastructure to create greener computing environment. E-waste (Management and Handling) Rules 2011 is in place in India since May 2012. The research questions which arise now are as follows: How does it help e-waste recycling? Which factors hinder the rules to be properly implemented? To what extent the e-waste (Management and Handling) Rules 2011 helps in implementing green computing approach?

17.2 Objective

This research tries to find the issues and challenges in green computing specifically the e-waste generation, e-waste recycling, its impact on the environment and the implementation of the e-waste rules in India. The work focuses on the computers peripherals and IT products, the largest contributor to the e-waste. Analysis is carried out to identify the issues through a Quality Function Deployment (QFD) analysis, considering the environmental aspects and the customer requirements within the computers and IT products users.

17.3 Literature Review on Green Computing

Green Computing has been popularized in the past few years. Several researches were carried out on green computing with different approaches. Malviya and Singh [4] present a brief study on green computing and discussed about different approaches of green computing. Jindal and Gupta [5] identified the key issues of green computing and evaluated different approaches to these problems. Jain et al. [6] showed implementation of Green Computing and IT together, can be cost effective. Anam and Syed [7], Roy and Bag [8] and Kiruthiga and Vinoth Kumar [9], presented e-waste minimization as an eco-friendly approach of green computing.

The QFD process uses matrices to help organizations to satisfy their customer requirements. The first of these matrices is called the house of quality (HOQ).

It displays the customers' wants and needs along the left side of the matrix and the technical requirements to meet these wants along the top of the matrix [10]. There were lots of application of QFD in various fields of research but literature in these fields is hardly available. Baidya [11] presented a model for predictive maintenance technology selection based on AHP and QFD. Pun [12] has presented the determinants of environmentally responsible operations and suggested green QFD as one of the tools for environmentally responsible operations. Hsu et al. [13] presented a methodology for data mining cycle in QFD to forecast future customer requirements for green design of life cycle.

17.4 Methodology

The methodology adopted in the study involves the following steps:

1. The environmental criteria and consumers (computer and the IT products users) criteria were identified through literature review and field study for proper implementation of green computing methodology by reducing the e-waste of computer peripherals and IT products.
2. The relationship ratings through field survey were obtained.
3. A standard statistical tool median was used to proceed to the calculation with a single crisp value. The final value was also rescaled with respect to the closest fixed rating value. Then the QFD analysis was performed.

QFD analysis basically considers the customer requirements and engineer's requirement. In this case, customer requirements represent the environmental requirements and the engineer's requirements are the requirements of the computer and the IT products users. Following steps were followed:

a. Identify the WHATs and HOWs and preparation of the relationship matrix.
b. The weights of the HOWs are placed at the base of the quality matrix. These weights are one of the main outputs of the QFD analysis and determine the overall importance considering the customer's and the environmental criteria.

$$\text{Weight (HOW)}_i = \text{V(HOW)}_{i1} \times \text{imp}(\text{WHAT}_1) + \cdots + \text{V(HOW)}_{in} \times \text{imp}(\text{WHAT}_n)$$

Where V(HOW)_{in} is the correlation value of HOW_i with WHAT_n, and imp (WHAT_n) represents the importance or priority of WHAT_n.

Table 17.1 QFD relationship matrix

Environmental/consumer (computer and the IT products users)	Degree of importance	Price of e-waste component	Awareness	Carrying/ transportation cost	Authorized collector	EPR	Effectiveness of the collector
E-waste cannot be dumped	5	1	9	3	9	9	3
Crude method must not be used for extraction of metals	4	1	9	0	3	9	0
Open burnings of cable should not be done	4	0	9	1	3	3	0
Recovery of materials	3	3	3	3	9	9	3
Recycling of materials	4	3	9	1	3	9	1
Extraction of heavy metals	2	1	3	1	3	3	1
Trans-boundary transfer of e-waste	4	0	1	0	0	3	0
Remanufacturing	3	3	3	3	9	9	0
Batteries to be segregated from the e-waste	3	1	3	1	3	9	3
Final disposal of the waste	4	1	3	3	1	3	3
Over all importance		48	202	58	154	240	51
Rank		6	2	4	3	1	5

17.5 Results and Discussion

In the primary research in preparing QFD analysis the main criteria considered under the environmental aspects are, (a) e-waste cannot be dumped, (b) crude method must not be used for extraction of metals, (c) open burning of cable should not be done, (d) recovery of materials, (e) recycling of materials, (f) extraction of heavy metals, (g) trans-boundary transfer of e-waste, (h) remanufacturing, (i) batteries to be segregated from the e-waste and (j) final disposal of the waste. The criteria considered for the user's requirements are, (a) price of e-waste component received, (b) awareness, (c) carrying/transportation cost, (d) authorized collector, (e) Extended Producer Responsibility (EPR) and (f) effectiveness of the collector. To find out the relationships importance factors a questionnaire was prepared based on the scale—none (=0), weak relationship (=1), medium relationship (=3) and strong relationship (=9). This questionnaire was used to collect ratings through a primary survey. The data thus obtained were used to prepare a QFD chart using Microsoft Excel and analysis was done as per the stated algorithms. The analysis provides an important view regarding the issues and challenges of e-waste management systems. The QFD analysis shows that EPR is the most important aspect and needs better implementation. The second important criterion is the awareness among the stakeholders that has to be increased. The third important criterion is the authorized collector which falls under legislative purview. This is because the implementation of the legislation is weak and the government has significant scope to look into the matter of e-waste dumping. It is observed (Table 17.1) that in terms of environmental factors, the degree of importance of e-waste dumping is the most important issue. The analysis showed that the least important of the criteria was the price of the e-waste component received.

17.6 Conclusion

It has been observed that EPR has not been implemented in the areas of e-waste recycling properly. This research work concludes that the responsibility of computer and electronic goods producers must be strengthened with respect to EPR aspect in the rules. Also poor awareness of the stakeholders on e-waste recycling and active participation of the authorized collector and recyclers are the stumbling blocks towards achieving green computing approach through e-waste recycling. Hence, these three are the prioritized areas which are to be looked into by the stakeholders and the legislative body for strengthening the e-waste rule implementation in India leading to the concept of Green Computing.

Acknowledgments Authors acknowledge the support of the UKIERI project at Mechanical Engineering Department, Jadavpur University, India and Aston University, UK.

References

1. W. Feng (ed.), Green computing: large-scale energy efficiency. (CRC Press, 2011)
2. P. Chakraborty, D. Bhattacharyya, Y. Sattarova Nargiza, S. Bedajna, Green computing: practice of efficient and eco-friendly computing resources. Int. J. Grid Distrib. Comput. **2**(3), 33–38 (2009)
3. B. Dougherty, J. White, D.C. Schmidt, Model-driven auto-scaling of green cloud computing infrastructure. Future Gener. Comput. Syst. **28**, 371–378 (2012)
4. P. Malviya, S. Singh, A Study about Green Computing. Int. J. Adv. Res. Comput. Sci. Softw. Eng. **3**(6), 790–794 (2013)
5. G. Jindal, M. Gupta, Green computing "future of computers". Int. J. Emerg. Res. Manage. Technol. 14–18, ISSN: 2278-9359 (2012)
6. A. Jain, M. Mishra, S.K. Peddoju, N. Jain, Energy efficient computing—green cloud computing. IEEE, 978–982 (2013)
7. A. Anam, A. Syed, Green Computing, E-waste management through recycling. Int. J. Sci. Eng. Res. **4**(5), 1103–1106 (2012)
8. S. Roy, M. Bag, Green computing—new horizon of energy efficiency and e-waste minimization—world perspective vis-à-vis Indian Scenario. *Emerging Technologies E-Government* (2013)
9. P. Kiruthiga, T. Vinoth Kumar, Green computing—an ecofriendly approach for energy efficiency and minimizing E-waste. Int. J. Adv. Res. Comput. Commun. Eng. **3**(4), 6318–6321 (2014)
10. L. Cohen, *Quality Function Deployment*, 1st edn. (Addison Wesley Publishing Company, New York, 1995)
11. R. Baidya, Some analysis of the application of condition based maintenance of some critical parts specific to power plant and engine using AHP and QFD, Master thesis, Mechanical Engineering Department, Jadavpur University, India (2014)
12. K. Pun, Determinants of environmentally responsible operations: a review. Int. J. Qual. Reliab. Manage. **23**(3), 279–297 (2006)
13. C.-H. Hsu, A.-Y. Chang, H.-M. Kuo, Data mining QFD for the dynamic forecasting of life cycle under green supply chain. WSEAS Trans. Comput. 11(1), 1–10 (2012)

Chapter 18
To Compare the Active Sites of a Series of Astacin Family Proteases by Multiple Sequence Alignment and Homology Modelling Methods

Indrani Sarkar

Abstract The astacins, a family of zinc-dependent endopeptidases have been widely detected in animal kingdom from humans to bacteria. They play a diverse role in both mature and developmental systems ranging from bone morphogenesis, tissue differentiation, hatching process, digestive function and in different diseases including cancer, inflammation, fibrosis and neurodegenerative diseases. In the absence of complete three-dimensional structure of the proteases, homology modelling has been carried to build at least the protease domain of this class of enzymes so that subsequent docking of substrates/inhibitors in future study will help to explore the differential substrate specificity of the astacins.

Keywords Bioinformatics · Molecular modelling · Astacin family · Zinc metalloendoproteases · Homology modelling · Sequence alignment · In silico drug design

18.1 Introduction

Metalloendopeptidases are zinc-dependent hydrolases that have many different roles in biological systems, ranging from connective tissue remodelling to removal of signal sequences in nascent proteins [1]. The "astacin" family of endopeptidases, found widely from bacteria to human, emerged in the 1990s. It was recognized as a consequence of cloning and sequencing of meprins; the metalloendopeptidases isolated from brush border membranes of rodents and humans. Astacins have been found to play a diverse role in both mature and developmental systems ranging from bone morphogenesis, tissue differentiation, hatching process, digestive function and

I. Sarkar (✉)
Department of Physics, Narula Institute of Technology, 81, Nilgunj Road, Agarpara, Kolkata 700109, West Bengal, India
e-mail: head_phy.nit@jisgroup.org

© Springer India 2015

K. Maharatna et al. (eds.), *Computational Advancement in Communication Circuits and Systems*, Lecture Notes in Electrical Engineering 335,
DOI 10.1007/978-81-322-2274-3_18

in different diseases including cancer. Among these meprins are the only astacin proteinases that function on the membrane extracellularly because they can be membrane bound or secreted. The first member of the astacin family was isolated and identified from the cray fish Astacus astacus L [2]. Subsequently, it was crystallized and the three-dimensional structure solved adding a new dimension to the understanding and exploration of these metalloendopeptidases [3]. All the members of this family have a protease domain containing approximately 200 amino acids, which shares an amino-acid sequence similarity of 29–99 %. Astacin family members are characterized by a unique 18 amino acid signature sequence HEXXHXXGFXHEXXRXDR containing the Zn binding motif HEXXH present in all metalloendopeptidases. Most of the known family members contain a COOH terminal to the protease domain. They are found to contain one or more copies of the EGF (epidermal growth factor) like E and/or CUB (complement subcomponents) domain [4, 6, 7, 9]. An attempt was made based on the astacin structure to model the three-dimensional structure of at least the protease domains of other members of this family by molecular modelling and then make a comparative study of the interactions of substrates and inhibitors with the active site of the astacins [8]. The substantial difference in the structure of the proteolytic domain of this group of metallopeptidase will definitely throw some light on the substrate specificity differences among the members of the same astacin family [5]. On the structural topology of their catalytic modules, the astacins are grouped in the 'metzincin superfamily' together with the serralysins, the adamalysins, reprolysins and the matrix metalloproteinases (MMPs).

18.2 Methodology

18.2.1 Sequence Alignment

It is a process by which two protein sequences are aligned to achieve maximum levels of identity between them. The comparisons of the sequence are made with the given sequence to that of the database sequence. This method helps to determine the structural, functional and evolutionary relationships among a group of proteins. Information generated from sequence alignment can help in predicting functions for proteins with unknown structure. It also determines the evolutionary relations of organisms and helps to predict the 3D structure of a protein. Homology is attributed to similarity due to a descent from a common ancestor, i.e., if two sequences that are from different organisms are similar they are termed as Homologous. Multiple sequence Alignment (MSA) performs the alignment of more than two protein sequences. The sequence conservation of proteins domains and structures are found from it. It is an extrapolation of the pair-wise sequence alignment which gives alignment of similar sequences and provides a better alignment score. Analysis like Homology modelling for prediction of protein structure, Phylogenetic analysis, motif detection, etc., is based on the results of MSA.

18.2.2 Homology Modelling

The sequences of 22 astacin sequences were retrieved from UniProtKB/SwisspProt [4] and were found to have good similarity in the catalytic Met-turn and zinc binding motif. MSA was performed with ClustalW. The GONNET matrix was applied to know the conserved residues among the sequences. The astacins contained four conserved Cysteine residues forming two disulphide bridges in their structure: Cys-42, Cys-198, Cys-64, Cys-84 as in astacin. Multiple sequence analysis of nine target proteins (Hcw1, Hce2, Qucam1, Drtlr1, Drtld, Xebmp, Hubmp, Humepa, Humepb) using CLUSTLAW show a sequence identity of 35–46 % in comparison to astacin. These proteins also contain the characteristic Met-turn containing the SXMHY sequence of the metzincin family. Ramachandran plot of each of the modelled protease domain of astacins show most residues in the favourable region. Comparison of astacin with other group of metalloproteases reveals that there are subtle differences in amino acids which contribute in their substrate binding capacity. The generated modelled protein was then checked and prepared to correct any improper bond lengths, bond angles, conformations as well as add missing atoms and correct connectivity within the protein. The stereo chemical quality of the models was validated by Ramachandran plot using PROCHECK.

18.3 Results and Discussion

The metalloproteases meprin α and meprin β shows unique structural and functional features among all extracellular proteases. They are found to be highly expressed in kidney and intestine and are responsible for angiogenesis, cancer, inflammation, fibrosis and neurodegenerative diseases. Though the protease domains of these evolutionary related proteins are 55 % identical at the amino acid levels, the substrate and peptide bond specificities vary markedly. The sequence alignment of the catalytic domain of both the meprins are shown in Fig. 18.1.

Zn was modelled by coordination bond and subsequent models of the two meprins were built. The amino acid differences in active site contributing in inhibitor/substrate binding are listed in Table 18.1. The subsites S1, S2, S3, S1′, S2′, S3′ have differences in amino acid residues which are to be noted. Meprin α subunits prefer small or aromatic amino acids proximal to the scissile bond while meprin β subunit likes acidic residues proximal to the scessile bond [5]. For example, gastrin which has five glutamine residues is an excellent substrate for meprin β while it is not hydrolyzed by meprin α. The binding mode of a hydroxamic acid inhibitor Z-Pro-Lys-Pheψ(PO_2CH_2)L,D Ala-Pro-OMe with different subsites of the protease is shown in Table 18.1.

Human meprin α and meprin β were modelled with a hydramic acid derivative, which experimentally was found to be a good inhibitor. The interaction of the

```
humep-α     NGLRDPNTRWT-FPIPYILADNLGLNAKGAILYAFEMFRLKSCVDFKPYEGESSYIIFQQ 59
humep-β     NSIIGEKYRWP-HTIPYVLEDSLEMNAKGVILNAFERYRLKTCIDFKPWAGETNYISVFK 59
astacin     AAILGDEYLWSGGVIPYTFAGVSGADQS-AILSGMQELEEKTCIRFVPRTTESDYVEIFT 59
             .: . :  *.   *** : .    : . .** .::  . *:*: * *   *:.*: .
humep-α     -FDGCWSEVGDQHVG-QNISIGQG-CAYKAIIEHEILHALGFYHEQSRTDRDDYVNIWWD 116
humep-β     -GSGCWSSVGNRRVGKQELSIGAN-CDRIATVQHEFLHALGFWHEQSRSDRDDYVRIMWD 117
astacin     SGSGCWSYVG-RISGAQQVSLQANGCVYHGTIIHELMHAIGFYHEHTRMDRDNYVTINYQ 118
              .**** ** :  * *::*:  . *    . : **::**:**:**::* ***:** * ::
humep-α     QILSGYQHNFDTYDDSLITDLNTPYDYESLMHYQPFSFNKNASVPTITAKIPEFNSIIG- 175
humep-β     RILSGREHNFNTYSDDISDSLNVPYDYTSVMHYSKTAF-QNGTEPTIVTRISDFEDVIG- 175
astacin     NVDPSMTSNFD--IDTYSRYVGEDYQYYSIMHYGKYSFSIQWGVLETIVPLQNGIDLTDP 176
            .: ..   **:    *       :.  *:* *:***    :*  :         . : :  .: .
humep-α     -QRLDFSAIDLERLNRMYNCTTTH 198α
humep-β     -QRMDFSDSDLLKLNQLYNCSSSL 198
astacin     YDKAHMLQTDANQINNLYTNECSL 200
```

Fig. 18.1 Multiple sequence alignment

Table 18.1 Differences in the active sites of human meprin α and human meprin β

Astacin	Human meprin α	Human meprin β	Subsite	Inhibitor site
Trp 65	Trp 64	Trp 64	S3	P3 Pro
Tyr 67	Glu 66	Ser 66		
Tyr 101	Tyr 99	Trp 100	S2 (hydrophilic pocket)	P2 Lys
Asp 131	Asp 131	Asp 132	Backbone of P2 interacts with ser via water	
Ser 66	Ser 65	Ser 65		
Asn 127	Asn 125	Asn 126	S1	P1(Phe)_−P2(Lys)
Trp 65	Trp 64	Trp 64		
Cys 65	Cys 63	Cys 63		
Glu 93	Glu 91	Glu 92	Zinc	Interaction with phosphinic group
Tyr 149	Tyr 149	Tyr 150		
His 92	His 90	His 91	S1′	P1′(Ala)
Cys 64	Cys 63	Cys 63		Small uncharged residues are preferred due to pro 176
Thr 89	Ile 87	Thr 88		
Pro 176	Gly175	Gly175		
Phe 154	Phe 154	Phe 155	S2′ (prefers Pro in P2′)	P2′ (Pro)
Pro 176	Find out	Find out		
Asp 175	Gly 175	Asp 172		
Thr 89	Ile 87	Thr 88	S3′	Empty in the inhibitor
Gly 83				
Tyr 177	deleted	deleted		
Tyr 67	Glu 66	Ser 66	S4	Benzoyl group
Ile 71				

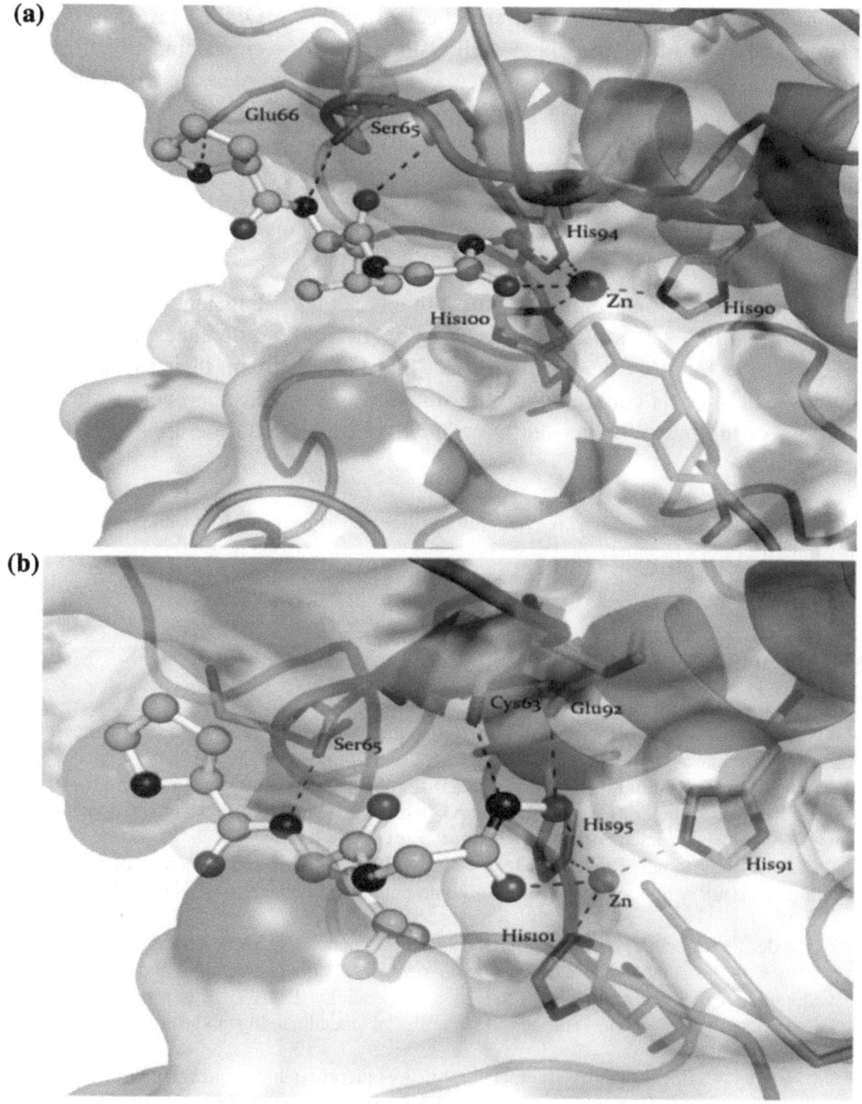

Fig. 18.2 Hydrogen bonding network and Zn coordination of human meprin-α (**a**) and -β (**b**). Inhibitor shown in *ball* and *stick* and important residues present in active site shown in stick. Inhibitor coordinates the catalytic zinc ion (*grey sphere*) via carbonyl oxygen of glycine and 'O' atom of hydroxamic acid. Figure created with PyMol

catalytic residues with inhibitor atoms are shown in Fig. 18.2 for both the meprins. It was seen from the models that the inhibitor made a network of hydrogen bonds, electrostatic interactions and coordinate bonds with the Zn atom to get stabilized inside the catalytic site of the meprins. The bond distances, bond angles and dihedral angles of the complex were checked by the MODELLER. The conformational changes of the protein as well as the inhibitor were calculated before and after binding. The value of binding free energy of the complex was also calculated by using Discovery Studio suit of programs.

18.4 Conclusion

The sequence analysis, alignment and homology modelling methodologies were applied to build a series of astacin proteases. The subtle differences in amino acid residues were found in the catalytic site which could explain their substrate specificities. The models could explain the function of certain amino acid residues on the drug or inhibitor binding the knowledge of which could be applied in further drug design.

Acknowledgments The author is thankful to the Department of Biotechnology, India for providing necessary facilities.

References

1. N. Abramowitz, I. Schechter, A. Berger, On the size of the active site in proteases. II. Carboxypeptidase-A. Biochem. Biophys. Res. Commun. **29**, 862–867 (1967)
2. Accelrys Discovery Studio Visualiser v 3.5.0.12158, Accelrys Software Inc, San Diego
3. J. Bella, M.J. Humphries, Calpha-H...O=C hydrogen bonds contribute to the specificity of RGD cell adhesion interactions. BMC Struct. Biol. **5**, 4 (2005)
4. W. Bode, F.X. Gomis-Rüth, R. Huber, R. Zwilling, W. Stöcker, Structure of astacin and implications for activation of astacins and zinc-ligation of collagenases. Nature **358**, 164–167 (1992)
5. W. Bode, P. Reinemer, R. Huber, T. Kleine, S. Schnierer, H. Tschesche, The X-ray crystal structure of the catalytic domain of human neutrophil collagenase inhibited by a substrate analogue reveals the essentials for catalysis and specificity. EMBO J. **13**, 1263–1269 (1994)
6. J.S. Bond, R.B. Beynon, The astacin family of metalloendopeptidases. Protein Sci. **4**, 1247–1261 (1995)
7. C.C. Broder, C. Becker-Pauly, The metalloproteases meprin α and meprin β: unique enzymes in inflammation, neurodegeneration, cancer and fibrosis. Biochem. J. **450**, 253–264 (2013)
8. A. Brunger, C.L. Brooks III, M. Karplus, Stochastic boundary conditions for molecular dynamics simulations of ST2 water. Chem. Phys. Lett. **105**, 495–500 (1984)
9. E. Dumermuth, E.E. Sterchi, W.P. Jiang, R.L. Wolz, J.S. Bond, A.V. Flannery, R.J. Beynon, The astacin family of metalloendopeptidases. J. Biol. Chem. **266**, 21381–21385 (1991)

Chapter 19
A New Way to Find Similarity/Dissimilarity of DNA Sequences on the Basis of Dinucleotides Representation

Subhram Das, Subhra Palit, Anindya Raj Mahalanabish and Nobhonil Roy Choudhury

Abstract In this paper, we have introduced a new way of representation of dinucleotides. They are directly represented by two-dimensional coordinates (x, y) in the Euclidean plane. By writing $z = x \times y$, ultimately the representation is given by three-dimensional coordinates (x, y, z). Based on the variances of cumulative values x', y', z' of x, y, and z respectively, a measure is determined to compare coding sequences of β-globin genes of 11 species. We have also used another measure for similar comparison of sequences. This is based on the probability distribution of dinucleotides occurring in the sequence. Finally, we have compared the similarity/dissimilarity matrix obtained under above two measures by the method of correlation coefficient.

Keywords Dinucleotide · Graphical representation · Coding sequence · Similarity/dissimilarity analysis · Correlation coefficient

S. Das (✉) · S. Palit · A.R. Mahalanabish · N.R. Choudhury
Department of Computer Science and Engineering,
Narula Institute of Technology, Kolkata, India
e-mail: subhram@gmail.com

S. Palit
e-mail: palitsubhra@gmail.com

A.R. Mahalanabish
e-mail: bitts.prince9@gmail.com

N.R. Choudhury
e-mail: nobhonil30390@gmail.com

© Springer India 2015
K. Maharatna et al. (eds.), *Computational Advancement in Communication Circuits and Systems*, Lecture Notes in Electrical Engineering 335,
DOI 10.1007/978-81-322-2274-3_19

19.1 Introduction

In Bioinformatics, the basic studying strategy for both DNA and proteins is the alignment of sequences. However, directly working with mononucleotides (A, G, C and T) leads to a lot of information loss. So we need to come up with different strategies to analyse the sequences in different possible ways. There have been many works to graphically and numerically represent DNA sequences. One such graphical representation was first given by Hamori and Ruskin in 1983 [1]. Graphs are more vivid and easier to understand, and hence a lot of other such work has also followed. Until now, graphical representation has been in a range of 2D to 6D of space, respectively [2].

Models based on mononucleotides include Z-curves and walking models [3, 4]. Z-curve representation has a special merit in the sense that it does not lose any biological information of the sequence. In fact it uses the classification of chemical structure based on purines/pyrimidines, amino/keto groups and strong/weak hydrogen bonds. Also the original sequence can be recovered from Z-curve. Nandy [5] had proposed a model by assigning A, G, T and C to the direction −x, +x, −y and + y, respectively of the Cartesian coordinate axes. But this led to overlapping and thereby loss of information. Another method was devised to prevent this loss, by using double vectors [6] but the results were not satisfactory. Anyway the mononucleotide models cannot represent the di and trinucleotides without complex calculations [7]. Trinucleotide (TN) models are rarely used due to problems with visualization [8]. So dinucleotide representation is an important choice.

What makes our representation better than the previous ones [9–13] based on dinucleotides (DN) is that it is a much easier method and hence more efficient. Variance shows the spread of the data rather than determining a theoretical centre, and the cumulative components reduce redundancy. We have taken up the information provided by the exon of β-globin genes of different species, which is essential for pharmaceutical purposes.

19.2 Construction of the DN Curve: Graphical Representation

There are four nucleotides: A, G, C and T. When their sequences are represented as dinucleotides, based on the position of the first bases and kinds of bases, there are 16 dinucleotides. Depending on the first base each is assigned a quadrant of the Cartesian coordinate axes (Fig. 19.1) [14].

Each dinucleotide is represented by a 2D coordinate (x, y). The signs of x and y are decided by the category of the base at the first site of dinucleotide, that is, {+, +} –> A, {−, +} –> G, {−, −} –> C and {+, −} –> T. The absolute values of x and y are given by the absolute (assumed) values of the bases |A| –> 1, |G| –> 2,

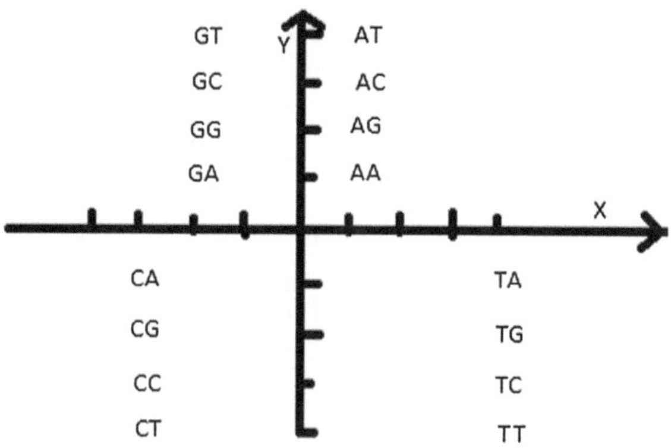

Fig. 19.1 The 16 dinucleotides placed in chosen quadrants

|C| -> 3, and |T| -> 4. For example, (−3, −2) indicates that the sign is {−, −}, i.e. first base is C and second base is G.

Now, let us consider all the dinucleotides of an arbitrary DNA sequence. Supposing S = ATCGT…CG is a DNA sequence with N bases, the number of dinucleotides is $N - 1$. We have plot sets (S) = (AT)(TC)…(CG)… to convert S into a curve, where, (base(n)base(n + 1)) corresponds to (xn, yn), the 2D coordinates of dinucleotide of base(n)base(n + 1) as introduced in Fig. 19.1, and n = 1, 2, 3,…, N − 1. For instance, in Table 19.1, we show how the calculation is done for a random sample DNA sequence ATGCCA.

From Fig. 19.1, we find that when $x > 0$, the first base of dinucleotide is A or T, while when $x < 0$, the first base is G or C, and then x divides the 16 kinds of dinucleotides into two groups, i.e. weak H–bond/strong H–bond groups. Similarly, $y > 0$ denotes the first base of dinucleotide as A or G, while $y < 0$ denotes the first base as C or T, and then the 16 kinds of dinucleotides are also divided into two groups by y, i.e. purine/pyrimidine groups. When $z > 0$, the first base of dinucleotide is A or C, while when $z < 0$, the corresponding base is G or T; hence z divides the 16 kinds of dinucleotides into two groups, i.e. amino/keto groups.

We define $z = x \times y$ and the respective cumulative sums of x, y and z as

$$x'_n = \sum_{i=1}^{n} x_i,\ y'_n = \sum_{i=1}^{n} y_i,\ z'_n = \sum_{i=1}^{n} z_i, \quad n = 1, 2, \ldots N - 1$$

Table 19.1 Values of dinucleotides and their corresponding coordinates for the sequence ATGCCA

Dinucleotide	X	Y	Z	X'	Y'	Z'
AT	1	4	4	1	4	4
TG	4	−2	−8	5	2	−4
GC	−2	3	−6	3	5	−10
CC	−3	−3	9	0	2	−1
CA	−3	−1	3	−3	1	2

Table 19.2 Classification of nucleotides based on division along *X*, *Y* and *Z* axes

Groups	Division at axis	Dinucleotides
Weak H-bond	*Y*	AA AG AC AT
		TA TG TC TT
Strong H-bond		GA GG GC GT
		CA CG CC CT
Purine	*X*	AA AG AC AT
		GA GG GC GT
Pyrimidine		TA TG TC TT
		CA CG CC CT
Amino	*Z*	AA AG AC AT
		CA CG CC CT
Keto		TA TG TC TT
		GA GG GC GT

where N is the number of nucleotides, and n is the number of dinucleotides in a given sequence.

Therefore, the 16 kinds of dinucleotides can be classified into two groups in three ways by x, y and z respectively, as presented in Table 19.2. Nevertheless, it is sufficient to derive a unique dinucleotide curve from it.

19.3 Analysis of Similarities/Dissimilarities Among Beta Globin Gene Sequences

Earlier many approaches to the coding sequences of the first exon of beta globin genes [6, 8–13, 15–19, 14, 20–28]. In order to have an unambiguous result, Nandy [5] suggested that the graphical techniques should be applied to complete coding sequences. In this segment, similarities analysis is performed on complete coding sequence of exon of beta globin genes among 11 species based on the dinucleotide curve discussed above in Sect. 19.2. Table 19.3 shows the information regarding these corresponding sequences.

19.3.1 Graphical Alignment

We first mention the percentages of different nucleotides in 11 β-globin genes in (Table 19.4).

We can obtain the information of dinucleotides from graphs of (x', n), (y', n) and (z', n) in (Fig. 19.2):

To establish our point we start with Bovine. We observe that the graph of (y', n) is always positive. Hence, the percentage of purine is more than that of pyrimidine.

Table 19.3 The complete coding sequences of β globin genes of 11 species

Species	NCBI ID	Location of each exon	Length of CDS
Human	U01317	62487...62278, 62409...62631, 62482...62610	444
Goat	M15387	279...364, 493...715, 1621...1749	438
Opossum	J03643	467...558, 672...894, 2360...2488	444
Gallus	V00409	465...556, 649...871, 1682...1810	444
Lemur	M15734	465...556, 649...871, 1682...1810	444
Mouse	V00722	275...367, 484...705, 1334...1,462	444
Rabbit	V00882	277...368, 495...717, 1291...1, 419	444
Rat	X06701	310...401, 517...739, 1377...1505	444
Gorilla	X61109	4538...4630, 4761...4,982, 5833...5881	364
Bovine	X00376	278...363, 492...714, 1613...1741	438
Chimpanzee	X02345	4189...4293, 4412...4633, 5484...5532	376

Table 19.4 Percentage of each nucleotide classification

Species	Strong H-bond	Weak H-bond	Purine	Pyrimidine	Keto	Amino
Bovine	54.01	45.99	51.72	48.28	56.98	43.02
Chimpanzee	56.27	43.73	50.67	49.33	56.53	43.47
Gallus	59.37	40.63	46.95	53.05	47.40	52.60
Goat	55.38	44.62	51.49	48.51	55.84	44.16
Gorilla	56.2	43.8	50.41	49.59	56.20	43.80
Human	56.43	43.57	50.34	49.66	54.63	45.37
Lemur	55.76	44.24	50.34	49.66	56.66	43.34
Mouse	55.76	44.24	49.89	50.11	51.69	48.31
Opossum	51.69	48.31	49.21	50.79	55.08	44.92
Rabbit	54.18	45.82	51.69	48.31	55.53	44.47
Rat	53.27	46.73	50.34	49.66	52.82	47.18

This is also observed for goat. Also, take chimpanzee for instance. We observe that (y', n) first becomes slightly negative and then continues as positive, (with a slight bend towards the line of origin) indicating dominance of purine over pyrimidine group. Similar nature is observed in rat, rabbit, lemur, human and gorilla. However, for opossum, gallus and mouse the (y', n) curve is more negative. Hence, the percentage of pyrimidine is more than that of purine.

Similarly, graph of (x', n) is always negative for all the 11 species, thereby meaning that the percentage of strong hydrogen bond is more than that of weak hydrogen bond. This is also observed from the chart given above.

Moreover, we observe that for all the species except gallus, the (z', n) graph is always negative. Hence, the percentage of keto group is more than that of amino. The (z', n) graph for gallus starts out as negative but then turns positive for approximately $n = 235$. We observe that for gallus amino group is more dominant

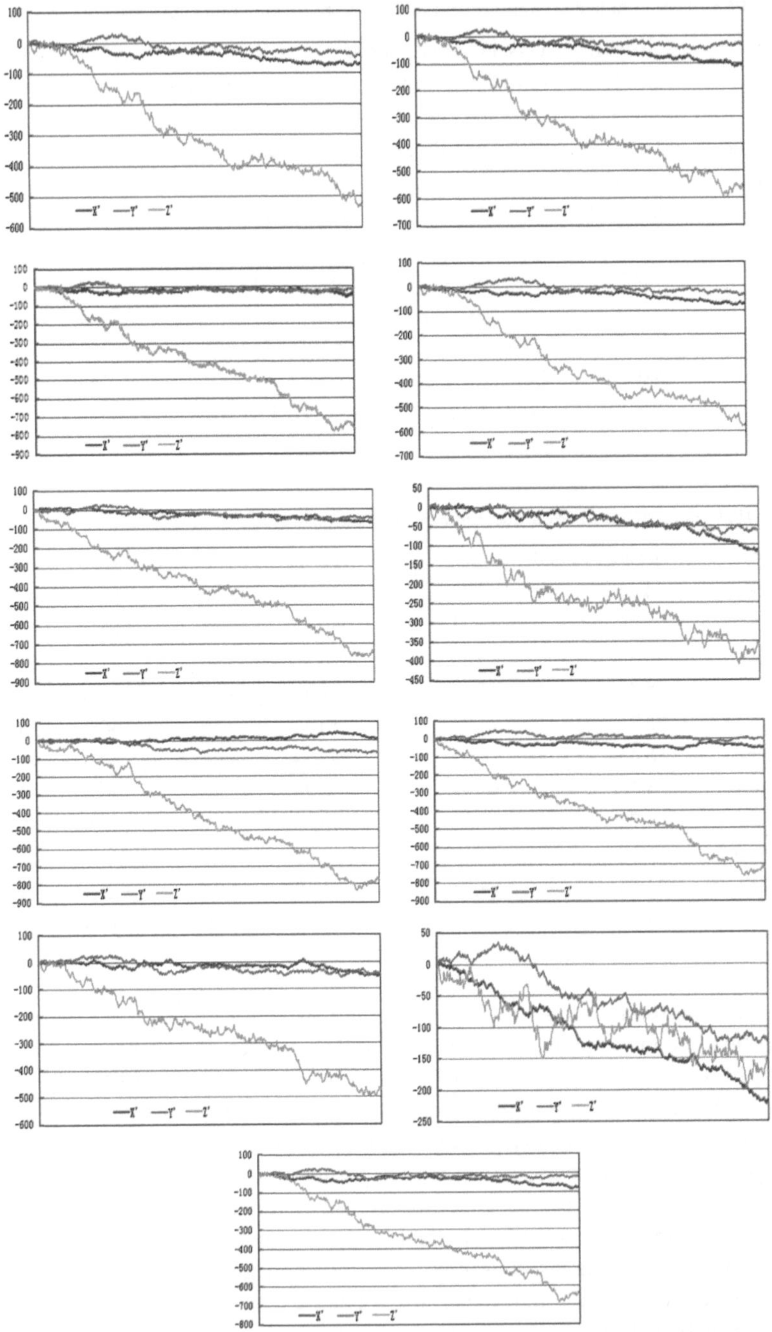

Fig. 19.2 2D curves on cumulative components x', y', z' of coding sequence of β-globin gene of bovine, chimpanzee, gallus, goat, gorilla, human, lemur, mouse, opossum, rabbit and rat, respectively

than the keto group. Thus, our method of representation of dinucleotides is able to describe the densities of the distributions of the individual dinucleotides.

19.3.2 Quantitative Measurement of Similarity/Dissimilarity of DNA Sequences

Mean is calculated as:

$$\mu_k = \sum_{i-1}^{n} k_i/n$$

where k is the component in $i = 1, 2, 3, \ldots, n$ and $n = N - 1$, where N is the number of nucleotides, and n is the number of dinucleotides in a given sequence.

Many authors have proposed E matrix, M/M matrix, L/L matrix, Lk/Lk matrix, and use their Eigen values as descriptors to do analysis of similarities or dissimilarities of DNA sequences [29]. However, the matrices become too large and complex to calculate Eigen values when DNA sequences are very long. Furthermore, there is some loss of information associated with these matrices [30]. Thus, we need some method that is more convenient and precise.

We have taken second-order moment/variance as our determining parameter. The variance is calculated as:

$$V_k = (\sum_{i-1}^{n} (k_i - \mu_k)^2)/n$$

Using Euclidean distance as the criterion of sequence similarities/dissimilarities, the smaller the Euclidean distance is, more similar the DNA sequences are. Therefore we can say that the distances between evolutionary closely related species are smaller, while those between evolutionary distant species are larger. The Euclidean distance between two sequences is defined as follows:

$$D(S_i, S_j) = \sqrt{\sum_{k-1}^{3} (V_k^{S_i} - V_k^{S_j})^2}$$

where, $V_k^{S_i}$ and $V_k^{S_j}$ are the kth component of the three-component vector V of sequences S_i and S_j, respectively. In this way, we obtain the matrix of similarities/dissimilarities of the 11 species presented in Table 19.5.

From Table 19.4, we find that gallus (the only non-mammal among them) and opossum (the most remote species from the remaining mammals) are most dissimilar to others among the 11 species. In case of humans, human–chimpanzee has the smallest distance, so they are more similar species pairs.

Table 19.5 Matrix of similarities/dissimilarities of the complete coding sequences of globin genes of 11 species based on 6 component vectors

	Bovine	Chimpanzee	Gallus	Goat	Gorilla	Human
Bovine	0	18786.24	47452.92	11953	24660.3	18983.8
Chimpanzee		0	28743.91	6833.89	5874.4	527.97
Gallus			0	35538.8	22928.4	28508.3
Goat				0	12708.2	7038.92
Gorilla					0	5710.54
Human						0
	Lemur	Mouse	Opossum	Rabbit	Rat	
Bovine	6086.16	38564.57	13754.39	6731.22	29365.6	
Chimpanzee	12728.4	19779.3	32532.74	12055	10584.8	
Gallus	41367.8	9297.82	61139.69	40740.6	18323.5	
Goat	5906.07	26611.97	25701.33	5221.94	17416.8	
Gorilla	18600.8	13907.71	38406.47	17929.2	4716.36	
Human	12914.9	19587.36	32722.16	12255.6	10422.7	
Lemur	0	32497.81	19808.56	817.51	23306.9	
Mouse		0	52306.32	31833.5	9225.33	
Opossum			0	20481	43110.1	
Rabbit				0	22635.5	
Rat					0	

19.4 Comparison of the Results of Two Types of Measures

The measure based on 16 components is naturally good enough as it deals with all the components of dinucleotides. Naturally, it is wise to check whether similar results based on parameters of our representation are in good match with the aforesaid results. For this purpose, we compare the results given by two matrices in the two cases by using the method of Pearson's correlation coefficient (PCC), whose formula is given as:

$$PCC = \frac{\sum XY - \frac{1}{N^2}\sum X \sum Y}{\sqrt{\left(\sum X^2 - \frac{1}{N^2}\left(\sum X\right)^2\right)\left(\sum Y^2 - \frac{1}{N^2}\left(\sum Y\right)^2\right)}}$$

where, X and Y stand for the entries in the columns of the matrices given in Table 19.5 and a table based on 16 components [14], respectively, and $N = 11$.

It is found that the correlation coefficient is nearly 0.9, which is with no doubt a good match.

19.5 Results and Discussion

In the earlier work on dinucleotide representation [14], mean of the cumulative values were used for calculating the distances between the sequences. But we have preferred variance in place of mean. The reason is that very often the mean value does not coincide with the given data values and may not remain very close to them. But variance has no such problem, as it simply tells us how much the data are spread away from the mean.

Evolutionary correlations among other species can also be obtained from general biological studies and these results coincide with Fig. 19.2 based on the implementation on a single link algorithm, and similar results are also obtained in recent papers using different methods [8, 11, 24].

19.6 Conclusion

The way of analysing DNA sequences depicted in this paper has a distinct advantage of being easier to perform and simple to understand. Any other slight discrepancies with actual closeness of species in reality may be due to the fact that, we have used only the information of the exon of the β-globin genes. Our results are at par with the earlier ones, if not better, for its simplicity in calculation.

References

1. E. Hamori, J. Ruskin, H curves, a novel method of representation of nucleotide series especially suited for long DNA sequences. J. Biol. Chem. **258**, 1318–1327 (1983)
2. A. Nandy, M. Harle, S.C. Basak, Mathematical descriptors of DNA sequences: development and applications. ARKIVOC **9**, 211–238 (2006)
3. C.T. Zhang, R. Zhang, Analysis of distribution of bases in the coding sequences by a diagrammatic technique. Nucl. Acids Res. **19**, 6313–6317 (1991)
4. R. Zhang, C.T. Zhang, Z curves, an intuitive tool for visualizing and analyzing the DNA sequences. J. Biomol. Struct. Dyn. **11**, 767–782 (1994)
5. A. Nandy, A new graphical representation and analysis of DNA sequence structure: I. methodology and application to globin genes. Curr. Sci. **66**, 309–314 (1994)
6. Z.J. Zhang, DV–Curve: a novel intuitive tool for visualizing and analyzing DNA sequences. Bioinformatics **25**, 1112–1117 (2009)
7. F.B. Guo, H.Y. Ou, C.T. Zhang, ZCURVE: a new system for recognizing protein–coding genes in bacterial and archaeal genomes. Nucl. Acids. Res. **31**, 1780–1789 (2003)
8. J.F. Yu, X. Sun, J.H. Wang, TN curve: a novel 3D graphical representation of DNA sequence based on trinucleotides and its applications, J. Theoret. Biol. (In press) doi:10.1016/j.jtbi.2009.08.005
9. X.Q. Liu, Q. Dai, Z.L. Xiu, T.M. Wang, PNN–curve: a new 2D graphical representation of DNA sequences and its application. J. Theoret. Biol. **243**, 555–561 (2006)
10. X.Q. Qi, J. Wen, Z.H. Qi, New 3D graphical representation of DNA sequence based on dual nucleotides. J. Theoret. Biol. **249**, 681–690 (2007)

11. Z.H. Qi, T.R. Fan, PN–curve: a 3D graphical representation of DNA sequences and their numerical characterization. Chem. Phys. Lett. **442**, 434–440 (2007)
12. Z. Cao, B. Liao, R.F. Li, A Group of 3D graphical representation of DNA sequences based on dual nucleotides. Int. J. Quantum Chem. **108**, 1485–1490 (2008)
13. Z.B. Liu, B. Liao, W. Zhu, G.H. Huang, A 2–D graphical representation of DNA sequence based on dual nucleotides and its application. Int. J. Quantum Chem. **109**, 948–958 (2009)
14. J.F. Yu, J.H. Wang, X. Sun, Analysis of similarities/dissimilarities of DNA sequences based on a novel graphical representation. State Key Laboratory of Bioelectronics
15. B. Liao, T.M. Wang, 3–D graphical representation of DNA sequences and their numerical characterization. J. Mol. Struct. (Theochem) **681**, 209–212 (2004)
16. R. Chi, K.Q. Ding, Novel 4D numerical representation of DNA sequences. Chem. Phys. Lett. **407**, 63–67 (2005)
17. Y.H. Yao, X.Y. Nan, T.M. Wang, A new 2D graphical representation—Classification curve and the analysis of similarity/dissimilarity of DNA sequences. J. Mol. Struct. (Theochem) **764**, 101–108 (2006)
18. B. Liao, K.Q. Ding, A 3D graphical representation of DNA sequences and its application. Theoret. Comput. Sci. **358**, 56–64 (2006)
19. B. Liao, T.M. Wang, Analysis of similarity/dissimilarity of DNA sequences based on non-overlapping triplets of nucleotide bases. J. Chem. Inf. Comput. Sci. **44**, 1666–1670 (2004)
20. W.Y. Chen, B. Liao, Y.H. Liu, W. Zhu, Z.Z. Su, A numerical representation of DNA sequence and its applications. MATCH Commun. Math. Comput. Chem. **60**, 291–300 (2008)
21. B. Liao, W. Zhu, Y. Liu, 3D graphical representation of DNA sequence without degeneracy and its applications in constructing phylogenic tree. MATCH Commun. Math. Comput. Chem. **56**, 209–216 (2006)
22. B. Liao, C. Zeng, F.Q. Li, Y. Tang, Analysis of similarity/dissimilarity of DNA sequences based on dual nucleotides. MATCH Commun. Math. Comput. Chem. **59**, 647–652 (2008)
23. M. Randic, J. Zupan, D. Vikic-Topic, D. Plavsic, A novel unexpected use of a graphical representation of DNA: graphical alignment of DNA sequences. Chem. Phys. Lett. **431**, 375–379 (2006)
24. Y. Guo, T.M. Wang, A new method to analyze the similarity of the DNA sequences. J. Mol. Struct. (Theochem) **853**, 62–67 (2008)
25. P.A. He, J. Wang, Characteristic sequences for DNA primary sequence. J. Chem. Inf. Comput. Sci. **42**, 1080–1085 (2002)
26. J. Wang, Y. Zhang, Characterization and similarity analysis of DNA sequences based on mutually direct–complementary triplets. Chem. Phys. Lett. **425**, 324–328 (2006)
27. Y.S. Zhang, W. Chen, Invariants of DNA sequences based on 2DD–curves. J. Theoret. Biol. **242**, 382–388 (2006)
28. J.F. Yu, X. Sun, J.H. Wang, TN curve: a novel 3D graphical representation of DNA sequence based on trinucleotides and its applications. Med. Eng. Phy. 459–460 (2009)
29. M. Randic, M. Vracko, N. Lers, D. Plavsic, Novel 2–D graphical representation of DNA sequences and their numerical characterization. Chem. Phys. Lett. **368**, 1–6 (2003)
30. B. Liao, T.M. Wang, Analysis of similarity/dissimilarity of DNA sequences based on 3–D graphical representation. Chem. Phys. Lett. **388**, 195–200 (2004)

Chapter 20
Single Person Hand Gesture Recognition Using Support Vector Machine

Sriparna Saha, Amit Konar and Jayashree Roy

Abstract A simple and easy-to-use system is designed for recognition of single person gestures using their movement of hands while expressing feelings. Here Kinect sensor is employed to generate the 20 body joint coordinates for a person. The gestures are comprised of six single hand gestures as well as four double hand gestures. In this paper, the authors have processed only one (right- or left-hand joint) coordinate while sculpting single hand gestures, whereas two (both right- and left-hand joints) coordinates are taken into account when double hand gestures are considered for each frame using the Kinect sensor. Once the coordinates are obtained, then normalization is carried out based on the coordinate of the hip centre for the first frame. The recognition procedure is based on support vector machine and produces high accuracy rate of 94.3 %.

Keywords Hand gesture · Kinect sensor · Trajectory · Variation

20.1 Introduction

Non-verbal communication between two persons is mainly conducted via body gestures. While expressing feelings, body movements, especially the motion of hands plays a pivotal role. Interpretation of gestures is a very crucial aspect of human–computer interaction.

S. Saha (✉) · A. Konar · J. Roy
Electronics and Tele-Communication Engineering Department,
Jadavpur University, Kolkata, India
e-mail: sahasriparna@gmail.com

A. Konar
e-mail: akonar@etce.jdvu.ac

J. Roy
e-mail: jayashree.jgec@gmail.com

© Springer India 2015

K. Maharatna et al. (eds.), *Computational Advancement in Communication Circuits and Systems*, Lecture Notes in Electrical Engineering 335,
DOI 10.1007/978-81-322-2274-3_20

Kinect sensor [1, 2] can be used to identify a person's gestures by noticing the hand movements. In this paper, we have developed an algorithm to identify an unknown person while portraying his/her emotional states by expressing hand gestures. Here totally ten hand gestures are taken into account. Single hand gestures can be displayed by the right or left hand, while double hand gestures are conveyed using both the hands. The proposed algorithm is also developed keeping in mind the utility of human–machine interaction.

Among various applications of Kinect sensor, it finds wide application in upper limb rehabilitation [3]. This proposed system is suitable for home-based motion capture by measuring finger joint kinematics. Oszust and Wysocki [4] have proposed a method for recognition of signed expressions observed using Kinect sensor. The proposed work helps hearing impaired people. The recognition phase consists of *k*-nearest neighbour classifier. Lai et al. [5] proposed a gesture-driven computer interface using Kinect. The method performs nearest-neighbour classification using the Euclidean distance metric. Arm swing, arm push, arm back, zoom in and zoom out are recognized in that paper. A system for monitoring of workers in industry is elaborated by Martin et al. [6]. The system is used to teach employees if their current lifting and carrying methods can be detrimental to their health.

In this paper, the Kinect sensor is employed to understand body gestures using two predefined features. Among twenty coordinates collected via Kinect sensor, only three coordinates are processed for this algorithm. These three coordinates are the hip centre, the left and right hands. To deal with the variation of a person's distance from Kinect sensor, the hand joint coordinates are normalized with the help of hip centre's coordinate of first frame. Here the normalization technique is used between two coordinate systems, one whose origin is at the IR projector of the Kinect sensor and another coordinate system's origin is at the hip centre of first frame of the person. Now the three-dimensional trajectories of the right and left hands with two body joints are examined for a specific gesture. The movement of the hand joints gives rise to data points per frame. For a total sequence, a trajectory is formed for a specific hand joint. This trajectory is utilized as a feature for identification of a gesture. Now for the purpose of recognition, we have acquired dataset from 20 different persons and each person has given data for 10 times for single as well as double hand gestures. The training to testing ratio is 4:1. High accuracy rate of 94.3 % is obtained using support vector machine.

The rest of the paper is organized as follows. The principals are introduced in Sect. 20.2. Experimental results are elaborated in Sect. 20.3. Section 20.4 concludes with future work.

20.2 Principals Used

The preliminary part of gesture recognition is to detect whether the gesture is enacted with single hand or double hands. For measuring this, a concept of threshold is brought forth. After getting the coordinates information from the Kinect sensor, the variation (v) of the left and right hands per frame is noted. The variation is calculated by the following equation

$$v = \sum_{i=1}^{M} f_{i+1} - f_i \tag{20.1}$$

where M is total number of frames, f_i depicts the coordinate for ith frame. If for a certain gesture, this variation value for the right-hand joint is greater than the threshold value, then the gesture is admitted as the right-hand gesture and vice versa. If the variations for both the hands are simultaneously greater than the threshold value (τ), then the gesture is confirmed as a double hand gesture.

After identifying the type of gesture, then the joint coordinate is normalized based on the hip centre's coordinate for the first frame. Let coordinate of hip centre for the first frame is (x_h, y_h, z_h) and the new and old coordinates for hand joint are (x^n_{feature}, y^n_{feature}, z^n_{feature}) and (x^o_{feature}, y^o_{feature}, z^o_{feature}). Coordinate normalization for each joint coordinate is done using the following equations:

$$x^n_{\text{feature}} = x^o_{\text{feature}} - x_h \tag{20.2}$$

$$y^n_{\text{feature}} = y^o_{\text{feature}} - y_h \tag{20.3}$$

$$z^n_{\text{feature}} = z^o_{\text{feature}} - z_h \tag{20.4}$$

The position from which the person starts gesturing, i.e. the distance of the person from the Kinect sensor differs widely. Thus coordinate normalization based on the coordinate of the hip centre of the first frame is performed. The initial hip centre is taken at the zero position and all the other coordinate values are transformed based on this.

The last part of the proposed work is application of support vector machine (SVM). The linear SVM [7] acts as a non-probabilistic binary linear classifier by taking the set of inputs and dividing them into two classes. It divides the data points on the basis of the gap between the data points using a hyper plane, such that the distance margin between the support vectors, and hence the two classes are maximized. However, linear SVM can be successfully used only where the data are linearly separable. This limitation can be overcome by mapping the data into a larger dimensional space using a kernel function [8–10]. The RBF or Gaussian kernel with the width of the Gaussian as 1 has been used in the present work. Binary SVM classification is carried out for every class of gesture in a one-against-all approach. The total procedure for single person hand gesture recognition is given in Fig. 20.1.

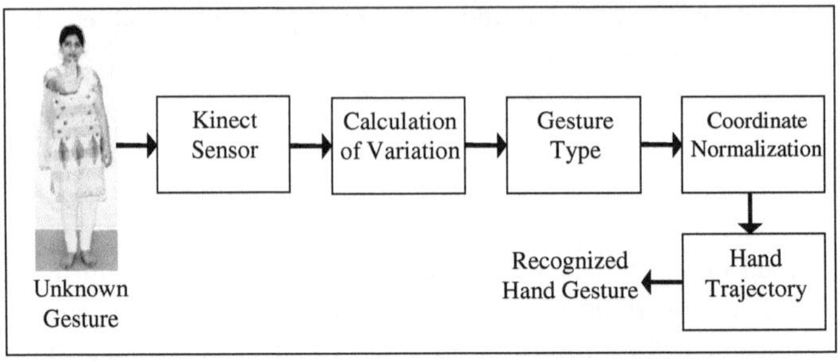

Fig. 20.1 Block diagram of the proposed work

20.3 Experimental Results

Figures 20.2 and 20.3 present the experimentally found single and double hand trajectories for subject 5, respectively. The single hand gestures are only shown for the right hand as the left-hand trajectories are easily obtained by flipping.

The McNemar's statistical test has been done on the comparative framework. Let f_A and f_B be two classifiers. Let n_{01} be the number of examples misclassified by f_A but not by f_B, and n_{10} be the number of examples misclassified by f_B but not by f_A. Then McNemar's statistic Z follows a χ^2 with degree of freedom equals to 1.

$$Z = \frac{(|n_{01} - n_{10}| - 1)^2}{n_{01} + n_{10}} \tag{20.5}$$

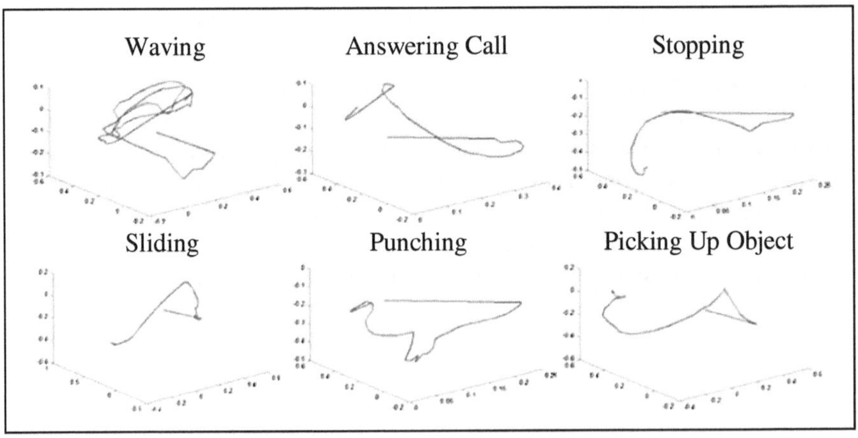

Fig. 20.2 Single hand trajectories for subject 5

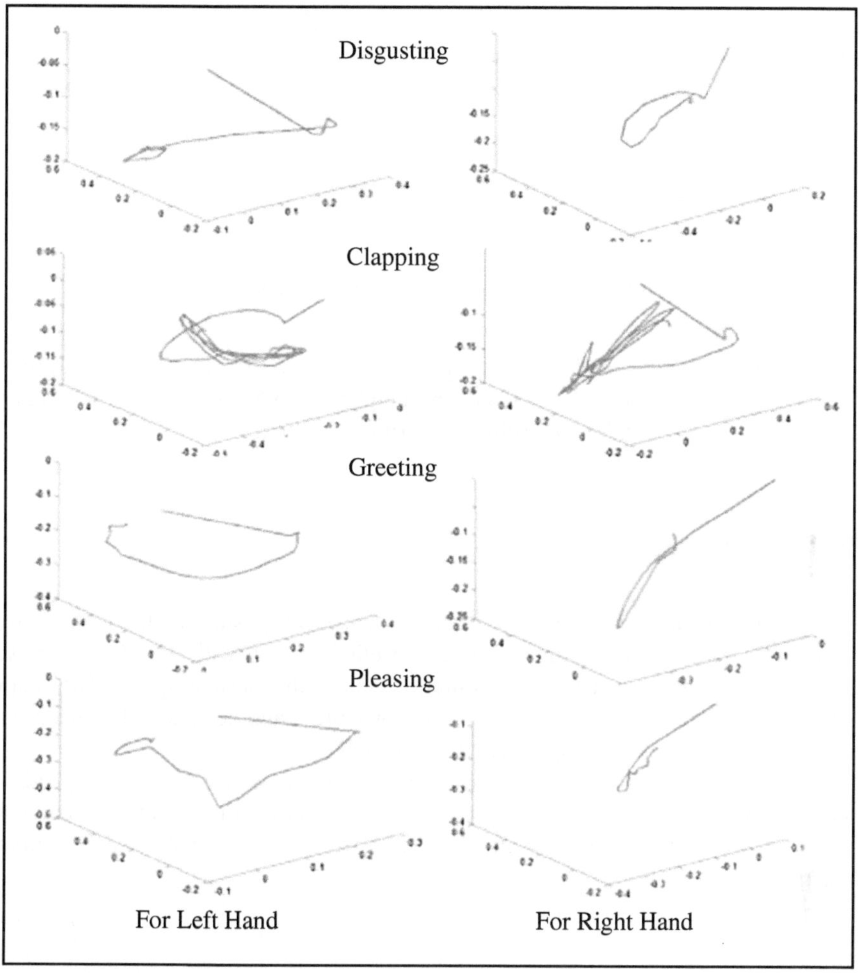

Fig. 20.3 Double hand trajectories for subject 5

Let A is the proposed SVM algorithm and B is one of the other three algorithms. In Table 20.1, the null hypothesis has been rejected, if $Z > \chi^2_{1,\alpha=0.05} = 3.84$, where 3.84 is the critical value for 1 degree of freedom at probability of 0.05.

Table 20.1 Performance analysis using McNemar's test

Algorithm *B*	Algorithm *A*			
	n_{01}	n_{10}	Z	Comment
*k*NN	6	17	4.3478	Reject
Perceptron	7	21	6.0357	Reject
FCM	1	18	13.4737	Reject

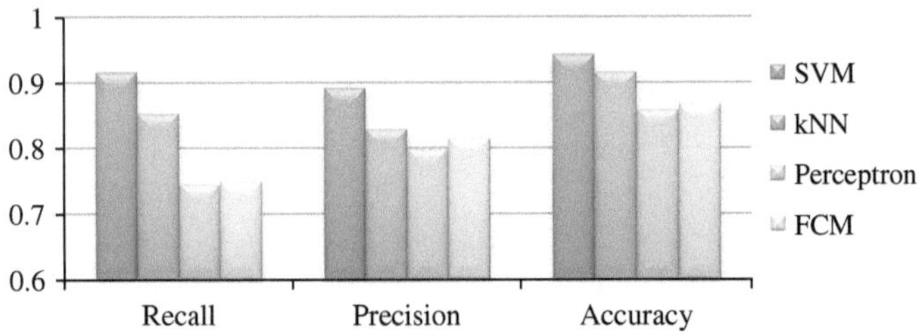

Fig. 20.4 Comparison between different gesture recognition algorithms

From Fig. 20.4, it is evident that the proposed method provides better result than support vector machine (SVM), *k*-nearest neighbor (*k*NN), perceptron and fuzzy *C*-means clustering (FCM) algorithms in terms of accuracy.

20.4 Conclusion

The novel approach explained in this paper deals with gesture of ten persons by calculating standard deviation in simple and efficient way with 94.3 % accuracy. Gestures are collected each of 3 s duration and with constant speed. The main advantages of the algorithm are reliability, robustness and accuracy. The methodology is wireless, easy to implement and cost effective.

One pitfall of the system is that during recognition of a gesture, the speed is nearly constant. The variation of speed may lead to erroneous result. The authors are trying to overcome this problem in future by recognizing the feature curve pattern.

References

1. T. Dutta, Evaluation of the Kinect™ sensor for 3-D kinematic measurement in the workplace. Appl. Ergon. **43**(4), 645–649 (2012)
2. R.A. Clark, Y.-H. Pua, K. Fortin, C. Ritchie, K.E. Webster, L. Denehy, A.L. Bryant, Validity of the microsoft kinect for assessment of postural control. Gait Posture **36**(3), 372–377 (2012)
3. C. Metcalf, R. Robinson, A. Malpass, T. Bogle, T. Dell, C. Harris, S. Demain, Markerless motion capture and measurement of hand kinematics: validation and application to home-based upper limb rehabilitation (2013)
4. M. Oszust, M. Wysocki, Recognition of signed expressions observed by Kinect Sensor, in *2013 10th IEEE International Conference on Advanced Video and Signal Based Surveillance (AVSS)* (2013), pp. 220–225
5. K. Lai, J. Konrad, P. Ishwar, A gesture-driven computer interface using Kinect, in *2012 IEEE Southwest Symposium on Image Analysis and Interpretation (SSIAI)* (2012), pp. 185–188

6. C.C. Martin, D.C. Burkert, K.R. Choi, N.B. Wieczorek, P.M. McGregor, R.A. Herrmann, P.A. Beling, A real-time ergonomic monitoring system using the Microsoft Kinect, in *2012 IEEE, Systems and Information Design Symposium (SIEDS)* (2012) pp. 50–55
7. T.M. Mitchell, Machine learning and data mining. Commun. ACM **42**(11), 30–36 (1999)
8. S. Saha, S. Ghosh, A. Konar, A.K. Nagar, Gesture recognition from indian classical dance using kinect sensor, in *2013 Fifth International Conference on Computational Intelligence, Communication Systems and Networks (CICSyN)* (2013), pp. 3–8
9. C. Cortes, V. Vapnik, Support vector machine. Mach. Learn. **20**(3), 273–297 (1995)
10. R.K. Sevakula, N.K. Verma, Support vector machine for large databases as classifier, in *Swarm, Evolutionary, and Memetic Computing*, (Springer 2012), pp. 303–313

Chapter 21
Ensemble Classifier-Based Physical Disorder Recognition System Using Kinect Sensor

Sriparna Saha, Monalisa Pal, Amit Konar and Jayahsree Roy

Abstract Gestures symbolizing body pain is a major challenging area in today's world. We have presented a system here using Microsoft's Kinect sensor for physical disorder recognition. Kinect sensor using its proprietary software development kit (SDK) approximates the human body in terms of 20 joint coordinates in three-dimensional space. We have taken into account 24 gestures related to physical disorder. For the training part, 3 datasets are formed where each dataset is constructed with the data acquired from 20 different subjects. For real time implementation, Kinect sensor is employed in laboratory for 24 h monitoring. The datasets are divided into 4:1 ratio for training and testing, respectively. After determining 171 features from each frame, ensemble classifier (in bagging framework) performs disorder recognition and yields 94.65 % accuracy in 0.0516 s.

Keywords Ensemble classifiers · Healthcare · Kinect sensor

21.1 Introduction

A gesture can be generated by any state or motion of a body or body parts. Gesture recognition aims at processing of the information that is ingrained in the gesture and is not expressed by speech or text. This work identifies 24 gestures involving pain in different parts of the body. After working in a fixed posture over a long duration,

S. Saha (✉) · M. Pal · A. Konar · J. Roy
Electronics and Telecommunication Engineering Department,
Jadavpur University, Kolkata, India
e-mail: sahasriparna@gmail.com

M. Pal
e-mail: monalisap90@gmail.com

A. Konar
e-mail: akonar@etce.jdvu.ac

J. Roy
e-mail: jayashree.jgec@gmail.com

© Springer India 2015

K. Maharatna et al. (eds.), *Computational Advancement in Communication Circuits and Systems*, Lecture Notes in Electrical Engineering 335,
DOI 10.1007/978-81-322-2274-3_21

people have been observed to perform certain typical gestures revealing pain at different muscles and body parts. These gestures are the subject of interest for our work and tracked using Kinect sensor [1–3]. The purpose of this work is to monitor the gesture of young persons while in sitting/standing posture, thereby sensing the early stage symptoms of certain disorders and making the subject aware about their health while suggesting an exercise as a cure. After taking guidance from several doctors, we have made this proposed system which is not only flexible but also provides medical guidance at low cost.

Kinect sensor consists of an IR camera, a visible RGB camera and an IR projector incorporated into it [1–3]. It eliminates the chances of error introduction owing to variation of appearance or colour of clothes. Other advantages of the device are optimum cost, portability and ability to perform throughout 24 h of the day. This work uses Kinect sensor at a frame rate of 30 fps during data acquisition.

Several works related to gesture recognition are found in the literature. Parajuli et al. have proposed a method on senior health monitoring [4]. The authors have detected the gestures when elders are likely to fall by measuring gait. The paper also deals with posture change during transition from sitting to standing and vice versa. Le et al. demonstrate scaling of angle features while using support vector machine as the classifier for health monitoring framework [5]. The author has extracted the skeletons with the help of Kinect sensor for the detection of lying, sitting, standing and bending postures while our work addresses more specific gestures related to healthcare. Kinect sensor has also found application of posture recognition in cases like children tantrum analysis [6]. The authors in [7] have proposed an elderly healthcare gesture recognition technique using a neural network. Also in [8], similar type of work has been done for gesture recognition for younger individuals with the help of principal component analysis and fuzzy C- means clustering. In both the two papers, the authors lack in preparing a large dataset addressing subjects belonging to different age groups.

In this proposed work, Kinect sensor is used to recognize human body gestures. Then 171 Euclidean distances are measured. This huge amount of information forms the feature space. We have created three datasets comprising gestures from different subjects. These datasets are divided in the ratio 4:1 for training and testing the classifier performance. For the recognition purpose ensemble classifier is applied. The overall accuracy obtained is 94.65 % with time complexity of 0.0516 s in Intel Core i3 processor and 4 GB RAM.

Rest of the paper is organized as follows. Section 21.2 provides a brief description of the concerned disorders related to muscle and joint pain and the ensemble classifier. Section 21.3 describes the experiment. Finally, the paper is concluded in Sect. 21.4.

21.2 Method of Study

This section includes the description concerned muscle and joint pains and ensemble classifier.

21.2.1 Muscle and Joint Pains Leading to Physical Disorders

Muscular and joint pains at the onset of these disorders give rise to 24 distinct gestures shown in Fig. 21.1. Early detection of these disorders prevents the progress of these disorders to chronic stages.

21.2.2 Ensemble Classifier

Ensemble classifiers are constructed on the basis of the fact that the decision accuracy of a group of classifiers is far more reliable than a single classifier. In our work, we have used 'tree' classifier as the base classifier [9, 10]. The base classifier creates binary tree where each node operates on one of the features from the dataset. The predictions of the individual base classifier are combined (weighted voting) to decide the class of the test samples. Two important criteria must be satisfied in selecting the individual classifiers: they must be accurate (error rate better than random guess; also called weak learners) and diverse (different error on new dataset). Bagging approach for multiclass classification is used while implementing ensemble classifier. In case of bagging, classifiers are trained by different datasets which are obtained from bootstrapping the original dataset, i.e. a subset of the

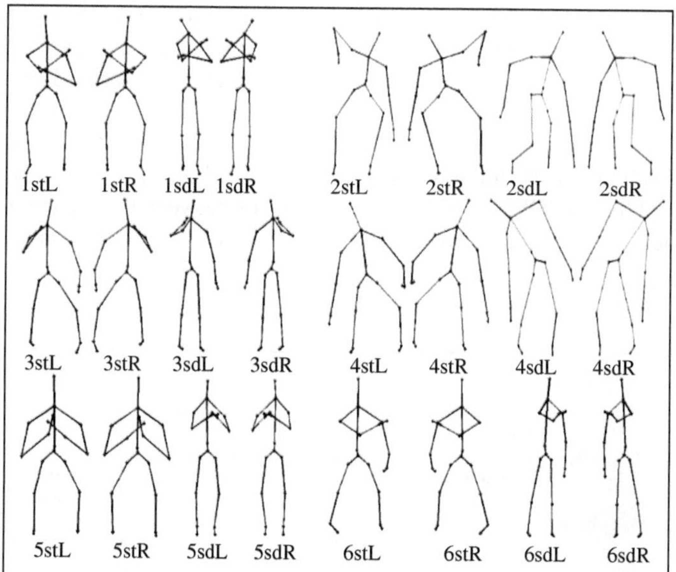

Fig. 21.1 Concerned disorders, *1* = Lumbosacral strain, *2* = Knee sprain, *3* = Prolapse intervertebral disc, *4* = Gastro-soleus muscle spasm, *5* = Spasmodic torticollis, *6* = Trapezius fibromyalgia, *7* = Plantar fasciitis, *st* = pain while sitting posture, *sd* = pain while standing posture, *L* = pain at *left side*, *R* = pain at *right side* and *B* = pain at *both sides*

dataset is created by randomly drawing (with replacement) n samples from the original dataset. The diversity among the weak learners is explored by this resampling procedure which is repeated T times. Finally, majority voting on the decision of the weak learners infer the class of an unknown sample. In this work, the value of T is considered to be 100 and the bootstrap size (n) is considered to be 30 % of the total dataset.

The working of the tree classifier can be stated as follows. Tree classifier [11] forms a binary tree on the training set T with N samples and n classes where each node operates on a single feature yielding the smallest Gini's diversity index [12] as given by (21.1) and splits the data into two sets $T1$ and $T2$ with $N1$ and $N2$ samples such that the condition in (21.2) is satisfied. The term $p(i)$ stands for the relative frequency of class i.

$$\text{gini}(T) = 1 - \sum_{i=1}^{n} p(i)^2 \tag{21.1}$$

$$\text{gini}(T) = \frac{N1}{N}\text{gini}(T1) + \frac{N2}{N}\text{gini}(T2) \tag{21.2}$$

21.3 Experimental Results

The proposed work involves gesture recognition from young persons. At first ($^{20}C_2$ − 19=) 171 features which are basically Euclidean distances between the different pairs of 20 joints, are extracted from each skeletal frame. Then ensemble classifier is applied for the classification part. This total procedure is picturized in Fig. 21.2. It is not feasible to state all the 171 features due to space constraint. The first five features for the unknown gesture shown in Fig. 21.2 are 0.3217, 0.4841, 0.2524, 0.2346 and 0.1989 which are the distances between the hip centre and five other joints viz. shoulder centre, head, shoulder left, elbow left and wrist left, respectively.

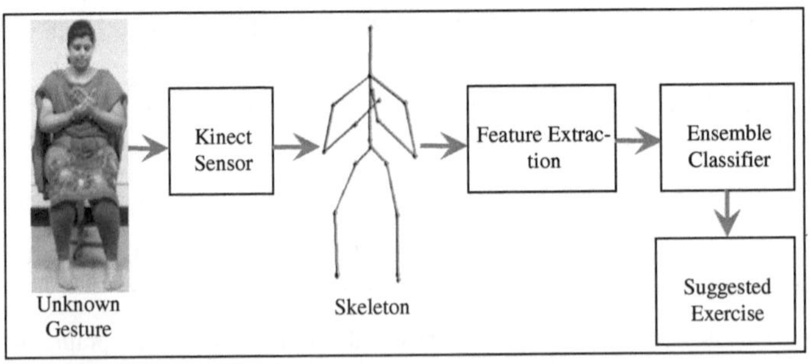

Fig. 21.2 Block diagram of the proposed work

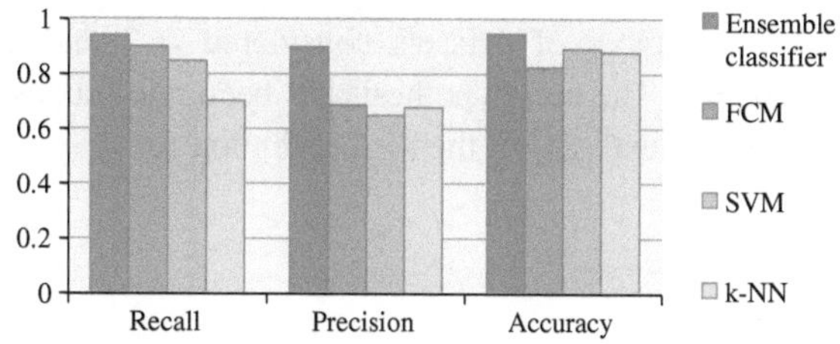

Fig. 21.3 Comparison of different physical disorder recognition algorithms

We have prepared three datasets from Jadavpur University research group scholars of age group 20–25, 25–30 and 30–35 years. In each dataset there are 20 subjects each imitating the gestures for 10 times. Each trial lasts for 2 s resulting in 60 frames from Kinect sensor. The 56th frame of each trial is selected as the final gesture which is expected to be achieved towards the end of a trial. The total size of each of the dataset is 24 (gestures) × 20 (subjects) × 10 (trials of each gesture) = 4,800. Now 3,840 gestures from each dataset are taken for training purpose and remaining are used for testing.

This paper compares the relative performance of the proposed algorithms with three traditional gesture recognition algorithms/techniques with respect to the three performance metrics viz. recall, precision and accuracy. This comparison is given in Fig. 21.3 from which we note that ensemble classifier outperforms the other algorithms. The comparative framework includes fuzzy *C*- means clustering (FCM) [8], Support Vector Machine (SVM) classifier [4] and *k*-Nearest Neighbour (*k*-NN) classification [13].

For statistical analysis, we have implemented Friedman test. Here the best of all the k algorithms, i.e. $i = [1, k]$, is assigned a rank of 1 and the worst is assigned the ranking k. Then the average ranking acquired by the i-th algorithm over all $j = [1, N]$ datasets is defined as R_i. Ranking of the classifiers is performed on the basis of the classification accuracy as shown in Table 21.1. Then χ_F^2 distribution with degree of freedom equals to $(k-1)$ [14] is measured using (21.3).

$$\chi_F^2 = \frac{12N}{k(k+1)}\left[\sum_{i=1}^{k} R_i^2 - \frac{k(k+1)^2}{4}\right] \tag{21.3}$$

Table 21.1 Ranks of classifiers for Friedman test

Dataset (age group)	Ensemble classifier	Fuzzy *c*-means algorithm	Support vector machine	*k*-nearest neighbour
1 (20–25)	1	2	3	4
2 (25–30)	1	2	4	3
3 (30–35)	1	2	3	4
Average ranks	1	2	3.33	3.67

In this paper, N = number of datasets considered = 3 and k = number of competitor algorithms = 4. The null hypothesis has been rejected, as $\chi_F^2 = 8.2040$ is greater than the critical value (7.81) of the χ_F^2 distribution for ($k-1$=) 3° of freedom at probability of 0.05.

21.4 Conclusion

The proposed hybrid system for gesture recognition is a simple and novel one. A large dataset of different disorder-related gestures are created and their diagnosis and treatment are implemented in daily life in home environment using ensemble classifier. The decaying nature of age leads to some disorders arising out of muscle fatigue and joint degradation.

Kinect detects only the skeleton of the subject thereby preserving the privacy of the subject. The scheme can be used for other applications as well like training in sports, learning different dance forms, teaching sign languages, etc. Rehabilitation areas involve early detection of disorders in elders like osteoporosis, rheumatism, arthritis, etc.

References

1. B.C. Munsell, A. Temlyakov, C. Qu, S. Wang, Person identification using full-body motion and anthropometric biometrics from kinect videos, in *Computer Vision–ECCV 2012. Workshops and Demonstrations*, (2012), pp. 91–100
2. P. Henry, M. Krainin, E. Herbst, X. Ren, D. Fox, RGB-D mapping: using Kinect-style depth cameras for dense 3D modeling of indoor environments. Int. J. Rob. Res. **31**(5), 647–663 (2012)
3. K. Khoshelham, Accuracy analysis of kinect depth data, in *ISPRS Workshop Laser Scanning*, vol. 38 (2011), p. 1
4. M. Parajuli, D. Tran, W. Ma, D. Sharma, Senior health monitoring using Kinect, in *Fourth International Conference on Communications and Electronics* (*ICCE*), (2012), pp. 309–312
5. T.-L. Le, M.-Q. Nguyen, T.-T.-M. Nguyen, Human posture recognition using human skeleton provided by Kinect, in *International Conference on Computing, Management and Telecommunications* (*ComManTel*) (2013), pp. 340–345
6. X. Yu, L. Wu, Q. Liu, H. Zhou, Children tantrum behaviour analysis based on Kinect sensor, in *Third Chinese Conference on Intelligent Visual Surveillance* (*IVS*) (2011), pp. 49–52
7. S. Saha, M. Pal, A. Konar, R. Janarthanan, Neural network based gesture recognition for elderly health care using kinect sensor, in *Swarm, Evolutionary, and Memetic Computing* (Springer, New York, 2013), pp. 376–386
8. M. Pal, S. Saha, A. Konar, A fuzzy C means clustering approach for gesture recognition in healthcare. Knee **1**, C7
9. R. Polikar, Ensemble based systems in decision making. IEEE Circuits Syst. Mag. **6**(3), 21–45 (2006)
10. T.G. Dietterich, An experimental comparison of three methods for constructing ensembles of decision trees: bagging, boosting, and randomization. Mach. Learn. **40**(2), 139–157 (2000)

11. S.C. Lemon, J. Roy, M.A. Clark, P.D. Friedmann, W. Rakowski, Classification and regression tree analysis in public health: methodological review and comparison with logistic regression. Ann. Behav. Med. **26**(3), 172–181 (2003)
12. L. Jost, Entropy and diversity. Oikos **113**(2), 363–375 (2006)
13. M. Oszust, M. Wysocki, Recognition of signed expressions observed by kinect sensor, in *10th IEEE International Conference on Advanced Video and Signal Based Surveillance* (*AVSS*), (2013), pp. 220–225
14. S. García, D. Molina, M. Lozano, F. Herrera, A study on the use of non-parametric tests for analyzing the evolutionary algorithms' behaviour: a case study on the CEC'2005 special session on real parameter optimization. J. Heuristics **15**(6), 617–644 (2009)

Chapter 22
Improved Prediction Accuracy with Reduced Feature Set Using Novel Binary Gravitational Search Optimization

Sankhadip Saha and Dwaipayan Chakraborty

Abstract Improvement of classifier prediction accuracy is a long run burning issue over the years in the field of data mining and machine learning application. Optimized feature set is the best strategy and feature selection is the only key to the optimization problem. Various heuristic search algorithms are proposed in the literature for the feature set selection task. In this context we have enlightened the feature set exploration capacity of gravitational search algorithm (GSA) which is based on the Newton's law of motion principle and the interaction of masses. Binary version of GSA with one modification is used for our application here. It is found that binary gravitational search algorithm (BGSA) is useful for finding only the relevant features while improving classifier accuracy from that with all features. We test our approach on six benchmark datasets from UCI machine learning repository.

Keywords Binary GSA · Feature selection · Stochastic algorithm · SVM · UCI

22.1 Introduction

In the history of development of various optimization techniques, it is greatly found that nature inspired philosophy plays a major role. It is because natural system is random in nature and random search often can fruitfully avoid the trap of local minima or maxima in the search or solution space. Various stochastic optimization methods are prescribed by researcher so far such as genetic algorithm [1], simulated

S. Saha (✉)
Department of Electrical Engineering, Netaji Subhash Engineering College, Kolkata, India
e-mail: imsankhadip@gmail.com

D. Chakraborty
Department of Electronics and Instrumentation, Netaji Subhash Engineering College, Kolkata, India
e-mail: dwaipayanchakraborty@hotmail.com

© Springer India 2015

K. Maharatna et al. (eds.), *Computational Advancement in Communication Circuits and Systems*, Lecture Notes in Electrical Engineering 335,
DOI 10.1007/978-81-322-2274-3_22

annealing [2], particle swarm optimization [3], ant colony optimization [4], fish swarm optimization [5], cuckoo search optimization [6], firefly optimization [7] and fruit fly optimization [8]. It has been proved that although these nature-inspired algorithms are based on different theoretical approaches they are computationally intelligent in searching route. The gravitational search algorithm (GSA) [9] is also nature inspired but different in nature in theoretical point of view. Unlike aforesaid optimization methods, gravity search algorithm has biological background rather it is Newton's physics related. This algorithm conceptualizes the fact that the whole universe is the total search space and position of a body mass is the searcher agent or solution vector attracts other body masses with attraction forces. This is the gravitational force that propagates the search area by navigating the masses all around. Original GSA is intended for continuous-valued vector space whereas many real optimization problems like feature selection and dimensionality reduction require discrete search space. In this regard many stochastic optimization algorithm is modified to the binary version like binary genetic algorithm [10], binary particle swarm optimization [11], etc. Rashedi et al. [12] first proposed the binary version of the gravity search algorithm where the binary search space is considered as a hypercube in which an agent may move to nearer and farther corners of the hypercube by flipping various numbers of the bits and gravitational forces gets converted into a probability value for each element of the binary vector, which guides whether that elements will take on the value 0 or 1. Though it cannot as surely be said that one specific optimization algorithm serves the feature selection task in all application of engineering. That means selection of proper optimization algorithm varies with different datasets, i.e. one optimization algorithm outperforms on a dataset when the other fails to perform well in terms of finding the best solution vector. In this paper we instigate the potential feature selection ability of GSA. We modify the algorithm for efficient convergence near best solution by giving adaptive weight to the velocity of masses so that body nearer to best solution can achieve fast movement.

22.2 Theory

22.2.1 Binary Gravitational Search Algorithm

The GSA is a metaheuristic optimization algorithm, introduced by Rashedi et al. [9], which is based on the law of gravity and the interaction between the masses in a system. It utilizes Newton's law of gravitational attraction as well as Newton's laws of motion. In GSA, each mass or particle represents a candidate solution and the system consists of many such particles which interact with and move towards each other, depending on their fitness. The solutions are vectors containing real-valued integers. Subsequently, a binary version of the algorithm was developed [12], in

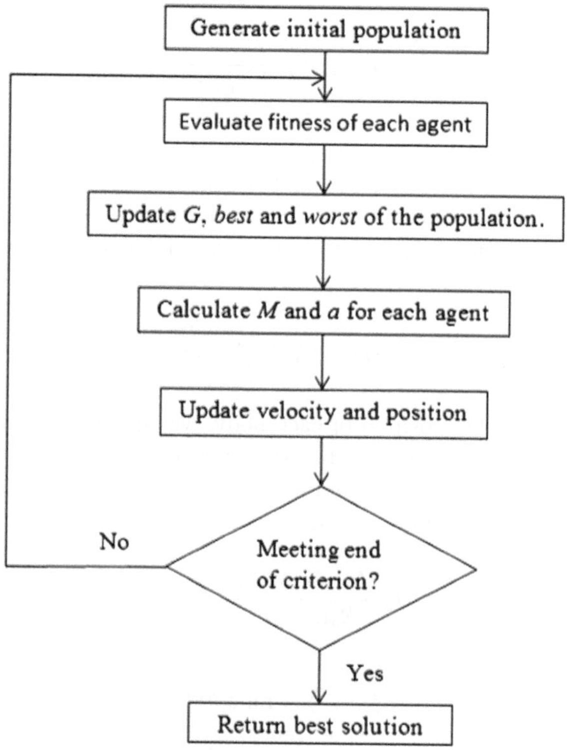

Fig. 22.1 Gravitational search algorithm flowchart

which the solution vectors are binary, i.e. they contain only 1 and 0. A brief description of the GSA, along with flowchart, is given in Fig. 22.1.

The GSA considers a system with N agents (masses) where each agent is a candidate solution and defined as:

$$X_i = \left(x_1^i, \ldots, x_d^i, \ldots x_n^i\right) \tag{22.1}$$

where x_i^d represents the position of the ith agent in the dth dimension and n is the total number of dimensions or parameters to be optimized. The initial population is generated by randomly choosing 0 or 1 for each dimension of each solution in the population.

After the fitness of the entire population is evaluated, the mass of candidate solution is calculated using the best and worst fitness of the population as well as the fitness of the respective agent, according to Eqs. (22.2) and (22.3).

$$q_i(t) = \frac{\mathrm{fit}_i(t) - \mathrm{worst}(t)}{\mathrm{best}(t) - \mathrm{worst}(t)} \tag{22.2}$$

$$M_i(t) = \frac{q_i(t)}{\sum_{j=1}^{N} q_j(t)} \tag{22.3}$$

where $\text{fit}_i(t)$ is the fitness of the ith agent and $M_i(t)$ is its corresponding weight, while best(t) and worst(t) are the best and worst fitness of the population, respectively. The gravitational constant G is a linear decreasing function which depends on both the initial value G_0 and the iterations of the algorithm, as shown in Eq. (22.4):

$$G(t) = G_0(t)\left(1 - \frac{t}{T}\right) \tag{22.4}$$

In order to update the position of each body, we need to calculate its velocity. This is done by first calculating the net gravitational force on a body, as in Eqs. (22.5) and (22.6), the acceleration of the body according to Eq. (22.7) and finally the velocity of the body, as shown in Eq. (22.8).

$$F_{ij}^d = G(t)\frac{M_i(t) \times M_j(t)}{R_{ij}(t) + \varepsilon}\left(x_j^d(t) - x_i^d(t)\right) \tag{22.5}$$

$$F_i^d = \sum_{j=1, j \neq i}^{N} \text{rand}_j F_{ij}^d(t) \tag{22.6}$$

$$a_i^d(t) = \frac{F_i^d(t)}{M_i(t)} \tag{22.7}$$

$$v_i^d(t+1) = \text{rand}_i \times v_i^d(t) + a_i^d(t) \tag{22.8}$$

In binary GSA, a probability function $S(v_i^d)$ is defined in Eq. (22.9) such that the probability of x_i^d is near zero for a small velocity $\left|v_i^d\right|$, and movement x_i^d is high for a large value of $\left|v_i^d\right|$. Once the probability $S(v_i^d)$ is calculated, the movement of the agents will follow the rule explained in Eq. (22.10).

$$S(v_i^d(t)) = \left|\tanh(v_i^d(t))\right| \tag{22.9}$$

$$\text{if rand} < S(v_i^d(t+1)) \text{ then } x_i^d(t+1) = \text{complement}(x_i^d(t))$$

$$\text{else } x_i^d(t+1) = x_i^d(t) \tag{22.10}$$

22.2.2 Proposed Modification

In binary GSA, the velocity of a particle is updated by using Eq. (22.8). In this paper, we introduce a new adaptive scheme for updating the velocity, given by Eq. (22.11).

$$v_i^d(t+1) = w_i \times v_i^d(t) + a_i^d(t) \tag{22.11}$$

where w_i is the Pearson's product moment coefficient between the agent with current best fitness and agent with the best fitness in the previous iteration, as shown in Eq. (22.12). Thus, the velocity of a mass, i.e. an individual solution, is lesser if the best solution in the current iteration is considerably different from the best solution in the previous iteration. On the other hand, the velocity is greater if the two best solutions are somewhat similar, and the body moves faster.

$$w_i = \text{corr}\{\text{best}(t),\, \text{best}(t-1)\} \tag{22.12}$$

22.2.3 Feature Subset Selection Using Binary Gravitational Search Algorithm

Binary gravitational search algorithm can be used effectively for feature selection [10, 13]. In such a case, the total number of dimensions in the solution vector is equal to the total number of features in the dataset. The candidate solutions each represent a different feature subset. The fitness function consists of a classifier which is trained and tested by the data corresponding to the feature subsets and the tenfold cross-validation error rate serves as the fitness of each contending solutions. The GSA aims to minimize this fitness, i.e. it tries to select the solution vector which provides the least amount of cross-validation error rate.

22.3 Results and Analysis

In order to validate the proposed modification, we have used six datasets with moderately large number of features from the UCI Repository [14]. All of the datasets are divided into training and testing subsets, where 70 % of the samples are used for training and the remaining 30 % for testing. At first, a K-nearest neighbours classifier, with $K = 5$, is used for finding the classification accuracy with all the features. Then, optimized feature subsets are selected from each dataset using the modified novel binary GSA algorithm with a total of 100 iterations and 30 agents is used. K-nearest neighbours classifier with $K = 5$ is implemented for evaluating the cross-validation error rate in the objective function. Again, the classification accuracy with the reduced feature subset is computed, and presented in Tables 22.1 and 22.2.

Table 22.1 Prediction accuracy with all features and reduced feature subset

Dataset used	Total number of features	Prediction accuracy with all features (%)	Reduced feature subset size	Prediction accuracy with reduced features (%)
Urban land cover (2014)	147	39.05	82	44.77
QSAR biodegradation (2013)	41	79.50	22	87.07
Theorem proving (2013)	51	51.44	25	52.04
Ionosphere (1989)	34	80.00	7	87.62
Sonar (1988)	60	70.97	29	72.58
Waveform (1988)	40	80.67	22	81.67

Table 22.2 Selected features for different datasets

Dataset used	Selected features
Urban land cover	[2,3,4,7,8,9,12,14,17,18,19,20,22,23,24,25,26,27,28,29,30,31,33,35, 38,39,40,42,43,45,46,47,48,49,50,51,52,53,57,58,59,61,62,65,67,68, 70,71,72,78,79,83,84,85,87,88,91,92,93,94,97,98,101,102,108,109, 111,112,113,114,117,120,122,123,127,130,132,133,134,135, 136,145]
QSAR biodegradation	[1,2,3,5,6,12,14,15,17,18,19,21,22,26,27,32,33,36,37,38,39,40]
Theorem proving	[1,2,3,4,5,6,7,16,17,18,19,21,23,24,26,27,29,30,31,36,37,41,42,46,49]
Ionosphere	[2,3,4,5,7,13,27]
Sonar	[1,3,4,6,8,9,10,11,12,17,18,22,26,29,31,32,36,39,41,43,44,45,47, 48,49,52,53,55,59]
Waveform	[2,3,4,5,6,7,8,9,10,11,12,13,14,15,16,17,18,22,30,31,34,37]

22.4 Conclusion

Stochastic optimization generally performs better than non-stochastic technique in finding the important features in the dataset. Hence classifier system accuracy and efficiency are improved with the dimensionality reduction. In this paper gravity search algorithm is applied as stochastic process in feature space to explore the relevant features. The GSA algorithm utilizes the theory of Newtonian physics and its searcher agents are basically the group of masses. Here binary masses are considered since feature selection is discrete problem. An appropriate modification is done to give adaptive weight to the body masses or searcher agent for efficient convergence of the optimization progression. Results confirm this algorithm's capability to improve the prediction accuracy of the classifier even if the number of features is considerably reduced.

References

1. J.H. Holland, *Adaptation in Natural and Artificial Systems* (The University of Michigan Press, Ann Arbor, 1975)
2. S. Kirkpatrick, C.D. Gelatto et al., Optimization by simulated annealing. Science **220**(4598), 671–680 (1983)
3. J. Kennedy, R.C. Eberhart, Particle swarm optimization, in *Proceedings of IEEE International Conference on Neural Networks*, vol. 4 (1995), pp. 1942–1948
4. M. Dorigo, M. Birattari, T. Stutzle, Ant colony optimization. IEEE Comput. Intell. Mag. **1**(4), 28–39 (2006)
5. X. Li, A new intelligent optimization-artificial fish swarm algorithm, Doctoral thesis, Zhejiang University, Zhejiang (2003)
6. X.-S. Yang, S. Deb, Cuckoo search via lévy flights, in *Proceedings of World Congress on Nature & Biologically Inspired Computing* (2009), pp. 210–214
7. X.-S. Yang, Firefly algorithm, stochastic test functions and design optimization. Int. J. Bio-Inspired Comput. **2**(2), 78–84 (2010)
8. W.-T. Pan, A new fruit fly optimization algorithm: taking the financial distress model as an example. Knowl. Based Syst. **26**, 69–74 (2012)
9. E. Rashedi, H. Nezamabadi-pour et al., GSA: a gravitational search algorithm. Inf. Sci. **179**(13), 2232–2248 (2009)
10. W. Siedlecki, J. Sklansky, A note on genetic algorithms for large-scale feature selection. Pattern Recogn. Lett. **10**(5), 335–347 (1989)
11. J. Kennedy, R.C. Eberhart, A discrete binary version of the particle swarm algorithm, in *IEEE International Conference on Computational Cybernetics and Simulation*, vol. 5 (1997), pp. 4104–4108
12. E. Rashedi, H. Nezamabadi-pour, S. Saryazadi, BGSA: binary gravitational search algorithm. Nat. Comput. **9**(3), 727–745 (2010)
13. X. Wang, J. Yang et al., Feature selection based on rough sets and particle swarm optimization. Pattern Recogn. Lett. **28**(4), 459–471 (2007)
14. K. Bache, M. Lichman, UCI machine learning repository (University of California, School of Information and Computer Science, Irvine, 2013), http://archive.ics.uci.edu/ml

Chapter 23
Proton Density Variation in Ionosphere Before Strong Earthquake Using GOES-15 Data

Pranab Hazra and Tamanna Islam

Abstract Seismic activity has some relationship with proton density which reveled from the observation of Scotia Sea Earthquake. In this paper, proton density variations are observed with the help of GOES-15 satellite data considering the density of proton in between 8:00 and 12:00 (I.S.T) for a few days before and after the earthquake. The earthquake time was at 9:04 a.m. on 17 November 2013 in Scotia Sea (with the magnitude of 7.5 M and depth of 10 km). In this paper, the variations of proton density and ion densities are statistically analysed from GOES-15. The GOES N Series of spacecraft continuously observe and measure the meteorological phenomena in real time, providing the meteorological community and the atmospheric scientist greatly improved observational and measurement data of the Western Hemisphere. Because of forecasting and space environmental monitoring, these enhanced operational services also improve support for atmospheric science research, numerical weather prediction models and environmental sensor design and development. The GOES satellites provides timely environmental information to meteorologists and their audiences alike—graphically displaying the intensity, path and size of storms. The result shows that the variation in the anomalies before the earthquakes have different sign from case to case and that there exists a strong relationship between the amplitude and the magnitude of the earthquake. It has also been found that the proton density at night detects variations significantly before the earthquakes. Due to the plate movements under Scotia Sea, the submicron aerosols are emerged by which ionosphere gets ionized and electric field becomes stronger; then electrons are accelerated but huge amount of protons comes from solar radiation. So there is a chance of decreasing proton density in ionosphere. Anomalies in the positive and negative pulses in GOES-15 data during

---

P. Hazra (✉) · T. Islam
Department of ECE, Narula Institute of Technology, Kolkata, India
e-mail: pranabhazra2007@gmail.com

T. Islam
e-mail: tamannaslm7@gmail.com

© Springer India 2015

K. Maharatna et al. (eds.), *Computational Advancement in Communication Circuits and Systems*, Lecture Notes in Electrical Engineering 335,
DOI 10.1007/978-81-322-2274-3_23

1–7 days before all studied earthquakes during quiet geomagnetic conditions indicates that these anomalous behaviours are highly regarded as seismo ionospheric precursors.

Keywords Sferics · Earthquake · Ionosphere · ELF · Proton density

23.1 Introduction

It was recognized that a seismic electromagnetic inconsistency is a climax of some course of action which begins a few days prior to the main event and reside until a few days after it. Earthquakes are a geophysical phenomena which involve processes that are irregular, non-linear and complicated. Therefore, it is necessary to find more reliable methods to closely study the process and its relevant parameters. A dynamic process involves an energy transfer due to crust displacement during earthquake preparation and at the time of the shock a break down between the source and the environment occurs. The diverse physical and chemical effects on the lithosphere, atmosphere and ionosphere also observed and it is possible to detect. These variations of lithosphere, atmosphere and ionosphere parameters before the main earthquakes are considered as hint of impending earthquakes (earthquake precursors). The effects of the preseismic activity on the ionosphere can be investigated using the ionospheric electron and proton densities. The solar and geomagnetic activities, season, latitude and longitude and other unknown parameters have a greater impact on the daily variations of the ionosphere. From the satellite measurements, the equatorial anomaly also appears during night time in high solar activity conditions [1].

The November 17, 2013 earthquake in the Scotia Sea took place due to the effect of either left-lateral strike slip faulting on an east-west oriented plane, or right-lateral faulting on a north-south plane. The site of the event adjacent to the east-west oriented plate boundary between the Antarctica and Scotia Sea plates implies the left-lateral faulting scenario is most expected. The November 17 earthquake is the latest in a series of moderate-to-large earthquakes to strike the same region over the past several days. Though the region surrounding the Scotia Sea is familiar with earthquakes, the greater part occur around the subduction zone adjacent to the South Sandwich Islands, to the east of the November 17 earthquake.

23.1.1 GOES Satellite Data

GOES 15, is a part of the Geostationary Operational Environmental Satellite (GOES) system operated by the US National Oceanic and Atmospheric Administration. GOES 15 was launched atop a Delta IV-M + (4,2) rocket flying from Space Launch Complex 37B at the Cape Canaveral Air Force Station [2, 3].

In this study, depending on the time, geographical position and proton density variations of each day earthquake the closest satellite approaches from about 8 days before to 5 days after the earthquake are chosen.

23.1.2 Geomagnetic Data

The geophysical conditions and geo-magnetic storms of the solar system have strong effects on the ionospheric parameters especially in the equatorial and polar regions. Also, auroral activity plays a vital role in the mid-latitude ionosphere perturbations. The ionospheric current, equatorial storm time, ring current in periods of solar-terrestrial disturbances produce considerable geomagnetic field instability observed on the ground. Consequently, the measured parameters may display variations even if the seismic activity is absent.

A global impact being observed all over the world due to the ionospheric effect of a geomagnetic storm, while the seismogenic effect is observed only by stations with the distance less than 2,000 km from the potential epicenter [1]. Data published in the USGS earthquake report recommend that the movement is relatively slow at a rate of, approximately, 6 mm per year.

23.2 Observations

In Figs. 23.1, 23.2, 23.3, 23.4, 23.5, 23.6, 23.7, 23.8, 23.9, 23.10, 23.11, 23.12, 23.13, 23.14, 23.15 and 23.16, we are considering the proton density variation on ELF signals as precursors of earthquake in between 8:00 and 12:00 (I.S.T) for a few

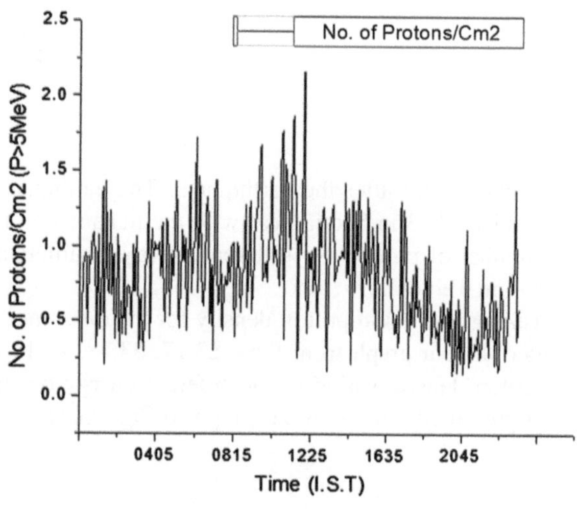

Fig. 23.1 Proton density (P > 5 MeV) in ionosphere on 08 November 2013

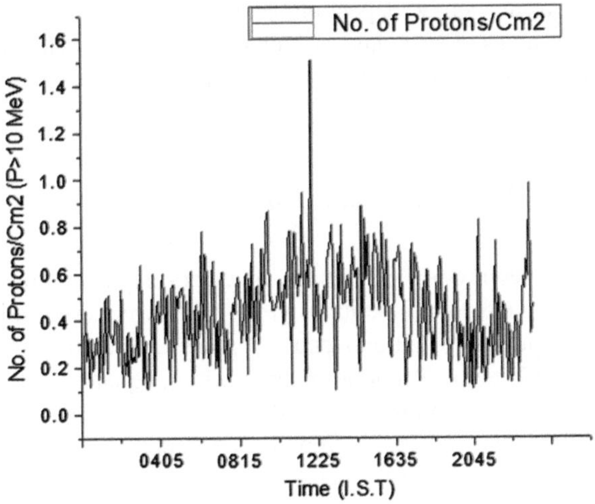

Fig. 23.2 Proton density (P > 10 MeV) in ionosphere on 08 November 2013

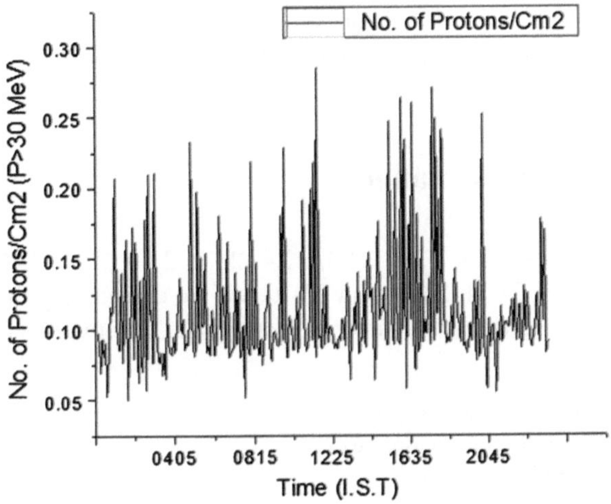

Fig. 23.3 Proton density (P > 30 MeV) in ionosphere on 08 November 2013

days before and after the earthquake. The earthquake time was 9:04 a.m. on 17 November 2013 in Scotia Sea (with the magnitude of 7.5 M and depth of 10 km).

Number of protons for different energy in different date is presented in tabular format in Table 23.1.

Taking maximum proton density for the different energy of the proton, we have plotted the bar graph from Figs. 23.17, 23.18, 23.19 and 23.20.

Proton density variation for different energy on the day of the earthquake has been presented through a bar graph in Fig. 23.21.

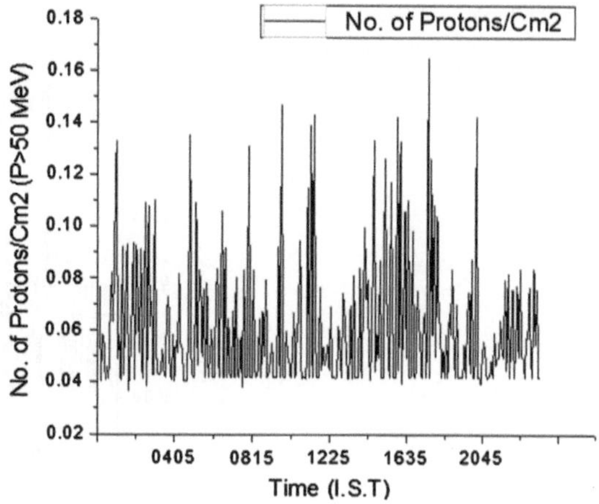

Fig. 23.4 Proton density (P > 50 MeV) in ionosphere on 08 November 2013

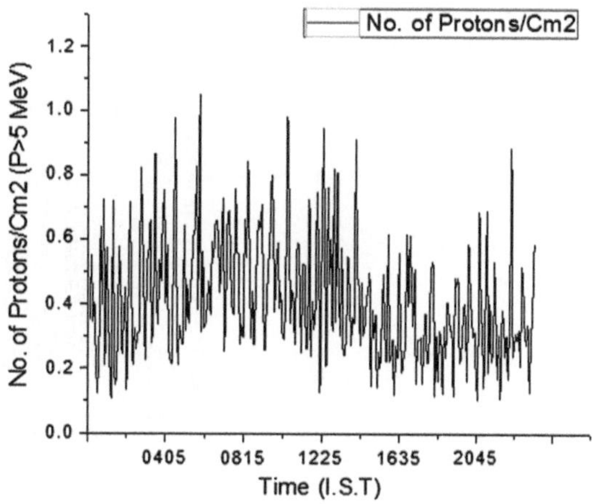

Fig. 23.5 Proton density (P > 5 MeV) in ionosphere on 15 November 2013

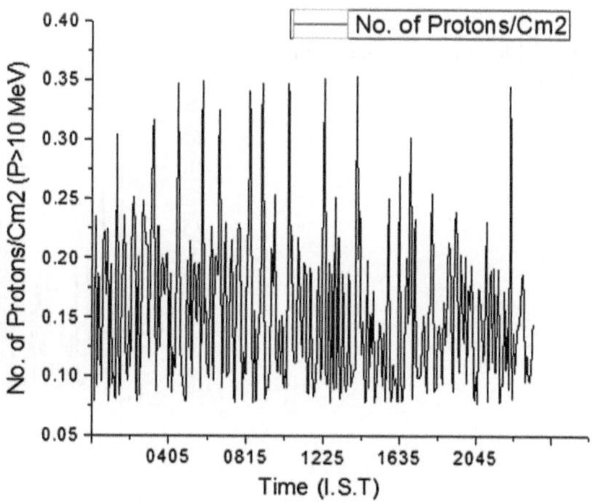

Fig. 23.6 Proton density (P > 10 MeV) in ionosphere on 15 November 2013

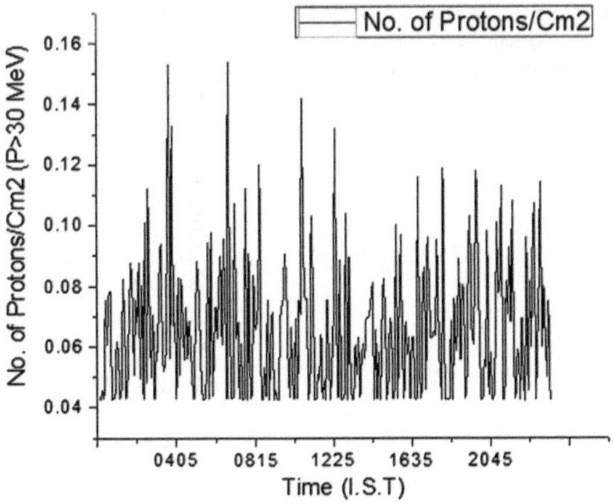

Fig. 23.7 Proton density (P > 30 MeV) in ionosphere on 15 November 2013

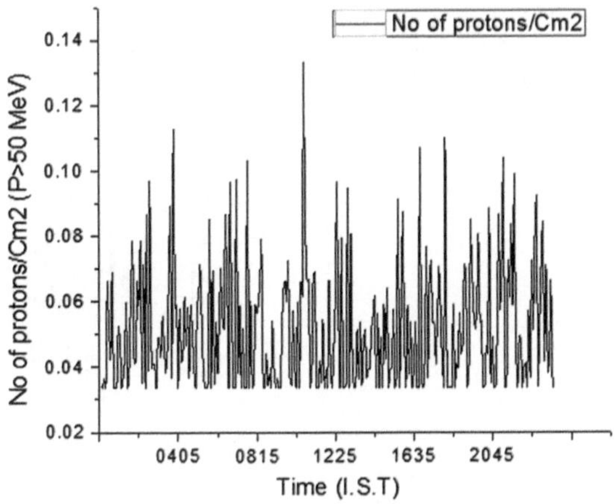

Fig. 23.8 Proton density (P > 50 MeV) in ionosphere on 15 November 2013

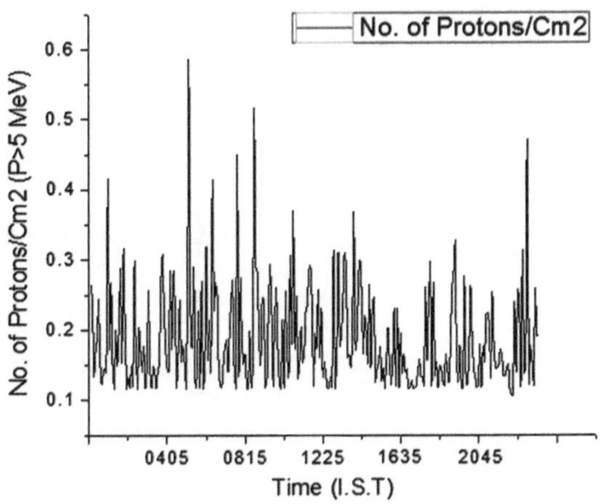

Fig. 23.9 Proton density (P > 5 MeV) in ionosphere on 17 November 2013

Fig. 23.11 Proton density (P > 30 MeV) in ionosphere on 17 November 2013

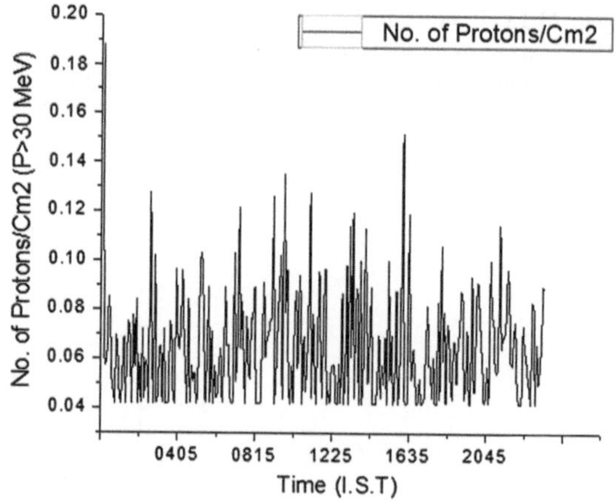

Fig. 23.10 Proton density (P > 10 MeV) in ionosphere on 17 November 2013

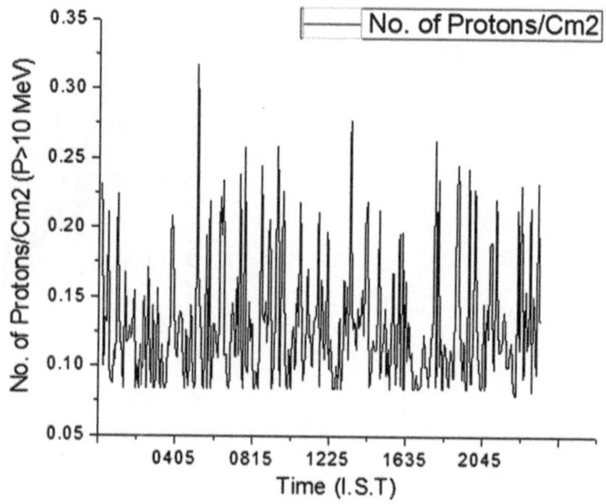

Fig. 23.12 Proton density (P > 50 MeV) in ionosphere on 17 November 2013

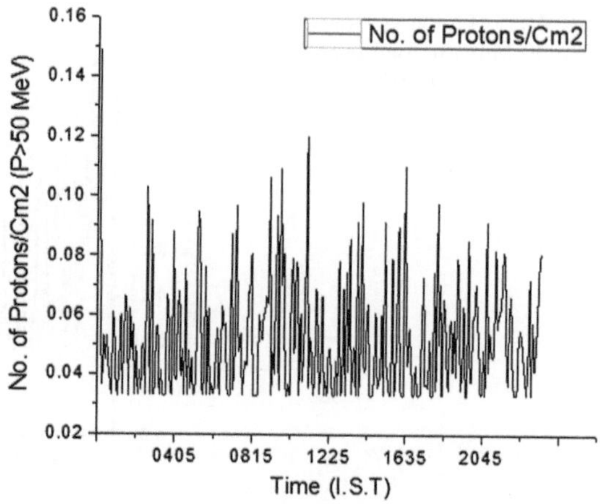

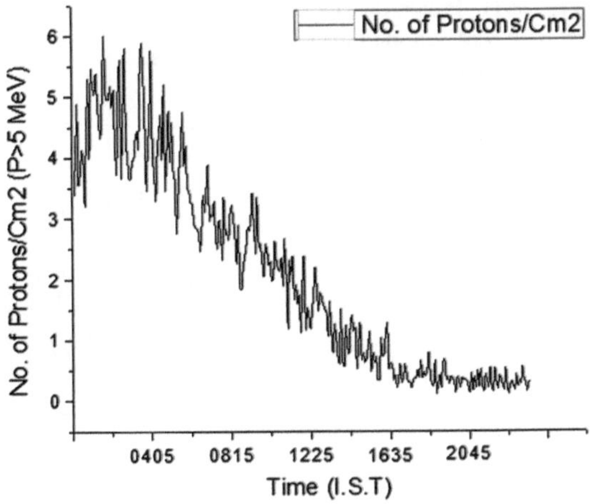

Fig. 23.13 Proton density (P > 5 MeV) in ionosphere on 20 November 2013

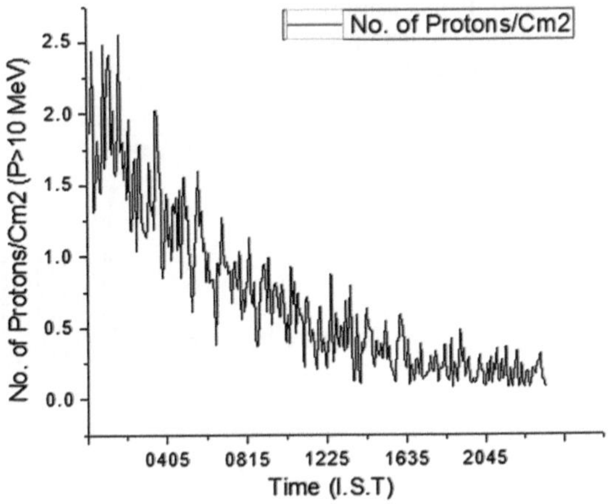

Fig. 23.14 Proton density (P > 10 MeV) in ionosphere on 20 November 2013

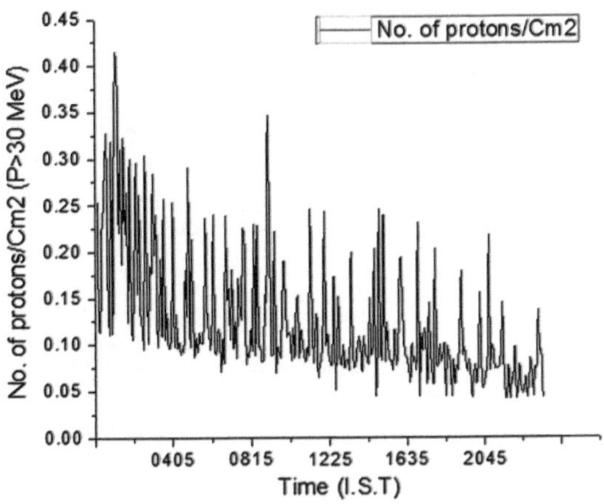

Fig. 23.15 Proton density (P > 30 MeV) in ionosphere on 20 November 2013

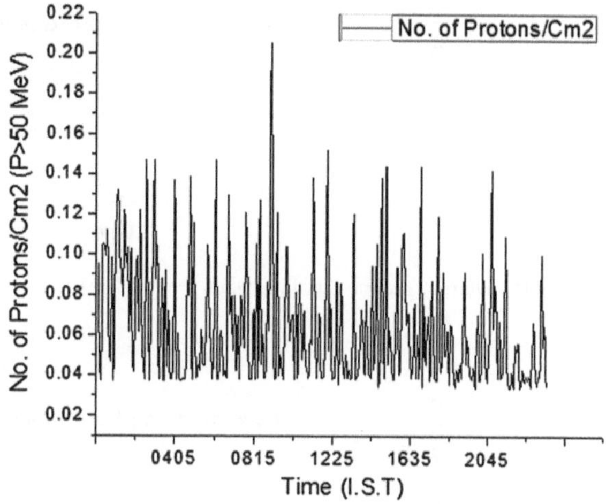

Fig. 23.16 Proton density (P > 50 MeV) in ionosphere on 20 November 2013

Table 23.1 Number of protons/cm^3 for different energy in different date

Date	Time (I.S.T)	Max. no of proton/cm^3 (p > 5 Mev)	Max. no of proton/cm^3 (p > 10 Mev)	Max. no of proton/cm^3 (p > 30 Mev)	Max. no of proton/cm^3 (p > 50 Mev)
8/11/13	8 a.m.–12 a.m.	2	1.5	0.255	0.16
9/11/13	8 a.m.–12 a.m.	4	1.6	0.325	0.18
11/11/13	8 a.m.–12 a.m.	3.5	1.2	0.25	0.16
12/11/13	8 a.m.–12 a.m.	1.9	0.85	0.275	0.135
15/11/13	8 a.m.–12 a.m.	1.4	0.35	0.15	0.135
16/11/13	8 a.m.–12 a.m.	0.85	0.375	0.15	0.135
17/11/13	8 a.m.–12 a.m.	0.6	0.375	0.15	0.135
18/11/13	8 a.m.–12 a.m.	0.55	0.275	0.17	0.12
19/11/13	8 a.m.–12 a.m.	5	3.5	0.8	0.325
20/11/13	8 a.m.–12 a.m.	6	2.5	0.40	0.21

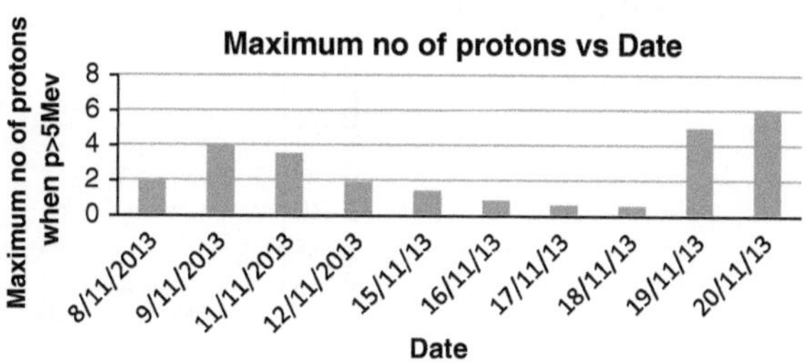

Fig. 23.17 Diurnal proton variation when P > 5 MeV

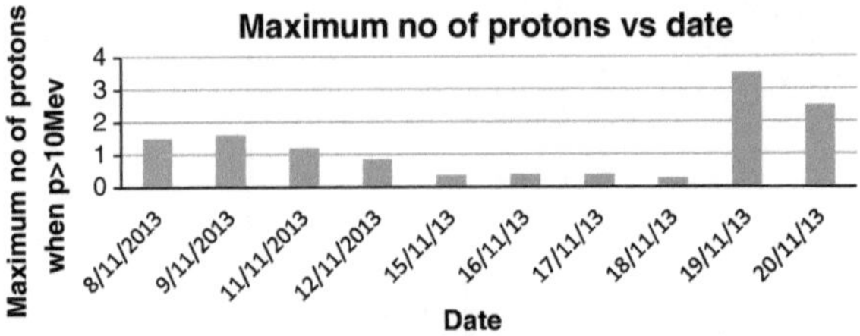

Fig. 23.18 Diurnal proton variation when P > 10 MeV

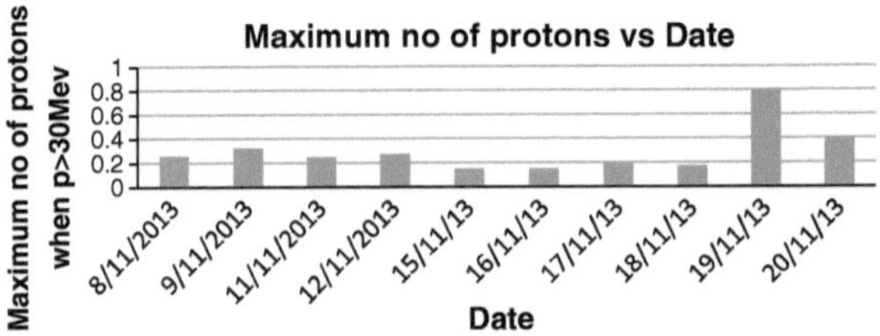

Fig. 23.19 Diurnal proton variation when P > 30 MeV

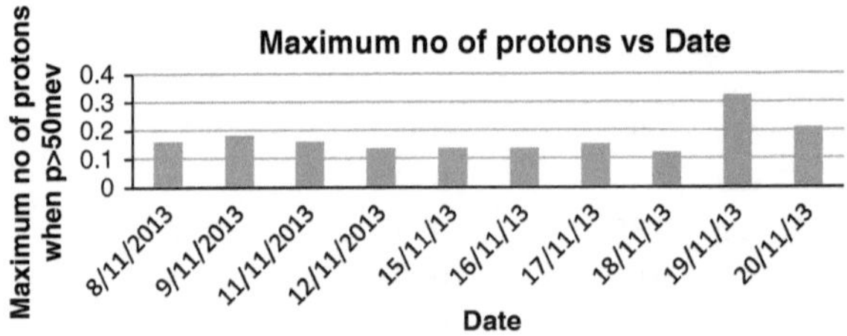

Fig. 23.20 Diurnal proton variation when P > 50 MeV

Fig. 23.21 Proton density variation for different energy on 17 November 2013

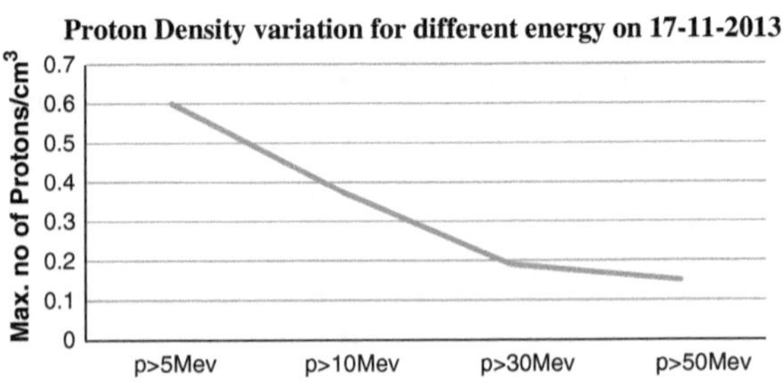

23.3 Conclusion

Some characteristic features of two large earthquakes have been analysed from the recorded data. The signal is characterized by spiky variations which indicates the proton density commencing several hours prior to the occurrences of earthquakes. The nearer the epicenters from the receiver, the higher is the amplitudes of spikes. The amplitude of spikes is very much dependent on the magnitude of earthquake. It should be pointed out that the pre-seismic ionospheric anomalies which appear 1–8 days before the earthquakes can be positive as well as negative. Here we consider the proton density as a parameter which is having a nature of decreasing towards the earthquake date, because geomagnetic activity was very quiet during the days of the studied earthquake's day. The detected anomalies can be interpreted as the pre-seismic ionospheric variations. The seismo ionospheric anomalies represented here are very significant for short-term earthquake prediction. It may be due to the plate movements under Scotia Sea the submicron aerosols are emerged by which ionosphere gets ionized and electric field becomes stronger, then electrons are accelerated but huge amount of protons comes from solar radiation. So there is a chance of decreasing proton density in ionosphere. During any large earthquake, there will be coupling between lithosphere–atmosphere and ionosphere through some probable channels like chemical channel and electromagnetic channel. From chemical channel, there will be a gas emanation, changes in geophysical parameters which introduce chemical/conductivity changes in air resulting in a modification of the atmospheric electric field perturbing the plasma density in the ionosphere. EM channel is supposed to introduce VLF emission, ionizations, electric charge redistribution above the surface of the earth by which anomalous electric field would be generated producing large-scale irregularities.

References

1. S.A. Pulinets, A.D. Legen, T.V. Gaivoronskaya, V.K. Depuev, Main phenomenological features of ionospheric precursors of strong earthquakes. J. Atmos. Sol.-Terr. Phys. **65**, 1337–1347 (2003)
2. M.B. Gokhberg, V.A. Morgounav, T. Yoshino, I. Tomizawa, Experimental measurement of electromagnetic emission possible related to earthquakes in Japan. J. Geophys. Res. **87**, 7884–7888 (1982)
3. Y. Fuzinawa, K. Takahashi, Electromagnetic radiations associated with major earthquakes. Phys. Earth Planet. Inter. **105**, 249–259 (1998)

Chapter 24
Analysis of Similarity Between Protein Sequences Through the Study of Symbolic Dynamics

Jayanta Pal, Anilesh Dey, Soumen Ghosh, D.K. Bhattacharya and Tarunima Mukherjee

Abstract Protein sequence analysis is an important tool to decode the logic of life. The rapid growth of protein sequences is constantly throwing several challenges to the bioscientists. So several methods are being improvised to assign mathematical descriptors to protein sequences, in order to quantitatively compare the sequences and determine similarities and dissimilarities between them. In this paper, for the analysis of protein sequences their 4-bit ranks and entropies are studied and compared with each other which yields satisfactorily convenient results.

Keywords Hydropathic scale · Symbolic dynamics · Rank · Entropy · Q–Q plot

J. Pal (✉) · T. Mukherjee
Computer Science and Engineering Department, Narula Institute of Technology, Kolkata 700109, India
e-mail: jayantapal1978@yahoo.com

T. Mukherjee
e-mail: tarunima91@gmail.com

A. Dey
ECE Department, Narula Institute of Technology, Kolkata, India
e-mail: anilesh.dey@gmail.com

S. Ghosh
Information Technology Department, Narula Institute of Technology, Kolkata, India
e-mail: soumenghosh.kolkata@gmail.com

D.K. Bhattacharya
Instrumental Music Department, Rabindra Bharati University, Kolkata, India
e-mail: dkb_math@yahoo.com

© Springer India 2015

K. Maharatna et al. (eds.), *Computational Advancement in Communication Circuits and Systems*, Lecture Notes in Electrical Engineering 335,
DOI 10.1007/978-81-322-2274-3_24

24.1 Introduction

The basic idea behind the numerical characterization of protein sequences is that specific sequences are generally unique and therefore possess a characteristic signature which conveniently helps us to recognize an unknown species and measure its similarity with others. In the past few years, considerable efforts are being made in order to find reliable, robust and fully automated methods to mathematically characterize sequence segments. Quite a few approaches are being proposed which are laborious and time consuming at the same time, and thus they are not efficient enough for serving our purpose in a short time.

For example, PSI-BLAST [1] and HMM [2] are regarded as very powerful procedures for detecting the remote homology between proteins. PSI-BLAST helps to look in detail into the database for sequences that match the query sequence by utilizing a scoring matrix exclusively formulated for the query. In HMM, the modelled system is assumed to be a Markov process with unobserved (hidden) states. A HMM can be considered as the simplest dynamic Bayesian network. But these methods do not work when query sequences lack significant similarity to the database sequences [3].

Later the protein sequences were expressed in 2D graphical representations [4–7]. But it resulted in an increased complexity, because the protein strings are built on 20 alphabets (representing the 20 amino acids) unlike the strings that are built on 4 alphabets only (representing DNA or RNA). A simple, convenient and fast approach was made through dynamic 2D protein sequences [8]; however, the results obtained were not so accurate. An attempt was made which dealt with characterization of protein sequences by extracting sequence features, rather than using the query strings directly for comparison among themselves. But the dimension of the vectors obtained was so small that it resulted in significant loss of information [9].

To eliminate a few of the above shortcomings, recently a new approach is made where the 20 amino acids are classified into 4 groups [10], in such a way that it does not affect the results in that method. But the subsequent steps are quite complex, as the method involves calculations among 56-dimensional vectors.

Due to the limitation of availability of convenient methods, a robust and efficient computational system is important. In this paper, we present a novel approach to analyse the similarity of protein sequences of nine ND5 (NADH Dehydrogenase subunit 5) proteins. Without loss of generality, the 20 amino acids are classified to 4 subgroups based on their hydropathy characteristics. Each of the subgroups is expressed by 2 bits (binary digits). Thus the protein sequences are expressed as binary sequences, and the subsequent calculations are done accordingly. We compute the rank and entropy of the sequences, and make a comparative study between each pair of species. The remainder of this paper is organized as follows: The methodology is described in Sect. 24.2. The experimental results and discussions are reported in Sect. 24.3 and the conclusions are drawn in Sect. 24.4.

Table 24.1 Database source

No.	Species name	ID	Database
1	Human (*Homo sapiens*)	AP-000649	NCBI
2	Gorilla (*Gorilla gorilla*)	NP-008222	NCBI
3	Common chimpanzee (*Pan troglodytes*)	NP-008196	NCBI
4	Pigmy chimpanzee (*Pan paniscus*)	NP-008209	NCBI
5	Fin whale (*Balaenoptera physalus*)	NP-006899	NCBI
6	Blue whale (*Balenoptera musculus*)	NP-007066	NCBI
7	Rat (*Rattus norvegicus*)	AP-004902	NCBI
8	Mouse (*Mus musculus*)	NP-904338	NCBI
9	Opossum (*Didelphis virginiana*)	NP-007105	NCBI

24.2 Methods

24.2.1 Data Acquisition

A protein sequence of length n can be described as a linear succession of n symbols from the 20-letter amino acid alphabet A, C, D, E, F, G, H, I, K, L, M, N, P, Q, R, S, T, V, W, Y. We work with nine ND5 protein sequences of nine different species, obtained from the NCBI database. The subjects are chosen in such a way that it includes a varied collection of mammals, ranging from human and chimpanzee to rat and mouse. The data used is given in Table 24.1. All sequences are represented as binary sequences and are processed pairwise in MATLABR2010a, with the help of Q–Q plotting.

24.3 Proposed Methodology

24.3.1 Classification of Protein Sequences

Clearly, computations involving as much as 20 different types of amino acids become troublesome. So recently, several attempts are made to simplify the protein sequences, which result in easier calculations and at the same time, ensuring preservation of data. Here we adopt the classification [10] based on hydropathic scale. A hydropathic scale measures the hydrophilic and hydrophobic properties of each of the 20 amino acid side-chains under consideration. According to hydropathic scale, the 20 amino acids can be grouped into three hydropathy states: Strongly hydrophilic (POL), strongly hydrophobic (HPO) and weakly hydrophilic or weakly hydrophobic (Ambi) [3]. Here the abbreviations, i.e. POL, HPO and Ambi, for hydropathy characteristics as used by Panek et al. are adopted [11]. Since Proline and Glycine are not classified into any hydropathy set because of their

Table 24.2 Subrouping of amino acids

Hydropathy characteristics	Abbreviation	Amino acids
Strongly hydropathic (polar)	POL	R, D, E, N, Q, K, H
Strongly hydrophobic	HPO	I, L, V, A, M, F
Weakly hydrophilic or weakly hydrophobic (ambiguous)	Ambi	S, T, Y, W
Special	None	C, G, P

unique backbone properties, and Cystine is excluded from any set because it has polarisable properties [12], these three amino acids are taken in one group. Finally, the 20 amino acids are classified into four groups (Table 24.2).

24.3.2 Similarity Index

Each of the group has to be represented by a unique *k*-digit binary number. As there are only four groups, it will suffice if we choose *k* = 2.

Suppose a protein sequence is given by $\{x_1, x_2, x_3, \ldots, x_n\}$, where x_i represents the amino acid in the *i*th position of the series. This sequence is converted into symbolic dynamics by a mapping $f{:}\{x_1, x_2, x_3, \ldots, x_n\} \rightarrow B$, which is defined by:

$$\begin{aligned} f(x_i) &= 00, \quad \text{when } x = \mathrm{R, D, E, N, Q, K, H} \\ &= 01, \quad \text{when } x = \mathrm{L, I, V, A, M, F} \\ &= 10, \quad \text{when } x = \mathrm{S, T, Y, W} \\ &= 11, \quad \text{when } x = \mathrm{C, G, P} \end{aligned}$$

Thus by the mapping f we see the sequence $\{x_1, x_2, x_3, \ldots, x_n\}$ of length *n* becomes the binary sequence *B* of length 2*n*.

For example, if we have a protein sequence segment as RDVACSTLIAAAA then by symbolic dynamics we get the binary sequence *B* as 000001011110-10010101010101

Calculation of Rank:

Let *Wk* denote the *k*-bit word that represents a unique pattern in the binary sequence *B*. Obviously there exists 2 k such words. Let *Wik* denote the *i*th *k*-bit word. Corresponding to each *Wik* the rank of *Wik*, denoted by *R*(*Wik*) is defined as

R(*Wik*) = number of times *Wik* appears in binary sequence *B*.
Thus, for the above sequence we obtain the ranks as follows:
R(0000) = 2, *R*(0001) = 1, *R*(0010) = 2 and so on.

It is claimed in [13] that the occurrence of these *k*-bit words may reflect the underlying dynamics of the original time sequence.

24.3.3 *Calculation of Entropy*

We use a slightly different approach in which in place of rank, Shanon's entropy is used as the parameter. In [14], to measure entropy of a protein sequence we first find out the rank of each *Wik* that occurs in the sequence. The formula for Shanon's Entropy *H*(*wk*) is given by

$$H(wk) = -p(wk)\log p(wk),$$

where *p*(*wk*) represent the probability of the rank of a specific word in the query sequence.

24.4 Q–Q Plot (Quantile–Quantile Plot)

In statistics, a Q–Q plot [15] ("Q" stands for quantile) is a probability plot which is a graphical method for comparing two probability distributions by plotting their quantiles against each other. Q–Q plot is a plot of the quantiles of two distributions against each other, or a plot based on estimates of the quantiles. The pattern of points in the plot is used to compare the two distributions. The points plotted in a Q–Q plot are always non-decreasing when viewed from left to right. If the two distributions under comparison are identical, the Q–Q plot follows the line $y = x$. If the two distributions agree after transforming the values linearly in one of the distributions, then also the Q–Q plot follows some line, but this line is not necessarily the line $y = x$.

24.4.1 *Our Methodology*

First we consider the 4-bit words, starting from 0000 to 1111, which can be expressed mathematically as *w*4*i* where $i = 0, 1, 2,\ldots,15$. This collection of 4-bit words over the whole binary sequence is obtained by shifting one data point at a time.

Now the rank is calculated for each such word. This actually produces a sequence of rank values for the binary sequence of the protein sequence under consideration. Finally, the probability distributions of the rank values for every pair of protein signals are compared by Q–Q plot.

For example, we obtain Figs. 24.1 and 24.2 by the Q–Q plotting of the ranks of Human–Common Chimpanzee and Human–Mouse protein sequences.

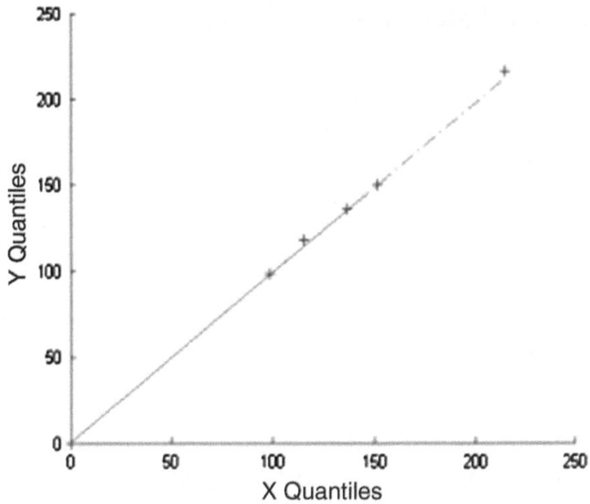

Fig. 24.1 Q–Q plot of RANK value of human–common chimpanzee protein sequences

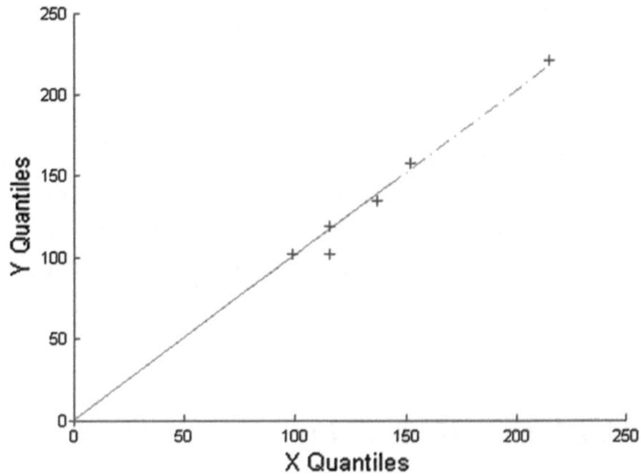

Fig. 24.2 Q–Q plot of RANK value of human–mouse protein sequences

It is clearly evident from the figure that there is a marked similarity between Human and Common Chimpanzee in terms of their protein sequences, whereas Human and Mouse have a great deal of dissimilarity.

Next we use the Entropy of species in place of Rank. Here we obtain Figs. 24.3 and 24.4 by Q–Q plotting Human–Common Chimpanzee and Human–Mouse protein sequences, respectively.

Fig. 24.3 Q–Q plot of entropy

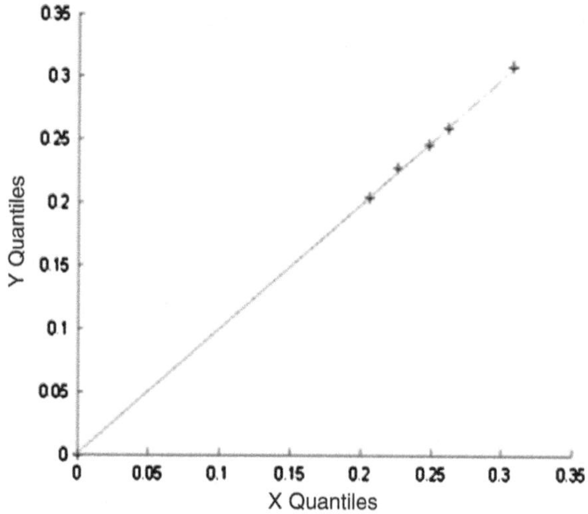

Fig. 24.4 Q–Q plot of entropy

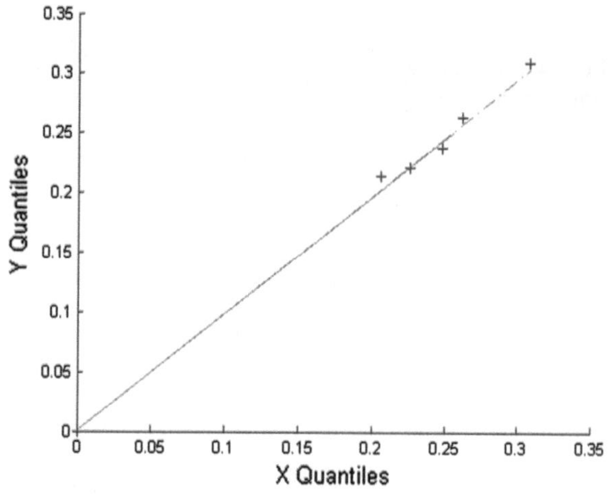

24.5 Results and Discussions

24.5.1 Q–Q Plot of the RANK Value Series for Protein Sequences of Different Species

An illustration of Q–Q plots are obtained for the RANK value sequences for different protein sequences represented in the following figures (Figs. 24.5, 24.6, 24.7, 24.8, 24.9, and 24.10).

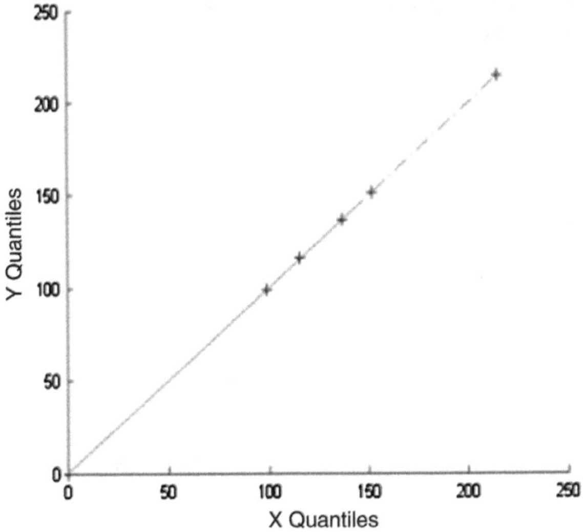

Fig. 24.5 Human versus gorilla

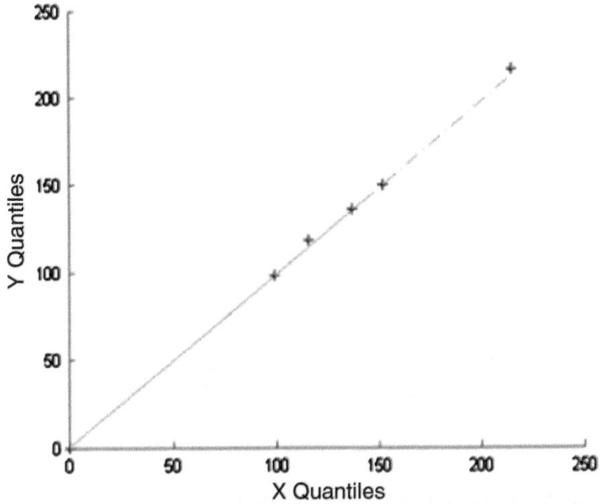

Fig. 24.6 Human versus common chimpanzee

Fig. 24.7 Human versus pigmy chimpanzee

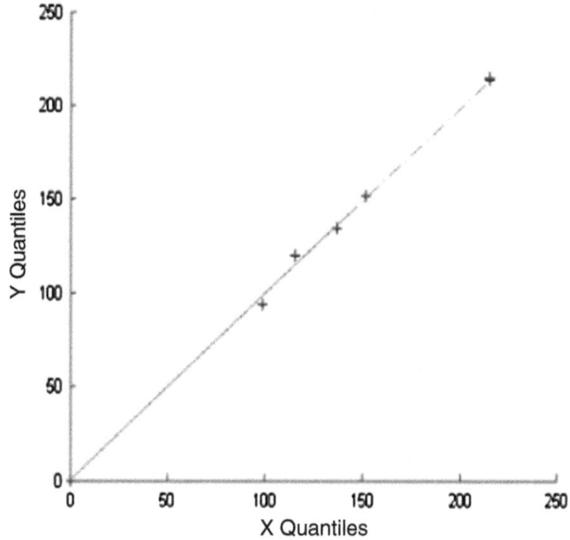

Fig. 24.8 Human versus fin whale

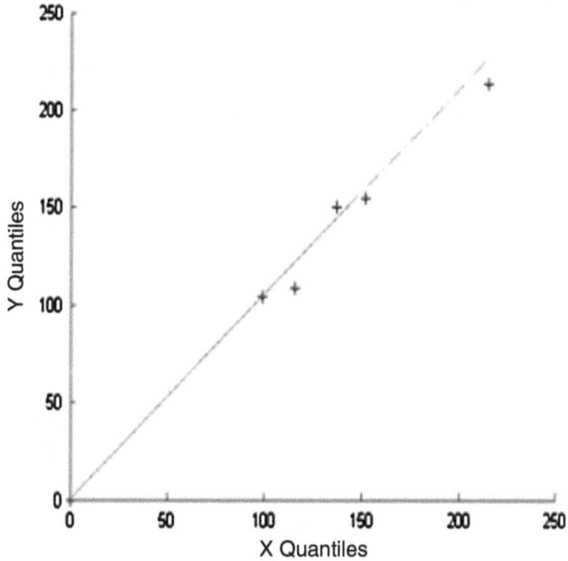

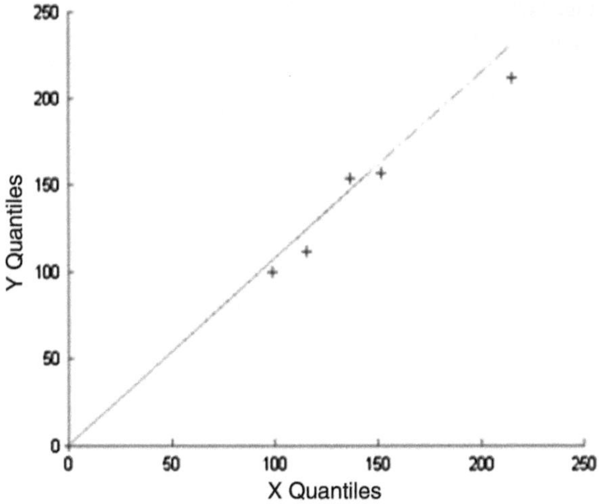

Fig. 24.9 Human versus blue whale

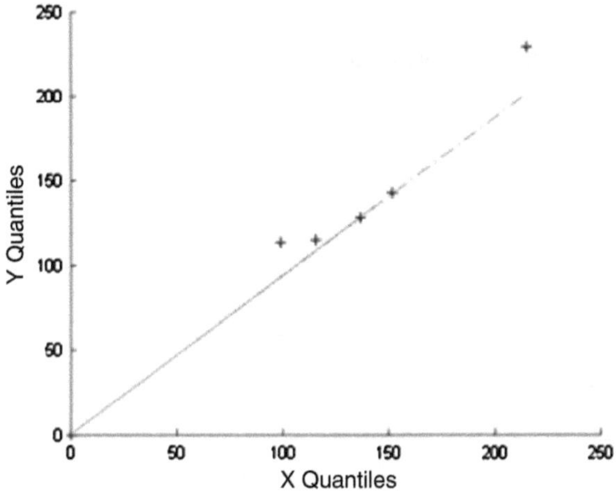

Fig. 24.10 Human versus rat versus opossum

24.5.2 *Q–Q Plot of the Entropy Value Sequences for Protein Sequences of Different Species*

As an illustration, Q–Q plots are obtained for the entropy value sequences for different protein sequences in Figs. 24.11, 24.12, 24.13, 24.14, 24.15, 24.16, 24.17, 24.18, 24.19, 24.20, 24.21 and 24.22.

Thus we have found out all the Q–Q plot graphs between each pair of species, using both rank and entropy as the parameters. We see both the result sets yield considerably satisfactory results. But on closer examination, we find in the human–gorilla pair, in Fig. 24.11, that all the points are on the dotted straight line, whereas

Fig. 24.11 Human versus gorilla

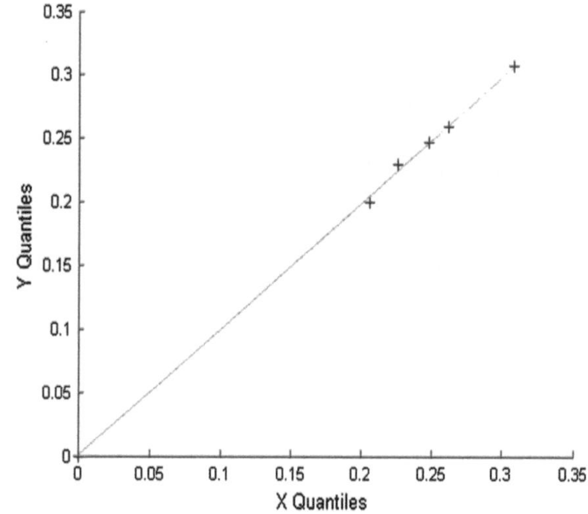

Fig. 24.12 Human versus common chimpanzee

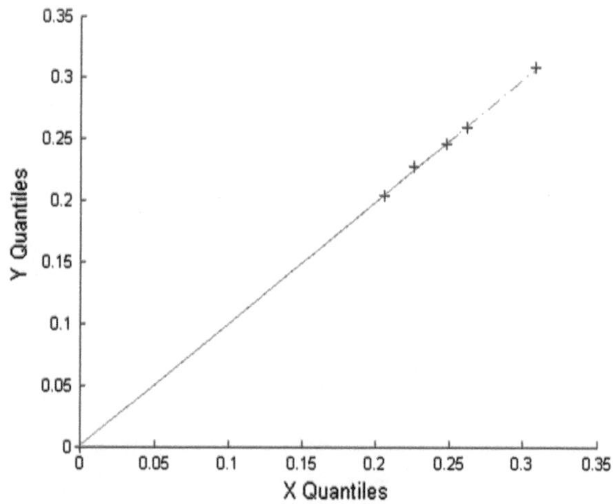

Fig. 24.13 Human versus pigmy chimpanzee

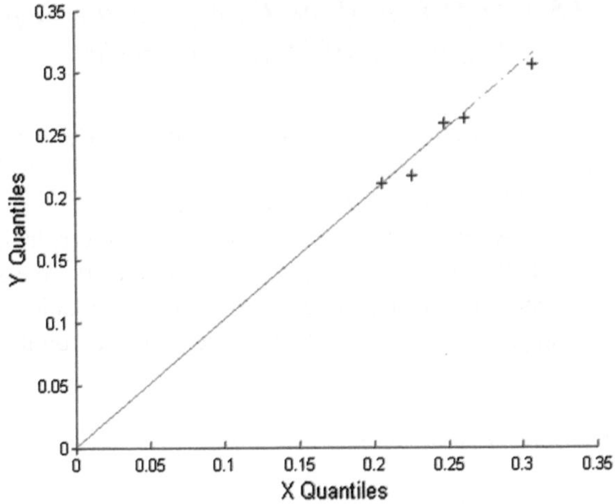

Fig. 24.14 Human versus fin whale

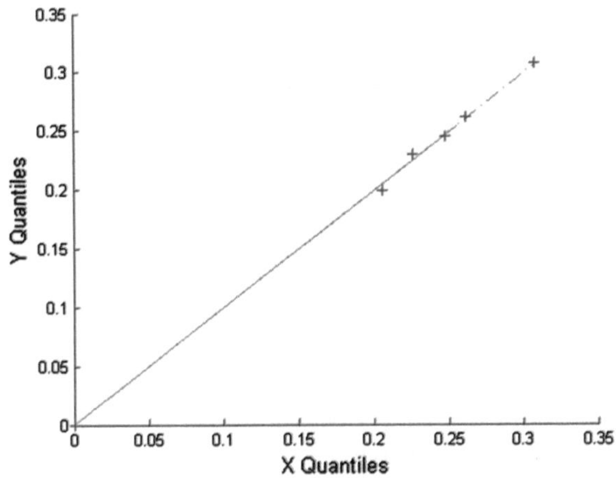

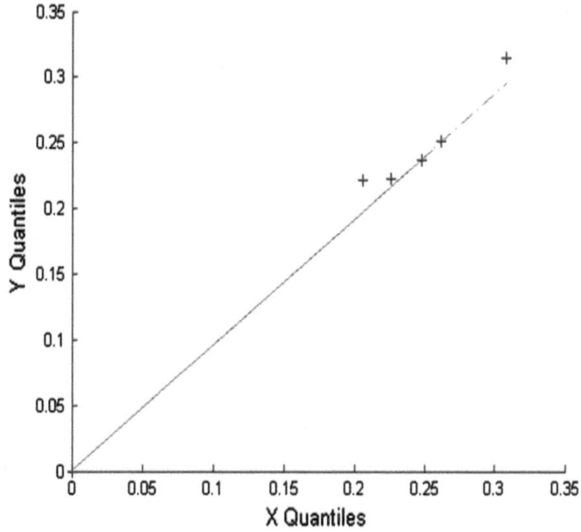

Fig. 24.15 Human versus blue whale

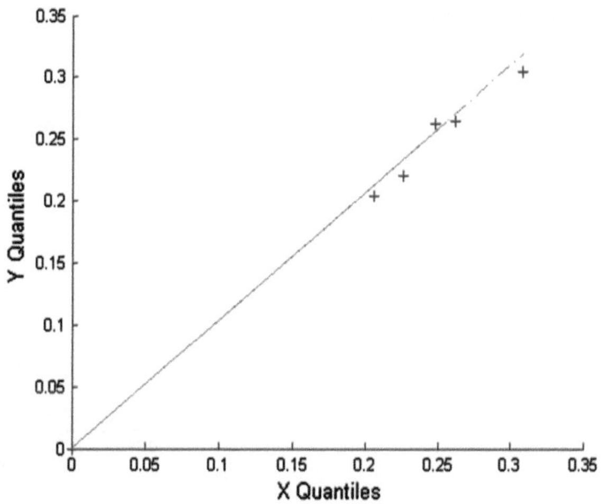

Fig. 24.16 Human versus rat

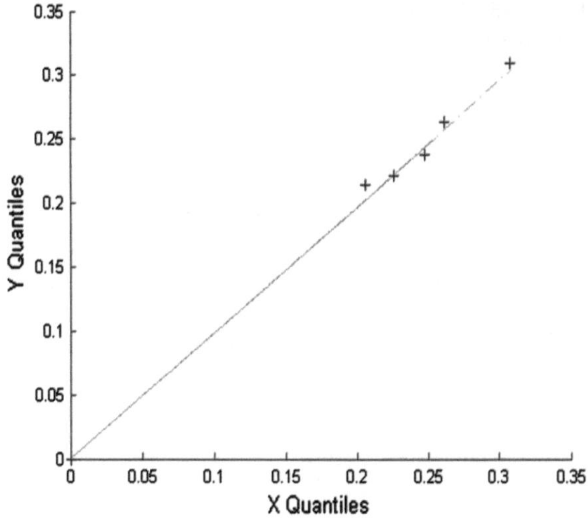

Fig. 24.17 Human versus mouse

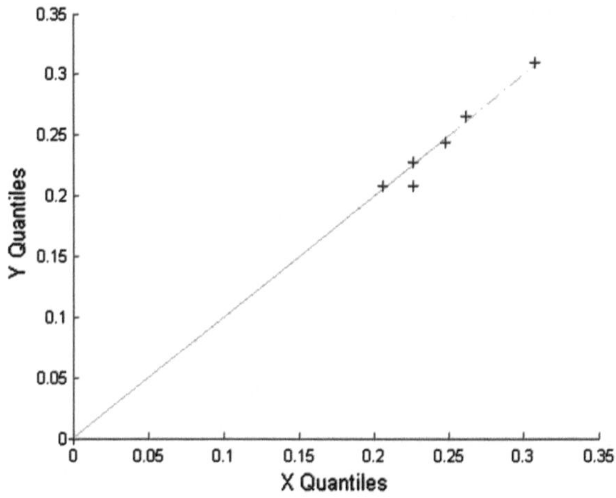

Fig. 24.18 Human versus opossum

Fig. 24.19 Gorilla versus common chimpanzee

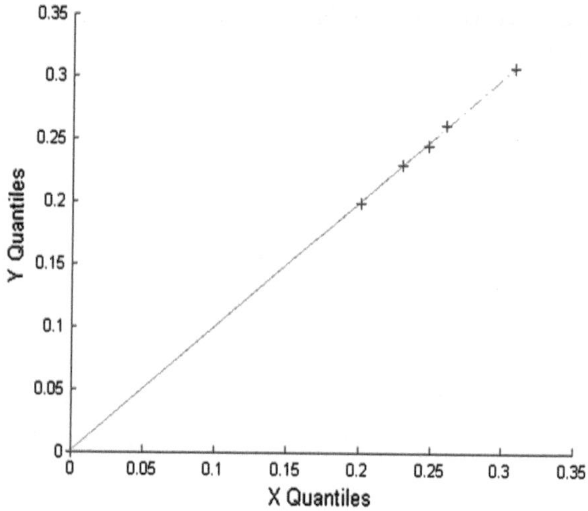

Fig. 24.20 Gorilla versus pigmy chimpanzee

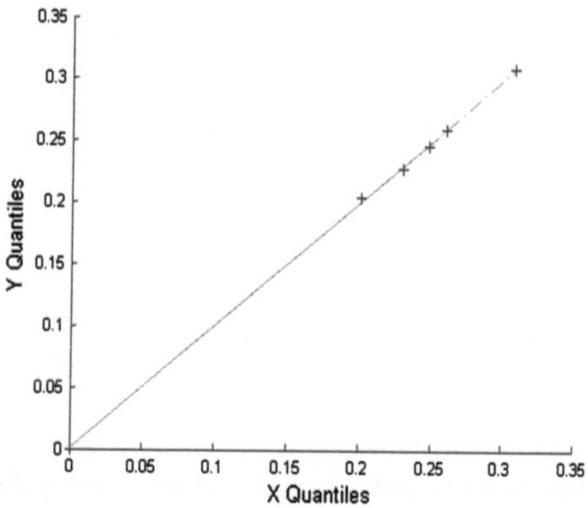

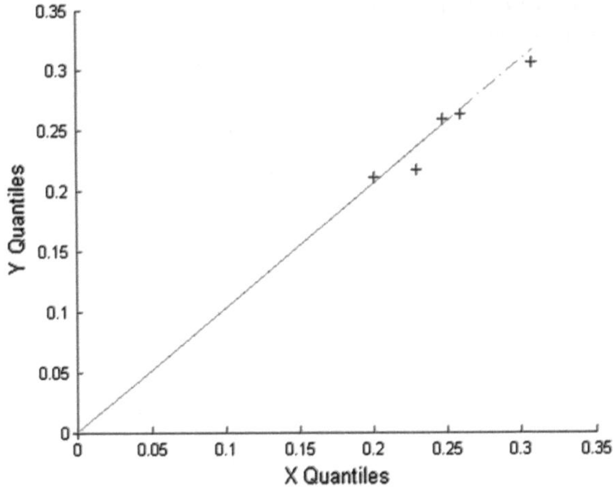

Fig. 24.21 Gorilla versus fin whale

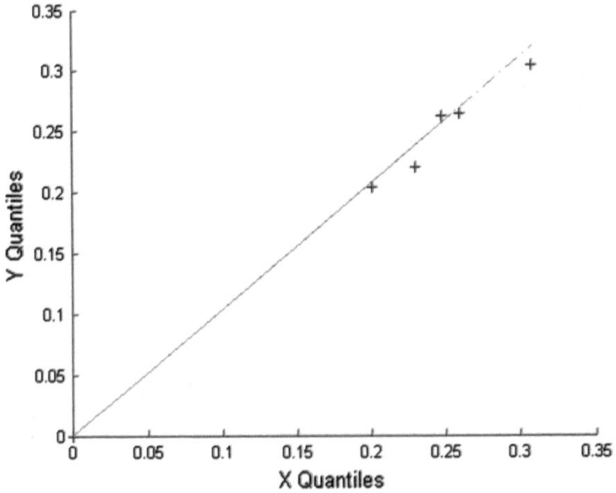

Fig. 24.22 Gorilla versus blue whale

they do not lie on the same straight line as in Fig. 24.6. So the distribution of the rank values is unable to distinguish between human and gorilla (Fig. 24.23), whereas the dissimilarity between human and gorilla is clearly visible in the plot using entropy values (Fig. 24.24).

Thus evidently, we are getting a more perfect similarity–dissimilarity graph by using Shannon's entropy as the parameter of protein sequence analysis, in place of rank.

Fig. 24.23 Q–Q plot of RANK value of human versus gorilla

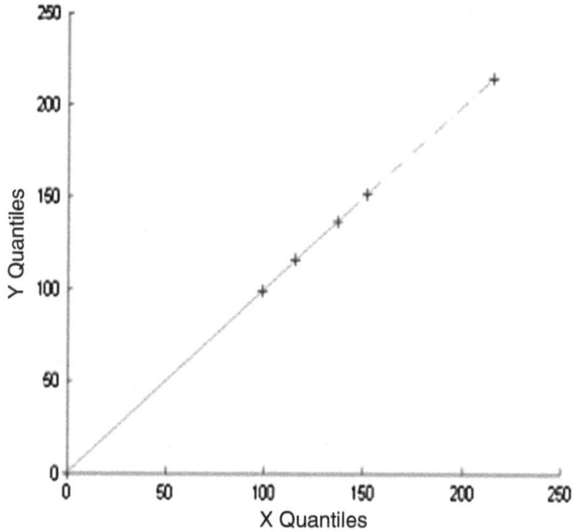

Fig. 24.24 Q–Q plot of entropy value of human versus gorilla

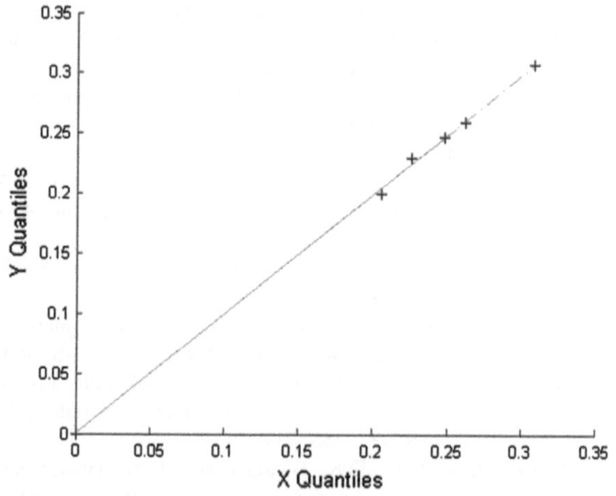

24.6 Conclusion

The binary representation of rank and entropy value sequences of the protein sequences yield satisfactory results when shown on Q–Q plot. From here, we can draw a few significant inferences: The subgrouping of amino acids to four groups is a convenient attempt to reduce the complexity of dealing with 20 alphabets. Binary representation of sequences with the help of symbolic dynamics eases the subsequent calculations. Both Rank and Entropy are, simple yet ,correct measures of measuring the similarity and dissimilarity of different protein sequences. This study yields more satisfactory results when entropy is used as the parameter. This method can even be used to find out the similarity and dissimilarity of an unknown query sequence with the sequence of a known species. This will give us an idea about what could be the species of the unknown sequence. However, it may be noted that the present study is carried out only on a smaller sample size of data and so further investigation is required on a larger sample size of data to substantiate the results of the present work.

References

1. S. Altschul, T. Madden, A. Schafer, J. Zhang, Z. Zhang, W. Miller, D. Lipman, Gapped blast and psi-blast: a new generation of protein database search programs. Nucleic Acids Res. **25**, 3389–3402 (1997)
2. A. Krough, M. Brown, I. Mian, K. Sjolander, D. Haussler, Hidden markov models in computational biology: applications to protein modelling. J. Mol. Biol. **235**, 1501–1531 (1994)
3. M. Bhasin, G.P.S. Raghava, Gpcrped: ansvm-based method for prediction of families and subfamilies of g-protein coupled receptors. Nucleic Acids Res. **32**, 383–389 (2004)
4. M. Randic, 2-D graphical representation of proteins based on virtual genetic code. SAR-QSAR Environ. Res. **15**, 147–157 (2004)
5. F.L. Bai, T.M. Wang, On graphical and numerical representation of protein sequences. J. Biomol. Str. Dyn. **23**, 537–545 (2006)
6. B. Liao, J.W. Luo, R.F. Li, W. Zhu, Novel method for analyzing proteome. Int. J. Quant. Chem. **107**, 1295–1300 (2007)
7. W. Zhu, B. Liao, J.W. Luo, R.F. Li, Numerical characterisation and similarity analysis of neurocan gene. MATCH Commun. Math Comput. Chem. **57**, 143–155 (2007)
8. Y.H. Yao, Q. Dai, C. Li, P.A. He, X.Y. Nan, Y.Z. Zhan, Analysis of similarity/dissimilarity of protein sequences. Proteins **73**, 864–871 (2008)
9. Y.Q. Liu, Y.S. Zhang, A new method for analysing H5N1 avian influenza virus. J. Math. Chem. (2009). doi:10.1007/s0-009-9641-2
10. Y. Zhang, X. Yu, Analysis of Protein Sequence Similarity
11. J. Panek, I. Eidhammer, R. Aasland, A new method for identification of protein (sub)families in a set of proteins based on hydropathy distribution in proteins. Proteins **58**, 923–934 (2005)
12. W. Taylor, Identification of protein sequence homology by consensus template alignment. J. Mol. Biol. **188**, 233–258 (1986)
13. A. Dey, S.K. Palit, D.K. Bhattacharya, D.N. Tibarewala, Effect of music on autonomic nervous system through the study of symbolic dynamics of heart rate variability signals, RTCCE-2013, ISBN: 978-981-07-6184-4. doi:10.3850/978-981-07-6184-4 08
14. C.C. Albert, Y.A.L. Goldberger, C.K. Peng, Information-Based Similarity Index. http://www.physionet.org/physiotools/ibs/doc/ibsi.pdf
15. W.L. Martinez, A.R. Martinez, *Computational Statistics Handbook with Matlab®*, 2nd edn. (Chapman & Hall/CRC, Boca Raton 2002)

Part IV
Advances in Computer and Network

Chapter 25
A Secure Group-Based Communication Scheme in Disaster Response Environment Using Delay Tolerant Network

Chandrima Chakrabarti, Ananya Banerjee and Sanchari Chakrabarti

Abstract When a major disaster strikes, large parts of the critical infrastructure can be completely debilitated for weeks. In such situations the principal need is to quickly re-establish minimal communication infrastructures for support of rescue operations. For this purpose smart phone-based ad hoc opportunistic network can be built to rescue people from different affected areas. Relief workers from different agencies divide into small groups based on different categories of needs. In such sensational environments secure and immediate group communication among small groups and to the remote monitoring system is urgent as there can be some malicious nodes that intend to intercept and alter sensitive data for the purpose of corruption. To implement group communication using Delay Tolerant Network (DTN) some authorized carrier nodes can be allocated to deliver messages and shared group keys can be assigned for encryption-decryption purposes. The performance of the proposed scheme is evaluated on ONE simulator (The Opportunistic Network Environment Simulator) version 1.4.1.

Keywords Ad hoc opportunistic network · Group communication · Delay tolerant network

C. Chakrabarti (✉) · A. Banerjee
Department of Computer Science and Engineering, Narula Institute of Technology, Kolkata, India
e-mail: chandrima.narula@gmail.com

A. Banerjee
e-mail: ananyabanerjee.narula@gmail.com

S. Chakrabarti
Brahmananda Keshab Chandra College, Kolkata, India
e-mail: dearsanchari@gmail.com

© Springer India 2015

K. Maharatna et al. (eds.), *Computational Advancement in Communication Circuits and Systems*, Lecture Notes in Electrical Engineering 335,
DOI 10.1007/978-81-322-2274-3_25

25.1 Introduction

Our modern communication infrastructure will be totally damaged or unavailable when a major disaster takes place. To build the minimal infrastructure for rescue operations will be a prime need in such situation. Such communication networks can be infrastructure-less with typical mobility pattern and delay tolerance. One example of such an "infrastructure-less" network is described by the perception of hastily formed networks [1–3], built-up using smart phones, carried by emergency responders and emergency vehicles working at the disaster sites to carry out post-disaster recovery operations. Relief workers from different agencies try to divide into small groups based on different categories of situational needs in a timely manner [4]. However, different group members can be located in different positions for the purpose of receiving or forwarding data. If we consider medical teams or military teams as groups, they are not working in a particular area but are scattered in different areas. Even group members may not know the actual location of its own group members. So, to build up proper communication within a group is very difficult. Also, in such sensational environments, secure and immediate group communication among small groups and to the remote monitoring station is very essential, as there can be some malicious node intending to intercept and alter sensitive data for the purpose of corruption.

In this paper, our objective is to ensure fast and reliable delivery of group messages in post-disaster communication networks by avoiding the possibility of data modification by malicious nodes as far as possible. We propose a fast and secured Group-based Communication Scheme where a group of reliable and authorized static and dynamic carrier nodes are employed to deliver messages in sparsely connected network. In our paper, group member nodes need to encrypt data with a shared group key, which is only available to group member nodes. Some authorized carrier nodes are assigned for the purpose of forwarding or receiving different group messages and delivering them appropriately on an urgent basis. We also compared our performance with the paper by Zhou et al. [5] using ONE simulator and we obtained better results using our scheme from the perspective of data delivery and overhead.

The rest of the paper is organized as follows. Section 25.2 describes the related work. Section 25.3 illustrates our system model. The simulation parameters and performance comparisons are discussed in Sect. 25.4. The paper is concluded with a discussion on the future work in Sect. 25.5.

25.2 Related Work

Nowadays, application of Delay Tolerant Network (DTN) in the post-disaster environment has gained the increasing attention of researchers. Substantial research has been carried out to facilitate group-based data exchange for the purpose of increasing reliability in a fragmented network like DTN.

In such an opportunistic set-up, a node must decide whether to forward packets to an encounter node based on contact history [6, 7]. But in such case authors [6, 7] did not consider the question of data integrity, which may be hampered. Li et al. [8] proposed an enhanced group-based routing protocol for DTN, in which the relay node is selected based on social group information obtained from historical encounters. We feel that in the disaster environment it is not possible to meet each group member within a zone, because they are scattered everywhere. So, encounter metric does not always work. Zhou et al. [5] discussed their Group Based Epidemic Routing (G-Epidemic) for propagating group information with the help of group members. However, we observed that in this scheme data delivery is very low as in a fragile environment like DTN, nodes contacts are opportunistic in nature. So a node needs to wait for a long time to find its own group member or friend. At the same time, authors of [5] did not consider secure delivery of data.

Inspired by existing research works, we aim to design a secure group-based communication scheme using shared group key and authorized carrier nodes for fast delivery in a disaster response communication scenario.

25.3 System Model

In the post-disaster environment each volunteer works in a group. Examples of such groups can be a medical team, a military team, NGOs, etc., which are mainly interest-based groups. These groups can be located in different locations of the disaster affected area. The total number of groups, type of groups, number of people allocated to each group may be varied based on the disaster type and situation. In disaster struck area, volunteer groups will choose some buildings, schools as local shelter for temporary residence of victims after rescuing them. Shelters can also be used as the local office of different volunteer groups, temporary storage of needs like rice, dry foods, medicine, etc., for survival. Updated information about all shelters (information like actual situation of the areas, demand of each shelter, number of victims) will be forwarded from time to time to the Control Station node, which is situated in a nearby city.

Now, from each shelter, volunteer groups will be divided into lesser number of groups and will work in different locations as shown in Fig. 25.1; but before that, each volunteer group needs to register at a shelter and get the shared secret key for ensuring secure message exchange.

As in Fig. 25.1, the topmost node, Control Station node controls all the shelter nodes and gets up-to-date information from them. Volunteer nodes are responsible for forwarding messages securely either directly in one hop or via carrier nodes which are used in group communication for delivering messages from time-to-time. Carrier nodes are also responsible for communication between Control Station node and Shelter nodes. There are two types of carrier nodes:

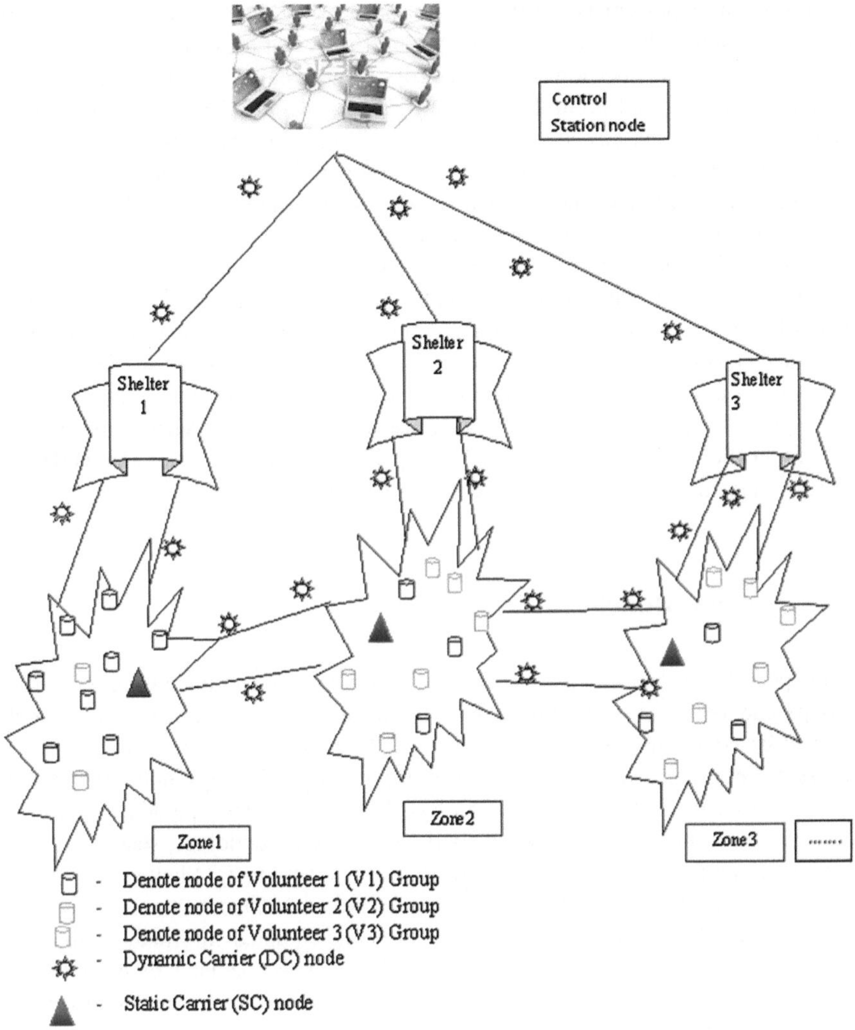

Fig. 25.1 Hierarchy of different nodes

(i) Static Carrier (SC) node
(ii) Dynamic Carrier (DC) node

SC nodes are fixed, situated in a zone (as in Fig. 25.1) and mainly serve as the repository for message exchange like the local post office.

DC nodes are roving from one zone to another for inter zone message exchange like mobile van of the post office. As we have used carrier nodes for data delivery, so, here, volunteer nodes need not wait much for forwarding a packet if no forwarder is found. As a result, delivery of messages will be faster than any other scheme.

Table 25.1 Example of neighbour table of a volunteer node

Node id	Group id	Location
s1 (shelter node)		(x, y)
v11s1 (volunteer node of group1)	g1	(x11, y11)
v12s1 (volunteer node of group1)	g1	(x14, y14)
v21s1 (volunteer node of group2)	g2	(x12, y12)
v31s1 (volunteer node of group3)	g3	(x13, y13)
Sc11 (static carrier node1)		(x1, y1)

25.3.1 Data Forwarding Using Volunteer Node

Suppose a volunteer node wants to send a message to its group member node; first, it encrypts the message with the shared group key. Then after searching the neighbour table, if it finds the destination node in its 1 hop neighbour, delivers it directly.

When a volunteer node within a particular zone needs to send some messages to another node, it first consults its neighbour node (Table 25.1).

But if it cannot find it in the neighbour table, then it needs the help of the carrier node. If that particular volunteer node finds any DC node within its vicinity, it forwards the message to that DC node, as using DC node a message can be forwarded fast. But if a volunteer node cannot find any DC node, then it forwards the message to the nearest SC node as it knows the SC node's location. The detail of this mechanism is discussed in Sect. 25.3.4, Algorithm 1.

25.3.2 Data Forwarding Using Carrier Node

In a scheduled time (after 3–4 h) when DC node visits the zone, it will forward (download) messages first, then it will collect (upload) new messages from SC nodes at first and then from other volunteer nodes.

These DC nodes always prefer to come in contact with SC node before meeting any volunteer node. When downloading and uploading of information between DC and SC nodes are completed, then both of them will come into contact with volunteer nodes for delivering messages. After collecting messages from DC nodes, SC nodes forward these messages to the appropriate volunteer node. We assume carrier node has the filtering capability to avoid redundancy of messages. The detail of this mechanism is discussed in Sect. 25.3.4, Algorithm 2 for DC and Algorithm 3 for SC node. An SC node has to maintain its neighbour table as (Table 25.2).

Table 25.2 Example of neighbour table of a static carrier node

Node id	Group id	Location
s1 (shelter node)		(x, y)
v11s1 (volunteer node of group1)	g1	(x11, y11)
v12s1 (volunteer node of group1)	g1	(x14, y14)
v21s1 (volunteer node of group2)	g2	(x12, y12)
v31s1 (volunteer node of group3)	g3	(x13, y13)
dc11 (dynamic carrier node1)		(x1, y1)
dc12 (dynamic carrier node2)		(x2, y2)

25.3.3 Communication with Shelter Node and Control Station Node via Carrier Node

If shelter node has to send some message to the affected zone or vice versa, which is not specific to a group, shelter node can broadcast the message. If any volunteer group has to send some message from shelter to the disaster affected zone or vice versa, then volunteer group or node needs to encrypt the message using shared group key. After collecting all information from the disaster struck zone, at shelter, different volunteer groups will decrypt the information and try to generalize the overall information. These volunteer groups (those who reside at shelter) will again encrypt that information using shared group key and send it towards Control Station node via carrier nodes.

Using this scheme Control Station (CS) node gets updated information about needs from different shelter nodes from time-to-time via carrier nodes. We assume CS node has the authority to get access of all shared group keys. CS node allocates relief materials for meeting the needs of different shelters and sends them via the carrier nodes as well.

25.3.4 Algorithms

ALGORITHM 1: DATA FORWARDING using Volunteer node

```
STEP1: Initialize current_node, source, destination, 1
hop neighbor node, receiver node, forwarder node, Dynamic
Carrier (DC) node, Static Carrier (SC) node, shared group
key for different groups available
STEP2: If source = current_node
i) Encrypt data using shared group key
ii) Search neighbor table for destination
```

```
If destination = 1 hop neighbor node
Deliver data
End if
Else Repeat the following steps (step 2.1 to 2.4) until
next forwarder is found
2.1: Check if any Dynamic Carrier node found within
transmission range
2.2: If found
2.3: Forward data to DC node
2.4: Else Check neighbor table for Static Carrier node
If found
Forward data to SC node
End if
End if
End if
// When current_node ≠ source
Step 3: If current_node = destination
Receive data
Decrypt data using shared group key
End if
```

ALGORITHM 2: DATA FORWARDING using DC node

```
STEP1: Initialize volunteer node, Dynamic Carrier (DC)
node, Static Carrier (SC) node, transmission range, data-
base of SC node, database of DC node
STEP2: If any volunteer node comes within the transmis-
sion range of any Dynamic Carrier (DC) node then DC node
will not accept data from it. DC node first comes with
the transmission range of a SC node within that zone
End if
STEP3: DC node checks if any SC node is found within
transmission range
If found
Forward data to SC node
Receive data from SC node
Else Store data to its database until a SC node found
End if
//After meeting with SC node within a zone, a DC node can
meet a volunteer node;
If any volunteer node found within the transmission range
of DC node
```

```
If found
DC node checks its database whether it has any data des-
tined for that volunteer node
If found
Deliver data to the volunteer node
Receive data from that volunteer node
End if
End if
```

ALGORITHM 3: DATA FORWARDING using SC node

```
STEP1: Initialize volunteer node, Dynamic Carrier (DC)
node, Static Carrier (SC) node, transmission range, data-
base of SC node, database of DC node
STEP2: If any volunteer node comes within the transmis-
sion range of any Static Carrier (SC) node then SC node
checks its own database whether it has some message
destined for that particular volunteer node; SC node
first forwards that data then receives data from that
volunteer node
End if
STEP3: SC node checks if any DC node is found within
transmission range
If not found
Store data to its database
Else
Download/receive data from DC node
Upload/forward data to DC node
End if
```

25.4 Simulation

We have implemented our scheme using Opportunistic Network Environment (ONE) simulator [9]. The nodes are positioned on default map. The details of our simulation parameters are shown in Table 25.3.

The performance of our proposed scheme and the scheme in [5] are evaluated in terms of packet delivery probability and overhead ratio. We have used the following formulae:

Delivery Probability = Number of Packets Delivered/Number of Packets Created and Overhead Ratio = (Number of packets relayed−Number of packets delivered)/Number of packets delivered.

Table 25.3 Parameters used for simulation

Parameter	Value
Number of groups of volunteer nodes	3
Number of volunteer nodes	Varies from 60 to 200
Number of shelter nodes (stationary)	3
Number of control station nodes (stationary)	1
Number of static carrier nodes (stationary)	10 % of total number of volunteer nodes
Number of dynamic carrier nodes	10 % of total number of volunteer nodes
Speed of volunteer node	0.5–1.5 m/s
Speed of dynamic carrier node	1.5–5.5 m/s
Transmission range	10 m
Transmission speed	2 Mbps
Buffer size of volunteer node	5 MB
Buffer size of carrier (both static and dynamic carrier) node	500 MB
Message size	500 kB–1 MB
Movement model	Shortest path map based movement
Routing protocol	Spray and wait routing
Simulation time	12 h

From Fig. 25.2 it is evident that the delivery probability improves in our proposed scheme compared to the G-epidemic scenario discussed in [5]. We also observed the performance of our proposed scheme and the scheme discussed in [5] in terms of Overhead Ratio as represented in Fig. 25.3 where overhead ratio decreases significantly in our proposed scheme compared to the G-epidemic scenario discussed in [5].

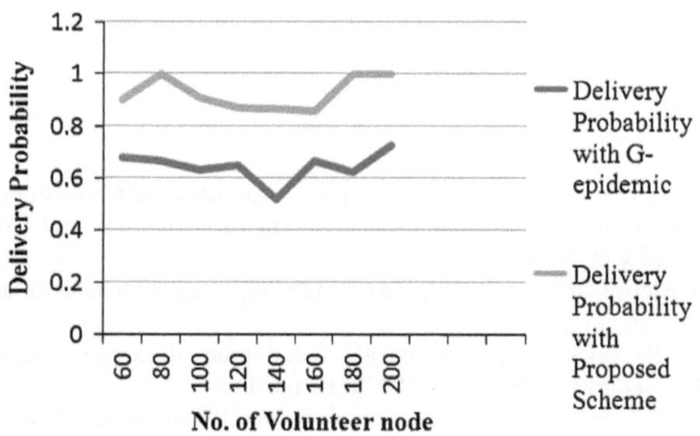

Fig. 25.2 Simulation result of number of volunteer nodes versus delivery probability with G-epidemic and our proposed scheme; our proposed scheme gives better results

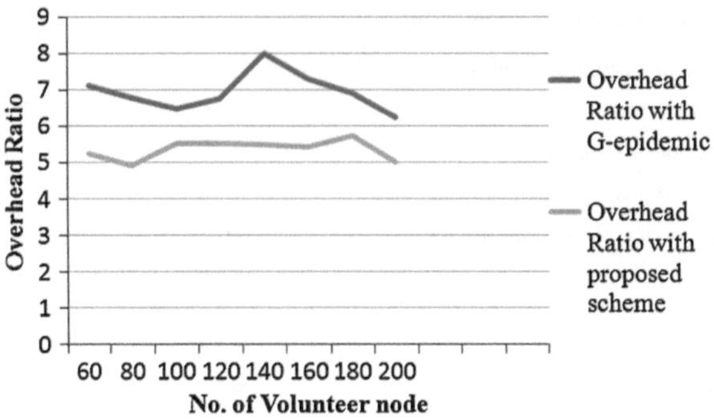

Fig. 25.3 Simulation result of number of volunteer nodes versus overhead ratio with G-epidemic and our proposed scheme; our proposed scheme gives better results

25.5 Conclusion and Future Work

This paper presents a secure group-based data delivery scheme using shared group key for data encryption–decryption by the group members and authorized carrier nodes for fast and reliable data dissemination. It is a novel approach due to the following aspects:

(i) We have tried to achieve data integrity, content of data remains intact.
(ii) Using our scheme, unnecessary dropping, non-forwarding, colluding attacks can be avoided.
(iii) Unnecessary delay in data dissemination can be avoided.
(iv) Our scheme gives better data delivery probability; overhead ratio is less compared to the scheme discussed in [5].

Here we assumed that carrier nodes are authorized and reliable. But if this situation varies, then our system will not give better delivery. In the near future we will try to solve this issue from a larger perspective.

References

1. M. Asplund, S.N. Tehrani, J. Sigholm, Emerging information infrastructures: cooperation in disasters, critical information infrastructure security, in *Third International Workshop, CRITIS*, Rome, Italy, 13–15 Oct 2008
2. The audit of disaster-related aid (Main), ISSAI 5520, http://eca.europa.eu/portal/pls/portal/docs/1/22006772.PDF
3. C. Chakrabarti, R. Chaki, Improved cluster based route discovery algorithm for ad-hoc networks, in *Proceedings of IEEE ICCIA* (2011), pp. 1–4
4. C. Chakrabarti, A. Banerjee, S. Roy, An observer-based distributed scheme for selfish-node detection in a post-disaster communication environment using delay tolerant network, in

Proceedings of IEEE International Conference on Applications and Innovations in Mobile Computing (2014), pp. 151–156
5. R. Zhou, Y. Cao, J. Jin, D. Zhu, Group based epidemic routing for delay and tolerant networks, wireless communications networking and mobile computing (WiCOM), in *Proceedings of 6th International Conference*, 23–25 Sept 2010
6. J. Wu, M. Xiao, L. Huang, Homing spread: community home-based multi-copy routing in mobile social networks, in *Proceedings of IEEE INFOCOM* (2013)
7. J.W. Chang, C. Chen, CROP: community-relevance-based opportunistic routing in delay tolerant networks, in *Proceedings of IEEE Wireless Communications and Networking Conference (WCNC)* (2013)
8. F. Li, C. Zhang, Z. Gao, L. Zhao, Y. Wang, Social feature enhanced group-based routing for wireless delay tolerant networks, mobile ad-hoc and sensor networks (MSN), in *Proceedings of Eighth IEEE International Conference*, 14–16 Dec 2012
9. The Opportunistic Network Environment Simulator (The ONE), www.netlab.tkk.fi/tutkimus/dtn/theone/Ver.1.4.0,2010 (2010)

Chapter 26
A New Approach to Trace the Behaviour Pattern of Nodes in the Delay Tolerant Network

Ananya Banerjee, Chandrima Chakrabarti and Angana Chakraborty

Abstract Due to lack of features of traditional network characteristics, DTN becomes one of the promising research paradigms of wireless network. As there are intermittent connectivity and store-carry-forward features, end-to-end acknowledgement is not possible. DTN is also named as 'opportunistic network'. But interestingly, intermediate nodes may always not behave fairly. Sometimes, due to limited resources (bandwidth and memory) intermediate nodes may behave 'selfish'. Intermediate nodes may also behave unfairly to next hop node intentionally. This behaviour is known as 'misbehaving behaviour'. Recent studies on 'selfish node' and 'misbehaving node' show us the negative impact on network performance (delivery ratio, delivery cost, delivery latency). So we design a slightly new concept inspired by real-life application in our paper showing the positive aspect of the involved nodes in the network. Apart from participating in data forwarding, node may behave as 'traffic node' trying to show the right path towards destination.

Keywords Reputation · Traffic · Forwarder

26.1 Introduction

With the recent proliferation of wireless communication devices, intermittently connectivity on the edge will quickly become a practical scenario. The disruption/delay tolerant networks which are not homogeneous (loosely connected by nature)

A. Banerjee (✉) · C. Chakrabarti
Department of Computer Science and Engineering, Narula Institute of Technology, 81, Nilgunj Road, Kolkata 700109, India
e-mail: ananyabanerjee.narula@gmail.com

C. Chakrabarti
e-mail: Chandrima.narula@gmail.com

A. Chakraborty
IIEST, 103, Subhas Nagar Road, Dum Dum Cantt, Kolkata 700065, India
e-mail: angana.chakraborty9@gmail.com

© Springer India 2015

K. Maharatna et al. (eds.), *Computational Advancement in Communication Circuits and Systems*, Lecture Notes in Electrical Engineering 335,
DOI 10.1007/978-81-322-2274-3_26

obtain structure from node interaction and node mobility, and generally includes distributed administrators instead of central administrator. This type of Ad-hoc infrastructure less network can be described by formation of networks where networks are made-up with social cooperation, dynamic topology, reliable and secured data delivery, limited resources, store-carry forward manner, intermittent connectivity and distributed authority [1]. Such networks are applicable in Inter-planetary Internets, sensor and ad hoc networks, rural and underwater communication networks and other challenging communication like disaster, military and any other emergency services. A trend has been shown that researcher is more interested in exploring the negative impact of selfish nodes and malicious nodes. They have also found many ways to detect and avoid above-mentioned nodes. But we believe that intermediate nodes also play proper roles as relay nodes. As in delay tolerant network (DTN), route establishment is not possible in advanced; we incorporate the concepts of traffic nodes in networks to help other participating nodes to get the view of the path towards destination. Moreover, based on traces pattern of nodes, we analyse the node's behaviour as cooperative and non-cooperative nodes. We also able to find out group-biasness or un-biasness of the nodes. We have simulated the scheme on one simulator and analysed the performance.

26.2 Related Work

Application of delay tolerant network in disaster management has been drawing a great attention among researchers in recent years. Adequate research works has been done to detect the selfish nodes or misbehaving nodes using reputation-based strategies. Furthermore, recent studies on the impact of selfish behaviour or misbehaving behaviour in DTNs show that the performance metrics (i.e. delivery ratio, delivery overhead and delivery latency) are rigorously affected if a large portion of the nodes is not corporative [2].

In [2], authors showed a survey on some representative strategies in the literature belonging to each category (barter based, credit based and reputation based), and finally compared the performance of these strategies for preventing different types of selfish behaviour.

In [3], source node is liable for giving reward to other nodes for successful bundle forwarding by showing proper relay evidence. In Wei et al. [4] demanded that intermediate nodes are given reward (good reputation) only after successful message delivery. This way authors are claimed to detect selfish nodes in the network. But this scheme lacks from collusion problem and group-biasness.

Inspired by the existing research works, we aim to design a new scheme that estimate global visited route with reputation and perception matrix to show the right path (selection of forwarder) towards destination and we are also able to find out group-biasness or un-biasness of node in a post-disaster communication scenario. Inspired by paper [5], we have also tried to detect the malicious activities like dropping, non forwarding, false token generation, colluding attacks and tried to stay away from those nodes during data forwarding.

26.3 System Model

In order to guarantee the reliable data communication in a post-disaster communication environment, we propose to engage a group of traffic nodes. These traffic nodes are trustworthy and are distributed in nature. Each traffic node is allocated to monitor a particular zone and move around in that zone randomly. The role of traffic node is to randomly check the nodes in its vicinity to get the information about recent visited routes and to finally assign a reputation to the nodes. Each traffic node periodically prepares a global visited route with reputation and perception matrix and broadcast it. As a consequence, nodes in its vicinity may aware of status (reputation) of nodes in the network and select the right path towards destination.

However, at the beginning, no traffic information is available to select the path. Thus, the nodes in any zone select the forwarders randomly without any support from the traffic nodes. A node selects a forwarder either (a) based on its own view about the nature of nodes in the network or (b) based on the visited route matrix with reputation and perception matrix published periodically by the traffic nodes in its vicinity. We have implemented our scheme based on Prophet routing algorithm [6]. In our model, we have proposed two schemes.

26.3.1 Forwarder Selection Based on a Node's Available Reputation and Higher Probability

Each node in the network initially selects the other node as forwarder node based on already available reputation and higher probability. Each node maintains routing tables. One is forwarder routing (fr) table and token given (tg) routing table. Token can be granted only by the destination to the previous node after successful message delivery to destination. We assume that message will reach from source (N_i) to destination (N_l) via intermediate nodes (N_j, N_k) (Tables 26.1, 26.2, 26.3 and 26.4).

According to our model, destination node (N_l) has sent token to the previous node (N_k) for successful message delivery (Tables 26.5 and 26.6).

Table 26.1 Forwarder routing table maintained at N_i node

Node ID	Source node	Message ID	Destination node ID	Timestamp	Token received	Intermediate nodes ID
N_i	N_i	M_i	N_l	t_1	–	N_j

Table 26.2 Forwarder routing table maintained at N_j node (assuming $t_1 < t_2$)

Node ID	Source node	Message ID	Destination node ID	Timestamp	Token received	Intermediate nodes ID
N_j	N_i	M_i	N_l	t_2	–	N_j, N_k

Table 26.3 Forwarder routing table maintained at N_k node (assuming $t_2 < t_3$)

Node ID	Source node	Message ID	Destination node ID	Timestamp	Token received	Intermediate nodes ID
N_k	N_i	M_i	N_l	t_3	–	N_j, N_k, -

Table 26.4 Forwarder routing table maintained at N_l node (assuming $t_3 < t_4$)

Node ID	Source node	Message ID	Destination node ID	Timestamp	Token received	Intermediate nodes ID
N_l	N_i	M_i	N_l	t_4	–	N_j, N_k, -

Table 26.5 Token given (tg) routing table maintained at N_l node (assuming $t_4 < t_5$)

Node ID	Token given to node ID	Message ID	Number of token given	Timestamp
N_l	N_k	M_i	1	t_5

Table 26.6 Forwarder routing table updated at N_k node (assuming $t_5 < t_6$) after getting token from destination (N_l)

Node ID	Source node	Message ID	Destination node ID	Timestamp	Token received	Intermediate nodes ID
N_k	N_i	M_i	N_l	t_6	1	N_j, N_k, -

26.3.2 Global Visited Route with Reputation and Perception Matrix Published by the Traffic

Traffic nodes generally move around its designated zone and may ask any node to share forwarder routing (fr) table and token given (tg) routing table. Inspired by the paper [5], traffic nodes are assumed to have access to all the public keys of the participating nodes in the network. Thus, they are able to authenticate of each token by decrypting it with the corresponding public key of the node. This feature supports strongly against false token generation. In our scheme, the traffic nodes tries to find out the information about the routes visited by the nodes by querying randomly any node in the network. We assume in our model that traffic node considers those nodes having tokens. Based on acquired information, traffic nodes maintain visited route (vr) tables. Based on token given by destination node, traffic node rewarded the node by giving the reputation to the node for successful delivery to destination. This visited route table is updated time to time by each traffic node.

We use same reputation schemes that we have used in paper [5]. Here, we assumed that the initial reputation value for each node is R, any non-zero positive number (representing non-selfish node). If a traffic node finds that the last known reputation of node was R and currently the node is still found as unbiased, i.e. both the Group-bias as forwarder and as receiver are ≤1, then the reputation will not change and will remain at R. However, if a node is found to have any kind of group-biasness, the traffic will reduce the last known reputation of that node by a reputation degradation factor (say, k) which is a non-zero positive integer (we assume $k > 1$), to assign its current reputation. Initially we assume the reputation value as R. For each type of bias (selfish activity), reputation will be degraded by a factor of k. So, if at some point of time, traffic nodes observes that a node has changed its selfish behaviour, i.e. the group-bias becomes ≤1 again, then, the last known reputation will be upgraded (multiplied) by a factor k (Tables 26.7, 26.8, 26.9 and 26.10).

Traffic nodes are scheduled to meet periodically at some special meeting points as in paper [5], called rendezvous points to exchange their visited route tables. A global node visited route with reputation matrix is thus formed by combining the information of all the traffic nodes. This global node visited route with reputation

Table 26.7 Forwarder routing table collected by traffic (T_p) node

Node ID	Source node	Message ID	Destination node ID	Timestamp	Number of token received	Intermediate nodes ID
N_k	N_i	M_i	N_l	t_3	1	N_j, N_k, -
N_m	N_p	M_j	N_r	t_7	1	N_s, N_m, -

Table 26.8 Category of biasness and corresponding reputation degradation factor

Sl. no.	Category of biasness	Reputation degradation factor
1	Group-bias as forwarder > 1 and group-bias as receiver > 1	$1/k^2$
2	Group-bias as forwarder > 1	$1/k$
3	Group-bias as receiver > 1	$1/k$

Table 26.9 Visited route (vr) table maintained at traffic (T_p) node

Node ID (from)	Route	Message ID	Timestamp
N_k	N_m–N_p–N_k–N_l	M_i	t_3
N_m	N_p–N_s–N_m–N_r	M_j	t_7
N_p	N_q–N_k–N_p–N_l	M_k	t_9

Table 26.10 Visited route (vr) table maintained at traffic (T_q) node

Node ID (from)	Route	Message ID	Timestamp
N_k	N_m–N_p–N_k–N_s	M_l	t_4
N_r	N_s–N_m–N_r–N_q	M_n	t_6
N_s	N_m–N_q–N_s–N_k	M_s	t_7

matrix is then published by the traffic nodes in their respective monitoring zone so that local nodes may use that to select trustworthy node to reach the destination (Table 26.11).

From the above sample table, traffic nodes and other nodes not only have the idea about visited routes but also group-biasness of a node. Our model says that if contact times are half or more than half of number of times routes found from source to destination, pair of nodes belongs to same group. Based on contact times, we opine that

$$\left(\mathrm{N}_m, \mathrm{N}_p, \mathrm{N}_k\right): \quad \text{same group.}$$
$$\left(\mathrm{N}_k, \mathrm{N}_p, \mathrm{N}_s\right): \quad \text{same group.}$$

Now, traffic nodes have the following perception matrix (Tables 26.12, 26.13, 26.14 and 26.15).

We assume that threshold value of reputation is *R*th [5]. If a node's reputation value is less than *R*th, node is declared as selfish node or misbehaving node by the

Table 26.11 Global visited route matrix created by traffic nodes T_p and T_q

Node ID (from)	Route	Message ID	Timestamp
N_k	N_m–N_p–N_k–N_l	M_i	t_3
N_m	N_p–N_s–N_m–N_r	M_j	t_7
N_p	N_q–N_k–N_p–N_l	M_k	t_9
N_k	N_m–N_p–N_k–N_s	M_l	t_4
N_r	N_s–N_m–N_r–N_q	M_n	t_6
N_s	N_m–N_q–N_s–N_k	M_s	t_7

Table 26.12 Perception matrix created by T_p and T_q

Nature of nodes	Node Ids
Group-bias	N_m, N_p, N_k, N_s
Group-unbias	N_q, N_r

Table 26.13 Reputation matrix created by T_p

Traffic node ID	Node Id			
	N_k	N_m	N_p	N_s
T_p	$R1_{t5}$	$R2_{t8}$	$R4_{t9}$	$R4_{t10}$

Table 26.14 Reputation matrix created by T_q

Traffic node ID	Node Id			
	N_k	N_m	N_s	N_q
Tq	$R1_{t3}$	$R2_{t9}$	$R4_{t9}$	$R3_{t7}$

Table 26.15 Global Reputation matrix created by T_p and T_q

Node Ids	N_k	N_m	N_p	N_s	N_q
Reputation	$R1_{t5}$	$R2_{t9}$	$R4_{t9}$	$R4_{t10}$	$R3_{t7}$

traffic nodes, will get a warning message to improve its behaviour. By using our scheme, we are also able to find out group-biasness or un-biasness of a node.

26.4 Simulation

We have implemented our scheme using opportunistic network environment (ONE) simulator [7]. The nodes are positioned on default map. We have implemented our scheme on Prophet routing. The movement we follow is the shortest path-based movement model. We assume that there are three groups of volunteer nodes: Two shelter nodes, one control node and one group of traffic nodes. Simulation is done for 12 h. In our scheme, we assume that a shelter node has the capability to generate the message. At the beginning, we have started simulation with 50 volunteers, 5 traffic nodes, 2 shelter nodes and one control node. 50 volunteer nodes are divided into 3 groups: group 1, group 2 and group 3 consist of 20 nodes, 20 nodes and 10 nodes, respectively. Now we have observed the impact on delivery ratio in presence of traffic nodes and also in absence of traffic nodes. After then, number of volunteer nodes are increased from 50 to 200 and also the impact on delivery ratio are observed in presence of traffic nodes and also in absence of traffic nodes. Traffic nodes are also increased from 10 to 20 % and up to 30 % of total volunteer nodes.

The performance of our scheme is evaluated in term of packet delivery ratio and overhead. We have used the formulae [5] to estimate packet delivery ratio and overhead.

Delivery Probability = Number of Packets Delivered/Number of Packets Created

Overhead Ratio = (Number of packets relayed − Number of packets delivered)/ Number of packets delivered

From Fig. 26.1, it is clearly evident that the delivery probability increases in presence of traffic nodes compared to absence of traffic nodes. If we increase the percentage of traffic nodes, delivery probability improves significantly.

We also noticed the performance of our proposed scheme in terms of Overhead Ratio as shown in Fig. 26.2 where overhead ratio degrades significantly in presence of traffic nodes compared to absence of traffic nodes in the scenario.

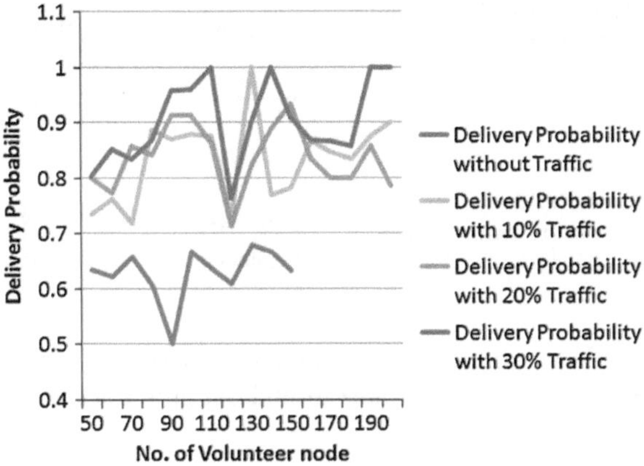

Fig. 26.1 Simulation result of number of volunteer node versus delivery probability without and with presence of traffic nodes

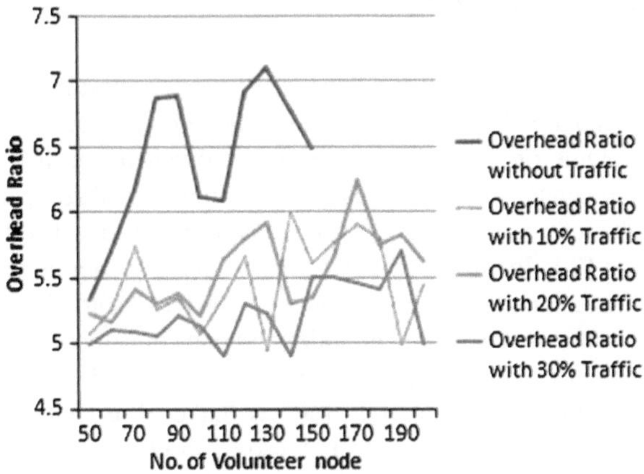

Fig. 26.2 Simulation result of a number of volunteer node versus overhead ratio without and with presence of traffic nodes

26.5 Conclusions

This paper presents post-disaster scenario based on traffic nodes which are aimed to help other nodes in the network to get the route view to reach the destination. The novelties of our traffic nodes based scheme are

- Traffic nodes are aimed to publish the global visited route with reputation and perception matrix. By which other nodes get the overview to find out the route (by selecting trustworthy forwarder) towards destination.
- In our scheme, routing tables are not exchanged among volunteer nodes. As a consequence we can detect colluding attack.
- According to our scheme no false token generation, dropping, non-forwarding attacks can be detected as we are using token concept. Token can be issued by the destination node only and decrypted by the public key of the destination. Interestingly, public key is possessed by the trustworthy traffic nodes only.

- We are able to find out group-biasness or un-biasness of a node in post disaster scenario. Node with bad reputation will get a chance to improve its reputation by participating in data forwarding in future.

In this paper, we have not addressed the issue of data integrity and mobility of traffic nodes. We intend to solve the problem in future in global perspective.

References

1. D.D. Deb, S. Bose, S. Bandyopadhyay, Coordinating disaster relief operations using smart phone/PDA based peer-to- peer communication. Int. J. Wireless Mobile Networks **4**(6), 27–44 (2012)
2. J. Miao, O. Hasan, S.B. Mokhtar, L. Brunie, K. Yim, An analysis of strategies for preventing selfish behavior in mobile delay tolerant networks, in *The Sixth International Conference on Innovative Mobile and Internet Services in Ubiquitous Computing* (*IMIS-2012*), Palermo, Italy, pp. 208–215. 4–6 July 2012
3. L. Wei, Z. Cao, H. Zhu, A user-centric reputation based incentive protocol for delay/disruption tolerant networks, in *IEEE Globecom* (2011), pp. 1–5
4. L. Wei, H. Zhu, Z. Cao, X. Shen, A user-centric and socialaware reputation based incentive scheme for delay/disruption tolerant networks, in *Ad-hoc, Mobile, and Wireless Networks*, ser. Lecture Notes in Computer Science, vol. 6811, ed. by H. Frey, X. Li, S. Ruehrup (Springer, Berlin, 2011), pp. 177–190
5. C. Chakrabarti, A. Banerjee, S. Roy, An observer-based distributed scheme for selfish-node detection in a post-disaster communication environment using delay tolerant network, in *IEEE, AIMoC* (2014), pp. 151–156
6. F. Li, J. Wu, A. Srinivasan, Thwarting black hole attacks in disruption-tolerant network using encounter tickets, in *Proceedings of IEEE INFOCOM* (2009), pp. 2428–2436
7. The Opportunistic Network Environment Simulator (The ONE), http://www.netlab.tkk.fi/tutkimus/dtn/theone/, Verision 1.4.0.

Chapter 27
A New Approach to Generate the RC4 S-Box

Suman Das, Hemanta Dey and Ranjan Ghosh

Abstract The RC4 stream cipher has two components: KSA and PRGA. Though this simple and fast cipher has proved itself as robust and is trusted by many organizations, a number of researchers claim that RC4 has some weaknesses and bias in its internal states. Researchers point to the *swap* function of RC4 as the main reason for its weakness, especially in KSA. In this paper, the authors eliminate KSA and use a mathematical process to generate the initial internal state array of RC4. The original RC4 and the modified RC4 are tested with the NIST Statistical Test Suite and it is found that the modified RC4 gives better security.

Keywords Modified RC4 · Modified KSA · Random S-Box · Irreducible polynomial · NIST test suite

27.1 Introduction

RC4 is one of the most popular stream ciphers, which is simple, efficient, fast and easy to implement. Based on a table-shuffling principle, RC4 is designed for fast software and hardware implementation and widely used in many commercial products and standards. It contains an initialization routine and a random number generator, where the random values are selected from an internal state array and two elements are *swapped* for every byte reported. The RC4 cryptanalysis has been mainly devoted to the statistical analysis of the output sequence, or to the initialization weaknesses.

S. Das (✉)
Department of Computer Science and Engineering, University of Calcutta, Calcutta, India
e-mail: aami.suman@gmail.com

H. Dey · R. Ghosh
Institute of Radio Physics and Electronics, University of Calcutta, 92, A.P.C. Road, Kolkata 700009, India
e-mail: hemantadey13@gmail.com

© Springer India 2015

K. Maharatna et al. (eds.), *Computational Advancement in Communication Circuits and Systems*, Lecture Notes in Electrical Engineering 335,
DOI 10.1007/978-81-322-2274-3_27

RC4 contains a secret array S of size N (generally, 256), in which integers (0 to $N-1$) are swapped, depending on two index pointers i and j in a deterministic (for i) and pseudo-random (for j) way. There are two components of the cipher: Key-Scheduling Algorithm (KSA) and Pseudo-Random Generation Algorithm (PRGA).

The KSA turns an identity permutation into a random-looking permutation and the PRGA generates the key-stream bytes, which are XORed with the plaintext bytes to generate ciphertext bytes. All additions in KSA and PRGA are additions modulo N.

There are several works on the strengths and weaknesses of RC4. It has been argued that there are biases in the PRGA due to propagation of biases in the KSA, which shows that there is a significant interest in the cryptographic community for RC4.

In this paper, we have replaced the KSA in a mathematical way of calculating and filling the S-array with the multiplicative inverses of the 256 bytes (0–255) in Galois Field 2^8 [GF(2^8)]. Also, the initial value of j in PRGA has been calculated from the key values, not from 0 as stated in the original algorithm—the main structure of the PRGA remains the same. We then compare and analyse this variant of RC4 statistically with the original RC4, following the guidelines given by the National Institute of Standards and Technology (NIST), USA in their Statistical Test Suite, coded by us. It has been found that though RC4 itself is quite secure to use; even after so many years of its primary design, the new variant is able to prove itself as more efficient (Table 27.1).

Table 27.1 The RC4 stream cipher

KSA	PRGA
Input: Secret Key K	Input: S-Box S – The o/p of KSA

`for i = 0, ... N - 1`
`  S[i] = i;`
`next i`
`j = 0;`
`for i = 0, ..., N - 1`
`{ j = j + S[i] + K[i]`
`  swap(S[i], S[j]);`
`}`
`next i`
	`i = 0; j = 0;`
	`while TRUE`
	`{i = i + 1`
	` j = j + S[i]`
	` swap(S[i], S[j]);`
	` z = S[S[i] + S[j]];`
	`}`

Output: S-Box S generated by K	Output: Random Stream Z

27.2 Motivation

RC4 has gone through tremendous analysis since it became public. Roos [1] showed some weakness in KSA and defined weak keys for RC4 with some important technical results. He showed strong correlation between the secret key bytes and the final key-stream generated. He also identified several classes of weak keys.

Paul and Preneel [2] presented a new statistical bias in the distribution of the first two output bytes of RC4 and proposed a new key-stream generator, namely RC4A with much less operations per output byte. They also described a new statistical weakness in the first two output bytes of RC4 key-stream and recommended to drop at least the initial 2 N bytes, where N is the size of the internal S-Box. They proposed to introduce more random variables in PRGA to reduce the correlation between the internal and the external states.

Maitra and Paul [3] revolved the non-uniformity in KSA and proposed additional layers over the KSA and the PRGA. They named the modified cipher as RC4+, which avoids existing weaknesses in RC4. They presented a three-layer architecture in a scrambling phase after initialization to remove weaknesses in KSA (KSA+). They also introduced some extra phases to improve the PRGA (PRGA+).

Akgün et al. [4] detected a new bias in the KSA and proposed a new algorithm to retrieve the RC4 key in a faster way. Their framework significantly increases the success rate of key retrieval attack. They showed that KSA leaks information about the secret key if the initial state table is known.

Tomasevic and Bojanic [5] introduced an abstraction in the form of general conditions about the current state of RC4. Strategy has been used to favour more promising values that should be assigned to unknown entries in the RC4 table. They proposed a new technique to improve cryptanalytic attack on RC4, which is based on new information from the tree representation of RC4.

Nawaz et al. [6] introduced a new 32-bit RC4 like faster key-stream generator. It has a huge internal state and offers higher resistance against state recovery attacks. This is suitable for high speed software encryption.

Sen Gupta et al. [7] implemented a hardware architecture to generate two key-stream bytes per clock cycle using the idea of loop unrolling and hardware pipelining. They thoroughly studied RC4 designing problem from the viewpoint of throughput.

Church [8] gave the complete list of irreducible polynomials for prime moduli (2, 3, 5, 7 and 11) of each degree. The determination of the exponents provided in itself a very satisfactory control of the irreducibility of the polynomials.

Daemen and Rijmen [9] defined a process of creating cryptographic S-Boxes by a mathematical process in GF(2^8) field. They showed the process of calculating multiplicative inverse of a byte (as an 8-bit binary string) in GF(2^8) with an irreducible polynomial as the modulus.

Das et al. [10] avoided the KSA and used a robust PRBG, BBS, to fill-up the internal state array, thus eliminating the swap function of KSA, which has been found to give better security.

27.3 A Modified RC4

Roos [1] and others strongly discussed the weakness of KSA and weak keys in RC4. Roos argued that in KSA, only the line of *swap* directly affects the state table S while exchanging two elements and hence the previous line $j = j + S[i] + K[i]$ is responsible for calculating the indexes. Here the variable i is deterministic and j is pseudo-random. Therefore, the swap between two elements may happen once, more than once, or may not happen at all—thus bringing a weakness in the KSA. He showed that there is a high probability of about 37 % for an element to be not swapped at all.

In this paper, we propose to introduce a modified RC4, where all the cells of the initial internal state array S will be filled up by calculating the multiplicative inverses of the 256 bytes (0–255) using any one of the 30 irreducible polynomials [8] in GF(2^8) as the modulus [11, 12]. In this way, the state array S can be filled up by 256 random values without using any swap function. A list of these irreducible polynomials, given in hexadecimal notations, is as follows [13, 14]:

11B	11D	12B	12D	139	13F	14D	15F	163	165	169	171	177	17B	187
18B	18D	19F	1A3	1A9	1B1	1BD	1C3	1CF	1D7	1DD	1E7	1F3	1F5	1F9

The outputs of RC4 (original and modified) have been tested statistically using the guidance of NIST, by the NIST Statistical Test Suite. For both the algorithms, the same text file has been encrypted 500 times by using 500 same encryption keys, generating 500 ciphertexts for each algorithm, each of which contains at least 1,342,500 bits, as recommended by NIST. The two sets of ciphertexts are then tested statistically to find out if the security varies for the original and the modified algorithms.

27.4 The NIST Statistical Test Suite

NIST developed a Statistical Test Suite, which is an excellent and exhaustive document consisting of 15 tests developed to test various aspects of randomness in binary sequences produced by cryptographic algorithms [15, 16]. The tests are as follows:

1. *Frequency (Monobit) Test* Number of 1s and 0s in a sequence should be approximately the same, i.e. with probability ½.
2. *Frequency Test within a Block* Whether frequency of 1s in an M-bit block is approximately M/2.

3. *Runs Test* Whether number of runs of 1s and 0s of various lengths is as expected for a random sequence.
4. *Test for Longest-Run-of-Ones in a Block* Whether the length of the longest run of 1s within the tested sequence (M-bit blocks) is consistent with the length of the longest run of 1s as expected.
5. *Binary Matrix Rank Test* Checks for linear dependence among fixed length sub-strings of the sequence, by finding the rank of disjoint sub-matrices of it.
6. *Discrete Fourier Transform Test* Detects periodic features in the sequence by focusing on the peak heights in the DFT of the sequence.
7. *Non-overlapping Template Matching Test* Occurrences of a non-periodic pattern in a sequence, using a non-overlapping m-bit sliding window.
8. *Overlapping Template Matching Test* Occurrences of a non-periodic pattern in a sequence, using an overlapping m-bit sliding window.
9. *Maurer's Universal Statistical Test* Whether or not the sequence can be significantly compressed without loss of information, by focusing on the number of bits between matching patterns.
10. *Linear Complexity Test* Finds the length of a Linear Feedback Shift Register (LFSR) to generate the sequence—longer LFSRs imply better randomness.
11. *Serial Test* Determines the number of occurrences of the 2^m m-bit overlapping patterns across the entire sequence to find uniformity—every pattern has the same chance of appearing as of others.
12. *Approximate Entropy Test* Compares the frequency of all possible overlapping blocks of two consecutive/adjacent lengths (m and $m + 1$).
13. *Cumulative Sums Test* Finds if the cumulative sum of a sequence is too large or small. Focuses on maximal excursion of random walks, which should be near 0.
14. *Random Excursions Test* Finds if the number of visits to a state within a cycle deviates from expected value, calculates the number of cycles having exactly K visits in a cumulative sum random walk.
15. *Random Excursions Variant Test* Deviations from the expected visits to various states in the random walk, calculates the number of times that a state is visited in a cumulative sum random walk.

In each test, for a bit sequence, NIST adopted different procedures to calculate the probability values (*P-values*) for different tests from the observed and expected values under the assumption of randomness. The Test Suite has been coded by us and used to study the randomness features of RC4 and the modified RC4.

27.5 Results and Discussions

After analysing the outputs of the original RC4 and the modified RC4, using the NIST Statistical Test Suite, as described above, it has been found that the modified algorithm creates a tweak in RC4 to increase its security. The final analysis and comparison is displayed in Table 27.2, where the Proportion of Passing (POP)

Table 27.2 Comparison of POP status and uniformity distribution generated by the 15 NIST tests for the modified RC4 and RC4

Test↓	POP status		Uniformity distribution	
	Modified RC4	RC4	Modified RC4	RC4
1	0.992000	0.988000	3.941953^{-01}	4.154218^{-01}
2	0.990000	0.992000	5.625911^{-01}	4.904834^{-01}
3	0.988000	0.992000	5.749035^{-01}	8.920363^{-01}
4	0.992000	0.982000	4.299224^{-01}	5.790211^{-01}
5	0.990000	0.984000	9.926704^{-01}	2.492839^{-01}
6	0.988000	0.980000	2.518375^{-01}	4.170881^{-02}
7	0.990000	0.990000	1.311223^{-01}	8.272794^{-01}
8	0.990000	0.992000	2.729770^{-01}	2.224804^{-01}
9	0.984000	0.982000	5.996926^{-01}	3.856456^{-02}
10	0.998000	0.992000	2.319564^{-01}	5.462832^{-01}
11	0.996000	0.982000	6.038408^{-01}	1.699807^{-01}
12	0.986000	0.992000	4.635119^{-01}	2.953907^{-01}
13	0.986000	0.995000	6.869553^{-01}	8.201435^{-01}
14	0.986500	0.983500	6.987989^{-01}	6.729885^{-01}
15	0.991333	0.985889	1.868185^{-02}	8.386675^{-02}
Total	10	6	8	7

status and Uniformity Distribution of the NIST tests for these two algorithms are displayed and compared. The best values of a particular test for each algorithm are shaded (in rows) and then the numbers of shaded cells for each are counted (in columns). The higher count (here for the modified RC4) gives a better result for that particular algorithm, which shows that this one has better security than the other, at least for this particular data set.

POPs and Uniformity Distributions generated by the two algorithms, the modified RC4 and RC4, for the 15 NIST tests, compared to the expected values [15], are displayed in Tables 27.3 and 27.4. Distributions of *P*-values generated by the algorithms for 15 NIST tests are displayed in Tables 27.5 and 27.6. Here, the interval between 0 and 1 is divided into 10 sub-intervals and the *P*-values lying within each sub-interval are counted, which should be uniformly distributed in each sub-interval [15].

Histograms on distributions of *P*-values for two NIST tests (4 and 8) for the modified RC4 and RC4 are displayed in Figs. 27.1a, b and 27.2a, b, respectively. Scattered graphs on the POP Status for the 15 tests for these two algorithms are displayed in Fig. 27.3a, b. If most of the values are greater than the expected values, the data is considered to be random. For a particular algorithm, the more number of POPs tend to 1 for the 15 tests, the more random will be the data sequence.

Finally, it has been observed that besides using the original KSA, a suitable mathematical model can also be used to generate a secured internal *S*-array for RC4, which may give better randomization in ciphertexts.

Table 27.3 POP status and uniformity distribution generated for the modified RC4

Test↓	Expected POP	Observed POP	Status	Uniformity distribution	Status
1	0.976651	0.992000	Successful	3.941953^{-01}	Uniform
2	0.976651	0.990000	Successful	5.625911^{-01}	Uniform
3	0.976651	0.988000	Successful	5.749035^{-01}	Uniform
4	0.976651	0.992000	Successful	4.299224^{-01}	Uniform
5	0.976651	0.990000	Successful	9.926704^{-01}	Uniform
6	0.976651	0.988000	Successful	2.518375^{-01}	Uniform
7	0.976651	0.990000	Successful	1.311223^{-01}	Uniform
8	0.976651	0.990000	Successful	2.729770^{-01}	Uniform
9	0.976651	0.984000	Successful	5.996926^{-01}	Uniform
10	0.976651	0.998000	Successful	2.319564^{-01}	Uniform
11	0.980561	0.996000	Successful	6.038408^{-01}	Uniform
12	0.976651	0.986000	Successful	4.635119^{-01}	Uniform
13	0.980561	0.986000	Successful	6.869553^{-01}	Uniform
14	0.985280	0.986500	Successful	6.987989^{-01}	Uniform
15	0.986854	0.991333	Successful	1.868185^{-02}	Uniform

27.4 POP status and uniformity distribution generated for RC4

Test↓	Expected POP	Observed POP	Status	Uniformity distribution	Status
1	0.976651	0.988	Successful	4.154218^{-01}	Uniform
2	0.976651	0.992	Successful	4.904834^{-01}	Uniform
3	0.976651	0.992	Successful	8.920363^{-01}	Uniform
4	0.976651	0.982	Successful	5.790211^{-01}	Uniform
5	0.976651	0.984	Successful	2.492839^{-01}	Uniform
6	0.976651	0.98	Successful	4.170881^{-02}	Uniform
7	0.976651	0.99	Successful	8.272794^{-01}	Uniform
8	0.976651	0.992	Successful	2.224804^{-01}	Uniform
9	0.976651	0.982	Successful	3.856456^{-02}	Uniform
10	0.976651	0.992	Successful	5.462832^{-01}	Uniform
11	0.980561	0.982	Successful	1.699807^{-01}	Uniform
12	0.976651	0.992	Successful	2.953907^{-01}	Uniform
13	0.980561	0.995	Successful	8.201435^{-01}	Uniform
14	0.98528	0.9835	Unsuccessful	6.729885^{-01}	Uniform
15	0.986854	0.985889	Unsuccessful	8.386675^{-02}	Uniform

Table 27.5 Distribution of *P*-values generated for the modified RC4

Test↓	1	2	3	4	5	6	7	8	9	10
1	59	47	47	40	54	41	48	63	52	49
2	58	60	51	48	44	48	46	58	46	41
3	58	46	51	55	48	45	36	52	56	53
4	55	51	57	60	38	57	45	46	44	47
5	56	53	48	49	53	49	47	45	49	51
6	53	64	51	39	51	40	56	56	48	42
7	53	47	46	48	40	72	46	47	55	46
8	62	56	38	54	46	39	47	51	58	49
9	58	60	45	54	45	41	46	46	50	55
10	50	47	62	35	46	58	58	53	43	48
11	109	98	117	98	105	88	94	93	92	106
12	55	47	62	44	58	45	46	49	40	54
13	104	88	109	92	103	93	102	108	90	111
14	406	378	405	378	403	399	385	396	429	421
15	804	851	899	917	888	917	913	908	954	949

Horizontal ranges for tables: 1: 0.0–0.1, 2: > 0.1–0.2, 3: > 0.2–0.3,…, 10: > 0.9–1

Table 27.6 Distribution of *P*-values generated for RC4

Test↓	1	2	3	4	5	6	7	8	9	10
1	69	49	49	48	53	47	45	46	50	44
2	50	48	54	38	52	44	63	47	48	56
3	56	58	49	43	55	47	49	48	50	45
4	39	60	51	50	47	59	52	43	52	47
5	59	55	49	49	48	38	45	65	50	42
6	57	46	47	44	60	37	48	69	39	53
7	42	44	47	48	53	51	57	49	50	59
8	47	56	41	56	43	49	65	57	41	45
9	61	54	57	54	58	53	33	41	54	35
10	52	43	57	47	44	45	64	46	53	49
11	100	114	123	103	99	103	83	97	89	89
12	50	56	66	49	55	50	40	49	41	44
13	91	105	106	97	100	103	103	111	99	85
14	439	399	415	381	398	383	400	384	402	399
15	956	853	902	890	949	902	896	934	833	885

Horizontal ranges for tables: 1: 0.0–0.1, 2: > 0.1–0.2, 3: > 0.2–0.3,…, 10: > 0.9–1

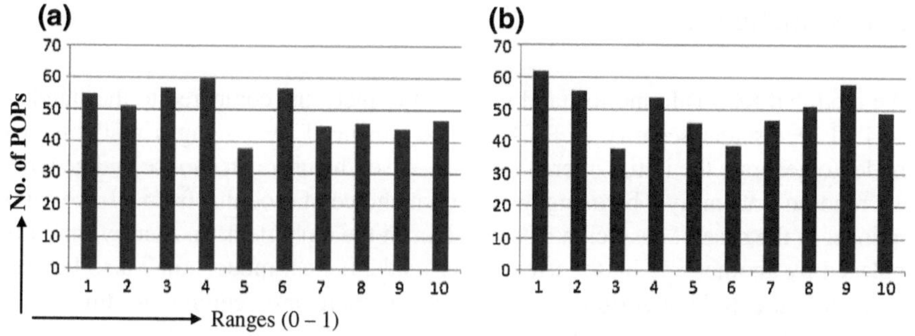

Fig. 27.1 **a**, **b** Histograms for POP distribution of tests 4 and 8 for modified RC4

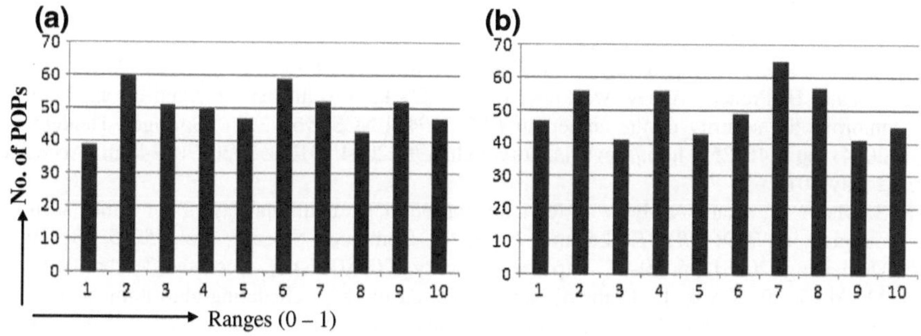

Fig. 27.2 **a**, **b** Histograms for POP distribution of tests 4 and 8 for RC4

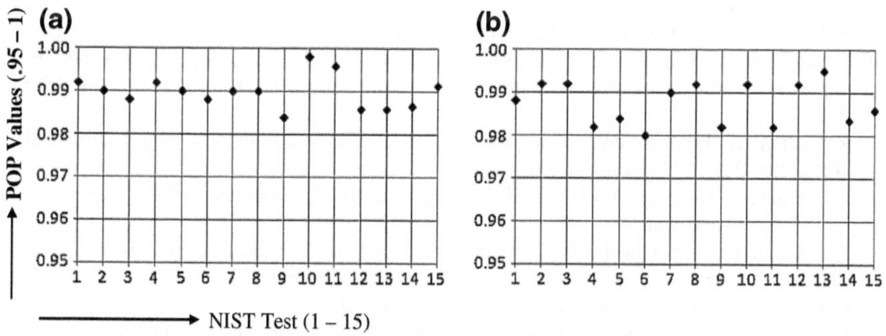

Fig. 27.3 **a** Scattered graph on POP status on 15 NIST tests for modified RC4. **b** Scattered graph on POP status on 15 NIST tests for RC4

27.6 Conclusion

The modified RC4 is found to stand in the better merit list compared to the standard RC4. It seems that security in RC4 will be enhanced by driving a mathematical model to generate the initial internal *S*-array. Also, the user can choose and generate any random state array *S* by using his/her own choice of modulus from a large set of options. In the case of suspicion of a trapdoor in the ciphertext, an *S*-array might be replaced by another one by the user. Other mathematical models may also be used to modify the RC4 algorithm and studies on them are required to find better opportunities to generate secure RC4 internal state arrays.

References

1. A. Roos, A class of weak keys in the RC4 stream cipher. Post in sci.crypt (1995)
2. S. Paul, B. Preneel, A new weakness in the RC4 keystream generator and an approach to improve the security of the cipher, in *FSE 2004,* LNCS, vol. 3017 (Springer, Heidelberg, 2004), pp. 245–259. http://www.iacr.org/archive/fse2004/30170244/30170244.pdf. Accessed 2 July 2014
3. S. Maitra, G. Paul, Analysis of RC4 and proposal of additional layers for better security margin, in *INDOCRYPT*, Lecture Notes in Computer Science, vol. 5365 (Springer, Heidelberg, 2008), pp. 40–52. http://eprint.iacr.org/2008/396.pdf. Accessed 2 July 2014
4. M. Akgün, P. Kavak, H. Demicri, New results on the key scheduling algorithm of RC4, in *INDOCRYPT*, Lecture Notes in Computer Science, vol. 5365 (Springer, Heidelberg, 2008), pp. 40–52. http://link.springer.com/content/pdf/10.1007/978-3-540-9754-5_4.pdf. Accessed 2 July 2014
5. V. Tomašević, S. Bojanić, Reducung the state space of RC4 stream cipher, in *ICCS 2004*, *LNCS*, vol. 3036, ed. by M. Bubak et al. (Springer, Berlin, 2004), pp. 644–647. http://link.springer.com/chapter/10.1007%2F978-3-540-24685-5_110#page-1. Accessed 2 July 2014
6. Y. Nawaz, K.C. Gupta, G. Gong, A 32-bit RC4-like Keystream Generator, IACR Eprint archive (2005) eprint.iacr.org/2005/175.pdf. Accessed 2 July 2014
7. S. Sen Gupta, A. Chattopadhyay, K. Sinha, S. Maitra, B.P. Sinha, High-performance hardware implementation for RC4 stream cipher. IEEE Trans. Comput. **82**(4), (2013). Accessed 2 July 2014
8. R. Church, Tables of irreducible polynomials for first four prime moduli. Ann. Maths. 2nd Sr. **36**(1), 198–209 (1935). http://www.jstor.org/stable/1968675
9. J. Daemen, V. Rijmen, AES Proposal: Rijndael, Version 2, Submitted to NIST (1999). http://csrc.nist.gov/encrytion/aes
10. S. Das, H. Dey, R. Ghosh, RC4 stream cipher with a modified random KSA, in *Natural Conferences on Emerging Trends on Computing and Communication* (*ETCC*), (B.P.Poddar Institute of Management and Technology, Kolkata, India, 2014)
11. B. Foruzan, *Cryptography and Network Security* (Tata McGraw-Hill, Special Indian Ed., New Delhi, 2007)
12. D.R. Stinson, *Cryptography—Theory and Practice* (Department of Combinatorics and Optimization, University of Waterloo, Ontario, 2002)
13. Federal Information Processing Standards Publication (FIPS), Announcing Advanced Encryption Standard (AES) (2001). http://csrc.nist.gov/publications/fips/fips197/fips-197.pdf
14. FIPS, PUB 197: The Official AES Standard, 2001-11-26, http://csrc.nist.gov/publications/fips/fips197/fips-197.pdf Accessed 29 April 2010

15. National Institute of Standard and Technology (NIST), Tech. Admin., U.S. Department of Commerce, A Stat. Test Suite for RNGs and PRNGs for Cryptographic Appls (2010). http://csrc.nist.gov/publications/nistpubs800/22rec1SP800-22red1.pdf
16. S.J. Kim, K. Umeno, A. Hasegawa, *Corrections of the NIST Statistical Test Suite for Randomness* (Communications Research Lab., Inc. Admin. Agency, Tokyo, 2004)

Chapter 28
An Improved Intellectual Analysis Precedence and Storage for Business Intelligence from Web Uses Access Data

S. Ganeshmoorthy and M.R. Bharath Kumar

Abstract With the growth of data mining web usage, user behaviour analysis is a useful area for business intelligence. There are several techniques to extract interesting pattern and knowledge which will be used for business intelligence from user's access records. However, analysis of large Web log files is a convoluted task not fully addressed by existing web access techniques. In order to provide better storage and user behaviour from huge datasets the proposed intellect storage, precedence analysis (ISPA) algorithm has been introduced. The user session and history are considered as feedback for clustering. The proposed system considers the total number of hits, time spent by the user on a particular page and links. Based on these parameters, personalization has been proposed. The implementation of an effective pruning technique and FP-growth algorithm has provided better results and performance. This also considers outlier detection in order to group the links effectively. Experimental results are presented using user click through logs to validate the effectiveness of the proposed methods.

Keywords Data mining · Web mining · Intrusion detection · Web usage mining · Association rules · Threshold · FP-growth algorithm · Minimum support threshold · Minimum confidence threshold

S. Ganeshmoorthy (✉)
Department of Computer Applications, Nehru Institute of Engineering and Technology, Coimbatore, Tamilnadu, India
e-mail: Ganeshmoorthy.mca@gmail.com

M.R. Bharath Kumar
Nehru Institute of Engineering and Technology, Coimbatore, Tamilnadu, India
e-mail: mrbharathee@gmail.com

© Springer India 2015

K. Maharatna et al. (eds.), *Computational Advancement in Communication Circuits and Systems*, Lecture Notes in Electrical Engineering 335,
DOI 10.1007/978-81-322-2274-3_28

28.1 Introduction

Data mining is a method of extracting hidden prognostic information from large or huge and high-dimensional databases. It is a powerful new technology to help companies focus on the most important information in their data warehouses. Data mining tools helps to predict future trends and behaviours allowing businesses to make proactive knowledge-driven decisions.

The proposed system overcomes the above drawbacks by proposing an effective optimal algorithm called 'Intellect Storage, Precedence Analysis (ISPA) algorithm'. The name of the algorithm itself denotes the advantages and features of the proposed system, i.e. providing effective, intellectual storage process and interest analysis of the user search goal.

The algorithm considers and provides search results based on the user interest. The user interest has been identified by the feedback clustering (clicked URL, time spent in the page and frequency). For effective storage FP-growth has been applied.

This maintains the data stream in tree format based on the frequency of clicks and maximum time spent by the user.

For fast searching, effective techniques have been proposed. Through the proposed technique the advantages of effective and intellectual storage are achieved, which eliminates unnecessary logs. Fast data retrieval based on the user interest needs effective clustering technique based on the click stream. Find effective outlier (abnormal activities of the user) from clustering finds individual interest mining as well as group mining.

28.2 Literature Study

In [1] a novel approach is proposed to infer user search goals for a query by clustering the proposed feedback sessions. In [1] the feedback session was defined as both clicked and unclicked streams of URL. The system then identifies the user need through the last URL. The main drawback of this approach is click stream alone does not provide better knowledge of the user search goals. The approach suffers from a lot of inability in data clustering and is also time-consuming. The system was considered for offline extraction than online. So difficulty in dynamic environment only provides popular keywords than the user's exact need and interest. In [2] is proposed access control (AC) architecture where every process of the users will be extracted and monitored. As in [1], the paper [2] also considered the clicked URL and its count for identifying user search goal. The impact and usage of [3] is identifying the outliers in the clustering phase. The WWW continues to grow at an amazing rate as an information gateway and as a medium for conducting business. Web mining is the extraction of interesting and useful knowledge and implicit information from artefacts or activity related to the WWW [4, 5]. Based on several research studies we can broadly classify Web mining into three domains: content, structure and usage mining [6, 7]. The discussions in this paper will be limited to Web usage mining. Web servers record and accumulate data about user interactions whenever requests for resources are received. Analysing the Web access logs can help understand the user behaviour and the Web structure. From the business and applications point of view, knowledge obtained from the Web usage patterns could be directly applied to efficiently manage activities related to e-business, e-services, e-education and so on [8, 9]. Accurate Web usage information could help to attract new customers, retain current customers, improve cross marketing/sales, effectiveness of promotional campaigns, tracking leaving customers and find the most effective logical structure for their Web space [10]. User profiles could be built by combining users' navigation paths with other data features, such as page viewing time, hyperlink structure and page content [11]. What makes the discovered knowledge interesting has been addressed by several works.

Results previously known are often considered as uninteresting. So the key concept to make the discovered knowledge interesting will be its novelty or unexpectedness appearance [12–14]. Mining evolving click streams is the subject of only a few recent research efforts [11, 15, 16]. In [15] is given an immune system inspired approach.

Login time	IP address	Host name	URL	Session ID	Login time	Pagel login time	Pagel logout time	User ID	Minimum and maximum time in every page

28.3 Developments of an Effective Intellect Storage, Precedence and Analysis from Web User Access Data

The proposed system generally includes the following several steps: Data collection, data preprocessing, session-based clustering and multi-parameter calculation and segmentation.

28.3.1 Data Collection

Data collection is the first step of the Web usage mining, and the data authenticity and integrality would directly affect the following works smoothly carrying on and the final recommendation of characteristic service's quality. At present, for Web usage in mining technology the main data origin has three kinds: Server data, client data and middle data (*agent server data and package detecting*). The first step in the

Web usage mining process consists of collecting the relevant Web data, which will be analysed to provide us useful information about users' behaviour.

<ip_addr><base_url>–< date><method><file><protocol><code><bytes><referrer> <user_agent>

Sample Web log format

The proposed systems Web log

Date/time	IP address	Host name	URL	UserID
7/17/2012 1:56:15 PM	127.0.0.1	Localhost	http://localhost:3366/webminingfinal/Home.aspx	Ram

There are two main sources of data for Web usage mining, corresponding to the two software systems interacting during a Web session: Data on the Web server side and data on the client side. In addition, when intermediaries are introduced in the client–server communication, they can also become sources for usage data, like proxy servers and packet sniffers. It will consider each of these sources in the following paper. Also it is trying to associate the data collection methods with the requirements imposed by different classes of personalization functions.

28.3.2 *Steps of Proposed Approach*

The system contains the following steps involved:

- Data preprocessing
- Session-based clustering
- Multi parameter-based clustering
- Segmentation.

28.3.2.1 Data Preprocessing

The purpose of data preprocessing is to cleanse the dirty/noise data, extract and merge the data from different sources, and then transform and convert the data into a proper format.

In this phase, the system applied in the preprocessing to find the frequent accesses clustered dataset of the Web log file. The data are preprocessed in data cleaning, user identification, session identification and path completion.

28.3.2.2 Session-Based Clustering

This proposed session-based clustering presents an approach for measuring similarity of interests among Web users from their past access behaviours. The similarity measures are based on the user sessions extracted from the user's access logs. A multi-level scheme for clustering a large number of Web users is proposed.

28.3.2.3 Cookies

Cookies can be used to track individual users thus making the sessionizer task easier. However, the use of cookies also raises the concern of privacy, and thus requires the cooperation of the users.

28.3.2.4 Session

All data that need to be available to the application across different requests within the same session are called session state or session state data. For storing such data between requests, there are in general two possibilities:

- Sending the state information back to the client. With the next request the current state is transmitted to the server again.
- Keeping the necessary data structures on the server. No session data (except the referencing identifier) is transmitted to the client.

The following mechanisms for session identification are common:

- Unique identifier in cookie.
- Unique identifier as URL parameter.
- Unique identifier in path portion of URL.

The task of user and session identification is to find out the different user sessions from the original Web access log.

28.3.2.5 Multi-parameter Clustering

In this paper, a multi-parametric clustering model for solving cluster analysis problems is presented. It shows how this model can be used to find optimal solutions for certain variations in the clustering problem or, in other cases, for an approximation of the general clustering problem.

In the proposed multiparameter methodology clustering, the common interest of web users will be grouped.

Process of ISPA.

Several existing systems have proposed an automatic user feedback session clustering, which is constructed from user click through logs. From the user click stream, the need and interest have been identified. All the existing approaches only concentrate on the click count rather than on measuring ‘time count’. Identifying user search goals based on the click stream does not provide the exact interest and need of the user.

The proposed system provides the following solutions for existing problems:

- Solution against Storage problem.
- Effective user search goal detection. The evaluation of user search goal inference is a big problem.
- Solution against Optimization problem in feedback clustering.

28.3.3 Algorithm Steps

Step 1: Read all Web log files (Wf) from Web server Ws
Step 2: Combine all the Web logs and perform Step 3
Step 3: Preprocess by cleansing unwanted transactions
Step 4: Apply divisive clustering technique
$D(Wf) = \sum(I \text{ to } n)$ {Trans $_i$(split(divisive pattern p))}
Step 5: Perform pattern discovery p
Step 6: Apply a priori algorithm for every discovery of pattern p
Step 7: Analyse the pattern Pa by applying FP-Growth

The main challenge of frequent pattern mining is how to select the items and transactions to find frequency. The session clustering algorithm is implemented to process the Web history and pattern mining mechanism initially. Then FP-growth will be applied.

28.4 Performance Evaluation

28.4.1 Implementation

The proposed ISPA has been designed for this purpose. The Business intelligence from Web log is composed by a C# program (History Analysis). In ISPA, there is a file called history.txt. This file contains information about the navigation history of the user. The system maintains another file called log.txt which contains the set of rules that have been retrieved from the user history.txt file from the server.

The proposed system keeps a queue with the latest visited links from this list and the rules contained in the log.txt file from the data file proposed system builds the proposed ISPA algorithm.

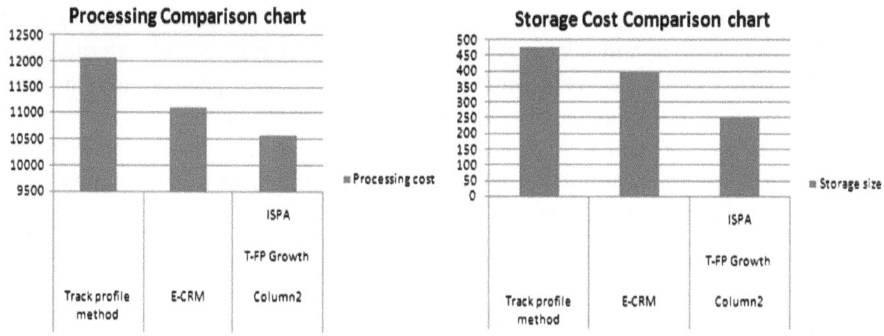

Fig. 28.1 Graph of processing and storage

28.4.1.1 Web log Analysis

The Web log which is known as the user search history analysis function is to update the set of rules according to the user history. Here the FP-growth algorithm has been implemented. That is, update the data.txt file according to the history.txt file content.

28.5 Result and Analysis

Using the methodology and metrics presented above, experiments were performed to evaluate the three cluster methods. The results presented in this section provide a detailed analysis and benefits of the proposed approach which has been created to personalize Web directories. The results define and show the proposed system is working well in the high-dimensional and huge dataset. Even the proposed system obtained good performance by all methods the use of session-based clustering and ISPA for the personalization of Web directories appears to be the most promising.

It helps to identify latent information in the users' choices and derives high-quality community directories that provide significant benefits to their users. The results presented here provide an initial measure of the benefits that this can obtain by personalizing the user Web directories to the needs and interests of user communities.

The result of the proposed algorithm is performing well in computation time. The log files and computation time is measured in this graph as shown in Fig. 28.1.

28.6 Conclusion

The system presented a framework for mining, tracking and validating evolving multifaceted user profiles with high and reliable performances on websites that have all the challenging aspects of real-life the system provided and proves that the experiments provide better performance on real datasets. Web usage mining include evolving user profiles and access patterns, dynamic webpages and external data describing ontology of the Web content. A multifaceted user profile summarizes a group of users with similar access activities and consists of their viewed pages, search engine queries and inquiring and inquired companies. In the future work, more semantic information will be introduced into mining system so that queries of similar meanings can be clustered and generalized.

References

1. R. Cooley, B. Mobasher, J. Srivastava, Web mining: information and pattern discovery on the world wide web, in *Proceedings of Ninth IEEE International Conference Tools with AI* (*ICTAI '97*) (1997), pp. 558–567
2. O. Nasraoui, R. Krishnapuram, A. Joshi, Mining web access logs using a relational clustering algorithm based on a robust estimator, in *Proceedings of Eighth International World Wide Web Conference* (*WWW '99*) (1999), pp. 40–41
3. T. Yan, M. Jacobsen, H. Garcia-Molina, U. Dayal, From user access patterns to dynamic hypertext linking, in *Proceedings of Fifth International World Wide Web Conference* (*WWW '96*) (1996)
4. R. Kosala, H. Blockeel, Web mining research: a survey. ACM SIGKDD Explor. **2**(1), 1–15 (2000)
5. R. Cooley, Web usage mining: discovery and application of interesting patterns from web data. Ph. D. thesis, Department of Computer Science, University of Minnesota, 2000
6. S. Chakrabarti, *Mining the Web: Discovering Knowledge from Hypertext Data* (Morgan Kaufmann Publishers, San Francisco, 2003)
7. G. Chang, M.J. Healey, J.A.M. McHugh, J.T.L. Wang, *Web Mining, Mining the World Wide Web*, Chapter 7 (Kluwer Academic Publishers, San Francisco, 2001), pp. 93–104
8. P.M. Chen, F.C. Kuo, An information retrieval system based on an user profile. J. Syst. Softw. **54**, 3–8 (2000)
9. D.W. Cheung, B. Kao, J. Lee, Discovering user access patterns on the world wide web. Knowl.-Based Syst. **10**, 463–470 (1997)
10. S.E. Jespersen, J. Thorhauge, T.B. Pedersen, A hybrid approach to web usage mining, in *Proceedings of 4th International Conference Data Warehousing and Knowledge Discovery, the* (*DaWaK'02*), *LNCS 2454* (Springer, Germany, 2002), pp. 73–82
11. O. Nasraoui, C. Rojas, C. Cardona, A framework for mining evolving trends in web data streams using dynamic learning and retrospective validation. Comput. Netw. Spec. Issue Web Dyn. **50**(14), (2006)
12. C. Aggarwal, J.L. Wolf, P.S. Yu, Caching on the world wide web. IEEE Trans. Knowl. Data Eng. **11**(1), 94–107 (1999)
13. R. Agrawal, R. Srikant, Fast algorithms for mining association rules, in *Proceedings of the 20th International Conference on Very Large Databases*, ed. by J.B. Bocca, M. Jarke, C. Zaniolo (Morgan Kaufmann, Santiago, 1994), pp. 487–499
14. F. Coenen, G. Swinnen, K. Vanhoof, G. Wets, A framework for self adaptive websites: tactical versus strategic changes, in *Proceedings of the Workshop on Web Mining for E-commerce: Challenges and Opportunities* (*KDD'00*) (2000), pp. 75–78

15. O. Nasraoui, C. Cardona, C. Rojas, F. Gonzalez, Mining evolving user profiles in noisy web clickstream data with a scalable immune system clustering algorithm, in *Proceedings of Workshop Web Mining as a Premise to Effective and Intelligent Web Applications* (*WebKD' 03*), (2003), pp. 71–81
16. P. Desikan, J. Srivastava, Mining temporally evolving graphs, in *Proceedings Workshop Web Mining and Web Usage Analysis* (*WebKDD' 04*), (2004)

Chapter 29
Algorithms for Road Network Range Queries in Location-Based Services

Prosenjit Gupta

Abstract Consider a set of mobile users moving along a road network. There are points of interest (POI), possibly of different types at fixed locations on the network. Users may be interested in "pull-type" information/advertisements from facilities within a range. Stores and establishments associated with the facilities may be interested in sending "push-type" offers/advertisements to users in their vicinity. In this work, we provide a theoretical framework for such range queries in location-based mobile advertising.

Keywords Location-based services · Query algorithms · Mobile advertising

29.1 Introduction

29.1.1 Location-Based Services

Advances in mobile devices and positioning technologies have given rise to several location-based services (LBSs). Mobile applications which combine spatial coordinates with other attributes to provide value-added services are becoming extremely popular. LBSs are interesting not only from an economic point of view but are being used for emergency, navigation, travel, information, billing, tracking and advertising services. These services provide customized information for the users by considering their locations. In this paper, we consider location-dependent queries wherein the query depends on the query issue location [1] and the retrieved objects are assumed to be stationary.

P. Gupta (✉)
Department of Computer Science and Engineering,
Heritage Institute of Technology, Kolkata, India
e-mail: prosenjit_gupta@acm.org

© Springer India 2015

K. Maharatna et al. (eds.), *Computational Advancement in Communication Circuits and Systems*, Lecture Notes in Electrical Engineering 335,
DOI 10.1007/978-81-322-2274-3_29

29.1.2 Push- and Pull-Type Services

LBS can be classified as push-type or pull-type services. In push-type services, the user receives information based on his or her location without actually having to request it. It may be received with prior consent as in a subscription-based system. It may also be received without prior consent, (if local regulations allow) for instance as a special offer sent by a Chinese restaurant to all users within a 1 km radius. In pull-type services, the user pulls information from the network which may be location enhanced. Finding all Italian restaurants within a 1 km radius of the user is an example.

29.1.3 Queries on Road Networks

As pointed out by Papadias et al. [2], earlier work in spatial databases assumed distances between points as Euclidean distances. However, since users move along pre-defined road networks, the true distance between a pair of points is the *network distance* measured along the road network.

29.1.4 Range Queries on Road Networks

In Computational Geometry [3], the *range searching* problem involves finding points in a query range. The *point enclosure* problem involves reporting objects covering a query point.

Consider mobile users moving along a road network. There are *facilities* or points of interest (POI), possibly of different types at fixed locations on the network. Users may be interested in "pull-type" information/advertisements from facilities within a range. This motivates the *network range query* problem. Stores and establishments associated with the facilities may be interested in sending "push-type" offers/advertisements to users in their vicinity. This motivates the *network point enclosure* problem. In this work, we provide an algorithmic framework for such range queries in location-based mobile advertising.

29.2 Preliminaries

We are given a weighted road network graph RNG $= (V, E, W)$, where E is the set of directed edges (roads); V is the set of vertices of RNG; W is the weight function defined on the edges, the weight of edge e being $W(e)$. S is a set of sites located on members of E or V and q is a query point.

V, S, q are all points but with different roles in the network. (Points in S may also have facility type and other associated attributes). Let (p_x, p_y) be coordinates of point p.

For any point p on the road network, p is located on some member of V or E. If p is on the directed edge $e = (u, v)$, of E, we define left(p) $= u$ and right(p) $= v$. If p is on a vertex v of V, left(p) $=$ right(p) $= v$.

For points a and b on the RNG, we let $d(a, b)$ be the shortest road network distance between a and b. If $e = (u, v)$ is an edge, $d(u, v) = W(e)$. If either a or b is not on the road network, $d(a, b) = \infty$. Facilities are POI located on edges or vertices.

A facility f located on directed edge (u, v) is associated with two parameters: $r(f)$ being the radius of influence and $s(f) = d(u,f)$. The quantity $r(f)$ denotes the maximum distance at which an user is of interest to the facility. For, e.g. a restaurant may have offers for users within a radius of 1 km, in which case for restaurant f, $r(f) = 1$ km. For any facility f on a directed edge from u, we also define $\pi(u,f) = r(f) - d(u,f)$. The quantity $\pi(u,f)$ denotes the influence of f at u. For an edge (u, v), let $\tau(u, v) = \max_{f \text{ on}(u,v)} \pi(u,f)$. The quantity $\tau(u, v)$ is the maximum influence at u of any facility located on edge (u, v).

For vertices u, v, $w \in V$ and edge (v, w), and u, v not necessarily adjacent, let $\beta(u, v, w) = \tau(v, w) - d(u, v)$. Then $\beta(u, v, w) = (\max_{f \text{ on}(v,w)} \pi(u,f) - d(u, v)$. $= (\max_{f\_\text{on}\_(v,w)}(r(f) - (d(v,f) + d(u, v)))$. Thus $\beta(u, v, w)$ is the maximum influence at u of the facilities on edge (v, w).

Let $\mu(u, v) = \max_{w \in \mathcal{N}(v)} \beta(u, v, w)$, where $\mathcal{N}(v) = \{w \in V : (v, w) \in E\}$ and u and v are adjacent. Thus $\mu(u, v)$ is the maximum influence at u of the facilities in the neighbourhood of v, i.e. on edges emanating from v.

A network disk $D(q, r)$ with centre q and radius r is the set of all points p on the RNG such that $d(q,p) \leq r$. $D(q, r)$ is a set of road segments all connected to q.

29.3 The Data Structures

Let $\mathcal{F}(u, v)$ be the list of facilities on edge (u, v) stored in non-increasing order of their $\pi(u,f)$ values. Let $\mathcal{N}(u)$ be the list of neighbours v of vertex u sorted in non-increasing order of $\tau(u, v)$. Let $\mathcal{D}(u)$ be the list of vertices $v \in V - \{u\}$ such that $\mu(u, v) > 0$ sorted in non-increasing order of $\mu(u, v)$.

29.4 Influence Regions Containing Query User

During pre-processing, suppose we have pre-computed all distances $d(u, v)$ where $u \in V$ and $v \in V$. With each vertex $u \in V$ we store lists $\mathcal{N}(u)$ and $\mathcal{D}(u)$ as defined above. We also store the list of facilities on each edge (u, v), i.e. $\mathcal{F}(u, v)$. It is important to note here that facilities are stored only once, on the edge they are located.

Given a query point q located on edge $e = (u, v)$, we would like to find the sites s such that $q \in D(s, r(s))$. First we would like to locate vertices w such that there is at least one edge (w, z) which has a facility s with $q \in D(s, r(s))$. We call such a vertex w *relevant* to q. Recall that $\mathcal{D}(v)$ is the list of vertices $w \in V - \{v\}$ such that $\mu(v, w) > 0$ sorted in non-increasing order of $\mu(v, w)$.

Let vertex w be relevant to q. Then there is a facility s on an edge (w, z) such that $q \in D(s, r(s))$. Then,

$$
\begin{aligned}
& d(q,v) + d(v,w) + d(w,s) \leq r(s) \\
& \Rightarrow d(q,v) + d(v,w) \leq r(s) - d(w,s) \\
& \Rightarrow d(q,v) + d(v,w) \leq \pi(w,s) \\
& \Rightarrow d(q,v) + d(v,w) \leq \max_{s \text{ on}(w,z)} \pi(w,s) \\
& \Rightarrow d(q,v) + d(v,w) \leq \tau(w,z) \\
& \Rightarrow d(q,v) \leq \tau(w,z) - d(v,w) \\
& \Rightarrow d(q,v) \leq \beta(v,w,z) \\
& \Rightarrow d(q,v) \leq \max_{z \in N(w)} \beta(v,w,z) \ \text{ where } N(w) = \{z \in V : (w,z) \in E\}. \\
& \Rightarrow d(q,v) \leq \mu(v,w)
\end{aligned}
$$

Thus to find vertices w relevant to q, it is sufficient to walk down list $\mathcal{D}(v)$ and find vertices w with $\mu(v, w) \geq d(q, v)$.

We conclude:

Lemma 29.1 *If there are m vertices relevant to a query point q, we can find and report them in $O(m)$ time.*

Once we have a list of relevant vertices, we can try to locate a list of relevant edges incident at these vertices. Given a vertex w relevant to a query point q on edge $e = (u, v)$, an edge $e' = (w, z)$ is relevant to q if edge e' has at least one facility s such that $q \in D(s, r(s))$. Recall that $\mathcal{N}(w)$ is the list of neighbours z of vertex w sorted in non-increasing order of $\tau(w, z)$.

Let edge $e' = (w, z)$ be relevant to q. Then there is a facility s on an edge (w, z) such that $q \in D(s, r(s))$. Then,

$$
\begin{aligned}
& d(q,v) + d(v,w) + d(w,s) \leq r(s) \\
& \Rightarrow d(q,v) + d(v,w) \leq r(s) - d(w,s) \\
& \Rightarrow d(q,v) + d(v,w) \leq \pi(w,s) \\
& \Rightarrow d(q,v) + d(v,w) \leq \max_{s \text{ on}(w,z)} \pi(w,s) \\
& \Rightarrow d(q,v) + d(v,w) \leq \tau(w,s)
\end{aligned}
$$

Thus to find such edges (w, z), it is sufficient to walk down list $\mathcal{N}(w)$ and report edges (w, z) with $\tau(w, z) \geq d(q, v) + d(v, w)$.

We conclude:

Lemma 29.2 *If there are e edges which are relevant to a query point q, we can locate and report them in $O(e)$ time.*

Once we have a list of relevant edges, we can try to locate a set of relevant facilities on these edges. Given an edge $e' = (w, z)$ relevant to q, a facility s on e' is relevant to q iff $q \in D(s, r(s))$. Recall that $\mathcal{F}(w, z)$ be the list of facilities s on edge (w, z) stored in non-increasing order of their $\pi(w, s)$ values.

Let facility s on e' be relevant to q. Then,

$$\begin{aligned} & d(q, v) + d(v, w) + d(w, s) \leq r(s) \\ & \Rightarrow d(q, v) + d(v, w) \leq r(s) - d(w, s) \\ & \Rightarrow d(q, v) + d(v, w) \leq \pi(w, s) \end{aligned}$$

Thus to find facilities s on an edge (w, z) relevant to a query point q, it is sufficient to walk down the list $\mathcal{F}(w, z)$ and report facilities s with $\pi(w, s) \geq d(q, v) + d(v, w)$.

We conclude:

Lemma 29.3 *If there are k facilities relevant to a query point q, they can be located and reported in $O(k)$ time.*

To summarize, given a query point q on edge e of the RNG, we find the vertices which are relevant to q. For each such relevant vertex, we find the edges incident at such vertices which are relevant to q. For each such relevant edge, we find facilities on these edges which are relevant to q. If for a query point q, there are m relevant vertices, e relevant edges and k relevant facilities, our algorithm runs in $O(m + e + k) = O(k)$ time.

We conclude:

Theorem 29.1 *The data structure for the network point enclosure problem occupies $O(|V|^2 + n)$ space where $|V|$ is the number of vertices in the RNG (i.e. the road intersections) and n is the number of facilities. The queries can be answered in optimal $O(k)$ time where k is the output size.*

29.5 Sites Within a Query Range

We consider the following problem:

Preprocess S and RNG $G = (V, E)$ into a data structure such that given a query point q on RNG and a distance r, all sites within $D(p, r)$ can be reported efficiently.

A facility f located on directed edge (u, v) is associated with one parameter: $s(f) = d(u, f)$. For an edge (u, v), let $\tau(u, v) = \min_{f \text{ on } (u,v)} d(u, f)$.

For vertices u, v, $w \in V$ and edge (v, w), and u, v not necessarily adjacent, $\beta(u, v, w) = \tau(v, w) + d(u, v)$. Let $\mu(u, v) = \min_{w \in N(v)} \beta(u, v, w)$, where $N(v) = \{w \in V : (v, w) \in E\}$. Let $\mathcal{F}(u, v)$ be the list of facilities on edge (u, v) stored in non-decreasing order of their $d(u,f)$ values. Let $\mathcal{N}(u)$ be the list of neighbours v of vertex u sorted in non-increasing order of $\tau(u, v)$. Let $\mathcal{D}(u)$ be the list of vertices $v \in V - \{u\}$ such that $\mu(u, v) > 0$ sorted in non-decreasing order of $\mu(u, v)$.

Given a query point q located on edge $e = (u, v)$, and a radius r, we would like to report facilities s so that $q \in D(s, r)$.

First we would like to locate vertices w such that there is at least one edge (w, z) which has a facility s with $q \in D(s, r)$. We call such a vertex w *relevant* with respect to q. Recall that $\mathcal{D}(v)$ is the list of vertices $v \in V - \{u\}$ such that $\mu(u, v) > 0$ sorted in non-decreasing order of $\mu(u, v)$. We show that to find relevant vertices w, it is sufficient to walk down list $\mathcal{D}(v)$ and find vertices w with $\mu(v, w) \leq r - d(q, v)$.

For each such relevant vertex w, we would like to find edges $e' = (w, z)$ so that e' has at least one facility s with $q \in D(s, r)$. Recall that $\mathcal{N}(w)$ is the list of neighbours z of vertex w sorted in non-decreasing order of $\tau(w, z)$. We show that to find relevant edges $e' = (w, z)$, it is sufficient to walk down $\mathcal{N}(w)$ and report edges (w, z) so that $\tau(w, z) \leq r - d(q, v) - d(v, w)$.

For each such relevant edge $e' = (w, z)$, we would like to find relevant facilities s on e' so that $q \in D(s, r)$. Recall that $\mathcal{F}(w, z)$ is the list of facilities s on edge (w, z) stored in non-decreasing order of their $d(u, s)$ values. To find such relevant facilities, it is sufficient to walk down $\mathcal{F}(w, z)$ and report facilities s with $d(w, s) \leq r - d(q, v) - d(v, w)$.

We conclude:

Theorem 29.2 *The data structure for the network range searching problem occupies $O(|V|^2 + n)$ space where $|V|$ is the number of vertices in the RNG (i.e. the road intersections)* and *n is the number of facilities. The queries can be answered in optimal $O(k)$ time where k is the output size.*

29.6 Conclusions and Further Research

We have provided a theoretical framework for answering location-based range queries. Clearly for a large number of facilities and relatively smaller number of road intersections (number of vertices in the road network graph), this is an efficient solution. The challenge will be to reduce the space while keeping the query time and the simplicity of the solution reasonable. Building a system that supports a wide range of LBS queries [4] around this theoretical framework will be part of our further research.

References

1. A.Y. Seydim, M.H. Dunham, V. Kumar, Location dependent query processing, in *Proceedings, 2nd ACM International Workshop on Data Engineering for Wireless and Mobile Access* (ACM Press, New York, 2001), pp. 47–53
2. D. Papadias, J. Zhang, N. Mamoulis, Y. Tao, Query processing in spatial network databases, in *Proceedings, VLDB* (2003), pp. 802–813
3. M. de Berg, M. van Kreveld, M. Overmars, O. Schwarzkopf, *Computational Geometry: Algorithms and Applications* (Springer, Berlin, 2000)
4. S. Ilarri, E. Mena, A. Illarramendi, Location-dependent query processing: where we are and where we are heading. ACM Comput. Surv. **42**(3), 12-1–12-67 (2010)
5. P.K. Agarwal, J. Erickson, Geometric range searching and its relatives, in *Advances in Discrete and Computational Geometry, Contemporary Mathematics*, vol. 23, eds. by B. Chazelle, J.E. Goodman, R. Pollack (American Mathematical Society Press, Providence, 1999), pp. 1–56

Chapter 30
Scalable Hierarchical Collaborative Filtering Using BSP Trees

Joydeep Das, Ankit Kumar Aman, Prosenjit Gupta, Ammad Haider, Subhashis Majumder and Subhranil Mitra

Abstract *Collaborative Filtering* (CF) technique is used by most of the *Recommender Systems* (RS) for formulating suggestions of item relevant to users' interest. However, CF algorithms using the large dataset sometimes become very expensive as the similarity computation among n users is an $O(n^2)$ process. In this work, we propose a decomposition-based recommendation algorithm using *Binary Space Partitioning* (BSP) trees. We divide the entire users' space into smaller regions based on the location, and then apply the recommendation algorithm separately to each of the regions. Our proposed system recommends item to a user in a specific region only using the rating data of that particular region. This reduces the quadratic complexity of the CF process as we avoid the similarity computation over the entire data. The primary goal of our work is to reduce the running time as well as maintain a good recommendation quality. This ensures scalability, allowing us to tackle bigger datasets. Empirical evaluation of our approach on the MovieLens dataset demonstrates that our method is effective while reducing the running time.

Keywords Collaborative filtering · BSP trees · Recommendation systems · Scalability

J. Das (✉)
The Heritage Academy, Kolkata, West Bengal, India
e-mail: joydeep.das@heritageit.edu

A.K. Aman · P. Gupta · A. Haider · S. Majumder · S. Mitra
Department of Computer Science and Engineering, Heritage Institute of Technology, Kolkata, West Bengal, India
e-mail: ankitkraman@gmail.com

P. Gupta
e-mail: prosenjit_gupta@acm.org

A. Haider
e-mail: amdhdr@gmail.com

S. Majumder
e-mail: subhashis.majumder@heritageit.edu

S. Mitra
e-mail: subhranil.mitra1992@gmail.com

© Springer India 2015

K. Maharatna et al. (eds.), *Computational Advancement in Communication Circuits and Systems*, Lecture Notes in Electrical Engineering 335,
DOI 10.1007/978-81-322-2274-3_30

30.1 Introduction

Recommender systems recommend items that fit a user's tastes, in order to help the user in selecting/purchasing items from an overwhelming set of choices. Such systems have great importance in applications such as e-commerce, subscription-based services, information filtering, etc. Collaborative filtering (CF) techniques [1, 2] that predict the likely preferences of a user based on the known preferences of other similar users have been used effectively for generating high-quality recommendations. Most of the existing CF methods based on correlation criteria, singular value decomposition (SVD) and non-negative matrix factorization (NNMF) provide highly accurate predictions of ratings. However, these CF techniques suffer from high computational complexity. Therefore, in order to improve the recommendation performance (in terms of complexity and accuracy), sophisticated data structures and scalable architectures are required. To address this scalability problem, we propose a decomposition-based CF algorithm. Our goal is to partition the entire users' space into smaller regions and apply the recommendation algorithm separately to each regions. However, without using any arbitrary partitioning method, we employ an intelligent partition technique using *Binary Space Partitioning* (BSP) trees [3]. In this work, we partition the users' space according to the location of the user.

The proposed work partitions the entire users' space into smaller regions, and while recommending only use the ratings of a particular region. This should reduce the complexity (often $O(n^2)$, where n is the number of users) associated with the CF process. However, it may compromise our recommendation quality, because only the ratings of a particular region are used, and not the entire rating data. Our goal is to reduce the running time as well as ensure a good recommendation quality. Experiments performed indicate that our method reduces the running time as well as maintain an acceptable recommendation quality.

The rest of the paper is organized as follows: In Sect. 30.2, we provide background information about BSP and CF as well as review some past works related to recommender systems and CF. In Sect. 30.3, we present our recommendation method. In Sects. 30.4 and 30.5, we present our decomposition and recommendation algorithms, respectively, while Sect. 30.6 reports our experimental results. Section 30.7 concludes our work and discusses future research directions.

30.2 Background and Related Works

Before presenting our system in detail, we first provide background knowledge about BSP and CF. Then, we discuss some of the existing literature on recommender systems and clustering CF models.

30.2.1 Binary Space Partitioning

Binary space partitioning[1] is a method for recursively subdividing a space into two half-spaces by hyperplanes. This subdivision technique represents the objects within the space by means of a tree data structure known as a *BSP tree*. Applications of BSP tree include rendering, performing geometrical operations with shapes in CAD, collision detection in robotics and other applications that involve spatial decomposition. BSP is a generic process of recursively dividing a space into two until the partitioning criterion are met. In BSP, the hyperplanes that partition the space may have any orientation, rather than being aligned with the coordinate axes.

30.2.2 Collaborative Filtering

Collaborative Filtering algorithm is a subfield of machine learning that tries to predict user preferences based on past user behaviour in purchasing or rating of items. CF systems generate recommendations based on a subset of users that are most similar to the active user. Each time a recommendation is requested, the algorithm needs to compute the similarity between the active user and all other users, based on their co-rated items. This similarity computation becomes very expensive with the growth of both the number of users and items in the database. We briefly discuss some of the past works that address this scalability problem in the remaining of this section.

A number of CF recommender systems have been proposed in the literature, of which the most popular ones are product recommendation in Amazon,[2] movie recommendation in MovieLens,[3] and music recommendation in Last.fm.[4] To deal with the scalability problem of the CF process, Sarwar et al. [4] clustered the complete user set on the basis of user–user similarity and used the cluster as the neighbourhood. In contrast, O'Connor and Herlocker [5] use clustering algorithms to partition the item set on the basis of user rating data. Breese et al. [6] utilize Bayesian network and clustering approaches in their recommendation process to address the scalability issue. Das et al. [7] used Voronoi Diagrams to tessellate the plane and applied the recommendation algorithm separately in each Voronoi cell.

---

1 http://en.wikipedia.org/wiki/Binary_space_partitioning.

2 http://www.amazon.com/.

3 http://www.movielens.org/.

4 http://www.last.fm/.

30.3 Our Contribution

In this paper, we present a method to deal with the scalability challenge without compromising recommendation quality. We propose a scalable CF recommendation algorithm using BSP tree-based space partitioning techniques. Our work uses the location of the users to partition the entire users' space into smaller clusters (or partitions). Once the clusters are formed, predictions for an individual can be made by averaging the opinions of the other users in that cluster. Our system, while recommending, tries to find the location of the target user, and then explores the preferences of her neighbourhood (cluster) using CF to generate the recommended items. We use MovieLens dataset to test the accuracy and performance of our algorithm. MovieLens is a CF recommender system developed by GroupLens Research Group. The dataset contains 1,000,209 anonymous ratings of approximately 3,900 movies made by 6,040 MovieLens users who joined MovieLens in 2000. Ratings are on a five star (integral) scale from 1 to 5.

30.4 The Decomposition Algorithm

In this work, we use the location of the users to decompose the users' space hierarchically on the basis of a BSP tree. MovieLens dataset has information about the user's location (zip code). We use the longitude and latitude of the centroids of the polygonal regions representing the zip codes as the user coordinates. Space partitioning is done on the basis of the zip codes of the users, that satisfy the following criteria. Let U represents the entire set of users. We partition the set U into p partitions $U_1, U_2, \ldots, U_p$, where $U_i \cap U_j = \phi$ for $1 \leq i,\ j \leq p$; and $U_1 \cup U_2, \ldots, \cup U_p = U$. For any user u, if $u \in U_i$ then the recommendation algorithm use the entire cluster U_i as the neighbourhood. Now we present our decomposition algorithm formally.

Algorithm: *BSP Decomposition*

Step 1: Count the number of users in each zip code and find two such zip codes that have the maximum number of users.

Step 2: Find the equation of the perpendicular bisector of the line joining the centroids of these two zip codes.

Step 3: Divide the users into two partitions along this perpendicular bisector.

Step 4: For each of the two newly formed partitions, repeat Steps 1–3 till the number of users in a partition is greater than a threshold value (500 in our case).

We depict our BSP based space decomposition method pictorially through Figs. 30.1, 30.2, 30.3 and 30.4.

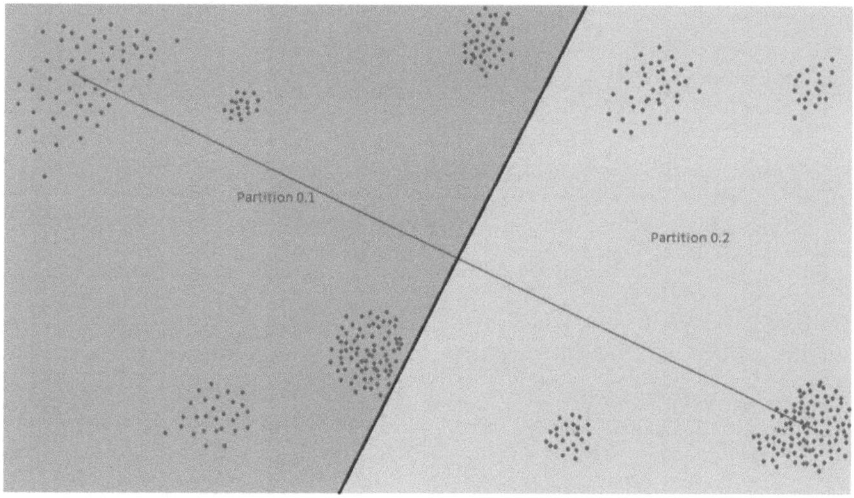

Fig. 30.1 BSP decomposition

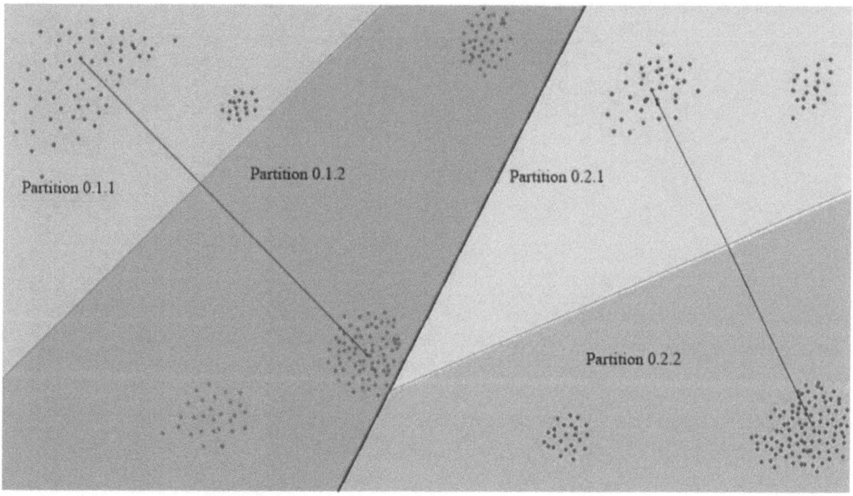

Fig. 30.2 BSP decomposition

In Fig. 30.1, we can see that the entire region 0 is partitioned into two partitions—partition 0.1 and partition 0.2 along the perpendicular bisector of the line joining the two zip codes that have the maximum number of users. Similarly in Fig. 30.2, partition 0.1 is divided into partition 0.1.1 and partition 0.1.2 while partition 0.2 gets divided into partition 0.2.1 and partition 0.2.2. This process is continued as long as the number of users in a partition is greater than a predefined threshold. Figure 30.4 depicts all the partitions resulting from our BSP decomposition process.

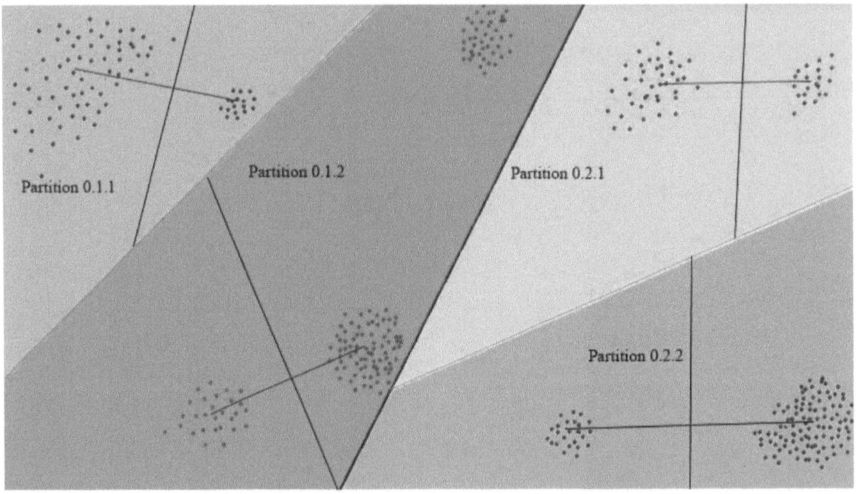

Fig. 30.3 BSP decomposition

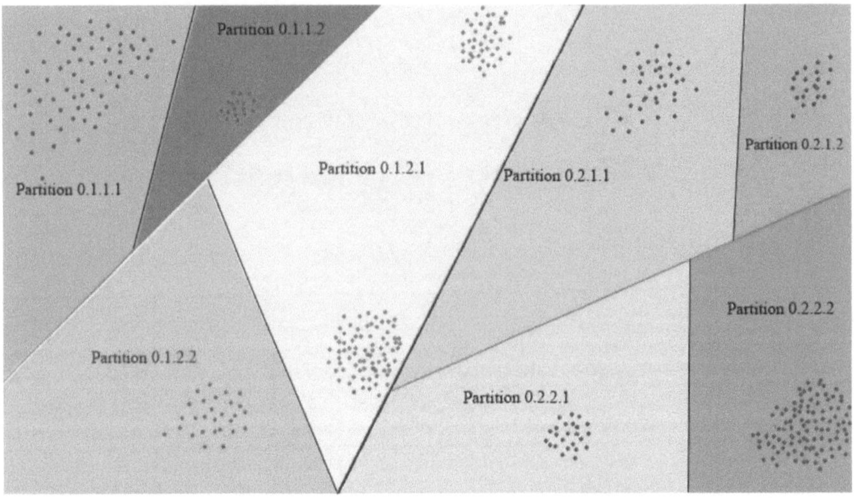

Fig. 30.4 BSP decomposition

30.5 The Recommendation Algorithm

In order to improve the scalability of our system, we apply the recommendation algorithm separately to the regions or clusters formed by the BSP decomposition algorithm. That is, recommendations for a user in a region are generated only using the ratings or preferences of the other users of that region. This helps us to reduce the complexity of the CF process as well as achieve a much faster recommendation time. In this work, we use a correlation-based CF algorithm which recommends item by computing user–user similarity among the users in a region. Pearson's correlation coefficient [8] is used to compute the similarity between the target user and all other users in the region. Our system provides recommendation for the following two categories of user.

- For an existing user in a cluster, the algorithm first finds her collaborative users based on the similarity score and then recommend *Top-N* items using the taste and preferences of the collaborative users.
- For a new user of the system, the algorithm use the IP address of the user to extract her location coordinates (longitude latitude), and accordingly place the user to her destined cluster (region). The *Top-N* movies highly rated by the users in the cluster are recommended to the user.

30.6 Experiments and Results

We conducted several experiments to evaluate the effectiveness of the proposed method. In this section, we describe the experimental settings in detail. We have tested our recommendation algorithm on the MovieLens dataset to validate our scheme. The user ratings of the dataset are randomly split into two sets—observed items (80 %) for training and held-out items (20 %) for testing. Ratings for the held-out items were to be predicted.

30.6.1 Evaluation Metric Discussion

In this work, we use mean absolute error (MAE) [8] to evaluate the prediction accuracy of the CF algorithm while recommendation quality is measured using *Precision* and *Recall* metric.

MAE: MAE is defined as the average of the absolute error. Absolute error is the difference between the predicted rating and actual rating. Let the actual user ratings be, $\{r_1, r_2, \ldots, r_n\}$, and predicted ratings are, $\{p_1, p_2, \ldots, p_n\}$, where n is the number of items. Then Absolute error,

$$E = \{e_1, e_2, \ldots, e_n\} = \{p_1 - r_1, p_2 - r_2, \ldots, p_n - r_n\}$$

and

$$\text{MAE} = \frac{\sum_{i=1}^{n} |e|}{n}$$

Table 30.1 Possible recommendations

	Customer likes (rating = 4 or 5)	Customer dislikes (rating = 1, 2 or 3)
Recommend	True positives	False positives
Do not recommend	False negatives	True negatives

Any prediction algorithm tries to minimize the MAE.

We have depicted the different combinations of recommendation that can be generated in a typical recommendation problem in Table 30.1. Note that a customer likes an item if he has given a rating of 4 or 5 to that item (in a scale of 1–5), otherwise dislikes it, i.e. his rating is 1, 2 or 3. A recommendation is positive if the recommended rating coincides with the actual rating given by the customer.

Precision: Precision measures the degree of accuracy of the recommendations produced by the algorithm. In our system, Precision measures what fraction of the recommended items are liked by the customers.

$$\text{Precision} = \frac{\text{True Positives}}{\text{True Positives} + \text{False Positives}}$$

Recall: The Recall metric is also known as the hit rate, which is widely used for evaluating *top-K* recommender systems. In our Recommender System, Recall measures what fraction of the items liked by the customers, has been recommended by the algorithm.

$$\text{Recall} = \frac{\text{True Positives}}{\text{True Positives} + \text{False Negatives}}$$

30.6.2 Recommendation Performance on MovieLens-1M Dataset

To test the quality of the overall recommendation, we tested the algorithm for all the regions formed by the BSP decomposition. We report and compare the results of the recommendation algorithm using different evaluation metrics in Table 30.2. In the Table, the column *time* reports the running time of our recommendation algorithm in the different regions. We compare the performance in the sub-regions formed by the BSP decomposition with that of the parent region. Note that, we present *Precision*@10 and *Recall*@10 to evaluate the quality of the *top*-10 recommended items. The bold numbers indicate that its value has an obvious improvement than the corresponding values in the parent region. Our experiments are run on a computer with Core i3—2100 @ 3.10 GHz × 4 CPU and 2 GB RAM.

Table 30.2 Results of recommendation on MovieLens-1M dataset

Region no.	MAE	Precision@10	Recall@10	Time (min)
1	0.463	0.856	0.782	1,023.5
1.1	0.482	**0.886**	**0.812**	170.54
1.2	**0.432**	**0.867**	**0.853**	130.78
1.1.1	0.498	0.878	**0.858**	31.61
1.1.2	0.501	**0.901**	**0.872**	23.21
1.2.1	0.452	**0.881**	**0.897**	10.62
1.2.2	0.462	0.842	**0.872**	11.75
1.1.1.1	**0.471**	0.823	**0.882**	2.23
1.1.1.2	0.512	0.837	**0.894**	3.33
1.1.2.1	**0.482**	0.887	**0.90**	0.5
1.1.2.2	0.523	0.8846	**0.913**	1.28
1.2.1.1	0.493	**0.902**	0.877	0.34
1.2.1.2	0.475	0.876	**0.912**	0.23
1.2.2.1	**0.443**	**0.864**	0.856	0.25
1.2.2.2	0.492	0.825	**0.913**	0.31

From Table 30.2, it is clear that our algorithm is working fine with high precision and recall values as well as lower MAE values. It is also clear from the table that the runtime of the algorithm is significantly less in the sub-regions than its parent region. As for example, the time required for recommending all the users in region 1 is 1023.5 min. However, when it is decomposed into two sub-regions 1.1 and 1.2, we have a total running time of 301.32 (170.54 + 130.78) min which is significantly less (by about 69 %) than the parent region 1. Analysing the results of the experiment performed, we can conclude that our approach is efficient in reducing the running time without sacrificing recommendation quality much. This ensures scalability and establishes that our method can be effective to deal with even bigger datasets.

30.7 Conclusion

In this paper, we propose a decomposition-based CF technique that deals with the scalability problem of the CF process by applying the recommendation algorithm separately to each of the clusters. Experimental analysis using real datasets show that our model is efficient and scalable. Online experimentation of the decomposition algorithm and the recommendation algorithm will be the focus of our future work. We will also try to use other metrics for finding the similarity between users with the aim of optimizing the decomposition technique and recommendation algorithm.

References

1. G. Adomavicius, A. Tuzhilin, Toward the next generation of recommender systems: a survey of the state-of-the-art and possible extensions, in *IEEE transactions on knowledge and data engineering* (2005), pp. 734–749
2. X. Su, T.M. Khoshgoftaar, A survey of collaborative filtering techniques. Adv. Artif. Intell. **2009**, 1–19 (2009)
3. T.M. Murali, P.K. Agarwal, J.S. Vitter, Constructing binary space partitions for orthogonal rectangles in practice, in *Proceedings of the European Symposium on Algorithms (ESA'98)*. LNCS, vol. 1461 (1998), pp. 221–222
4. B. Sarwar, G. Karypis, J. Konstan, J. Riedl, Recommender systems for large-scale ecommerce: scalable neighborhood formation using clustering, in *Proceedings of the Fifth International Conference on Computer and Information Technology* (2002), pp. 158–167
5. M. O'Connor, J. Herlocker, Clustering items for collaborative filtering, in *Proceedings of the ACM SIGIR Workshop on Recommender Systems* (1999)
6. J.S. Breese, D. Heckerman, C. Kadie, Empirical analysis of predictive algorithms for collaborative filtering, in *Proceedings of the UAI* (1998)
7. J. Das, S. Majumder, P. Gupta, Voronoi based location aware collaborative filtering, in *Proceedings of the IEEE Conference on Emerging Trends and Applications in Computer Science (NCETACS)* (2012), pp. 179–183
8. B. Bhasker, K. Srikumar, *Recommender Systems in e-Commerce* (Tata McGraw Hill, Noida, 2010)

Chapter 31
A Novel Approach for Non-cooperative Node Detection and Avoidance Using Reputation-Based Scheme in Mobile Ad hoc Network

Chandrima Chakrabarti, Ananya Banerjee, Sanchari Chakrabarti and Angana Chakraborty

Abstract A mobile ad hoc network (MANET) is a collection of wireless mobile nodes that form a dynamic network without the need for any infrastructure. Due to the dynamic nature of MANET, it is prone to different kinds of malicious attacks. In order to pursue secure communication in such networks, there are many research solutions proposed for detecting and avoiding such malicious activities. As we know, MANET works properly if participating nodes cooperate in routing and forwarding. However, a node may decide not to cooperate just to save its resources but still use network to relay its traffic. In this scenario, we propose a reputation-based strategy to detect non-cooperative or selfish nodes and to select proper forwarder node for improving overall packet delivery of the network. Moreover, our proposed solution ensures data integrity. We have also done survey and performance analysis on some existing malicious attacks detection and prevention techniques available. The entire simulation has been done using the Network Simulator (NS-2) (http://www.isi.edu/nsnam/ns/) and simulation results show better delivery compared to some of other existing techniques discussed in the papers by Khamayseh et al. (J Netw 7(1):116–125, 2012), Marti et al. (Proceedings of International

C. Chakrabarti (✉) · A. Banerjee
CSE Department, Narula Institute of Technology, Agarpara, Kolkata 700109, India
e-mail: chandrima.narula@gmail.com

A. Banerjee
e-mail: ananyabanerjee.narula@gmail.com

S. Chakrabarti
Brahmananda Keshab Chandra College, Baranagar, Agarpara, Kolkata 700109, India
e-mail: dearsanchari@gmail.com

A. Chakraborty
IIEST, Howrah 711103, India
e-mail: angana.chakraborty9@gmail.com

© Springer India 2015
K. Maharatna et al. (eds.), *Computational Advancement in Communication Circuits and Systems*, Lecture Notes in Electrical Engineering 335,
DOI 10.1007/978-81-322-2274-3_31

Conference on Mobile Computing and Networking (MOBICOM'OO), 255–265, 2000), Woungang et al. (Proceedings of IEEE Conference, 2012) and Hu et al. (IEEE J Sel Areas Commun 24(2):307–380, 2006).

Keywords Mobile ad hoc network · Malicious attacks · Reputation

31.1 Introduction

Due to the rapid evolution of wireless network, mobile ad hoc networks (MANETs) are becoming increasingly popular in various applications such as in the field of emergency preparedness and response, collaborative and distributed computing, mine site operations, battlefield military operations, electronic classrooms, conferences etc. [1]. Moreover, MANET becomes popular day by day because it requires no centralized administration or fixed network infrastructure and can be quickly and inexpensively form the set up as needed. In the dynamic MANET environment, nodes are assumed to cooperate among each other to provide routing service and forward packets. This requirement poses a security challenge when malevolent nodes are present in the network. Indeed, the existence of such nodes may not simply disrupt the normal network operations, but also generate severe security problems, like dropping, non-forwarding, authentication, data availability, confidentiality and integrity point of views. In MANETs, reputation-based strategies are considered for detection of selfish/malicious nodes as well as for determining the best forwarder node in case of data delivery. A reputation is a node's degree of cooperation in forwarding and receiving messages.

In this paper, we have done survey and performance analysis on some existing malicious attacks detection and prevention techniques. We have also proposed a model for detection and prevention of malicious attacks using a reputation-based technique. Our proposal is based on personal feedback table maintained by a node for calculating a node's reputation easily. After successful data packet exchange both the communicating nodes will also exchange credit packets; forwarder node will get forward credit from receiver node and receiver node will get receive credit from the forwarder node. Each node will be liable for showing the personal feedback table and credit tables to the Trusted Authority (TA) node for calculating each node's reputation perfectly. TA node is solely responsible for updating reputation values dynamically and broadcast these reputation values and status of a node time to time. After getting these values, a node can easily determine the best forwarder node and selfish/non-cooperative nodes in the network. We have simulated the scheme on Network Simulator (NS2) [2] and compared the performance with [3, 4–6].

The rest of the paper is organized as follows. Section 31.2 presents the related work in this domain. Section 31.3 illustrates our proposed system model. Results and Discussions are included in Sect. 31.4. The paper is concluded with a discussion on future work in Sect. 31.5.

31.2 Related Work

A considerable amount of research has been carried out in the area of MANETs security [7–10]. Some proposed schemes can detect and deal with malicious nodes [11]. The following sections discuss some of the techniques that have been proposed to achieve security in MANETs.

The authors in [12] enhanced the security of DSR protocol by enhancing the trust-based route selection mechanism. The authors in [13] proposed a protocol that calculates the reputation of a node by observing the node's behaviour, the observed value is later altered based on additional observations from other nodes. But node authentication is not mentioned here. In [3] the authors proposed a Trust Scheme for observing the behaviour of mobile nodes. Using this scheme, malicious nodes can be detected and avoided. But which particular node in the route is misbehaving cannot be identified. Node authentication is not used in [3].

In [7] authors proposed a secure Intrusion Detection (EAACK) System for MANETs. In this scheme there are three parts for detection of malicious activities. They are ACK, secure ACK (S-ACK) and misbehaviour report authentication (MRA). It is effective in detecting black hole attacks. Receiver collision problem can be solved. False acknowledgment is not possible.

In [4] authors designed the monitoring scheme to prevent black hole attacks. The solution is that each node in the network consists of two components. One is watchdog and second one is pathrater. Watchdog observes the behaviour of every neighbour by putting itself in promiscuous mode. Pathrater gives the rating of each node in the network. It is possible to detect black hole attack. But false acknowledgement is possible.

The authors in [5] made an intrusion detection-based solution known as anti-black hole mechanism (ABM). In this method, the difference between route request and route reply packets is calculated for every node. This difference between route request and route reply keeps changing due to the forwarding and broadcasting nature of nodes and is stored by every node. But in case of malicious node, difference is high. If the estimated value (difference) goes above threshold value, a block message is broadcasted by the detected node to other nodes to isolate the malicious node. The scheme suffered from high overhead. Node authentication is also not mentioned here.

The authors in [5] recommended a scheme for detecting and avoiding black hole attacks before the actual routing mechanism started by using fake RREQ packets to catch the malicious nodes. These fake RREQ packets are like the normal RREQ but they live only for certain time; fake destination address is used for identifying malicious nodes. But at the time of routing if any node behaves malicious then this mechanism will fail.

In [6] authors proposed a worm hole detection technique. First node is verified and its authorization is done using symmetric key cryptography. If authentication is true, sender can send the data to that particular node; otherwise, next node is selected to carry the data packet. This process continues until data packet reaches to destination. Here, malicious node cannot pretend to be authorized user. Overhead is huge in this case.

The authors in [14] suggested a method of detecting malicious node during route discovery and if malicious node is found, data packet is sent via another route. This is based on existing protocols like DSR, AODV and DSDV. In this approach encryption techniques, acknowledgement and principles of flow conservation are used to prevent attacks in network layer.

Inspired from these novel works, we have implemented one reputation-based scheme for selfish node detection and avoidance of those selfish nodes in case of further data transmission.

31.3 Proposed System Model

Nowadays, researchers are showing their interest in developing a secured ad hoc network. Many research works are done in detecting the selfish behaviour of node [7, 13]. Motivated from those works, we have implemented one reputation-based selfish node detection scheme based on Ad hoc On demand Distance Vector Routing (AODV) routing protocol. As this selfishness affects tremendously the packet delivery and efficiency of a network, we focused our work on improving the delivery ratio as well as the performance of the overall network.

A node's reputation is determined in terms of its forwarding and receiving characteristics in the recent past by the Trusted Authority (TA) node as in Fig. 31.1. In our proposed model, when a node gets some packet from another node, it needs to maintain one personal feedback table. Suppose, node A is the source of a message and node C is the destination. As node A cannot send the message directly to node C, it needs some intermediate node (here node B) to relay the message to node C which has the higher probability to reach destination node C as in Fig. 31.1. So, A first sends the message to B and updates its personal feedback table as shown in Table 31.1.

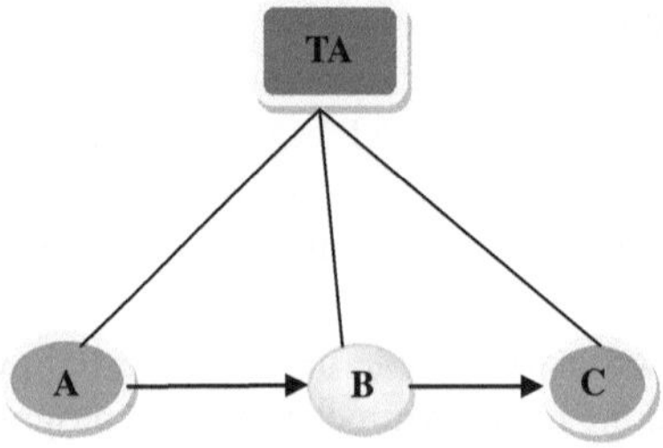

Fig. 31.1 Data exchange from source (A) to destination (C), where B is intermediate node; Trusted Authority (TA) node for calculating each node's reputation

After exchanging messages, node A and node B will exchange credit messages with each other. As node A is the sender so, it will get forward_credit from node B, as well as, node B will get receive_credit by node A. In this scheme, all the messages including data and credit messages are all encrypted messages, encrypted by the owner's private key. So, these messages can be decrypted by any node in the network which knows the public key of node A or node B. So, any node can decrypt the messages, but cannot encrypt it again. In this way, the modifications can be avoided.

From Table 31.1, it is cleared that at time T11, node A had given messages to node B and updated its Personal Feedback table. Here, based on node B's behaviour and previous performance, node A can assign any receive_credit (rc) value of node B up to 1 and node B can assign any forward_credit (fc) value of node A up to 1.

Now, both node A and B can show these credit values or exchange this personal feedback table only to the Trusted Authority for further communication, as reference.

Personal Feedback table maintained by node B is as follows (Table 31.2).

Node B, after getting the message, delivers it to the destination node C. After exchanging messages, node B and node C will exchange credit messages as discussed previously. Based on node C's behaviour and previous performance, node B will assign rc value of node C as 0.7 and node C will also assign fc value of node B as 0.8.

Now, both nodes B and C update their personal feedback tables as shown in Tables 31.3 and 31.4, respectively. We consider T12 > T11.

A sample Forward Credit Table is as shown in Table 31.5.

A sample Receive Credit Table is as shown in Table 31.6.

As shown in Table 31.6, a sample forward credit packet at a particular point of time (say T1) may have fc_id, forwarder node id, receiver node id, number of exchanged packets, exchanged messages' id and exchange time. Similarly, a sample

Table 31.1 Personal feedback table maintained by node A

Source_node id	Forwarder_node id	Time	Forward_credit	Receive_credit
A	B	T11	A-0.9 (by B)	B-0.8 (by A)

Table 31.2 Personal feedback table maintained by node B

Source_node id	Forwarder_node id	Time	Forward_credit	Receive_credit
A	B	T11	A-0.9 (by B)	B-0.8 (by A)

Table 31.3 Updated personal feedback table maintained by node B

Source_node id	Forwarder_node id	Time	Forward_credit	Receive_credit
A	B	T11	A-0.9 (by B)	B-0.8 (by A)
A	C (via B)	T12	B-0.8 (by C)	C-0.7 (by B)

Table 31.4 Personal feedback table maintained by node C

Source_node id	Forwarder_node id	Time	Forward_credit	Receive_credit
A	B	T12	B-0.8 (by C)	C-0.7 (by B)

Table 31.5 Sample forward credit table

Forward credit (fc)_id	Forwarder node id	Receiver node id	No. of exchanged packet	Exchanged msg id	Time
$fc1_{N1}$	A	B	1	M101	T1

Table 31.6 Sample receive credit table

Receive credit (rc)_id	Forwarder node id	Receiver node id	No. of exchanged packet	Exchanged msg id	Time
$rc1_{N2}$	A	B	1	M101	T2

receive credit packet at a particular time (say T2) may have rc_id, forwarder node id, receiver node id, number of exchanged packets, exchanged messages' id and exchange time.

These credit tables are exchanged as a sign of successful completion of data packet exchange. Receiver node will give this to its current forwarder node as a record of how many data packets it receives from current forwarder node and at the same time a forwarder node will give this to its receiver node as a record of how many data packets it forwards to the receiver node (we consider T2 > T1).

Based on forward credit and receive credit, a node will select a forwarder or receiver node but never exchange this personal feedback values to other nodes. So, it definitely reduces overhead.

Each node will also maintain as discussed in Gao et al. [15],

1. Delegation Evidence (D) = {M, A, B, Dst, TS, Exp}-Node A will maintain D.
2. Forwarding History Evidence (F) = {M, B, C, Dst, TS, Exp}-Node B will maintain F.
3. Contact History Evidence (C) = {M, B, C, TS}-Both node B, C will maintain C.

Here M—message, A—node A, B—node B, C—node C, Dst—destination, TS—timestamp and Exp—expiration time of the message.

However, a node will exchange personal feedback table, evidences and credit tables with the Trusted Authority (TA) node for gaining reputation.

31.3.1 Cooperative (Good, Medium) and Non-cooperative/ Selfish Node Detection by TA Node

Algorithm1: Cooperative, Non-Cooperative/Selfish Node Detection

```
STEP1: Initialize node under assessment (NA), TA node,
Delegation Evidence (D), Contact History Evidence (C),
Neighbor Table (N), Forwarding History Evidence (F)
STEP2: TA node demands all the nodes in the network to
submit their evidences and Neighbor tables
STEP3: Based on D, C, N; TA node will compare with F val-
ues
STEP4: If F = = Find (D, C, N)
STEP5: NA is cooperative
STEP6: Else
STEP7: NA is non-cooperative or Selfish
STEP8: End if
```

After determining cooperative and non-cooperative/selfish node, TA node will blacklist selfish nodes; but cooperative nodes will go through the reputation process for determining whether the node is good or medium.

Trusted Authority (TA) is the authorized node solely responsible for determining the reputation of a node based on a node's forward credit (fc) and receive credit (rc) values. From the forward credit table and receive credit table, reputation_value at time t is calculated as:

$$\text{Reputation\_value}(t) = \sum \text{fc\_value}(t) \,/ \sum \text{rc\_value}(t)$$

After getting the reputation of a node, TA node will decide whether that particular node is a good node or medium node.

If reputation_value <= Rth, the node is medium node.

If reputation_value > Rth, the node is good; where Rth = reputation threshold value.

TA node will maintain the reputation table as shown in Table 31.7.

From Table 31.7 it is cleared that TA node maintains the reputation table and reputation values with timestamp for each node and periodically updates this table.

TA node will broadcast this reputation table time to time so that all other nodes in the network can choose best forwarder node for data delivery.

Here also, selfish nodes will get another chance to gain their reputation. If any node fails to increase its reputation even after getting second chance, it will be temporarily avoided from future communication to improve packet delivery.

Table 31.7 Sample reputation table maintained by TA node

	N_1	N_2	…	N_i	…	N_j
TA	$R1_{t0}$	$R2_{t2}$	…	Rj_{tj}	…	Rk_{tk}

31.4 Results and Discussions

31.4.1 Simulation Details

We have implemented our scheme using Network Simulator (NS2) [2]. We have also made the comparison with other schemes discussed in the papers [3, 4–6]. The details of the simulation parameters are discussed in Table 31.8.

The performance of our proposed scheme and other schemes are evaluated in terms of packet delivery probability and routing overhead.

Delivery Probability: Delivery Probability defines the ratio of the number of packets received by the destination node to the number of packets sent by the source node.

Routing overhead: Routing overhead defines below formula

$$\text{Routing overhead} = \frac{(\text{No. of packets relayed} - \text{No. of packets delivered})}{\text{No. of packets delivered}}$$

31.4.2 Discussion About the Output

We have done our simulation on a basic packet dropping attack. Malicious nodes simply drop all the packets that they receive. We considered two scenarios.

In scenario 1, we have observed delivery probability in Fig. 31.2 where malicious nodes are varied from 0 to 40 %.

In scenario 2, we have observed routing overhead in Fig. 31.3 where malicious nodes are varied from 0 to 40 %.

We have compared the performance of our proposed scheme with

(i) Malicious nodes detection scheme discussed in paper [3]
(ii) Mitigating misbehaviour scheme discussed in paper [4]
(iii) Prevention of black hole scheme discussed in paper [5]
(iv) Wormhole attacks detection scheme discussed in paper [6].

Table 31.8 Parameters used

No of nodes	100
Terrain range	1000 × 1000 m^2
Speed of mobile node	0, 5, 10, 15, 20 m/s
Pause time	300 s
Packet size	512 Bytes
Simulation time	3,000 s
Packet transmission rate	4 packets/s
Routing protocol	Ad hoc on demand distance vector (AODV) routing

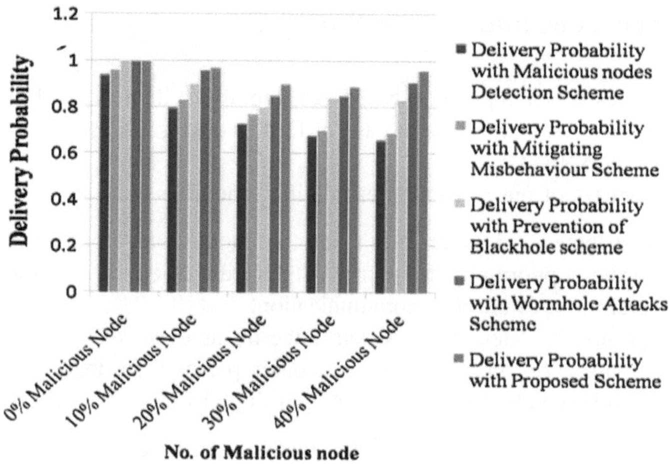

Fig. 31.2 Simulation result of delivery probability where malicious nodes are varied from 0 to 40 %

From Fig. 31.2, it is cleared that using our proposed scheme maximum delivery probability can be achieved compared to other schemes.

We can easily notice from Fig. 31.3, that in our proposed scheme routing overhead is minimum compared to other schemes.

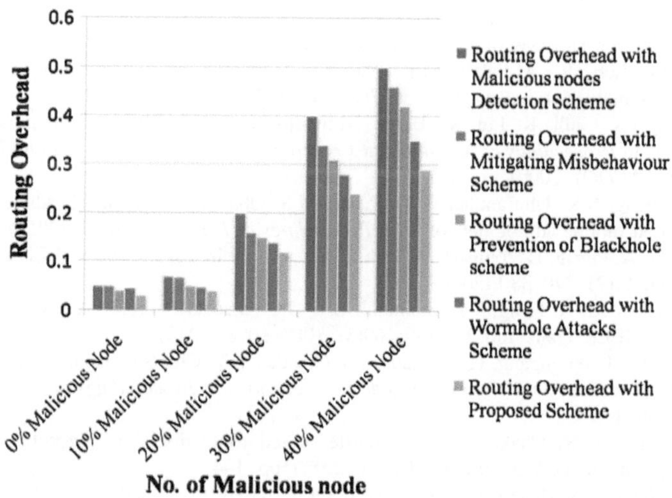

Fig. 31.3 Simulation result of routing overhead where malicious nodes are varied from 0 to 40 %

31.5 Conclusions and Future Work

This paper presents a reputation-based scheme for selfish node detection and avoidance of those selfish nodes in case of further data transmission in mobile ad hoc network (MANET).

The uniqueness of our reputation estimation scheme is that

(i) It is distributed and dynamic in nature, i.e. nodes with a bad reputation may get further opportunity to improve their reputation by participating and cooperating in future data communication.
(ii) Our proposed system ensures data integrity as every data packets are sent using public key cryptosystem and other personal feedback tables, credit tables and reputation tables are sent after digitally signed by the source node.

In this paper, we can easily detect the malicious activities like dropping, non-forwarding, false token generation, colluding attacks and we have made efforts to avoid those nodes during data forwarding. We have evaluated the performance in terms of delivery probability and routing overhead. However, the impact of mobility and scalability is not evaluated yet, which is included in our future work.

References

1. P. Michiardi, R. Molva, CORE: A Collaborative reputation mechanism to enforce node cooperation in mobile ad hoc networks, in *Proceedings of IFIP-Communication and Multimedia Security Conference* (2002)
2. NS2 simulation package, http://www.isi.edu/nsnam/ns/
3. Y. Khamayseh, R. Al-Salah, M.B. Yassein, Malicious nodes detection in MANETs: behavioral analysis approach. J. Netw. **7**(1), 116–125 (2012)
4. S. Marti, T.J. Giuli, K. Lai, M. Baker, Mitigating routing misbehavior in mobile ad hoc networks, in *Proceedings of International Conference on Mobile Computing and Networking (MOBICOM'OO)* (2000), pp. 255–265
5. I. Woungang, S.K. Dhurandher, R.D. Peddi, M.S. Obaidat, Detecting blackhole attacks on DSR-based mobile ad hoc networks, in *Proceedings of IEEE Conference* (2012)
6. Y.-C. Hu, A. Perrig, D. Johnson, Wormhole attacks in wireless networks. IEEE J. Sel. Areas Commun. **24**(2), 370–380 (2006)
7. E.M. Shakshuki, N. Kang, T.R. Sheltami, EAACK—a secure intrusion-detection system for MANETs. IEEE Trans. Ind. Electron. **60**(3), 1089–1098 (2013)
8. J.W. Huang, I. Woungang, H.C. Chao, M.S. Obaidat, T.Y. Chi, S.K. Dhurandher, Multi-path trust-based secure AOMDV routing in ad hoc networks, in *Proceeedings of IEEE Globecom* (2011), pp. 1–5
9. C. Chakrabarti, R. Chaki, Improved cluster based route discovery algorithm for ad-hoc networks, in *Proceedings of IEEE ICCIA* (2011), pp. 1–4
10. C. Chakrabarti, A. Banerjee, S. Roy, An observer-based distributed scheme for selfish-node detection in a post-disaster communication environment using delay tolerant network, in *Proceedings of IEEE AIMoC* (2014), pp. 151–156
11. A. Nadeem, M.P. Howarth, A survey of MANET intrusion detection and prevention approaches for network layer attacks. IEEE Commun. Surv. Tutorials **15**(4), 2027–2045 (2013)

12. M. Alsaadi, Y. Qian, Performance study of a secure routing protocol in wireless mobile ad hoc networks, in *2nd International Symposium on Wireless Pervasive Computing* (IEEE Computer Society, New York, 2007), pp. 425–430
13. G. Bella, G. Costantino, S. Riccobene, Evaluating the device reputation through full observation in MANETs, J. Inf. Assur. Secur. **4**, 458–465 (2009)
14. S. Arya, C. Arya, Malicious nodes detection in mobile ad hoc networks, J. Inf. Oper. Manag. **3** (1), 210–221 (2012) ISSN: 0976–7754 & E-ISSN: 0976–7762
15. Z. Gao, H. Zhu, S. Du, C. Xiao , R. Lu, PMDS: a probabilistic misbehavior detection scheme in DTN, in *IEEE ICC 2012—Wireless Networks Symposium* (2012)

Part V
Advances in Control System

Chapter 32
Performance Comparison of Brain Emotional Learning-Based Intelligent Controller (BELBIC) and PI Controller for Continually Stirred Tank Heater (CSTH)

Manoj Kumar Sharma and Anmol Kumar

Abstract In a process industry, the control of temperature is a significant requirement. The various conventional control schemes using PID controllers and soft computing-based controllers such as neural networks and fuzzy logic controllers have been implemented for the temperature control system in the industry. In this paper, the emotional behaviour-based controller called brain emotional learning-based intelligent controller (BELBIC) is implemented for continually stirred tank heater (CSTH) and its performance is compared with that of conventional proportional integral (PI) controller for various test conditions. The simulations are performed for: set point tracking, disturbance rejection and multi-set point tracking. It is observed that BELBIC gives better response as compared to PI controller in terms of performance parameters for different test conditions.

Keywords Emotional learning · BELBIC · PI controller · CSTH

32.1 Introduction

The classical controllers are widely used in the process control industry as it is easy to implement and tune them easily for given operating points. These are generally proportional (P), proportional integral (PI) and proportional integral-derivative

M.K. Sharma (✉) · A. Kumar
Department of Electrical and Electronics Engineering, University Institute of Engineering and Technology, Panjab University, Chandigarh, India
e-mail: mk07sharma@rediffmail.com

A. Kumar
e-mail: anmol_k97@hotmail.com

© Springer India 2015

K. Maharatna et al. (eds.), *Computational Advancement in Communication Circuits and Systems*, Lecture Notes in Electrical Engineering 335,
DOI 10.1007/978-81-322-2274-3_32

(PID) controllers. However, the performance of classical controllers shows degradation when there is variation in the plant parameters and set point.

The drawbacks of the classical controllers have been overcome by the different adaptive control schemes developed in last decades [1, 2]. With advancements in the soft computing techniques, various intelligent control techniques based on artificial neural network (ANN) [3], fuzzy logic [4] and genetic algorithm [5] have been successfully applied to various applications. The emotional behaviour of the human has been modelled in [6, 7]. It is a type of conditional learning associated with external emotional stimuli such as rewards and punishments received from various real-life situations. These emotional stimuli can cause various internal emotional states such as joy, sadness and fear affecting future decisions. BELBIC was proposed in [8], which was further modified in [9]. It was successfully utilized in several applications such as controller design of a path tracking problem [10], intelligent control of washing machines [11], etc. The process control systems are often nonlinear and the parameters are varied due to disturbance, noise etc. and thus it becomes tough to apply the control method. In this paper, continually stirred tank heater (CSTH) has been considered which is widely used in the process and chemical industries [12]. The conventional PI controller and BELBIC are implemented for the temperature control of the CSTH and results are compared.

In the second section, the computational model of limbic system of mammalian brain is described. The modelling of CSTH has been discussed in Sect. 32.3. The BELBIC design and implementation has been explained in Sect. 32.4. The simulations and results obtained for BELBIC and PI controllers are compared in Sect. 32.5. The conclusion is obtained in the Sect. 32.6.

32.2 Model of Limbic System

The emotional behaviour of brain is due to the limbic system which consists of: amygdala, orbitofrontal Cortex (OFC), thalamus and sensory cortex. A computational model of limbic system has been developed [13, 14] which is shown in Fig. 32.1.

The signals from the thalamus and from cortical areas are fed to amygdala while the cortical areas and the amygdala provide input to the orbito frontal. The reinforcement signal R_w is also given to the system. The difference between the output of amygdala and OFC is the output of the model. Each node A, is connected through the weight V and the output of this node is weighted input.

$$A_i = S_i * V_i \tag{32.1}$$

$$A_{th} = \max(S_i) \tag{32.2}$$

The weights V_i are adjusted as per Eq. (32.3) where α_A is learning rate parameter.

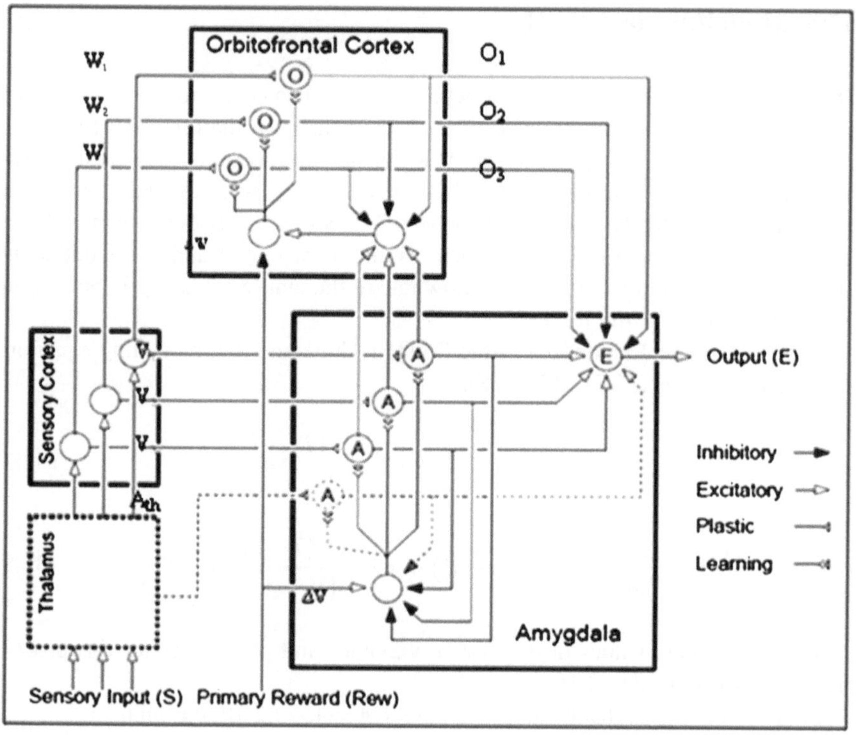

Fig. 32.1 Diagram of a model of brain limbic system

$$\Delta V_i = \alpha_A \cdot \max\left[0, R_w - \sum A_i\right] \tag{32.3}$$

It is observed that the conditioning in the amygdala is permanent. The term max in (32.3) implies that the amygdala gain can never be decreased and making the learning monotonically increasing. For each O node, there is a connection weight W. The output of node is given as

$$O_i = S_i * W_i \tag{32.4}$$

The connection weights in OFC are adjusted proportionally to the difference between the model output (MO) and the reinforcement signal R_w, α_O is the learning parameter and is adjustable between 0 and 1.

$$\Delta W_i = \alpha_o \cdot (\mathrm{MO} - R_w) \tag{32.5}$$

$$\mathrm{MO} = \sum A_i - \sum O_i \tag{32.6}$$

The externality of emotion formation means that through choosing the emotional cues or R_w signal, the control goals can be implicitly decided.

32.3 Modelling of CSTH

The CSTH system consists of tank whose fluid is heated up to the desired temperature. The fluid inside the tank receives the heat from the other fluid which is circulated in the jacket. To keep the heat transfer uniform, the tank fluid is continually stirred. The heat transfer rate is to be controlled so as to attain the desired temperature [15, 16]. The block diagram of CSTH which is MIMO system is shown in Fig. 32.2. The outlet flow rate is same as inlet flow rate. T and T_i are outlet and inlet temperature of the tank and F is flow rate of the tank whereas T_j, T_{ji} and F_{ji} are respective values for the jacket.

The mathematical modelling of the CSTH is done and two modelling equations are obtained as per (32.7) and (32.8).

$$\frac{\mathrm{d}T}{\mathrm{d}t} = \frac{F}{V} * (T_i - T) + UA * \frac{(T_j - T)}{(V * \rho * C_p)} \tag{32.7}$$

$$\frac{dT_j}{dt} = \frac{F_j}{V_j} * (T_{ji} - T_j) + UA * \frac{(T_j - T)}{(V_j * \rho * C_p)} \tag{32.8}$$

The steady-state values of the system variables and parameters for this process are as shown in Table 32.1:

From $\mathrm{d}T/\mathrm{d}t = 0$ at steady-state, we have $UA = 157.62$ Btu/°F min.

From $\mathrm{d}T_j/\mathrm{d}t = 0$ at steady-state, we have $F_{js} = 1.44$ ft^3/min.

The transfer function for process one, i.e. G_{p1} is transfer function between tank temperature T and jacket temperature T_j and G_{p2} is transfer function between jacket flow rate F_j and jacket temperature T_j and are given as:

$$G_{p1} = \frac{T(s)}{T_j(s)} = \frac{0.0174(s + 18.82)}{(s + 0.4017)} \tag{32.9}$$

$$G_{p2} = \frac{T_j(s)}{F_j(s)} = \frac{50(s + 0.3516)}{(s + 0.1844) * (s + 4.1842)} \tag{32.10}$$

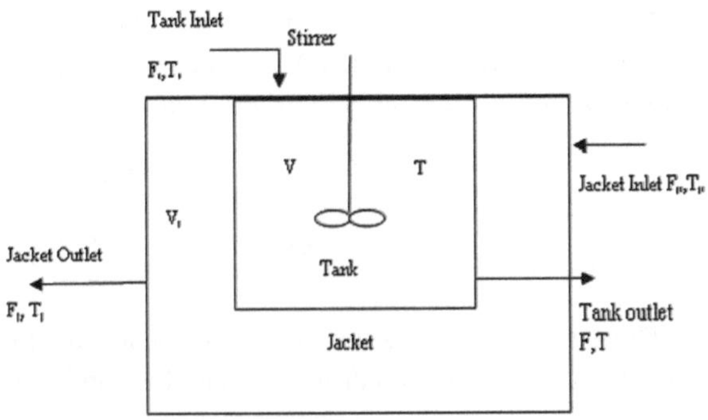

Fig. 32.2 Schematic diagram of CSTH

Table 32.1 Parameters and steady-state values

Parameters	Steady-state values
F_s	60 ft^3/h
T_{js}	50 °F
T_s	122 °F
V	10 ft^3
T_{jis}	200 °F
T_{js}	150 °F
V_j	1 ft^3
ρC_p	61.3 Btu/°F ft^3
ρC_{pj}	61.3 Btu/°F ft^3

32.4 Controller Design and Implementation

For the CSTH process which is a MIMO system, model reference control structure and cascade control scheme are used for better performance. In the BEL controller, the reinforcing signal, R_w is given from the outside, which in turn improves flexibility of the system. The cascade control scheme for CSTH is as shown in Fig. 32.3. The most important aspect of this controller is the selection of sensory and R_w signals associated with it. The sensory signal selected for both the primary and secondary controllers is:

$$S_i = k_e \cdot e \tag{32.11}$$

k_e = tunable constant, e = error.

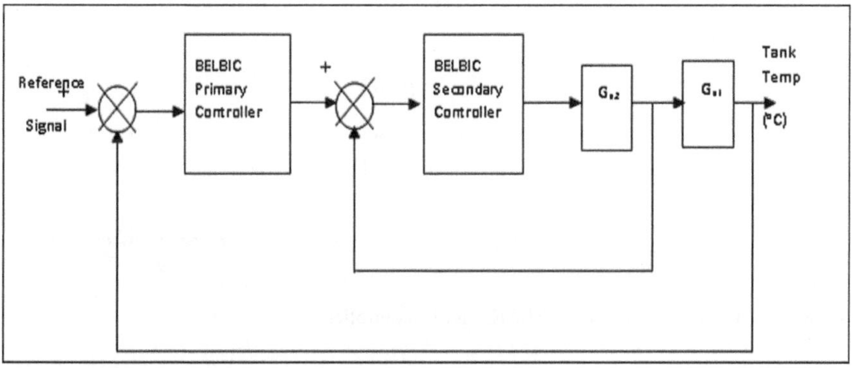

Fig. 32.3 BELBIC based cascade control scheme for CSTH

The R_w selected for both the controller is also identical, i.e.

$$R_w = k_p \cdot e + k_i \cdot \int e + k_u \cdot \int \mathrm{MO} \tag{32.12}$$

k_p, k_i, k_u are gains that are to be tuned to obtain the suitable controller.

32.5 Simulations and Results

This section presents simulation of the cascaded CSTH system. The simulations are done in MATLAB® and Simulink®. A comparison is made between BELBIC and conventional PI controller-based CSTH. The different tests have been performed to compare the performance of two controllers: set point tracking, disturbance rejection and multi-set point tracking.

32.5.1 Set Point Tracking

In this test, the performance of conventional PI controller and BELBIC controller for set point tracking is observed. The set point to be followed is 50 °C for 600 samples. It is seen that, BELBIC is able to reach the set point within 23 samples and tracked the same thereafter whereas the PI controller took 70 samples to reach the set point as shown in Fig. 32.4.

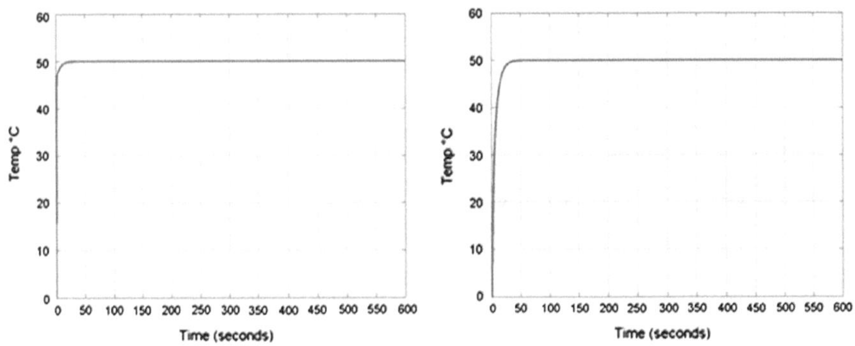

Fig. 32.4 Set point tracking for BELBIC and PI controller

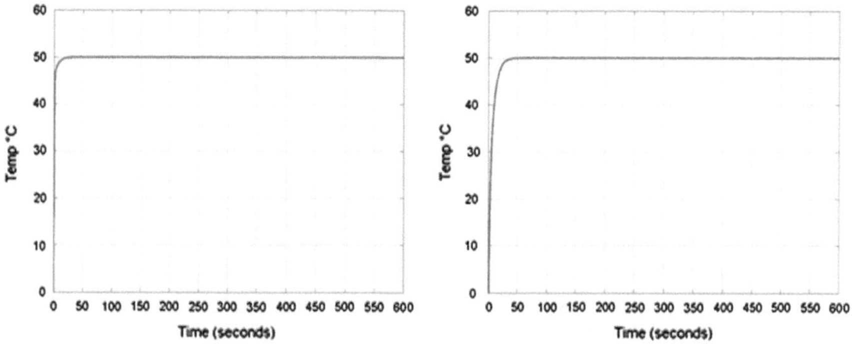

Fig. 32.5 Disturbance rejection comparison for BELBIC and PI controller

32.5.2 Disturbance Rejection

In the second test, the performance of the controller is assessed under disturbance condition. A Gaussian white noise is added in the forward path of the control loop with the set point of 50 °C for 600 samples. It is observed that both the controllers are able to reject the effect of disturbance introduced and followed the set point of 50 °C throughout the signal. Though, both PI and BELBIC controllers rejected the disturbance affectively, however, the BELBIC controller took 25 samples to reach the set point as compared to 75 samples taken by the PI controller as shown in Fig. 32.5. Thus, the BELBIC is better in responding to the disturbance.

32.5.3 Multi-Set Points Tracking

In this test, the three different set points are considered whereas the system is trained for one set point only. The total response is of 600 samples.

For the each 200 samples, set points of 40, 50 and 60 °C are set. The BELBIC and PI controllers successfully followed the set points for three different durations, although these were tuned for set point at 50 °C. The time taken for these controllers to reach the multiple set points is as shown in Fig. 32.6.

The results of the three tests are tabulated in Table 32.2.

The performance parameters namely ITAE, IAE and ISE have been calculated for both the controllers and these has been observed as shown in Table 32.3 and with white noise in Table 32.4.

It has been seen that the performance of BELBIC is much better than conventional PI controller.

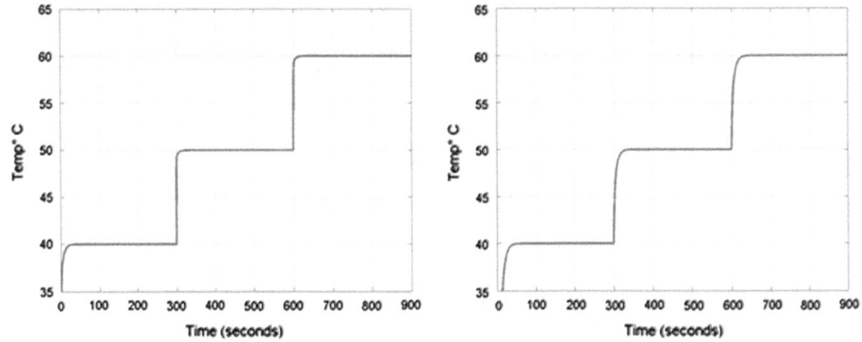

Fig. 32.6 Multi-set point tracking of BELBIC and PI controller

Table 32.2 Comparison of BELBIC and PI controller simulation results

Controller	Test-I	Test-II	Test-III		
			40 °C	50 °C	60 °C
BELBIC	23 samples	25 samples	35 samples	10 samples	20 samples
PI controller	70 samples	75 samples	50 samples	25 samples	50 samples

Test-I Set point Tracking (50 °C for 600 samples) (Time required to reach set point)
Test-II Disturbance rejection (Gaussian White Noise in forward path) (Time required to reach set point)
Test-III Multi-set points Tracking (Time required to reach different set point)

Table 32.3 Performance parameters for BELBIC and PI controller

Controller	ITAE	IAE	ISE
BEL	3,287	107.8	3,883
PI	3,971	607.4	3.067e + 04

Table 32.4 Performance parameters with white noise for BELBIC and PI controller

Controller	ITAE	IAE	ISE
BEL	2,382	105.6	3,897
PI	7,650	618.6	3.066e + 04

32.6 Conclusion

The biologically inspired controller, BELBIC with emotional learning and PI controller has been implemented for the continually stirred tank heater (CSTH). As the CSTH is an MIMO (multi-input multi-output) system so cascade control scheme has been used for implementing both the controllers. The simulations are done in MATLAB. The controller performance is observed for three different tests, i.e. tracking of set point, disturbance rejection and multi-set point tracking. The results obtained are compared. It is observed that both the controllers achieve the targets but the time taken for BELBIC is less considerable as compared to the PI controller. Thus, it can be concluded that for better enhanced control, the BELBIC can be a good option for CSTH temperature control.

References

1. B. Bequette, Wayne, *Process Dynamics—Modelling, Analysis and Simulation* (Prentice-Hall, Inc. Upper Saddle River, 1998)
2. D.R. Coughnowr, *Process Systems Analysis and Control* (Mcgraw Hill international editions, Singapore, 1991)
3. K.S. Narendra, K. Parthasarathy, Identification and control of dynamical systems using neural networks. IEEE Trans. Neural Netw. **1**(1), 4–27 (1990)
4. T.M. Takagi, M. Sugeno, Fuzzy identification of systems and its applications to modeling and control. IEEE Trans. Syst. Man Cybernet. **15**(1), 122–132 (1985)
5. R.J. Wai, J.D. Lee, K.H. Su, Supervisory enhanced genetic algorithm control for indirect field-oriented induction motor drive, in *Proceedings IEEE International Joint Conference on Neural Networks*, vol. 2, (2004), pp. 1239–1244
6. M. Fatourechi, C. Lucas and A. Khakisedigh, Reducing control effort by means of emotional learning, in *Proceedings of 19th Iranian Conference on Electrical Engineering*, Tehran, Iran (2001), pp. 41.1–41.8
7. C. Moren, Balkenius. Comput. Model Emotional Learn. Amygdala Cybern. Syst. **32**(6), 611–636 (2000)
8. J. Moren, Emotion and Learning—A Computational Model of the Amygdala. Ph.D dissertation, Lund University, Lund, Sweden (2002)
9. C. Lucas, R. Mohammadi, B.N. Araabi, Intelligent modeling and control of washing machine using LLNF modeling and modified BELBIC. Asian J. Control **8**(4), 393–400 (2006)
10. S. Jafarzadeh, R. Mirheidari, M.R.J. Motiagh, M. Barkhordari, Designing PID and BELBIC Controllers in Path Tracking Problem. Int. J. Comput. Commun. Control **3**, 343–348 (2008)
11. M.R. Mohammdi, C. Lucas, B.N. Araabi, Intelligent modeling and control of washing machine using LLNF modeling and modified BELBIC, in *International Conference on Control and Automation*, (Budapest, Hungary, 2005), pp. 812-817
12. K. Gaurav, S. Mukherjee, Design of Artificial Neural Network Controller for continually stirred tank heater, in *IECON 2012—38th Annual Conference on IEEE Industrial Electronics Society*, (Montreal, QC, 2012), pp. 2228–2231
13. C. Lucas, D. Shahmirzadi, N. Sheikholeslami, Introducing BELBIC: brain emotional learning based intelligent controller. Int J. Intell. Autom. Soft Comput. **10**(1), 11–22 (2004)
14. Z. Beheshti and S.Z.M. Hashim, A review of emotional learning and its utilization in control engineering. Int. J. Adv. Soft Comput. Appl. **2**(2) (2010)
15. G. Stephanopoulas, in *Chemical Process Control—An introduction to Theory and Practice* (Prentice-Hall of India, Upper Saddle River, 1983)
16. A.K. Goel, M.K. Sharma, Neeraj Gill, ANN controllers for industrial temperature control systems. J. Instrum. Soc. India **35**(1), 48–53 (2002)

Chapter 33
Discrete Design Approach Along with Performance Analysis by Controllability and Observability Testing for a Standard Prosthetic Arm Model

Swati Barui, Santu Gharai, Moupali Roy and Biswarup Neogi

Abstract Recently, within the scientific community, there has been consideration given to advanced prostheses to replace healthy body parts with artificial mechanisms and systems to improve function. Although no such 'enhancement prosthetics' have yet been created and demonstrated to work for healthy individuals, the morality and desirability of such technologies is debated. So in the point of view of welfare in society the Artificial Limb/Arms system is considered. The development of prosthetic control advances the research area with deterministic response analysis considering the dexterity of Prosthetic Arm Model. In this study, the standard prosthetic arm model is considered for performance analysis by controllability and observability testing. Then simulation procedure implies to build up the discrete domain transfer function. Stability analysis with all simulation aspect is required to expand the discrete domain transfer function. Jury test approach is further attended for this.

Keywords Standard prosthetic model · Controllability · Observability · Discrete domain transfer function · Jury test

---

S. Barui (✉) · S. Gharai · M. Roy
Narula Institute of Technology, Kolkata, India
e-mail: swati.engg2006@gmail.com

S. Gharai
e-mail: santu.gharai@gmail.com

M. Roy
e-mail: moupali.roy.29@gmail.com

B. Neogi
JIS College of Engineering, Kalyani, India
e-mail: biswarupneogi@gmail.com

© Springer India 2015

K. Maharatna et al. (eds.), *Computational Advancement in Communication Circuits and Systems*, Lecture Notes in Electrical Engineering 335,
DOI 10.1007/978-81-322-2274-3_33

33.1 Introduction

The "Prosthetic" word is originated from Prostheses mean anaesthetics. It is an artificial extension that replaces a missing body part [1]. Prostheses are typically used to replace parts lost by injury (traumatic) or missing from birth (congenital) or to supplement defective body parts. Prosthetic and dexterity are the related things for designing a standard arm model. Dexterity, adroitness in using the hands or body, is the prime or essential aspect in modelling dexterous hand [2]. The first practical prosthesis system was demonstrated in Hanover, Germany—the developments were slow. In twenty-first century the design and construction of the Hand prosthesis has shown great progress. Among this kind of prosthesis some are commercially available like well-known Otto Bock Sensor Hand, the Utah Arm and the I-Limb Hand from touch bionics are among the others [3]. Nowadays light and mobile arm design gets satisfactory solution for providing dexterity. The technology is based on electromyographic (EMG) signals and error control system. In the present work one reference continuous domain transfer functions corresponds to standard dexterous model is taken for controllability and observability testing and further analysis in discrete domain. Discrete domain analysis permits the reduction of characteristic roots from infinite to finite numbers and the analysis and design is consequently simplified. Firstly, optimum transfer function in discrete domain is obtained through simulation. Then Jury Test is verified for this model to prove the system stability in discrete domain.

33.2 Prosthetic Arm System

Prosthetic is the science that deals with realization of artificial limb to restore the originality of concerned for overall function of the living system. Dexterity helps fingers and hands to coordinate for completing fine tasks. Whenever some body part or limb is replaced artificially that replaced part needs to be well connected to the original connecting organ for faithful function of overall system. Hence the essential requirement of prosthetic limb is that it should be capable of processing various intelligences as obtainable [3]. A myoelectric prosthesis uses electromyographic signals or potentials from voluntarily contracted muscles within a person's residual limb on the surface of the skin to control the movements of the prosthesis, such as elbow flexion/extension, wrist supination/pronation (rotation) or hand opening/closing of the fingers and dexterities like griping and holding capabilities [4] (Fig. 33.1).

It is a simple feedback control system. An error signal is generated to the nervous system by comparing the desired position and the sensed position. The electromyographic (EMG) signal impulses are picked up by special sensor and an amplifier produce a voltage to drive the motor. The output of the motor circuit is the velocity of the limbs in one dimension [5].

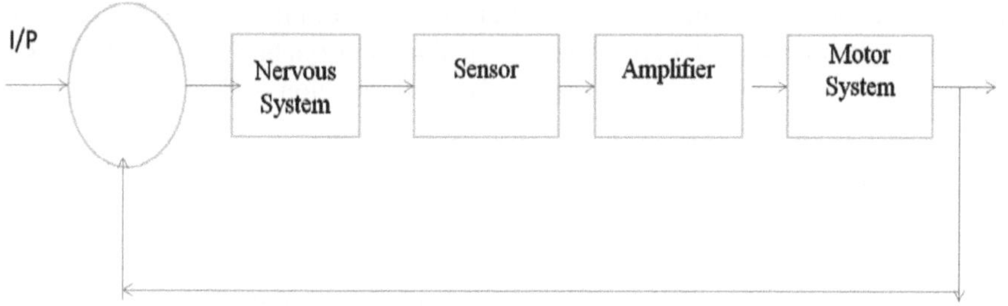

Fig. 33.1 Block diagram of prosthetic arm system [5]

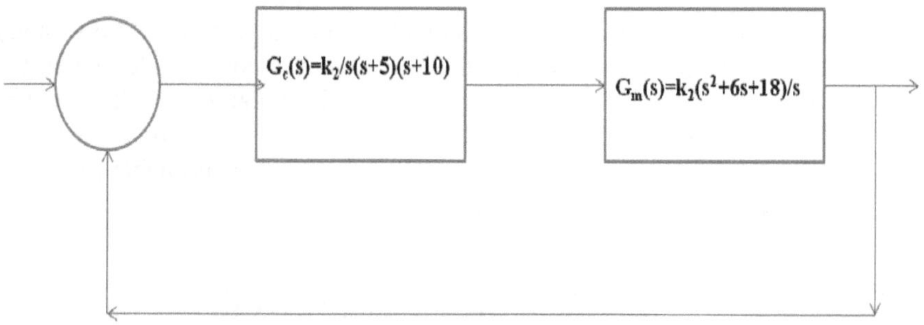

Fig. 33.2 Block diagram of the standard model [6]

33.3 Control Block Representation of the Standard Model

The standard transfer function is represented in block diagram form as shown in Fig. 33.2 [6]. To analyse the model, the control block and the transfer function of dexterous hand have been developed. Here $G_c(s)$ represents the controller and $G_m(s)$ is defined as Motor and Joint.

So, $T(s) = (s^2 + 6\,s + 18)k2/(s^4 + 15\,s^3 + (50 + k2)\,s^2 + 6\,k2\,s + 18\,k2)$ [6]

The system is stable for $k2 = 194.615$ [7].

Then, $T(s) = (194.6\,s^2 + 1168\,s + 3503)/(s^4 + 15\,s^3 + 244.6\,s^2 + 1168\,s + 3503)$

33.4 Performance Analysis with Controllability and Observability Testing

Suppose in a system initial state is $x(t_0)$ and desired state is $x(t)$, the system will be completely controllable if in a finite time ($t_0 \leq t \leq$ T) $x(t_0)$ can be transferred to $x(t)$ by an unconstrained control $u(t)$. To make a system flexible, i.e. all the closed loop system poles are placed arbitrarily, the need of completely controllable and observable system emerges. In order to determine whether the system is controllable or not, we need to deduce the state differential equation from the transfer function and hence calculating a sort of matrices we get the final controllability matrix P_c [8]. The canonical form of controllability with the dynamic equation of the system is

$$x'(t) = Ax(t) + Bu(t) \tag{33.1}$$

$$y(t) = Cx(t) + Du(t) \tag{33.2}$$

For either type of model, the output equation may be written as

$$y(t) = Cx(t);$$

where A = system matrix, B = input matrix and C = output matrix [8]. So to study the performance of this prosthetic arm that the system is controllable or not, the controllability matrix is produced. The system is defined controllable if the determinant of the controllability matrix is non-zero. The rank of the matrix n is equal to the system transfer function order, comments the system is controllable [9].

Here, continuous-time transfer function

$$G = \frac{194.6\ s^2 + 1168\ s + 3503}{s^4 + 15\ s^3 + 244.6\ s^2 + 1168\ s + 3503}$$

$$A = \begin{matrix} -15.0000 & -15.2884 & -9.1226 & -6.8419 \\ 16.0000 & 0 & 0 & 0 \\ 0 & 8.0000 & 0 & 0 \\ 0 & 0 & 4.0000 & 0 \end{matrix}$$

$$B = \begin{matrix} 4 \\ 0 \\ 0 \\ 0 \end{matrix}$$

$$C = 0 \quad 3.0409 \quad 2.2806 \quad 1.7105$$

$$D = 0$$

$$P_c = 1.0\text{e} + 04 * \begin{matrix} 0.0004 & -0.0060 & -0.0078 & 1.1183 \\ 0 & 0.0064 & -0.0960 & -0.1255 \\ 0 & 0 & 0.0512 & -0.7680 \\ 0 & 0 & 0 & 0.2048 \end{matrix}$$

Now, determinant of controllability matrix $P_c = 268435456 \neq 0$

The determinant of the P_c is non-zero as well as the rank of the system is equal to the system order so the designed system is controllable.

The observability properties of a plant have important practical consequence in analysis and, more importantly, designed of modern feedback control system [9].

If $x(0)$ be the initial state and $u(t)$ be the given control, then the system will be completely observable if $x(0)$ can be determined from observation history $y(t)$ for a finite time T. Observability is the ability of state variable estimation. The determinant of observability matrix is non-zero for maintaining the observability of the system. If the rank of the observability matrix P_O is equal to the order of the system transfer function then the system is observable [9].

$$P_O = 1.0\text{e}+03 * \begin{matrix} 0 & 0.0030 & 0.0023 & 0.0017 \\ 0.0487 & 0.0182 & 0.0068 & 0 \\ -0.4379 & -0.6891 & -0.4438 & -0.3329 \\ -4.4574 & 3.1438 & 2.6631 & 2.9960 \end{matrix}$$

Now, determinant of observability matrix $P_O = 1265488.217 \neq 0$

The determinant of the observable matrix is non-zero. The generated rank is equal to the order of the system. Hence the system is observable.

33.5 Analysis in Discrete Domain

The discrete domain is comparatively more effective than continuous domain for designing aspect. Here, as the model is analysed in discrete domain and the optimum case for the transfer functions from continuous domain to discrete domain are being developed through simulation using MATLAB.

$$T(z) = \frac{0.07919\,z^3 - 0.07182\,z^2 - 0.06555\,z + 0.06042}{z^4 - 3.45\,z^3 + 4.564\,z^2 - 2.75\,z + 0.6376}$$

Respective step response of the transfer function is shown in Fig. 33.3. Optimized results are found for sampling time 0.03 for this case.

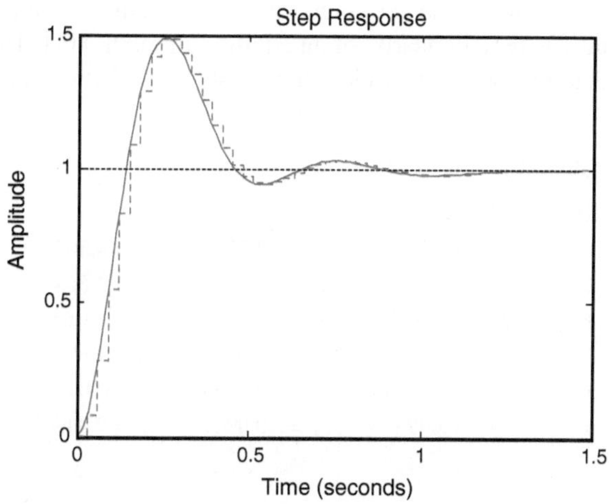

Fig. 33.3 Step response in z domain for the model

33.6 Stability Analysis by Jury Test

In case of discrete time system the Jury stability criterion is a method to determine the stability by analysing the coefficients of its characteristic polynomial. It has special application for stability analysis of biological system. Here, an algorithm is developed through jury test to verify the stability of the optimized discrete domain transfer function [10] (Table 33.1).

$$\text{Here, } T(z) = \frac{0.07919\, z^3 - 0.07182\, z^2 - 0.06555\, z + 0.06042}{z^4 - 3.45\, z^3 + 4.564\, z^2 - 2.75\, z + 0.6376}$$

So, the Characteristic polynomial $F(z) = z^4 - 3.45\, z^3 + 4.564\, z^2 - 2.75\, z + 0.6376$

$$F(1) = 0.0016[F(1) > 0 \text{ satisfied}]$$
$$(-1)4F(-1) = 12.4016[(-1)4F(-1) > 0 \text{ satisfied}]$$

where,

$$b_k = \begin{vmatrix} a_0 & a_{n-k} \\ a_n & a_k \end{vmatrix}, \ c_k = \begin{vmatrix} b_0 & b_{n-1-k} \\ b_{n-1} & b_k \end{vmatrix}$$

From the Jury format table of the standard dexterous arm transfer function, the satisfactory conditions for stability are found as

$$\begin{array}{ll} |a_0| < |a_5|, & \text{Satisfied} \\ |b_0| > |b_4|, & \text{Satisfied} \\ |c_0| > |c_3|, & \text{Satisfied} \end{array}$$

So the transfer function is proper as it proves the system stability. A stable system holds the set point without oscillating around it and all nearby initial conditions remain nearby of an equilibrium point [11]. This analysis will help for non-deterministic test of biological system and better modelling of the arm.

Table 33.1 Table of Jury test

Row	z^0	z^1	z^2	z^3	z^4
1	$a_0 = 0.6376$	$a_1 = -2.75$	$a_2 = 4.564$	$a_3 = -3.45$	$a_4 = 1$
2	$a_4 = 1$	$a_3 = -3.45$	$a_2 = 4.564$	$a_1 = -2.75$	$a_0 = 0.6376$
3	$b_0 = -0.5935$	$b_1 = 1.697$	$b_2 = -1.654$	$b_3 = 0.5503$	
4	$b_3 = 0.5503$	$b_2 = -1.654$	$b_1 = 1.697$	$b_0 = -0.5935$	
5	$c_0 = 0.04941$	$c_1 = -0.0969$	$c_2 = 0.0478$		
6	$c_2 = 0.0478$	$c_1 = -0.0969$	$c_0 = 0.04941$		

33.7 Conclusion

This work implies the performance analysis of a standard prosthetic arm control model with the support of control technology. The need of completely controllable and observable system appears considering the flexibility of the system. The focus is also made for discrete domain analysis for the optimum case. Jury stability criteria are applied for this case and the system stability is proved. Discrete domain stability is required to establish the sampled-data control representation for further analysis by considering the non-linear aspect. The work will build a shape for the welfare of society with the development aspect of handicap human beings.

References

1. R.T. Stefain, C.J. Savant, B. Shahian Jr., G.H. Hostetter, *Design of Feedback Control System* (Saunders College Publishing, Orlando,1994), pp. 259–263
2. D. Datta, N. Brain, Clinical application of myo electrically-controlled prostheses. Crit. Rev. Phys. Rehabil. Med. **4**(3–4), 215–239 (1992)
3. B. Neogi, S. Ghosal, R. Darbar, S. Ganguly, B. Brahma, Approach towards the characteristic interpretation of Electromyography (EMG) signal with clinical study, in *Recent Advances in Intelligent Computational Systems (RAICS)*. IEEE (Sept. 2011), pp. 788–791
4. R.N. Scott, P.A. Parker, Myoelectric prostheses: state of the art. J. Med. Eng. Technol. **12**(4), 143–151 (1988)
5. B. Neogi, S. Mandal, S. Banerjee, A. Das, D.N. Tibarewala, Design of prosthetic arms control transfer function with the analysis in digital domain introducing simulation aspect. Int. J. Eng. Res. Ind. Appl. **3**(I), 260–274 (2010). ISSN 0974-1518
6. S. Banerjee, S. Bhattacharjee, A. Nag, S. Bhattacharyya, B. Neogi, Discrete domain analysis of dexterous hand model by simulation aspect. C3IT-2012, Proc. Technol. **4**, 878–882 (2012). 2212-0173 © 2012 Published by Elsevier Ltd. doi: 10.1016/j.protcy.2012.05.144
7. S. Barui, P. Biswas, S. Gharai, Comparative response analysis through optimized transfer function generation for two standard dexterous arm model. Int. J. Adv. Res. Electr. Electron. Instrum. Eng. **3**(3), 2320–3765, 2278–8875 (2014)
8. M.N. Bandyopadhyay, *Control engineering theory and practice*. (Prentice-Hall of India Private Limited, New Delhi, 2003), p. 112
9. J.J. Distefano, A.R. Stubberud, I.J. Williams, *Theory and Problem of Feedback and Control System*. (Tata McGraw-Hill, New Delhi, 2003), p. 480
10. E.I. Jury, *Theory and Application of z-Transformation Method* (Willey, New York, 1946)
11. F. Szidarovszky, A.T. Bahill, S. Molnar, On stable adaptive control systems. Pure Math. Appl. **1B**(2–3), 115–121 (1990)

Part VI
Advances in Devices and Circuit

Chapter 34
Material Composition Dependence on Optimization of Small-Signal Properties of Si_xGe_{1-x} DDR IMPATT Diode

Arpan Deyasi and Swapan Bhattacharyya

Abstract Small-signal properties of Si_xGe_{1-x} DDR IMPATT diode is numerically computed using double iterative method by simultaneously solving Poisson equation, carrier diffusion equation and continuity equation in addition with mobile space charge effect for different material compositions. Conductance–susceptance profile is calculated at 94 GHz for two different values of x, and impedance is also determined as a function of operating frequency. Dependence of DC-to-RF conversion efficiency is calculated on Si mole fraction in the device. Results are important in designing the device for millimetrewave communication systems.

Keywords Conductance · Susceptance · Negative reactance · Operating frequency · Conversion efficiency

34.1 Introduction

This age of satellite communication demands high power source in microwave and millimetrewave frequency range, where transit time solid state devices become a natural candidate due to their possibility of operation in THz zone [1] with suitable window frequency, greater flexibility and reliability, improved performance, reduced size, cost and power requirements. Among the transit time family, IMPATT

A. Deyasi (✉)
Department of Electronics and Communication Engineering,
RCC Institute of Information Technology, Kolkata 700015, India
e-mail: deyasi_arpan@yahoo.co.in

S. Bhattacharyya
Department of Electronics and Communication Engineering,
JIS College of Engineering, Kalyani, India
e-mail: swapanbhattacharyya@ieee.org

© Springer India 2015

K. Maharatna et al. (eds.), *Computational Advancement in Communication Circuits and Systems*, Lecture Notes in Electrical Engineering 335,
DOI 10.1007/978-81-322-2274-3_34

is one of the natural choices because of capability of generating high RF power both in CW and pulse mode. It is a p-n junction diode, worked at reverse bias condition in avalanche breakdown region, which acquire high-frequency negative resistance due to an appropriate phase lag between external current and RF voltage applied across the diode; generated due to the combined effect of avalanche build-up delay and transit time delay [2]. DDR diodes provide higher conversion efficiency than SDR device, and different materials [3–5] are already used to fabricate DDR devices.

DC and RF properties of DDR IMPATT structure are investigated by several researchers in the last decade. Effect of optical communication on electric field and admittance of GaN and InP diodes are studied by Banerjee and Banerjee [6]. Electric field, breakdown voltage and power for DDR IMPATT is also computed for different compound semiconductors [7]. Effect of junction temperature is also analysed [5] recently for DDR structure. GaN SDR diode is especially investigated [8] in different frequency bands. Series resistance is also calculated for DDR IMPATT [9]. SiGe-based heterostructure devices are fabricated much earlier [10], and near optimum operating frequency is also calculated [11]. It is also suggested the heterojunction IMPATT exhibits less phase noise [12]. Novel materials [13] are also suggested to enhance power output and simultaneously reducing noise performance.

In present paper, susceptance–conductance profile along with negative resistance and reactance are computed for SiGe DDR structure, which is designed at 94 GHz. Conversion efficiency is also calculated as a function of material composition. Results are important for millimetrewave performance of the device.

34.2 Mathematical Modelling

Theoretical investigation of DDR IMPATT structure begins with double iterative technique where computation is initialized from field maximum near the metallurgical junction. DC field and current profiles for the diode that involves simultaneous solution of Poisson's equation, current density equations and continuity equations incorporating mobile space charge effecting the depletion layer; where iteration is carried out over magnitude of field maximum and its location in the depletion layer. Potential distribution across the junction of a p^+pnn^+ diode is given by the well-known Poisson's equation

$$\frac{\partial}{\partial x}E(x) = \frac{q}{\varepsilon}[N_D - N_A + p(x) - n(x)] \tag{34.1}$$

Continuity equations for electron and hole current are given by

$$\frac{\partial}{\partial t}(n,p) = \frac{1}{q}\frac{\partial J_{n,p}}{\partial x} + g - U_{n,p} \tag{34.2}$$

If the diffusion component of current caused by carrier concentration gradient in the space charge layer is considered, then expressions of J_n and J_p are given by-

$$J_{n,p} = q(n,p)v_{n,p} + qD_{n,p}\frac{\partial}{\partial x}(n,p) \tag{34.3}$$

Considering contributions of both diffusion current and drift current, former has been considered as a perturbation term over the major drift current in the avalanche region. Using perturbation technique, the expressions for hole and electron concentration in the space charge layer (considering both diffusion and drift) can be written as-

$$\begin{aligned}(n,p)(x) = &\left(\frac{J_{n,p}}{qv_{n,p}}\right) - \left(\frac{D_{n,p}}{qv_{n,p}^2}\right)\left(\frac{\partial J_{n,p}}{\partial x}\right) \\ &+ \left(\frac{D_{n,p}^2}{qv_{n,p}^3}\right)\left(\frac{\partial^2 J_{n,p}}{\partial x^2}\right) - \left(\frac{D_{n,p}^3}{qv_{n,p}^4}\right)\left(\frac{\partial^3 J_{n,p}}{\partial x^3}\right) + \cdots\end{aligned} \tag{34.4}$$

Hence the mobile space charge density due to both drift and diffusion components is obtained for the equation-

$$\rho(x) = \left[\left(\frac{J_p}{v_p}\right) - \left(\frac{J_n}{v_n}\right) + \left\{\left(\frac{D_p}{v_p^2}\right)\left(\frac{\partial J_p}{\partial x}\right)\right\} + \cdots \tag{34.5}$$

Boundary conditions for the electric field in the depletion region can be written as

$$E(-x_1) = E(x_2) = 0 \tag{34.6}$$

where $-x_1$ and x_2 define the edges of p and n layers, respectively. The boundary conditions for normalized current densities are given by

$$k(-x_1) = {}^{2}\!/\!_{M_p} - 1 \tag{34.7a}$$

And

$$k(x_2) = 1 - {}^{2}\!/\!_{M_n} \tag{34.7b}$$

where $M_{n,p} = {}^{J_0}\!/\!_{J_{ns.ps}}$ are computed considering $|J_p - J_n| = 0.95J_0$, extension of avalanche zone has been defined as the distance between two points (x_1, x_2) on either side of avalanche centre. Computation starts from the position of field maximum (x_0) near metallurgical junction and then proceeds towards edges of depletion layer. Double iterations are then carried out over field magnitude (E_0) and position of field maximum (x_0) in depletion region to satisfy boundary conditions for field and current densities. Simultaneous satisfaction of boundary conditions at depletion layer edges gives appropriate solution and DC field and current profiles are obtained.

Diode negative resistance and negative reactance can be computed using modified Runge–Kutta method. It is obtained by performing integration over the entire active space charge region as

$$-Z_{R(/X)} = \int_{x_1}^{x_2} -R(/X)\mathrm{d}x \tag{34.8}$$

Extension of avalanche zone has been defined as the distance between two points (x_1, x_2) on either side of avalanche centre. Once the values of maximum electric field (E_m) and its position (x_0) for which both the boundary conditions are simultaneously satisfied are computed at a given current density, this computer program provides more accurate profiles of electric field and carrier current.

The diode admittance is expressed in the form

$$Y = -G + jB = \frac{1}{-Z_R + jZ_X} \tag{34.9}$$

where

$$G = \frac{Z_R}{Z_R^2 + Z_X^2} \tag{34.10a}$$

$$B = \frac{-Z_X}{Z_R^2 + Z_X^2} \tag{34.10b}$$

It is assumed that both $-G$ and B are normalized with the diode area.

34.3 Results and Discussion

Two different IMPATT devices are considered for simulation with appropriate doping profiles involving exponential and error functions. Figure 34.1a shows the *G–B* plot for $Si_{0.9}Ge_{0.1}$ device, where highest value of negative conductance is achieved at 92 GHz. Further increasing operating frequency increases the susceptance with decreasing negative conductance. Its quality factor Q_P is 3.07. By increasing the percentage of Ge, we see a shift in peak of the *G–B* plot towards higher operating frequency. This is shown in Fig. 34.1b. From the graph, it can be seen that the peak occurs at 102 GHz for $Si_{0.7}Ge_{0.3}$ DDR device. Quality factor for the device is now 0.65. Thus a decrease of quality factor speaks for the increase in stability.

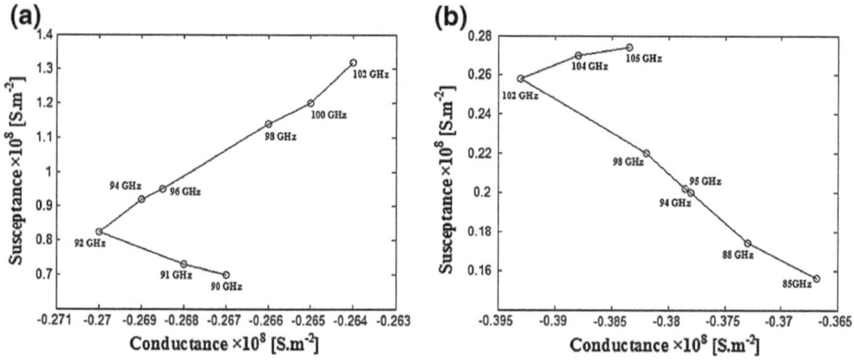

Fig. 34.1 **a** Negative conductance–susceptance plot for $Si_{0.9}Ge_{0.1}$ DDR IMPATT. **b** Negative conductance–susceptance plot for $Si_{0.7}Ge_{0.3}$ DDR IMPATT

Figure 34.2 shows the negative resistance and negative reactance profiles with operating frequency of the device. In Fig. 34.2a, parameters are plotted for $x = 0.1$, whereas in Fig. 34.2b, they are plotted for $x = 0.3$. From both the plots, it is observed that resistance and reactance increase with frequency. The rate of increment is quite higher for resistance than the reactance. One noticeable feature can be observed from the graphs that the magnitude of negative impedance is higher for lower percentage of Ge. Hence increase of Ge mole fraction not only stabilizes the device, but also increases the negative impedance.

DC-to-RF conversion efficiency is computed and plotted with Ge mole fraction to estimate the device performance. Result is shown in Fig. 34.3. From the plot, it can be seen that with increase of Ge percentage, conversion efficiency linearly increases. For $Si_{0.9}Ge_{0.1}$ DDR structure, efficiency is 5.5 %, whereas it becomes 8.4 for $Si_{0.7}Ge_{0.3}$ device. Hence for higher stable device, efficiency is quite moderate, and higher than SDR diode.

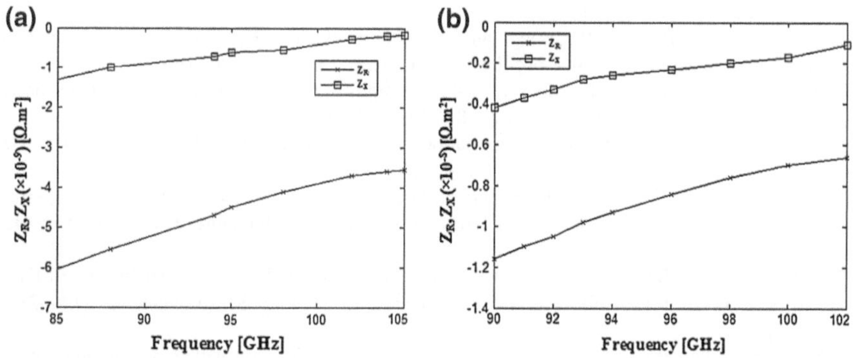

Fig. 34.2 **a** Negative resistance and negative reactance with operating frequency for $Si_{0.9}Ge_{0.1}$ DDR IMPATT. **b** Negative resistance and negative reactance with operating frequency for $Si_{0.7}Ge_{0.3}$ DDR IMPATT

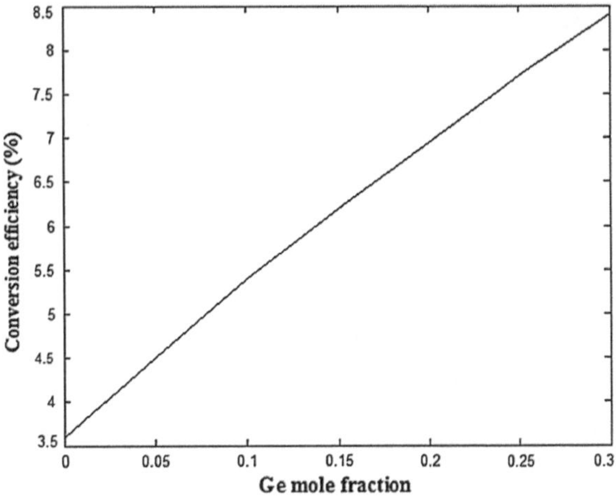

Fig. 34.3 DC-to-RF conversion efficiency with Ge mole fraction for $Si_{1-x}Ge_x$ DDR IMPATT

34.4 Conclusion

The Comparative study between two different DDR diodes show that peak of G–B plot appears at higher frequency for $Si_{0.7}Ge_{0.3}$, and quality factor also becomes lower than that obtained in $Si_{0.9}Ge_{0.1}$ diode. Conversion efficiency is also higher for the former, and is quite moderate. Simulated result suggests that for better efficient device with higher stability and higher frequency of operation, mole fraction of Ge should be higher up to an optimum value.

References

1. M. Mukherjee, S.K. Roy, Wide band gap III–V transit time diode in terahertz regime: studies on the effects of punch through on high frequency characteristics and series resistance of the device. Curr. Appl. Phys. **10**, 646–651 (2010)
2. C.A. Lee, R.L. Batdrof, W. Wiegmann, G. Kamnisky, The Read diode—an avalanching transit time negative resistance oscillator. Appl. Phys. Lett. **6**, 89–90 (1965)
3. A.K. Panda, D. Pavlidis, E. Alekseev, DC and high-frequency characteristics of GaN based IMPATT. IEEE Trans. Electron. Dev. **48**, 820823 (2001)
4. M. Mukherjee, Effects of optical illumination on 4H-SiC DDR IMPATT diodes at 140 GHz. CODEC-06, Kolkata, India, (Dec 2006), pp. 298–301
5. A. Acharyya, S. Banerjee, J.P. Banerjee, Dependence of dc and small-signal properties of double drift region silicon IMPATT device on junction temperature. J. Electron. Dev. **12**, 725–729 (2012)
6. S. Banerjee, J.P. Banerjee, Studies on optical modulation of III–V GaN and InP based DDR IMPATT diode at sub-millimeter wave frequency. Int. J. Eng. Sci. Technol. **2**, 2790–2801 (2010)
7. S. Banerjee, Dynamic characteristics of IMPATT diodes based on wide bandgap and narrow bandgap semiconductors at W-band. Int. J. Eng. Sci. Technol. **3**, 2143–2153 (2011)

8. B. Chakrabarti, D. Ghosh, M. Mitra, High frequency performance of GaN based IMPATT diodes. Int. J. Eng. Sci. Technol. **3**, 6153–6159 (2011)
9. A. Acharyya, J.P. Banerjee, Numerical modeling of series resistance of millimeter-wave DDR IMPATTs. Int. J. Electron. Electr. Eng. **2**, 9–18 (2012)
10. E.J. Brailovskii et al., Si_xGe_{1-x}GaAs heterojunction IMPATT-diodes. Cryst. Res. Technol. **21**, 413–421 (1986)
11. P.R. Tripathy, M. Mukherjee, S.P. Pati, Possible realization of near optimum efficiency from n-Si-Ge/p-Ge-Si DDR heterostructure IMPATT diode. Int. J. Mater. Eng. **2**, 17–22 (2012)
12. M.J. Bailey, Heterojunction IMPATT diodes. IEEE Trans. Electron. Dev. **39**, 1829–1834 (1992)
13. S.R. Pattanaik, J.K. Mishra, G.N. Dash, A new mm-wave GaAs ~ Ga0.52In0.48P heterojunction IMPATT diode. IETE J. Res. **57**, 351–356 (2011)

Chapter 35
Analytical Computation of Absorption Coefficient for Intersubband Transition in MQW Structure

Pratyush Kundu, Prameet Ghosh and Arpan Deyasi

Abstract Absorption coefficient of the multiple quantum well structure is analytically calculated for intersubband transition between the lowest three energy states in presence of an electric field applied along the direction of quantum confinement. Transition energies are obtained by solving time-independent Schrödinger equation subject to the appropriate boundary conditions, and corresponding absorbance is computed considering Lorentzian lineshape function. Suitable tuning of structural parameters along with axial electric field (applied along the direction of quantization) makes blueshift/redshift of absorption peaks. Results are computed considering the first-order band nonparabolicity of Kane-type, and are compared with that obtained for parabolic overestimation. Corresponding oscillator strength is also computed. Results are important for designing the MQW-based optical detector.

Keywords Absorption coefficient · Oscillator strength · Multiple quantum well · Intersubband transition · Lorentzian lineshape

35.1 Introduction

Research on emerging nanophotonic devices has gained momentum in the last decade due to the possibility of exploiting their unique features in diverse applications [1–3]. Thanks to the rapid progress in material science and microelectronics technology, theoretical and experimental investigations on semiconductor

P. Kundu (✉) · P. Ghosh · A. Deyasi
Department of Electronics and Communication Engineering,
RCC Institute of Information Technology, Kolkata 700015, India
e-mail: pkundu.001@gmail.com

P. Ghosh
e-mail: pramitg1@gmail.com

A. Deyasi
e-mail: deyasi_arpan@yahoo.co.in

© Springer India 2015

K. Maharatna et al. (eds.), *Computational Advancement in Communication Circuits and Systems*, Lecture Notes in Electrical Engineering 335,
DOI 10.1007/978-81-322-2274-3_35

nanostructures lead to the possible realization of novel electronic [4] and photonic [5] devices. Quantization of energy states in low-dimensional semiconductors along different directions leads the formation of quantum well, wire or dot, which are characteristically different from their bulk counterparts. These structures can be used as infrared detectors [6, 7], which has immense applications in medical [8], defence [9] and communication [10] fields. Characteristics of these IR devices may be studied from absorption coefficient profile, and thus its accurate determination becomes very important from the device design point of view.

In the present paper, absorption coefficient of MQW structure is computed for the intersubband transition between ground states and lowest two excited states in the presence and absence of bias for different well width. For realistic result, first-order band nonparabolicity is taken into account, and Lorentzian lineshape function is considered. Results are important for designing the IR detector.

35.2 Mathematical Modelling

Kindly Optical absorption coefficient for quantum well can be written as

$$\alpha(\omega) = \frac{q^2 m^*}{2\varepsilon_0 \varepsilon_r c n_r \hbar L m_0{}^2} \frac{P^2}{\hbar\omega} \sum_{n,m} \langle g_v^m \mid g_c^n \rangle \Theta(E_{mn} - \hbar\omega), \tag{35.1}$$

where $\langle g_v^m \mid g_c^n \rangle$ is the overlap integral between z-dependent envelope functions of conduction band and valence band, L is the width of quantum well, Θ is the Heavyside step function, n_r is the refractive index of the well material and 'P' is defined as

$$P = N \int_{\text{cell}} \psi_{k_c}^*(\vec{r}\,') p_A \psi_{k_v}(\vec{r}\,') d^3 r' \tag{35.2}$$

The overlap integral defined in Eq. (35.1) provides the selection rule for transition. It may be written as

$$\langle g_v^m \mid g_c^n \rangle = \delta_{k,k'} \int_{-L/2}^{L/2} \chi_m^h(z) \chi_n^e(z) \mathrm{d}z \tag{35.3}$$

Substituting the values for wavefunction for quantum well, Eq. (35.1) may be modified as

$$\alpha(\omega) = \frac{2\pi q^2 \hbar}{\varepsilon_0 \varepsilon_r c n_r m^{*2} V} \frac{1}{\hbar\omega} \sum_{i,j} \left| \langle i | p_z | j \rangle^2 \right| \delta(E_j - E_i - \hbar\omega)(f_{\text{FD}}^i - f_{\text{FD}}^j), \tag{35.4}$$

where '*i*' and '*j*' are the initial and final states, and f_{FD} is the Fermi-Dirac distribution function. The factor 2 is added for spin degeneracy, and the δ function is introduced to conserve the momentum.

For two consecutive states, Eq. (35.4) may be written as

$$\alpha(\omega) = \frac{2\pi q^2 \hbar}{\varepsilon_0 \varepsilon_r c n_r m^{*2} V} \frac{1}{\hbar\omega} \left[\frac{8\hbar}{3L}\right]^2 \sum_{i,j} \delta(E_j - E_i - \hbar\omega)(f^i_{FD} - f^j_{FD}) \tag{35.5}$$

If the quantum well is doped and Fermi energy level is above the ground state and first excited state is completely empty, then

$$\alpha(\omega) = \frac{n_s \pi q^2 \hbar}{\varepsilon_0 \varepsilon_r c n_r m^{*2} L} \frac{1}{\hbar\omega} \left[\frac{8\hbar}{3L}\right]^2 \delta(\Delta E - \hbar\omega), \tag{35.6}$$

where n_s is sheet charge density. Defining oscillator strength between 1st state to 2nd state as

$$f_{21} = \frac{2}{m^* \hbar\omega} \left[\frac{8\hbar}{3L}\right]^2 \tag{35.7}$$

Equation (35.6) can be put into the form

$$\alpha(\omega) = \frac{n_s \pi q^2 \hbar}{2\varepsilon_0 \varepsilon_r c n_r m^* L} f_{21} \delta(\Delta E - \hbar\omega) \tag{35.8}$$

Experimental work shows the broadening of peaks can be theoretically obtained by introducing Lorentzian lineshape function replacing the ideal delta function. Hence final expression of absorption coefficient may be written as

$$\alpha(\omega) = \frac{n_s \pi q^2 \hbar}{2\varepsilon_0 \varepsilon_r c n_r m^* L} f_{21} \frac{\Gamma}{\pi\left[(\hbar\omega - \Delta E)^2 + \Gamma^2\right]}, \tag{35.9}$$

where Γ is half-width at half of the maximum.

35.3 Results and Discussion

Using Eq. (35.9), absorption coefficient is calculated as a function of operating wavelength for different well width, as shown in Fig. 35.1. For lower well dimension, it is found that the coefficient is less compared to that obtained for higher well width. This is due to the fact that with decrease of quantum well length, confinement increases, which makes the increase of energy difference between the bands, i.e. intersubband transition energy increases. This reduces the peak of

Fig. 35.1 Absorption coefficient with wavelength for different well dimensions in the absence of electric field

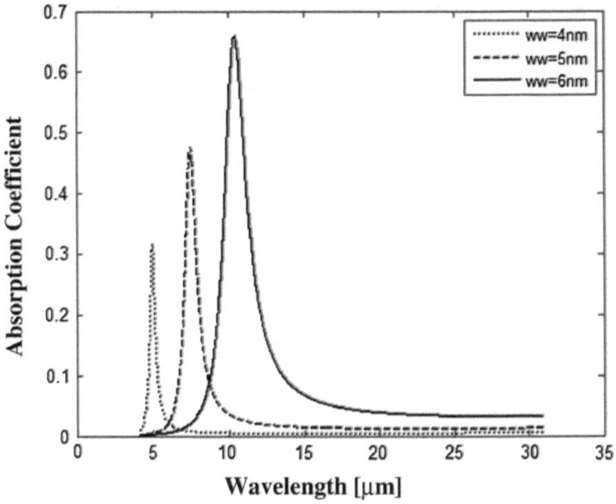

absorbance amplitude, provided half-width at half-maximum is kept constant throughout the simulation. With the increase of well dimension, redshift is observed for all the profiles corresponding to the transition energies. Peak value of the absorption coefficients also increases for all the cases, which can be verified from the plot. Application of electric field along the direction of confinement reduces the eigenenergies, and also the difference between them. Thus the decrease of inter-subband transition makes a redshift of the absorption peak for a precise geometry. This is plotted in Fig. 35.2. Also the spacing between the peaks increases though the magnitude of peak remains invariant.

Figure 35.3 shows the absorption coefficient for nonparabolic band structure and is compared with overestimated parabolic assumption. Absorption is plotted as a function of operating wavelength, and the plot suggests that a redshift appears when

Fig. 35.2 Absorption coefficient with wavelength for different well dimensions in presence of electric field

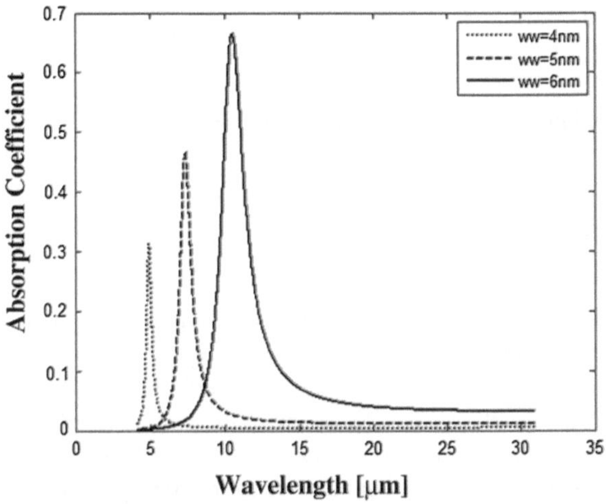

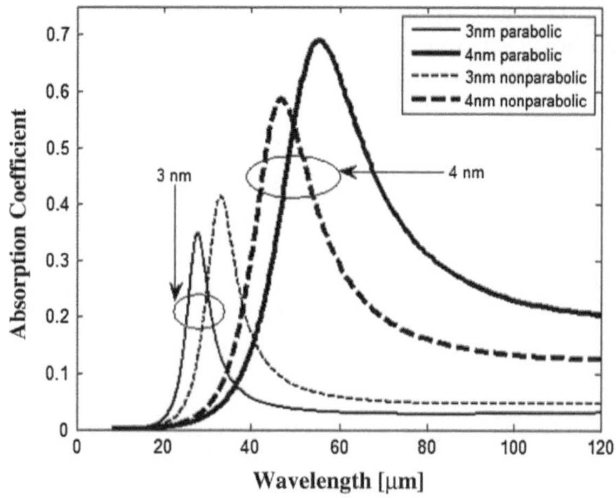

Fig. 35.3 Absorption coefficient with wavelength for different well widths considering both parabolic and nonparabolic dispersion relations

nonparabolicity is taken into account. This can be explained as follows: Band nonparabolicity consideration reduces the eigenenergy than that obtained for parabolic structure, and hence the separation between two consecutive states also reduces. This lowering of subband energy makes a shift of the peak of absorption coefficient profile towards the higher wavelength, and magnitude of the peak also increases. For two different well thicknesses, result is plotted.

Keeping the layer dimensions and material compositions constant, absorption coefficient is computed and plotted for first three intersubband transitions considering the first three eigenstates. Figures 35.4 and 35.5 show the absorption profiles for first three possible transitions when computed in the absence and presence of axial electric field, respectively. From the plots it may be noted that ΔE_{21} appears at the extreme right, whereas ΔE_{31} appears at the left with ΔE_{32} as intermediate. This

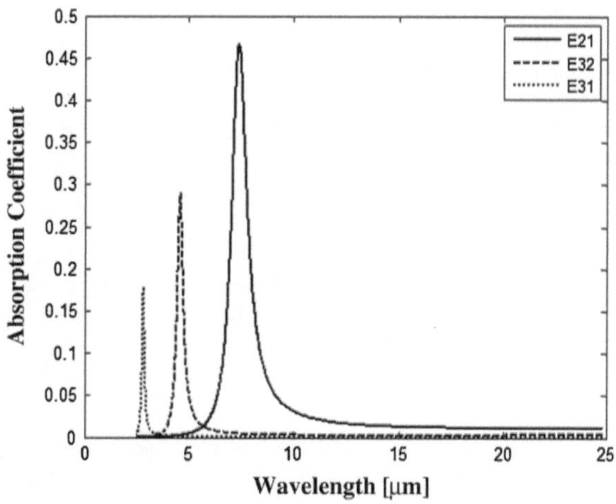

Fig. 35.4 Absorption coefficient with wavelength for first three intersubband transitions with 5 nm well width in the absence of electric field

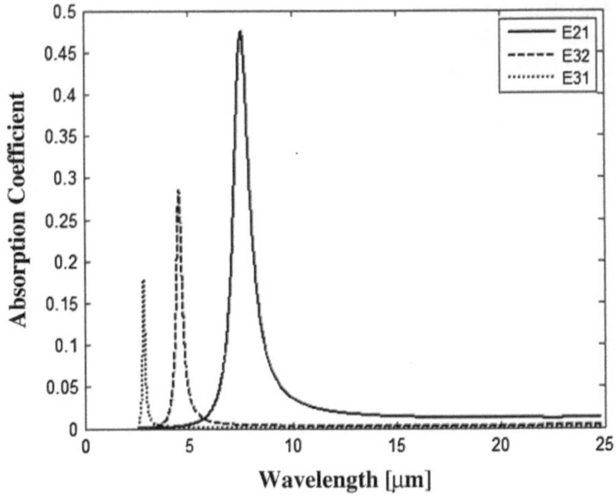

Fig. 35.5 Absorption coefficient with wavelength for first three intersubband transitions with 5 nm well width in the presence of electric field

is because ΔE_{21} has the lowest magnitude and ΔE_{31} attains highest value. With increasing the value of intersubband transition energy, magnitude of absorption peak also decreases.

Figure 35.6 shows the absorption coefficient profile with wavelength for different material compositions in a barrier layer. It is seen from the plot that with a gradual increment in InAs percentage in barrier region (barrier material is $In_{0.47}Ga_{0.53}As$), redshift is observed. The nonparabolic dispersion relation is also taken into consideration for this simulation. It is also observed that the peak value of absorption coefficient is increased with the increment to material composition value.

The result obtained for Fig. 35.6, is compared with the data obtained for parabolic dispersion relation, and comparative study is plotted in Fig. 35.7. For parabolic band structure, due to higher eigenenergies, and thus subband transition energies,

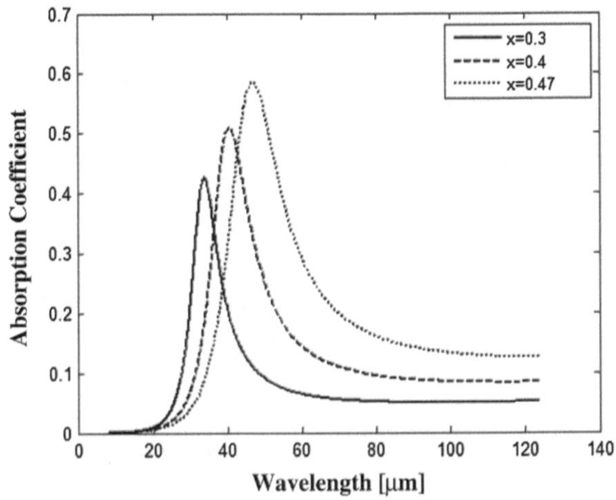

Fig. 35.6 Absorption coefficient with wavelength for different material compositions in barrier layer considering the nonparabolic dispersion relation

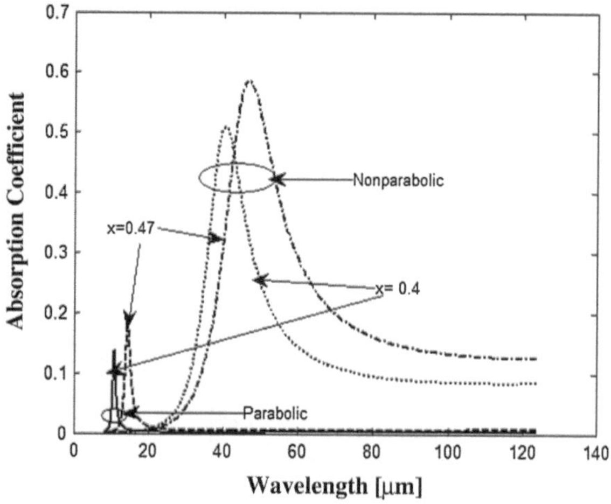

Fig. 35.7 Absorption coefficient with wavelength for different material compositions in barrier layer for both parabolic and nonparabolic dispersion relations

absorption profiles shows a blueshift w.r.t their nonparabolic profiles, and magnitude of the peaks decreases. Also for higher value of x, the shift in absorption value is considerable. Figure 35.8 shows the tuning of absorption coefficient by application of external electric field in axial direction. With increase of bias, eigenenergy decreases, this reduces the energy separation between the states. Hence intraband transition energy also decreases. This makes a redshift of the absorption profiles, as shown in the plot.

Using Eq. (35.7), oscillator strength is calculated as a function of wavelength. For different well dimensions, oscillator strength is calculated varying the wavelength accordingly. With the increase of wavelength, it is observed that the oscillator strength values are decreased. This is depicted in Fig. 35.9. Also for a given value of wavelength, oscillator strength decreases with the increasing well thickness.

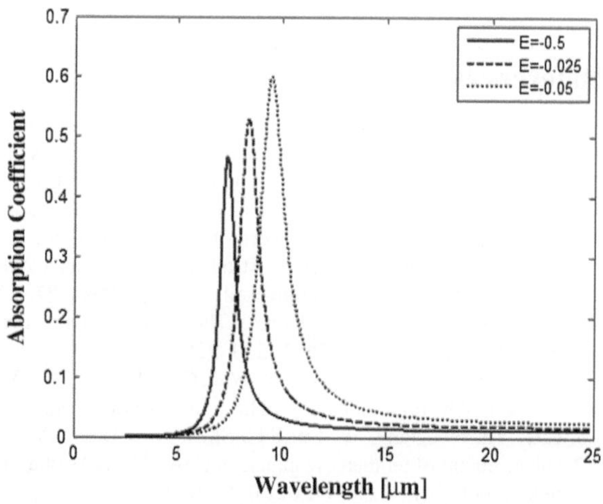

Fig. 35.8 Absorption coefficient with wavelength for different values of electric field

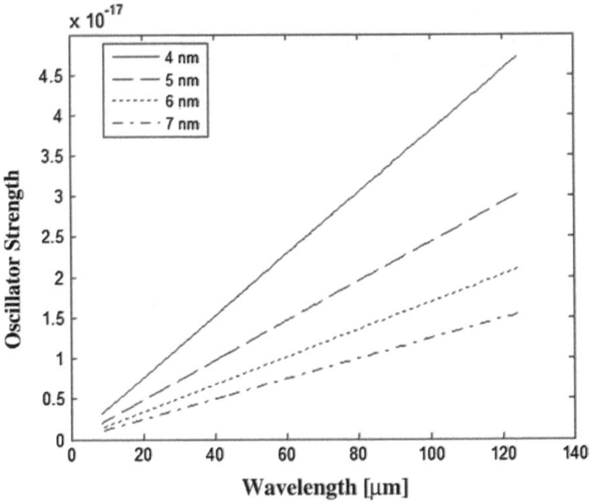

Fig. 35.9 Oscillator Strength with wavelength for different well dimensions

35.4 Conclusion

Absorption coefficient has been calculated and plotted as a function of wavelength for intersubband transition between lowest three eigenstates of multiple quantum well structure. Lorentzian lineshape function is considered to obtain the realistic result. Kane-type first-order band nonparabolicity has been taken into account, and results are compared with that obtained for the parabolic overestimation. With different well dimensions and material composition of barrier layer, absorbance and oscillator strength are computed. Effect on absorption spectra by tuning the electric field is also analysed, along with the zero bias effect. Redshift/blueshift is observed by a suitable variation of structural parameters and applied bias. Results are important in designing the quantum well-based optical detector in IR range.

References

1. T.K. Parashar, R.K. Pal, Modeling of GaAs/$Al_{0.2}Ga_{0.8}As$ quantum well gas detector for LWIR region. MIT Int. J. Electron. Commun. Eng. **1**, 97–100 (2011)
2. S. Fara, P. Sterian, L. Fara, M. Iancu, A. Sterian, New results in optical modeling of quantum well solar cells. Int. J. Photoenergy 810801 (2012). doi:10.1155/2012/810801
3. I. Friel, C. Thomidis, T.D. Moustakasa, Ultraviolet electroabsorption modulator based on AlGaN/GaN multiple quantum wells. J. Appl. Phys. **97**, 123515 (2005)
4. D. Urban, M. Braun, J. König, Theory of a magnetically controlled quantum-dot spin transistor. Phys. Rev. B **76**, 125306 (2007)
5. F. Qian, S. Gradecak, Y. Li, C.Y. Wen, C.M. Lieber, Core/Multishell nanowire heterostructures as multicolor, high-efficiency light-emitting diodes. Nano Lett. **5**, 2287–2291 (2005)
6. C.H. Yu, B. Zhang, W. Lu, S.C. Shen, H.C. Liu, Y.Y. Fang, J.N. Dai, C.Q. Chen, Strong enhancement of terahertz response in GaAs/AlGaAs quantum well photodetector by magnetic field. Appl. Phys. Lett. **97**, 022102 (2010)

7. S. Kalchmair, H. Detz, G.D. Cole, A.M. Andrews, P. Klang, M. Nobile, R. Gansch, C. Ostermaier, W. Schrenk, G. Strasser, Photonic crystal slab quantum well infrared photodetector. Appl. Phys. Lett. **98**, 011105 (2011)
8. A.G.U. Perera, Quantum structures for multiband photon detection. Opto-Electron. Rev. **14**, 103–112 (2006)
9. M.C. Phillips, M.S. Taubman, B.E. Bernacki et al., Design and performance of a sensor system for detection of multiple chemicals using an external cavity quantum cascade laser. Proc. SPIE **7608**, 76080D (2010)
10. R. Wei, N. Deng, M. Wang, S. Zhang, P. Chen, L. Liu, J. Zhang, Study of self-assembled Ge quantum dot infrared photodetectors, in *1st IEEE International Conference on Nano/Micro Engineered and Molecular Systems* (2006), pp. 330–333

Chapter 36
Analytical Computation of Band Structure of 1D Photonic Crystal Under Normal Incidence of Electromagnetic Wave

Arpan Deyasi, Sourangsu Banerji, Sayan Bose and Abhishek Halder

Abstract Band structure of one-dimensional photonic crystal is analytically computed under the normal incidence of electromagnetic wave inside first Brillouin zone. Plane wave expansion method is applied on Maxwell's equations, and set of eigenvalues are computed for different momentum values. Helmholtz theorem is used along with Fourier transform method to obtain the direction of wave propagation inside the lattice, and size of the Hamiltonian is reduced due to the identical behaviour of TE and TM modes. Result is computed for AlGaN/GaN composition, and is compared with conventional SiO_2/air material system. Result will be helpful to determine the dispersion relation of 1D photonic crystal, allowing us to design periodic photonic structures with required photonic bandgap. Knowledge may be utilized to design photonic crystal-based bandpass filter for use in photonic integrated circuit.

Keywords Photonic crystal · Photonic bandgap · Semiconductor heterostructure · Brillouin zone · Normal wave incidence

---

A. Deyasi (✉) · S. Banerji · S. Bose · A. Halder
Department of Electronics and Communication Engineering,
RCC Institute of Information Technology, Kolkata 700015, India
e-mail: deyasi_arpan@yahoo.co.in

S. Banerji
e-mail: sourangsu.banerji@gmail.com

S. Bose
e-mail: bose_sayan@yahoo.com

A. Halder
e-mail: abhi.abhishek.halder@gmail.com

© Springer India 2015

K. Maharatna et al. (eds.), *Computational Advancement in Communication Circuits and Systems*, Lecture Notes in Electrical Engineering 335,
DOI 10.1007/978-81-322-2274-3_36

36.1 Introduction

One-dimensional photonic crystal is the multilayer alternating periodic arrangement of dielectric materials along the direction of electromagnetic wave propagation [1]. Because of the periodicity, the structure exhibits a few windows in otherwise prohibited wavelength spectra, due to the formation of photonic bandgap, which may be exhibited in one, two or three dimensions [1, 2]. This unique property helps to block the propagation of some wavelength, and allows other wavelength in the spectra; thus can effectively be considered as an optical bandpass filter [3]. This phenomenon can be explained by the principle of Bragg's reflection [4], where wavelength of light is of the order of layer dimensions [5]. Materials exhibiting photonic bandgap can be used in designing photonic crystal fibre [6], which may replace the conventional optical fibre due to its highly improved performance from communication point of view [7].

Bandgap of 2D photonic crystal had previously been studied by varying column roundness by Hillebrand [8] using plane wave expansion method. Zhao calculated the width of bandgap [9] using Bragg's principle of reflection. Men optimized the computational problem using semidefinite programming and subspace methods [10]. Evolutionary algorithm [11] and level set method [12] have also been used for design of large bandgap crystal. In the present paper, band structure of one-dimensional photonic crystal is calculated by using plane wave expansion method technique for normal wave incidence, and AlGaN/GaN material composition is considered as a unit cell of the periodic arrangement. The use of semiconductor heterostructure adds novelty of the work in the respect that the structure can be treated as photonic multiple quantum well for electronic applications [13], and also it provides better performance as optical transmitting/emitting device [14]. Use of AlGaN/GaN material composition is taken up because as shown in [15], the predominant effect of carrier localization in undoped AlGaN alloys enhances with the increase in Al contents and is related to the insulating nature of AlGaN of high Al contents. Usage of high Al-content AlGaN layer also increases the overall figure of merit of the AlGaN/GaN due to the combined advantages of enhanced band offset, lattice mismatch induced piezoelectric effect. Comparative results speak for the advantage of the use of semiconductor heterostructure-based fabrication in photonic devices. Moreover, the study also reveals the ease of engineering a photonic bandgap in the region of interest when the crystal is fabricated with heterostructures.

36.2 Mathematical Modelling

Theoretical Lying on the fundamental principles of the Bloch function and the Fourier transform, the plane wave method (PWM) applies to a periodic structure of a photonic crystal in the wave-vector space. On solving the Maxwell equations, eigenvalues can be obtained to extract the dispersive feature of the photonic energy band. On the assumption that both *E*(*r*, *t*) and *H*(*r*, *t*) have sine surge modes with respect to real space r and time t, then Maxwell equations can be written as:

$$\nabla \cdot \varepsilon(r) \cdot E(r) = 0 \tag{36.1}$$

$$\nabla \cdot B(r, t) = 0 \tag{36.2}$$

$$\nabla \times H(r) = i\omega\varepsilon_0\varepsilon(r)E(r) \tag{36.3}$$

$$\nabla \times E(r) = -i\omega\mu_0 H(r) \tag{36.4}$$

where ω is the oscillation frequency of the electromagnetic field, $\varepsilon(r)$ is the corresponding dielectric constant of the crystal and is a function of space *r*, also ε_O and μ_O denote the dielectric constant and permeability in vacuum, respectively. On mathematically solving these equations in addition to the consideration of the harmonic mode, a simple important equation obtained can be written as:

$$\nabla \times \frac{1}{\varepsilon(r)} \nabla \times \vec{H}(r) = \frac{\omega^2}{C^2} \vec{H}(r) \tag{36.5}$$

Using Bloch's theorem, essentially in the case of an infinite periodic photonic crystal, a mode in a periodic structure can be expressed as a sum of infinite number of plane waves:

$$H(r) = \sum_{\vec{G}_i,\lambda} h_{G_i,\lambda} e^{i(\vec{k}+\vec{G}_i).\vec{r}} \hat{e}_\lambda \tag{36.6}$$

where $\lambda = 1, 2$, also κ denotes the wave vector of the plane wave, $\vec{G}$ is the reciprocal lattice vector, $\hat{e}_\lambda$ is used to denote the two unit axis perpendicular to the propagation direction $\overrightarrow{k} + \vec{G}$. $h_{G_i,\lambda}$ is used to represent the coefficient of the *H* component along the axes $\hat{e}_\lambda$. One thing to note here is that $(\hat{e}_1, \hat{e}_2, \overrightarrow{k} + \vec{G})$ are perpendicular to each other. Now, using the Fourier transform, the dielectric function can also be written as,

$$\varepsilon(r) = \sum_{\vec{G}} \varepsilon(\vec{G}) \exp(i\vec{G} \cdot r) \tag{36.7}$$

$$\varepsilon(\vec{G}) = \frac{1}{V} \oiiint_{\Omega} \varepsilon(r) \exp(-i\vec{G} \cdot r) \tag{36.8}$$

where Ω is the unit cell and *V* is the volume of the unit cell.

Eventually, Helmholtz's equation can be expressed in a form which is a standard eigenvalue problem,

$$\sum_{\vec{G}'} |k+G||k+G'|\varepsilon^{-1}(G-G')\begin{bmatrix} \hat{e}_2\hat{e}'_2 & -\hat{e}_2\hat{e}'_1 \\ -\hat{e}_1\hat{e}'_2 & \hat{e}_1\hat{e}'_1 \end{bmatrix}\begin{bmatrix} h'_1 \\ h'_2 \end{bmatrix} = \frac{\omega^2}{C^2}\begin{bmatrix} h_1 \\ h_2 \end{bmatrix} \tag{36.9}$$

Here, $\begin{bmatrix} \hat{e}_2\hat{e}'_2 & -\hat{e}_2\hat{e}'_1 \\ -\hat{e}_1\hat{e}'_2 & \hat{e}_1\hat{e}'_1 \end{bmatrix}$ matrix gives us the direction of the wave-vector propagation when it strikes a lattice site; i.e. causes diffraction on hitting a Bragg plane in the first Brillouin zone. Certain simplifications exist for both 1D and 2D case as well as for both in-plane and off-plane light propagation. $\begin{bmatrix} h'_1 \\ h'_2 \end{bmatrix}$ is an NX1 matrix which satisfies the number of plane waves that is being considered. Coming to the one-dimensional normal incidence case, both the TE and TM modes behave in the same way. Hence both the modes can be coupled and we reduce the number of equations from 2N to N. Noting the fact that both k and G have only two directions, $(-x$ and $+x)$ or $(-y$ and $+y)$ or $(-z$ and $+z)$ and accordingly we select $\hat{e}_1$ and $\hat{e}_2$. If $(-z$ and $+z)$ is so chosen then $\hat{e}_1, \hat{e}_2$ can be set to

$$\hat{e}_1 = \overrightarrow{x}, \quad \hat{e}_2 = \overrightarrow{y}$$

The eigen equation can be approximated to a N × N equation group

$$\sum_{G'} |k+G||k+G'|\varepsilon^{-1}(\vec{G}-\vec{G}') = \frac{\omega^2}{C^2}h(\vec{G}) \tag{36.10}$$

Here, D and H are in the x and y plane respectively and TE and TM mode are same.

36.3 Results and Discussion

The band structure of a periodic dielectric structure is given by the k–ω relation. As we know that according to the Bloch's theorem, the relation in a periodic structure is also periodic, so it suffices that it is enough to calculate only the k-points in the so-called first Brillouin zone. All other k's outside the first Brillouin zone are the same mode as the k in first Brillouin zone. Making use of the symmetry properties, we only need to calculate the k–ω relation in a zone with no symmetry property called irreducible Brillouin zone.

In the irreducible Brillouin zone given a k point, we will obtain a series of eigenvalues, i.e. a set of ω points. On changing the corresponding k points, continuously changed ω points for the same band is what we will get. So usually, we only need to calculate those points on the irreducible Brillouin zone edge, i.e. the boundary value, since the other frequencies for those k-points inside the irreducible Brillouin zone will fall inside the band region. In addition to these, the entire band structure of the dielectric periodic array is also plotted. Moreover it

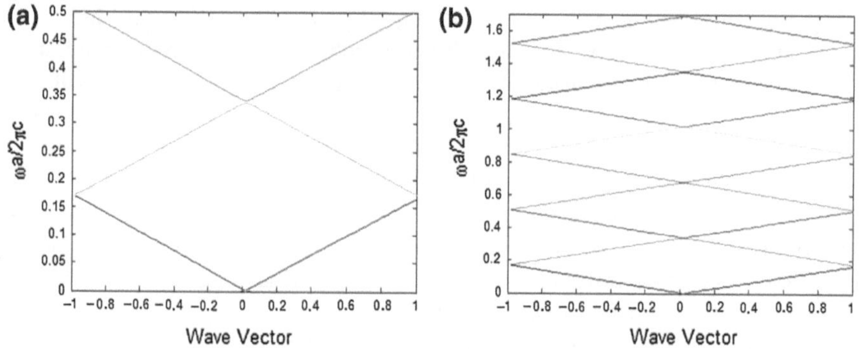

Fig. 36.1 **a** Band structure of a photonic crystal with AlGaN/GaN material composition in the first Brillouin zone. **b** Complete band structure of a photonic crystal with AlGaN/GaN material composition

turns out that the bands above and below the photonic bandgap can certainly be distinguished by where the power lies—in the high ε regions or in the low ε regions.

Using Eq. (36.10), the photonic band structure of a semiconductor heterostructure-based photonic crystal is obtained. Figure 36.1a, b depicts the band structure of the crystal within the first Brillouin zone and the band structure of the entire periodic structure, respectively, for AlGaN/GaN material composition. Similarly, Fig. 36.2a, b depicts the band structure of the crystal with SiO_2/Air material composition within the first Brillouin zone and the band structure of the entire periodic structure, respectively. The results are matched and analysed.

The lattice parameters are chosen in such a way in order to maximize the lowest bandgap. Size of the band as well as their thickness is essentially a function of the dielectric constants of the two materials. Figure 36.1a shows that the small splitting is observed at the central point, whereas large splitting is found out as we move

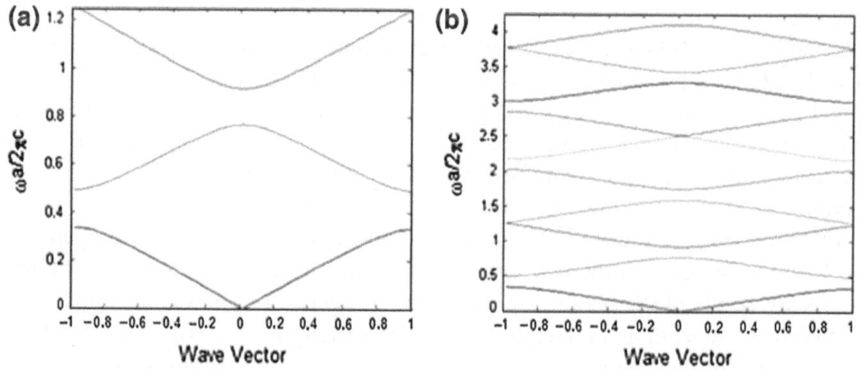

Fig. 36.2 **a** Band structure of a photonic crystal with SiO_2/Air material composition in the first Brillouin zone. **b** Complete band structure of a photonic crystal with SiO_2/Air material composition

away from the central X point towards the Γ point. This is exactly due to the consequence of band repulsion, a well-known physics for electronic band structures. For perfectly linear material, bending of the curves is negligibly small.

On similar lines in Fig. 36.1b, the complete band structure is plotted. It shows the existence of all the bands present in the dielectric periodic structure. Here the key feature observed is the absence of bandgap. The possible explanation for such a result is the fact that the displacement vector $D(r)$ is oriented in the plane, which ideally gives $\vec{D}(\vec{r}) \cdot \hat{k} = 0$. The fields at the zone-edge states have opposite phases in each unit cell in the direction of wave-vector k. Constant phases of the field have the direction perpendicular to k in each unit cell.

Fill factor (f) between any two bands can be written as

$$f = \frac{\int\limits_{V\varepsilon 1} \vec{E}^*(\vec{r}) \cdot \vec{D}(\vec{r}) \mathrm{d}v}{\int \vec{E}^*(\vec{r}) \cdot \vec{D}(\vec{r}) \mathrm{d}v} \tag{36.11}$$

It measures the amount of energy located inside the high dielectric permittivity material.

A seen from Fig. 36.2a, b, the nature of the behaviour of the band structure in case of SiO_2/Air material composition is similar to the case of AlGaN/GaN-based fabrication. No such anomaly is observed. Here also the maximization of the bandgap was performed through the effective utilization of the lattice parameters. The characteristic difference in the band structure obtained from SiO_2/Air material composition in comparison to AlGaN/GaN composition stems from the fact that the contrast in refractive index for both the cases differ which results in the difference of the size of the band in addition to its thickness. Due to the phenomenon of band repulsion here too a similar case of small splitting of the photonic bands is observed around the central point. However, the extent to which this splitting occurs is not as much as we see in case of AlGaN/GaN material composition case.

In Fig. 36.2a, we see the presence of bandgaps occurring between the first and the second band. The function f is calculated using Eq. (36.11) for the first and second bands is 0.1–0.5, respectively. Such a high contrast difference between the two bands is responsible for existence of a complete photonic bandgap. The first band has its power in the dielectric region and is characterized with low frequency as compared to which the second band has much of its power in the air regions and is associated with high frequency. Such difference creates a well distinct and prominent bandgap in the structure. In Fig. 36.2b, the complete band structure is plotted for SiO_2/Air material composition. On close observation, bandgaps are found to exist between the first and the second bands as well as between the fifth and sixth bands of the photonic crystal. The presence of a third bandgap is found between the seventh and the eight bands. The striking feature is that the bandgaps are uniform in their width. Under normal incidence of the electromagnetic wave, bandgaps are found in the structure fabricated with SiO_2/Air material composition as opposed to the AlGaN/GaN composition.

36.4 Conclusion

Under normal EM wave incidence, comparative analysis has been carried out for both conventional material as well as semiconductor heterostructure in designing photonic crystals. With AlGaN/GaN semiconductor heterostructure composition, the absence of a complete photonic bandgap is essentially a striking observation we had. Results indicate that AlGaN/GaN material composition-based crystals yield good results when smaller photonic bandgap is desired. Comparative study also reveals that for SiO_2/Air-based material composition, presence of complete photonic bandgap, i.e. between the first and the second, fourth and the fifth as well as the seventh and the eighth bands is well observed, which results a wider stopband in terms of filter applications. Such predominance of photonic bandgaps as observed from the plots obtained lead us to conclude that under normal incident wave it is preferable to go with the proposed heterostructure configuration which allows smaller separation of passbands compared to its conventional counterpart in photonic filter in higher frequency domain.

References

1. E. Yablonovitch, T.J. Gmitter, Photonic band structure: the face-centered-cubic case. Phys. Rev. Lett. **63**, 1950–1953 (1989)
2. R. Loudon, The propagation of electromagnetic energy through an absorbing dielectric. J. Phys. A. **3**, 233–245 (1970)
3. D. Mao, Z. Ouyang, J.C. Wang, A photonic-crystal polarizer integrated with the functions of narrow bandpass and narrow transmission angle filtering. Appl. Phys. B. **90**, 127–131 (2008)
4. A. Maity, B. Chottopadhyay, U. Banerjee, A. Deyasi, Novel band-pass filter design using photonic multiple quantum well structure with p-polarized incident wave at 1550 nm. J. Electron Devices **17**, 1400–1405 (2013)
5. K. Bayat, G.Z. Rafi, G.S.A. Shaker, N. Ranjkesh, S.K. Chaudhuri, S. Safavi-Naeini, Photonic-crystal based polarization converter for terahertz integrated circuit. IEEE Trans. Microw. Theory Tech. **58**, 1976–1984 (2010)
6. R.L. Wang, J. Zhang, Q.F. Hu, Simulation of band gap structures of 1D photonic crystal. J. Korean Phys. Soc. **52**, S71–S74 (2008)
7. I.S. Fogel, J.M. Bendickson, M.D. Tocci, M.J. Bloemer, M. Scalora, C.M. Bowden, J.P. Dowling, Spontaneous emission and nonlinear effects in photonic bandgap materials, pure and applied optics. J. Eur. Opt. Soc. Part A **7**, 393–408 (1998)
8. R. Hillebrand, W. Hergert, W. Harm, Theoretical band gap studies of two-dimensional photonic crystals with varying column roundness. Phys. Status Solidi B. **217**, 981–989 (2000)
9. J. Zhao, X. Li, L. Zhong, G. Chen, Calculation of photonic band-gap of one dimensional photonic crystal. Journal of Physics: Conference Series (Dielectrics 2009: Measurement Analysis and Applications, 40th Anniversary Meeting), (2009), 183 012–018
10. H. Men, N.C. Nguyenz, R.M. Freundx, P.A. Parrilo, J. Perairek, Band gap optimization of two-dimensional photonic crystals using semidefinite programming and subspace methods. ArXiv: 0907.2267 V1, (2009)
11. S. Preble, M. Lipson, Two-dimensional photonic crystals designed by evolutionary algorithms. Appl. Phys. Lett. **86**, 061111 (2005)

12. C.Y. Kao, S. Osher, E. Yablonovitch, Maximizing band gaps in two-dimensional photonic crystals by using level set methods. Appl. Phys. B. **81**, 235–244 (2005)
13. J. Chen, Y. Chen, Y. Shen, W. Zhou, J. Ren, Y. Zheng, L. Chen, Study of the band-gap structure of a 1D-photonic crystal by using different numerical approaches. J. Korean Phys. Soc. **56**, 1319–1324 (2010)
14. D.G. Popescu, P. Sterian, Photonic crystal fiber mode characterization with multipole method. J. Adv. Res. Phys. **2**, 021105 (2011)
15. M.D. Hodge, R. Vetury, J.B. Shealy, Fundamental failure mechanisms limiting maximum voltage operation in AlGaN/GaN HEMTs, in *IEEE International Reliability Physics Symposium* (2012)

Chapter 37
Frequency Response of Si/SiGe Heterojunction Bipolar Transistor

Arnima Das, Maitreyi Ray Kanjilal and Payel Biswas

Abstract Figure of merits are key performance indicators of any RF device and can be used to assess the performance of the heterojunction transistors operating at very high frequency range. For HBT, stability and power gain are the key figures of merits to consider. The Agilent Advanced Design System has been used here to simulate the frequency response of Si/SiGe HBT with Mextram model. The results of the simulation are used further to establish the figure of merits for the Si/SiGe HBT in the GHz frequency ranges that can be useful for operations at wireless and satellite communication systems.

Keywords Heterojunction bipolar transistor · Figure of merit · Rollett's stability factor · Maximum available gain · Maximum stable gain · Unilateral power gain

37.1 Introduction

The study and inventions in the field of radio frequency (RF) electronic devices was mostly related to the military projects and other exotic science projects until 1980 when satellite communication and wireless communication were started being used for civil purposes. Later on in the early twenty-first century, the rise of mobile communication in the civil world generated an increasing interest in this field. The latest communication system now is able to process and receive a huge amount of data in the RF frequency ranges from ultra high frequency (UHF—up to 3 GHz) to tremendously high frequency (THF—up to 3 THz).

A. Das (✉) · M.R. Kanjilal · P. Biswas
Narula Institute of Technology, Kolkata 700109, India
e-mail: arnimaz@gmail.com

M.R. Kanjilal
e-mail: mr.kanjilal@gmail.com

P. Biswas
e-mail: biswas.payel@gmail.com

© Springer India 2015

K. Maharatna et al. (eds.), *Computational Advancement in Communication Circuits and Systems*, Lecture Notes in Electrical Engineering 335,
DOI 10.1007/978-81-322-2274-3_37

However, the transistors still remain as the backbone of the communication system albeit the research has moved on the heterojunction transistors such as heterojunction bipolar transistor (HBT) and high electron mobility transistor (HEMT). These devices are capable of handling the frequency ranges required in the mobile communication era. Especially, HBT is a more interesting option as it can handle larger current flow and better amplification for a given smaller area than field effect transistors. This is made possible as sub-micron level base width can be fabricated, which is independent of the photolithography.

The heterojunction bipolar transistor is made by using two different semiconductor materials for the emitter and base regions. This creates an emitter-base heterojunction and in some cases also a base-collector heterojunction. In a heterojunction, the potential barrier in valence band is much higher than the potential barrier of conduction band. This limits the injection of holes or electrons from base to emitter allowing high concentration gradient to be achieved in the base region. This in turn reduces the base resistance while maintaining higher gain.

37.2 Figure of Merits of RF Transistors

The operation and characteristics of a particular device determine its area of application. Also each device should be judged by its performance under various external factors before practical application.

There are various figure of merits used for the assessment of RF electronic devices such as transistors. Figure of merits (FOM) gives a quantitative measurement of performance of an electronic device. This paper considers the stability and power gain of the transistor as principal FOM.

37.2.1 Stability

Stability over wide frequency range can be measured by using Rollett's stability factor (K) which is mathematically derived from Y parameters [1] as

$$K = \frac{2Re(y_{11})Re(y_{22}) - Re(y_{12}y_{21})}{|y_{12}y_{21}|} \tag{37.1}$$

For $K < 1$, the device is conditionally stable with unintended oscillations due to external reasons, while for $K > 1$ the transistor is unconditionally stable [2].

37.2.2 Power Gain

Power gain is considered as a key feature of a transistor and there are different types of power gain used to measure performance [3].

The transistor can obtain maximum power gain when the source and load impedances are matched, i.e. load impedance is a complex conjugate of the source impedance. The gain is expressed as

$$G_A = \frac{P_{\text{load,max}}}{P_{\text{source,max}}} \tag{37.2}$$

where, $P_{\text{load,max}}$ is the maximum available average power at the load and $P_{\text{source,max}}$ is the maximum power available from the source.

This is also known as maximum available gain (MAG) and is expressed as [4]

$$\text{MAG} = \left|\frac{y_{21}}{y_{12}}\right|\left(K - \sqrt{K^2 - 1}\right) \tag{37.3}$$

Again, under the condition of obtaining maximum gain, the maximum stable gain (MSG) can be measured, which is defined as [2]

$$\text{MSG} = \left|\frac{y_{21}}{y_{12}}\right| \tag{37.4}$$

It is the maximum gain obtainable from a transistor when it is combined with external matching impedance networks at load and source under the condition $K = 1$, at any operating frequency. From the mathematical definitions, it is found that MAG is always less than MSG except when stability factor is unity.

The other power gain FOM is the Mason's invariant, also called the unilateral power gain (U). This is also calculated over the entire range of frequency irrespective of the stability factor. This is defined as the gain obtained from a two-port network when there is zero feedback from output to input and load impedance is matched to the source impedance. Mathematically expressed as [3]

$$\text{U} = \frac{|y_{21} - y_{12}|^2}{4[Re(y_{11})Re(y_{22}) - Re(y_{12})Re(y_{21})]} \tag{37.5}$$

Although all the FOM are calculated using small signal parameter and applicable for small signal model, yet the same FOM are used to measure the performance in large signal operations also.

All the power gains in this paper are represented in Decibel-milliwatts (dBm) as this is the most common representation of power gain in RF devices. Also it can express both very large and very small values with relative ease.

$$\text{Gain (dBm)} = 10\log_{10}\left(\text{Gain}/0.001\right) \quad (37.6)$$

37.3 Results

Agilent advanced design system is a high-frequency/high-speed device simulation tool with inbuilt and custom editable integrated circuits (IC), packages, modules and boards. The small signal Mextram model of transistor (with product default parameters) available in the Agilent ADS 2009 has been used for the following simulations [5] with a matching impedance network design. The *Y* parameters were recorded from frequency 100 MHz–1 THz with 5 GHz interval for better response.

Figure 37.1 shows the plot of frequency versus Rollett's stability factor (*K*). From this figure it can be understood that the transistor has a very good stability at higher frequencies. The variation of stability in the low frequency region of Fig. 37.1 has been expanded in Fig. 37.2, to determine the critical frequency of

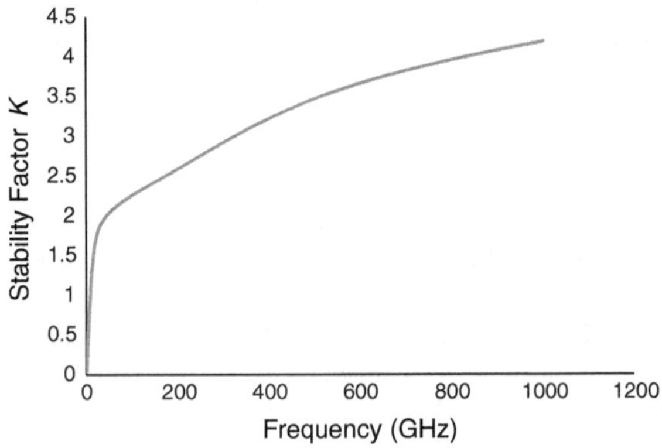

Fig. 37.1 Rollett's stability factor (*K*) versus frequency (GHz)

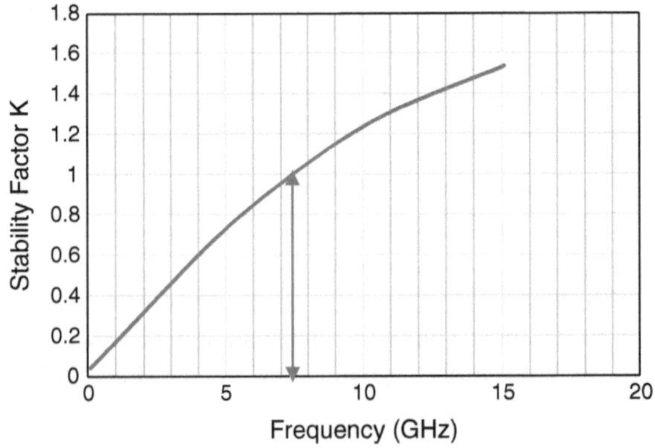

Fig. 37.2 Determining the critical frequency from the curtailed plot of Fig. 37.1

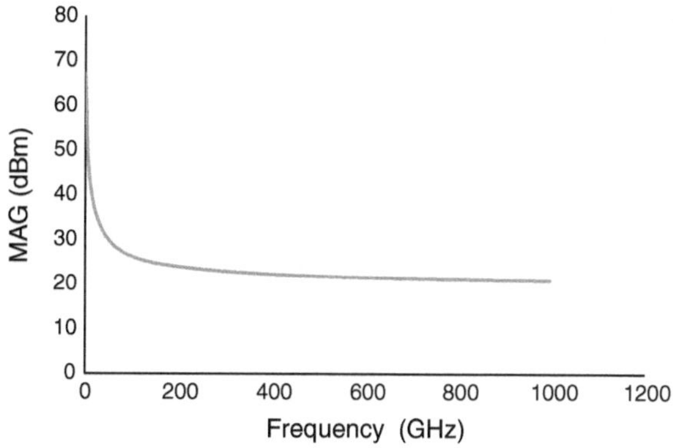

Fig. 37.3 Maximum available gain (dBm) versus frequency (GHz)

HBT under simulation. This is approximated at about 7.5 GHz. This frequency range completely agrees with that used in the satellite and microwave communication, hence justifying the use of SiGe HBT in the UHF and THF range.

All the three types of power gains are shown in the Figs. 37.3, 37.4 and 37.5 where the power gains are given in dBm. The theoretical understanding of MAG always being lower than that of MSG [2] can also be proved from the Figs. 37.3 and 37.4.

Also the gain as found from the results of the simulation are in agreement with the reported values of power level for wireless and mobile communication systems at such high radio frequency ranges [6]. Also it is noted that unilateral power gain is excellent for wireless operations.

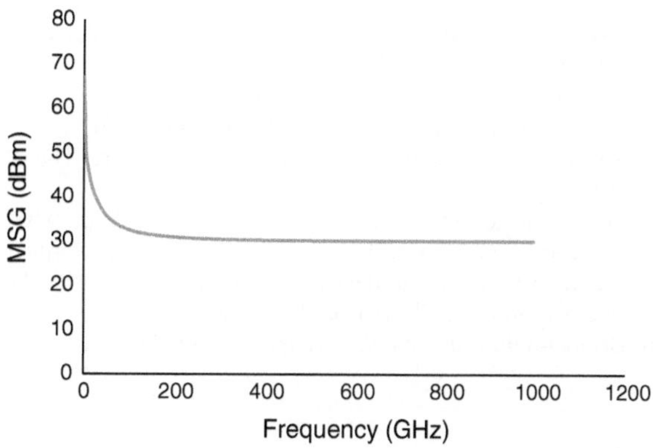

Fig. 37.4 Maximum stable gain (dBm) versus frequency (GHz)

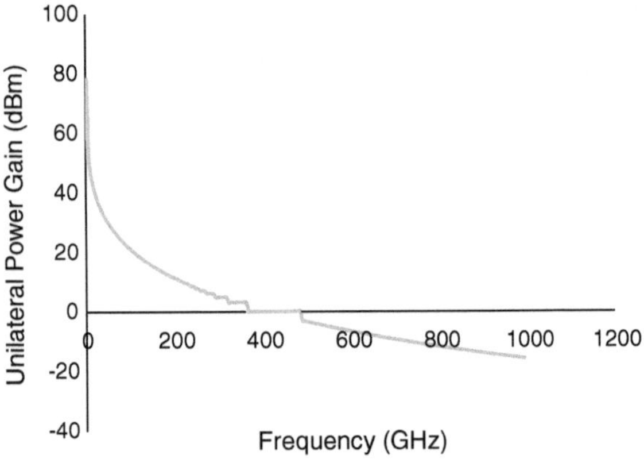

Fig. 37.5 Unilateral power gain (dBm) versus frequency (GHz)

37.4 Conclusion

It is essential to understand and analyse the figure of merits of any RF device in order to continue its future research for obtaining the optimum level of performance. In case of HBT, it is of utmost importance to take advantage of the size factor while trying to improve the performance by better fabrication techniques. While Si/SiGe HBT is very stable at higher frequencies and also shows a very good current gain, the thermal capacity of the same is a cause of concern at higher frequencies. Therefore, this is a wide open area of future research and development.

References

1. J.M. Rollett, Stability and power-gain invariants of linear two ports. Circ. Theor. IRE Trans. IEEE **9**(1), 29–32 (1962)
2. F. Schwierz, J.J. Liou, Semiconductor devices for RF applications: evolution and current status. Microelectron. Reliab., Elsevier **41**, 145–168 (2001)
3. S.J. Mason, Power gain in feedback amplifier. Circ. Theor. IRE Trans. IEEE **CT-1**(2), 20–25 (1954)
4. Stability factor and available gain of amplifiers. http://www.microwaves101.com
5. O. Esame, G. Yasar, I. Tekin, A. Bozkurt, Performance comparison of state-of-the-art heterojunction bipolar devices (HBT) based on AlGaAs/GaAs, Si/SiGe and InGaAs/InP. Microelectron. J., Elsevier **35**, 901–908 (2004)
6. Decibel-milliwatts. http://en.wikipedia.org/wiki/DBm

Chapter 38
An Approach for Designing an Optimized Reversible Parallel Multiplier by Reversible Gates

Shefali Mamataj, Biswajit Das and Saravanan Chandran

Abstract In today's electrical or electronic designing, power dissipation becomes dominant. The promising alternative to these technologies is the reversible computing. Because for irreversible circuits, information is lost and it dissipates energy as heat. The dissipated energy is proportional to the number of bits lost during computation. In case of reversible circuits, they do not dissipate energy in a large amount as irreversible circuits. So information is not lost in reversible circuits during the execution of operation. Thus generation of unique outputs from the particular inputs and reconstruction of inputs from the particular outputs can be achievable. So in the view point of designing issues, reversible logic is a promising field of research that finds applications in low power computing, quantum computing, optical computing and other emerging computing technologies, bioinformatics and nanotechnology-based systems. Here, an approach for the designing of optimized reversible parallel multiplier (ORPM) by reversible gates has been followed. In this regard, this paper represents a new partial reversible gate RPPG for partial product generation for the designing of ORPM. Also, two designing methods are proposed to design ORPM and a comparison is also made between these two methods and the existing reversible multiplier circuits reported in the literature in perspective of the number of reversible gates, constant input, garbage output and total logical calculation.

S. Mamataj (✉)
Department of Electronics & Communication Engineering,
MCET, Berhampore 742102, India
e-mail: smamataj@gmail.com

B. Das
Department of Computer Science & Engineering,
MCET, Berhampore 742102, India
e-mail: dasbiswaji@gmail.com

S. Chandran
Computer Centre, NIT, Durgapur 713209, India
e-mail: dr.cs1973@gmail.com

© Springer India 2015

K. Maharatna et al. (eds.), *Computational Advancement in Communication Circuits and Systems*, Lecture Notes in Electrical Engineering 335,
DOI 10.1007/978-81-322-2274-3_38

Keywords Reversible logic · Reversible gate · Reversible parallel multiplier · Garbage output · Constant inputs

38.1 Introduction

In order to talk about new trends and projects in the area of reversible logic, understanding of the reversible logic is very necessary. Reversible computation is generally treated as an unconventional type of computation. Mainly two types of reversibility can be shown: physical and logical. A logical reversible function can be achieved by a discrete, deterministic computational process where the transition function that maps old computational states to new states is a one-to-one mapping function; it means the output logical states uniquely define the input logical states of the computational operation. Power dissipation is a considerable issue in VLSI design as modern logic circuits offer a great deal of computing power in a small footprint. Reversible computation preserves information and prevents the lost of energy. Landauer has shown that every time when losing a bit of information, kTln2 joules of energy is lost to heat, where k is Boltzmann's constant and T is the absolute temperature at which the computation is performed [1]. In 1973, Bennett showed that to remove this energy loss it is necessary and sufficient to perform the computation in a reversible way [2]. It has been shown that reversible logic helps in saving this energy using the charge recovery process [3]. Multiplication is an important arithmetic operation which is widely used in a lot of computational units. It is necessary for the processors to have high-speed multipliers with less hardware complexity.

38.2 Reversible Logic

A reversible logic function is defined as a function for which each input vector maps to a unique output vector. From the given outputs, it is always possible to determine back its input in case of a reversible function, because there is a one state-to-one state relationship among the input and output states. For an $N \times N$ reversible logic gate, the inputs are denoted by I_1, I_2, I_3, …, I_N and the outputs are denoted by O_1, O_2, O_3, …,O_N. Using of the minimum number of reversible gates, garbage outputs and constant inputs are the main features that a reversible logic circuit should have [4]. The inputs, outputs, garbage outputs and constant inputs of a reversible gate are related as follows. Input + constant input = output + garbage.

The total logical operation of a reversible circuit [5] is calculated together with the number of AND operations, number of EX-OR operations and number of NOT operations. If α represents the number of EX-OR operations, β represents the number of AND operations and δ represents the number of NOT operations, then

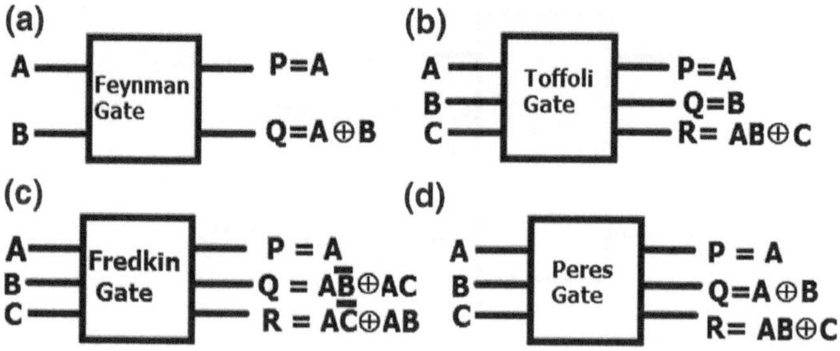

Fig. 38.1 Reversible gates. **a** Feynman, **b** toffoli gate, **c** fredkin gate, **d** peres gate

the total logical operation T is expressed as the sum of EX-OR, AND and NOT operations necessary for a specified circuit and can be expressed in terms of α, ß and δ. One of the most important 2 × 2 basic reversible logic gates is Feynman gate [6]; Toffoli gate [7], Fredkin gate [8] and Peres gate [9] are 3 × 3 reversible gates that can be used to establish all types of important combinational functions shown in Fig. 38.1.

38.3 Proposed Gate

A 5 × 5 partial reversible gate RPPG proposed here is shown in Fig. 38.2. The inputs (A, B, C, D, E) mapped to outputs ($P = A$, $Q = A.B$, $R = A.C$, $S = A.D$, $T = A.E$). The RPPG gate is mainly useful in generation of the partial product for multiplier circuit design.

In [10] another DKFG reversible gate was proposed as shown in Fig. 38.3. For the reversible DKFG gate, the input vector is given by $I_V = (0, A, B, C)$ which maps

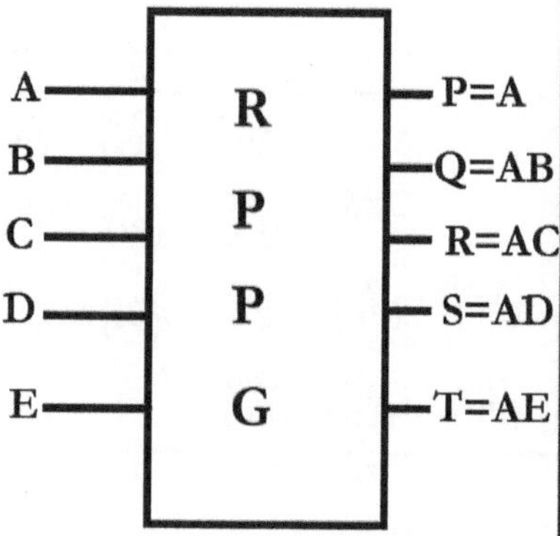

Inputs					Outputs				
A	B	C	D	E	P	Q	R	S	T
0	X	X	X	X	0	0	0	0	0
1	0	0	0	0	1	0	0	0	0
1	0	0	0	1	1	0	0	0	1
1	0	0	1	0	1	0	0	1	0
1	0	0	1	1	1	0	0	1	1
1	0	1	0	0	1	0	1	0	0
1	0	1	0	1	1	0	1	0	1
1	0	1	1	0	1	0	1	1	0
1	0	1	1	1	1	0	1	1	1
1	1	0	0	0	1	1	0	0	0
1	1	0	0	1	1	1	0	0	1
1	1	0	1	0	1	1	0	1	0
1	1	0	1	1	1	1	0	1	1
1	1	1	0	0	1	1	1	0	0
1	1	1	0	1	1	1	1	0	1
1	1	1	1	0	1	1	1	1	0
1	1	1	1	1	1	1	1	1	1

Fig. 38.2 Reversible RPPG gate and its truth table

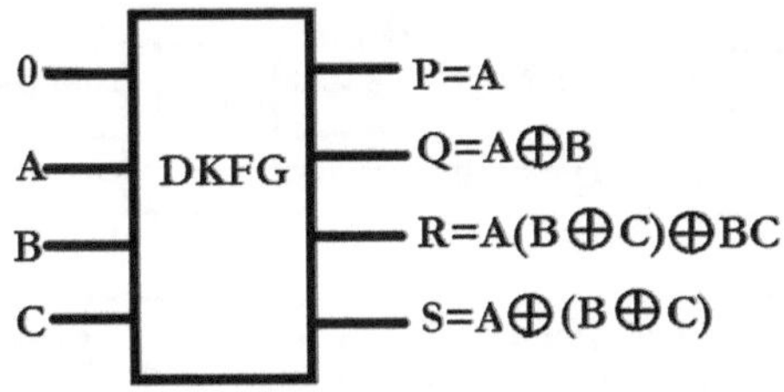

Fig. 38.3 DKFG reversible gate

the corresponding output vector is $O_V = (P, Q, R, S)$ and the gate has been used here for designing the multi-operand addition circuit (MOPAC) to implement the multiplier circuit.

38.4 Design Approaches

A reversible 4 × 4 multiplier circuit has two parts: partial product generation circuit (PPGC) and multi-operand addition circuit (MOPAC). The detailed information of these two parts is discussed in the subsequent sections:

The basic operation of a 4 × 4 reversible multiplier is shown in Fig. 38.4. The multiplier produces 16 partial product bits of the *X* and *Y* inputs in the form (Xi.Yi), where *i* vary between 0 and 3 to perform 4 × 4 multiplications. However, it can be applicable to design any other n × n reversible multiplier. Here an optimized reversible parallel multiplier (ORPM) has been designed using reversible gates. Here, basically, the reversible gates RPPG and PERES are used to design the multiplier and to realize full adder, reversible DKFG gates have been used.

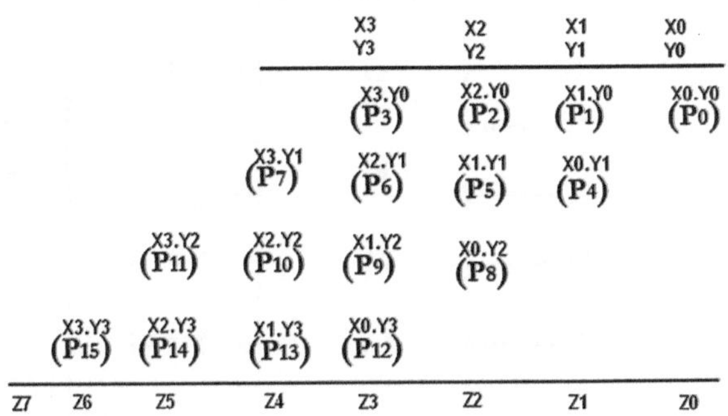

Fig. 38.4 The operation of the 4 × 4 parallel multiplier

38.4.1 Partial Product Generation

To design a multiplier, partial product generation is the main important part. For designing of a 4 × 4 reversible multiplier, making of the 16 partial product terms the proposed PPGC shown in Fig. 38.5 is designed by using RPPG gate. The PPGC

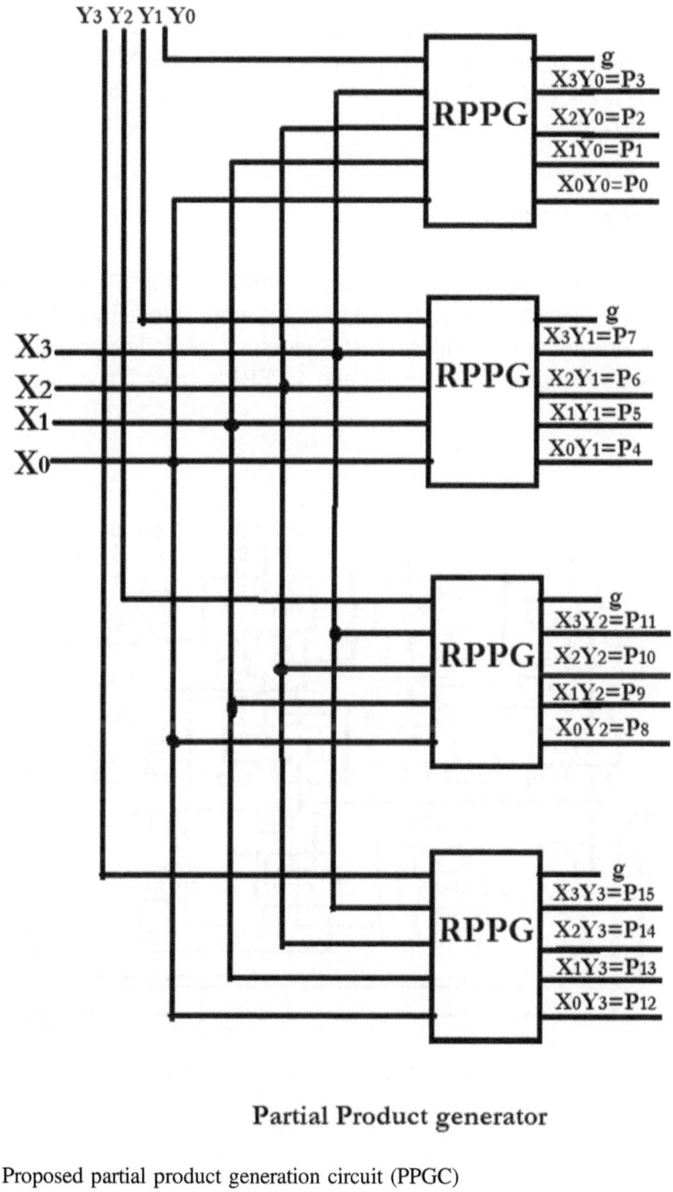

Partial Product generator

Fig. 38.5 Proposed partial product generation circuit (PPGC)

requires four numbers of RPPG gate which generates 16 product terms P_0, P_1, P_2, P_3, P_4, P_5, P_6, P_7, P_8, P_9, P_{10}, P_{11}, P_{12}, P_{13}, P_{14} and P_{15}. Also four garbage outputs are produced.

38.4.2 *Multi-operand Addition Circuit*

To implement multi-operand addition circuit (MOPAC), two different types of circuits have been followed here. The main purpose of this MOPAC is to add the product terms generated by the PPGC in a particular manner for getting the proper multiplication result.

38.4.2.1 MOPAC TYPE-I

The proposed four-operand TYPE-I adder is shown in Fig. 38.6. In this circuit, addition of three variables means full adder realization is done by DKFG reversible gates and for designing half adder, PERES gates are used. The proposed MOPAC TYPE-I uses 8 reversible DKFG gates and 4 PERES gates. There are 19 garbage outputs.

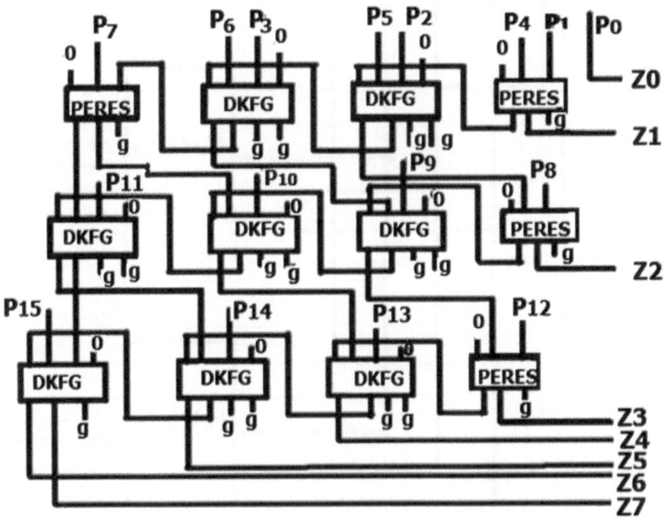

Fig. 38.6 Proposed multi-operand addition circuit (MOPAC TYPE-I)

Fig. 38.7 Proposed multi-operand addition circuit (MOPAC TYPE-II)

38.4.2.2 MOPAC TYPE-II

The proposed four-operand TYPE-II adder is shown in Fig. 38.7. DKFG gate is used here for designing reversible full adder and PERES gate is used for the realization of half adder. The proposed MOPAC TYPE-II uses 8 reversible DKFG gates and 2 PERES gates. There are 18 garbage outputs.

38.4.3 Proposed Optimized Reversible Parallel Multiplier (ORPM) Circuits

By using these two types of proposed MOPAC, implementation of two types of reversible multiplier can be made.

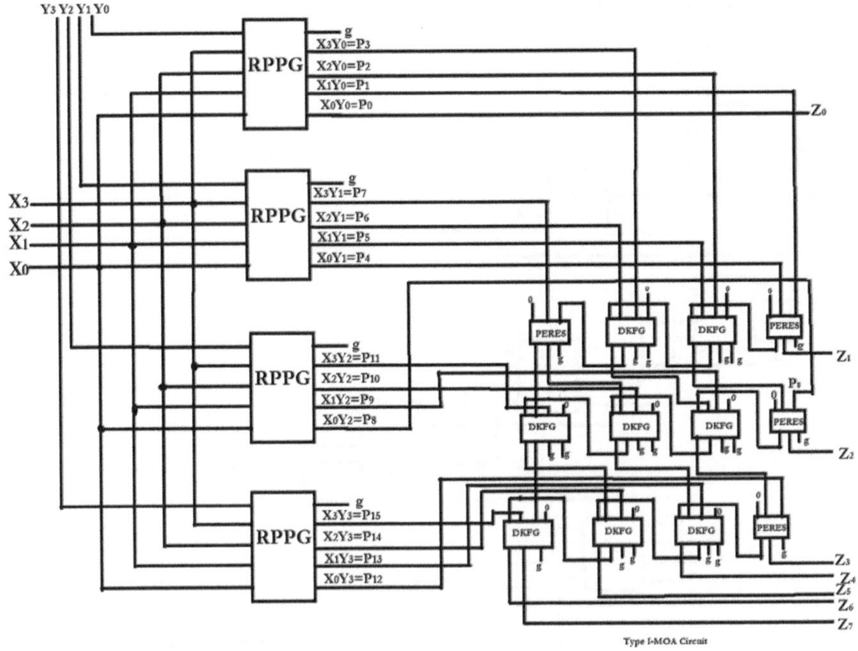

Fig. 38.8 Proposed multiplier (ORPM) TYPE-I

38.4.3.1 ORPM TYPE-I

This type of multiplier shown in Fig. 38.8 consists of the proposed partial product generation circuit (PPGC) and the proposed multi-operand addition circuit (MOPAC TYPE-I). Thus the circuit requires 16 number of gates altogether, 12 constant inputs and generates 23 garbage outputs. The multiplication outputs are Z_0, Z_1, Z_2, Z_3, Z_4, Z_5, Z_6 and Z_7. The total logical calculation is $(48\alpha + 36\beta)$.

38.4.3.2 ORPM TYPE-II

This type of multiplier shown in Fig. 38.9 consists of the proposed partial product generation circuit (PPGC) and the proposed multi-operand addition circuit (MOPAC TYPE-II). Thus the circuit requires 14 number of gates altogether, 10 constant inputs and generates 22 garbage outputs. The multiplication outputs are Z_0, Z_1, Z_2, Z_3, Z_4, Z_5, Z_6 and Z_7. The total logical calculation is $(44\alpha + 34\beta)$.

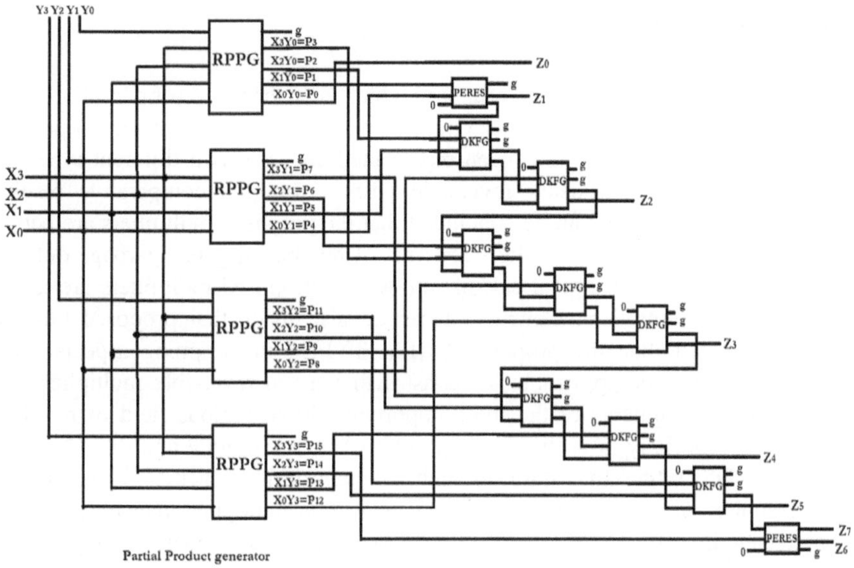

Fig. 38.9 Proposed multiplier (ORPM) TYPE-II

Table 38.1 Comparison between the proposed and the existing reversible multiplier designs

Multiplier	Gate count	Constant inputs	Garbage outputs	Total logical calculation
Bhagyalakshmi and Venkatesha [11]	40	52	52	$112\alpha + 36\beta$
Zhou et al. [12]	42	42	49	Not specified
Islam et al. [13]	44	44	52	$80\alpha + 36\beta$
Rangaraju et al. [14]	32	40	40	$112\alpha + 36\beta$
Basha and Amitha [15]	28	16	23	Not specified
Thaplyal et al. [16]	40	31	56	$80\alpha + 100\beta + 68\delta$
Thaplyal and Srinivas [17]	29	34	58	$110\alpha + 103\beta + 7\ 1\delta$
Shams et al. [18]	28	32	56	$92\alpha + 52\beta + 36\delta$
Nasar and Subbarao [19]	28	28	52	$80\alpha + 36\beta$
Proposed (ORPM TYPE-I)	16	12	23	$48\alpha + 36\beta$
Proposed (ORPM TYPE-II)	14	10	22	$44\alpha + 34\beta$

38.5 Comparison

A comparison result between the existing 4 × 4 reversible multipliers and the proposed multipliers is shown in Table 38.1. The comparison is done with respect to the number of gates, garbage output, constant input and total logical calculation.

38.6 Conclusion

Herein, a partial reversible gate is proposed which is further used for partial product generations to implement multiplier circuit. Two different types of multi-operand addition circuit (MOPAC) have been represented and also two different types of multipliers are designed using reversible gates. From the comparative results revealed in Table 38.1, it can be concluded that the proposed multipliers are better in respect to the existing multipliers in terms of number of gates, garbage output, constant input and total logical calculation. So the proposed multipliers are to be said optimized multipliers. Conclusion can be also made that proposed ORPM TYPE-II is better than the proposed ORPM TYPE-I in designing aspects. The designing can also be applicable for constructing $n \times n$ reversible multiplier circuits. As the multipliers are the most important part in various field of real-life applications especially in arithmetic logic unit (ALU) designing and digital signal processing (DSP), thus for future research, efficient design schemes for reversible more complex circuits in nanotechnology using this multiplier may be an interesting area to investigate. Alternate optimization methods are under investigation as a future work. Simulation of these proposed circuits may be another future work.

References

1. R. Landauer, Irreversibility and heat generation in the computational process. IBM J. Res. Dev. **5**, 183–191 (1961)
2. C.H. Bennett, Logical reversibility of computation. IBM J. Res. Dev. **17**(6), 525–532 (1973)
3. M.P Frank, Introduction to reversible computing: motivation, progress, and challenges, in *Proceedings of the 2nd Conference on Computing Frontiers* (2005), pp. 385–390
4. M. Perkowski, P. Kerntopf, Reversible logic invited tutorial, in *Proceedings*, EURO-MICRO, Warsaw, Poland, Sept 2001
5. M. Saiful Islam, Synthesis of fault tolerant reversible logic, in *IEEE* (2009)
6. R.P. Feynman, Quantum mechanical computers. Found. Phys. **16**(6), 507–531 (1986)
7. T. Toffoli, "Reversible computing", automata languages and programming, in *7th Colloquium of Lecture Notes in Computer Science*, vol. 85 (1980), pp. 632–644
8. E. Fredkin, T. Toffoli, Conservative logic. Int. J. Theor. Phys. **21**, 219–253 (1982)
9. K.A. Peres, Reversible logic and quantum computers. Phys. Rev. **32**, 3266–3276 (1985)
10. S. Mamataj, B. Das, A. Rahaman, An approach for realization of 2's complement adder subtractor using DKG reversible gate. IJETAE **3**(12), 205–209 (2013)
11. H.R. Bhagyalakshmi, M.K. Venkatesha, An improved design of a multiplier using reversible logic gates. IJEST **2**(8), 3838–3845 (2010)
12. R. Zhou, Y. Shi, J. Cao, H. Wang, Comment on design of a novel reversible multiplier circuit using HNG gate in nanotechnology. World Appl. Sci. J. **10**(2), 161–165 (2010)
13. M.S. Islam, M.M. Rahman, Z. Begum, M.Z. Hafiz, Low cost quantum realization of reversible multiplier circuit. Inf. Technol. J. **8**(2), 208–213 (2009)
14. H.G. Rangaraju, A. Babu, K.N. Muralidhara, Design and optimization of reversible multiplier circuit. IJCA **52**(10), 44–50 (2012)
15. M. Basha, G. Amitha, A reversible multiplier circuit design by using sleepy keeper technique. Int. J. Sci. Technol. **1**(1), 16–22 (2013)

16. H. Thaplyal, M.B. Srinivas, H.R. Arabnia, A reversible version of 4 × 4 bit array multiplier with minimum gates and garbage outputs, in *International Conference on Embedded System and Applications* (USA, 2005), pp. 106–114
17. H. Thaplyal, M.B. Srinivas, Novel reversible multiplier architecture using reversible TSG gate, in *IEEE International Conference on Computer Systems and Applications* (2006), pp. 100–103
18. M. Shams, M. Haghparast, K. Navi, Novel reversible multiplier circuit in nanotechnology. World Appl. Sci. J. **3**(5), 806–810 (2008)
19. S. Nasar, K. Subbarao, Design and implementation of MAC unit using reversible logic. IJERA **2**(5), 1848–1855 (2012)

Chapter 39
Analysis of Ambipolar Intrinsic Resistance of PIN Diode for Different Semiconductors Suitable for Power Devices

Shrabanti Das, Chiradeep Mukherjee, Saradindu Panda and B. Maji

Abstract The literature of power device must be aware of the fact of proper tradeoff between the choice of semiconductor material and the proper oxide along with it. With the growing semiconductor technology, the traditional semiconductor now has the promising competitor like silicon carbide and gallium nitride. The simulation considers Shockley-Read-Hall recombination model along with ambipolar diffusion mechanism. The analysis proves SiC-3C and SiC-6H as the most promising material for PIN diode manufacturing industries. The intrinsic region width and forward current are varied keeping another parameter constant while determining the resistance. The purpose of this work is to find out the ambipolar intrinsic resistance of PIN diode which will play key role in determining total ambipolar resistance of PIN diode.

Keywords Shockley-Read-Hall recombination · Ambipolar diffusion · Intrinsic resistance of power diodes · SiC power diode equations

S. Das (✉)
Jadavpur University, Kolkata, India
e-mail: shrabanti09@gmail.com

C. Mukherjee
Department of Electronics and Communication Engineering,
University of Engineering and Management, Jaipur, India
e-mail: chiradeep.1234321@gmail.com

S. Panda
Electronics and Communication Engineering Department,
Narula Institute of Technology, Kolkata, India
e-mail: saradindupanda@gmail.com

B. Maji
Electronics and Communication Engineering Department,
National Institute of Technology, Durgapur, India

© Springer India 2015

K. Maharatna et al. (eds.), *Computational Advancement in Communication Circuits and Systems*, Lecture Notes in Electrical Engineering 335,
DOI 10.1007/978-81-322-2274-3_39

List of Symbols

D_a Ambipolar diffusion coefficient in cm^2/s
D_n Electron diffusion coefficient in cm^2/s
τ_a Ambipolar lifetime in the base in s
τ_n Electron lifetime in s
τ_p Hole lifetime in s
p Hole concentration in/cm^3
N_B Base concentration in/cm^3
μ_n Mobility of electron in cm^2/s
μ_p Mobility of hole in cm^2/s
I_F Forward current
L_a Ambipolar diffusion length in micron
R_i Ambipolar intrinsic resistance in Ω

39.1 Introduction

The PIN diodes are used in phase shifter, limiter and attenuator circuits. The intrinsic resistance of PIN diode plays key role in finding out the total resistance of the PIN diode. Good estimation of diode impedance is required in microwave frequency, high frequency and low frequency operations. At low frequency intrinsic region acts as short circuit and in high frequency, impedance of the diode is dominated by conductivity of the intrinsic region [1]. The diffusion of positive and negative charge in the presence of an electric field is considered and implemented in this work. The ambipolar diffusion length is needed to govern the physics in intrinsic region [2]. PIN diode can operate in high reverse bias voltage and high switching frequency [3]. The application of PIN diodes is appreciable in optical physics as it is very sensitive to light waves. PIN diode was first proposed by Kleinmann [4] which has improved characteristics due to its smaller capacitance and lesser reverse leakage current [5]. Several parameters characterize PIN diode but there is still a lack of literature which describes the effects of intrinsic region width and forward current variation on intrinsic resistance.

This article analyses the impact of intrinsic region width and forward current on intrinsic resistance as it plays a key role to define the total ambipolar resistance of a PIN diode. Section 39.2 describes the basic equations which models the entire concepts using Shockley-Read-Hall recombination whereas Sect. 39.3 gives the detailed analysis of results obtained by the simulation using MATLAB. Modelling and analysis of intrinsic ambipolar resistance shows the path towards the total ambipolar resistance and hence Sect. 39.4 concludes with the findings obtained by the simulations which will help us in determining the impedance of PIN diode in small signal analysis.

39.2 Theory

39.2.1 PIN Diode Equation

Total impedance of PIN diode has two components; one of it is the resistance of intrinsic region. The intrinsic region of the PIN diode is the middle region amongst the p^+, n^- and n^+ subsequent regions as shown in Fig. 39.1.

Applying Shockley-Read-Hall [6, 7] recombination in the neutral base region, the ambipolar lifetime of the carrier is given as

$$\tau_a = \tau_p + \tau_n\left(\frac{p}{p + N_B}\right)$$

and the ambipolar diffusion coefficient is given as

$$D_a - D_n\left\{\frac{2p + N_B}{b(p + N_B) + p}\right\}$$

$$b = \frac{\mu_n}{\mu_p}$$

The intrinsic resistance of PIN diode is given as [8]

$$R_i = \left(\frac{V_t}{I_F}\right)\left(\frac{d_{\text{PIN}}}{L_a}\right)\tanh\left(\frac{d_{\text{PIN}}}{2L_a}\right)$$

The ambipolar diffusion length L_a is given as $\sqrt{\tau_a D_a}$ [9, 10]. The traps are located at the middle of the energy gap and the parameters $\tau_{a,}$ D_a and L_a are assumed to be completely injection dependent. The intrinsic resistance is casted by the width of intrinsic region d_{PIN} (as shown in Fig. 39.1) and the forward current I_F [11].

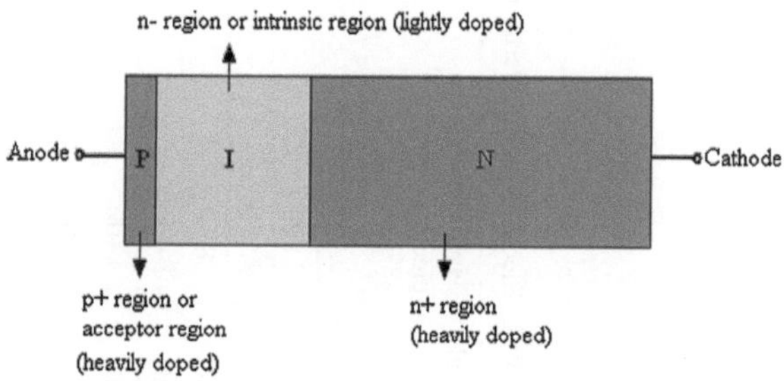

Fig. 39.1 Regions of PIN diode

Table 39.1 Electrical and material properties of different semiconductors

Material	Mobility of electron (cm^2 V^{-1} s^{-1})	Mobility of hole (cm^2 V^{-1} s^{-1})	τ_n (s)	τ_p (s)	Diffusion coefficients of electron (cm^2/s)	Diffusion coefficients of hole (cm^2/s)	Energy bandgap (eV)
Si	1,400	450	0.24×10^{-9}	0.09×10^{-9}	36	12	1.12
Ge	3,900	1,900	0.9×10^{-9}	0.9×10^{-9}	100	50	0.66
GaAs	8,500	400	4×10^{-8}	4×10^{-8}	200	10	1.424
SiC-3C	8,00	320	500×10^{-9}	500×10^{-9}	20	8	2.3
SiC-4H	900	120	10^{-9}	6×10^{-7}	22	3	3.3
SiC-6H	400	90	10^{-9}	4.5×10^{-7}	90	2	2.9
GaN-wZ	1,000	200	10^{-8}	10^{-8}	25	5	3.39
GaN-zB	1,000	350	10^{-8}	10^{-8}	25	9	3.2
InAs	40,000	500	3×10^{-8}	3×10^{-8}	103	13	0.35
InN	3,200	20	52×10^{-15}	52×10^{-15}	80	–	1.97

39.2.2 Semiconductor Materials for PIN Diode

The known semiconductor material hitherto involves Si, Ge and GaAs but silicon carbide (SiC) [12] claims the highest current operation. Not only the high forward current operation but also high dopant activation energy and strong impact of temperature coefficient of the device forces the involvement of SiC in the modelling of power device [13, 14]. The electrical and material properties of several semiconductors are given as in Table 39.1 [13, 15].

39.3 Analysis of Ambipolar Intrinsic Resistance for Different Width of Intrinsic Region and Forward Current

39.3.1 Variation of Intrinsic Region Width

The high injection of carriers from p^+ region to base region forces the analysis towards high current density in which intrinsic region width d_{PIN} becomes the determinant. The simulation bounds to 30–100 nm range in which other parameters are kept constant as shown in Table 39.2.

The intrinsic concentration is taken here as base concentration. Table 39.2 shows Si, Ge and Indium Nitride (InN) PIN diode possesses very high intrinsic resistance (even in the range of several mega-ohms) at 1 mA forward current. But SiC-3C and SiC-6H inhibit comparatively lower resistances of 766.1 and 670 Ω as obtained from the simulation shown in Fig. 39.2. Figure 39.3 shows the analysis graph which does not include Si, Ge and GaAs for their higher resistance values.

Table 39.2 Values of intrinsic resistance for different semiconductors

Sl. No.	Semiconductor	Ambipolar intrinsic resistance value (Ω)	Specifications
1	Si	1.245×10^5	Values taken at: T = 300 K Forward current $I_F = 1$ mA Acceptor Concentration: $10^{20}/cm^3$ Base concentration: $10^{14}/cm^3$ Reverse voltage: 5 kV
2	Ge	1.71×10^5	
3	GaAs	2,160	
4	SiC-3C	766.1	
5	SiC-4H	1,468	
6	SiC-6H	670.7	
7	GaN-wZ	6,344	
8	GaN-zB	5,087	
9	InAs	2,128	
10	InN	8.056×10^6	

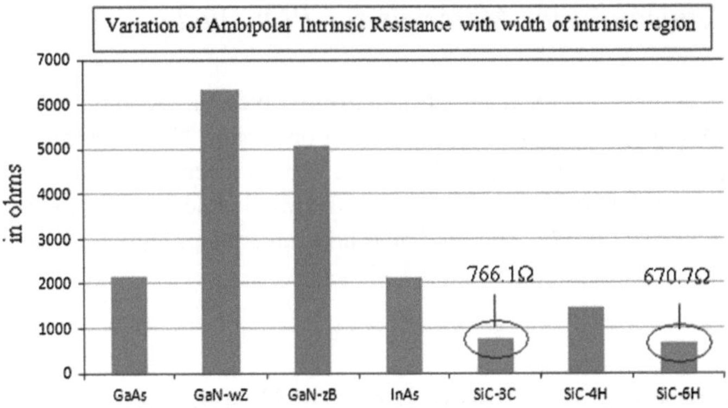

Fig. 39.2 Graph of ambipolar intrinsic resistance variation

Fig. 39.3 Results obtained by taking the variation in intrinsic region width

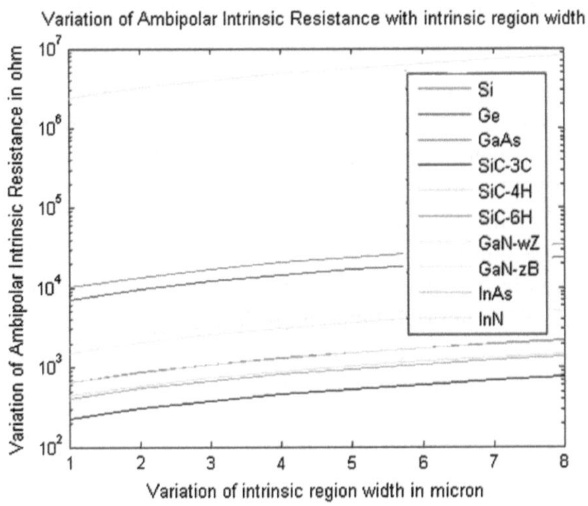

39.3.2 Variation of Forward Current

The forward current plays a pivotal role in determining the power rating of a power device [7]. Here, forward current is varied from 0.1 to 10 mA as shown in Table 39.3. The simulation as shown in Fig. 39.4, results in the value of 33.54 Ω intrinsic ambipolar resistance against SiC-6H PIN diode. The most promising semiconductors in designing of power device are hence silicon carbide and gallium nitride whose statistics are shown in Fig. 39.5.

Table 39.3 Values of intrinsic resistance for different semiconductors

Sl. No.	Semiconductor	Ambipolar intrinsic resistance value (Ω)	Specifications
1	Si	1,703	Values taken at: T = 300 K Width of *i*-region = 50 nm Acceptor concentration: $10^{20}/cm^3$ Base concentration: $10^{14}/cm^3$ Reverse voltage: 5 kV
2	Ge	377.1	
3	GaAs	108	
4	SiC-3C	38.31	
5	SiC-4H	73.42	
6	SiC-6H	33.54	
7	GaN-wZ	317.2	
8	GaN-zB	254.3	
9	InAs	156.4	
10	InN	4.028×10^6	

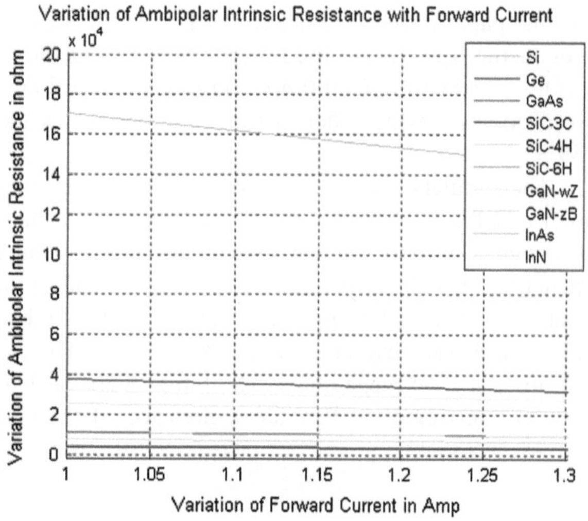

Fig. 39.4 Results of Ambipolar intrinsic resistance variation w.r.t. forward current for different semiconductors

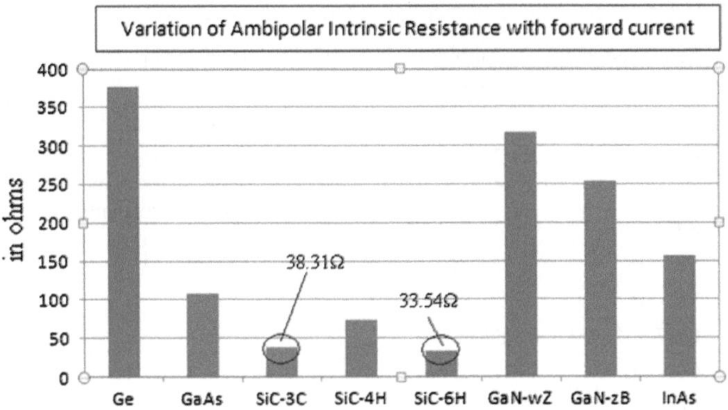

Fig. 39.5 Analytical graph of ambipolar intrinsic resistance variation for different semiconductors

39.4 Conclusions

In this paper intrinsic resistance of PIN diode were investigated for several semiconductors specifically used in power semiconductor devices. The intrinsic region width is varied in 30–100 nm and diode forward current is varied in 0.1–10 mA, respectively, taking other parameters constant as mentioned in Tables 39.2 and 39.3.

The simulation shows the value of ambipolar intrinsic resistances as 766.1 and 670.7 Ω for SiC-3C and SiC-6H, respectively, when the forward current remains constant. Similarly the variation of forward current increases the result as same materials which have the resistances of 38.31 and 33.54 Ω, respectively. Tables 39.2 and 39.3 show Si, Ge and InN PIN diode has higher ambipolar intrinsic resistances with respect to the SiC PIN diodes. The use of SiC lowers the resistance of up to several mega-ohms. The future work of determination of total ambipolar intrinsic resistance of PIN diode improves if the silicon carbide is chosen as the device material.

References

1. E. Gatard, R. Sommet, P. Bouysse, R. Quéré, An improved physics-based formulation of the microwave p-i-n Diode impedance. IEEE Microwave Wirel. Compon. Lett. **17**(3), 211–213 (2007)
2. L.H. González, E.B. Brito, S.N. Perez, M.A. Rodríguez, J.C. Yris, Proceedings modeling of SiC-PiN Diode with adjust of ambipolar diffusion length, in *SPEEDAM* 2010 *International Symposium on Power Electronics, Electrical Drives, Automation and Motion* (2010), pp. 541–544
3. N. Kim, W. Min, K. Sung, K. Suzuki, Y. Tanaka, H. Ohashi, Electric and physical characteristics of a SiC-PiN diode for high-power devices. J. Korean Phys. Soc. **52**(6), 1881–1885 (2008)

4. D.A. Kleinmann, Forward characteristics of the PIN diode. Bell Syst. Tech. J. **35**, 685–706 (1956)
5. F.J. Ramirez Jimenez, L. Mondragón-Contreras, P. Cruz-Estrada, Application of PIN diodes in physics research. in *Carretera México-Toluca S/N, La Marquesa, Ocoyoacac*. 57150, *MEXICO* (2006)
6. N.W. Ashcroft, N.D.Mermin, *Solid State Physics* (Brooks Cole, Boston, 1976)
7. D.T. Morisette, J.A. Cooper, Theoretical comparison of SiC PiN and Schottky diodes based on power dissipation considerations. IEEE Trans. Electron. Devices **49**(9), 1657–1664 (2002)
8. R. Caverly, G. Hiller, Microwave resistance of gallium arsenide and silicon p-i-n diode, in *IEEE MTT-S Digest* (1987)
9. M.P. Mikhailova, M. Levinshtein, S. Rumyantsev, M. Shur, *Handbook Series on Semiconductor Parameters*, vol. 1. (World Scientific, London, 1996), pp. 147–168
10. S.A. Jensen, J. Versluis, E. Canovas, J.J.H. Pijpers, I.R. Sellers, M. Bonn, Carrier multiplication in bulk indium nitride. Appl. Phys. Lett. **101**(22), 222113 (2012)
11. S.M. Sze, *Physics of Semiconductor Devices* (Wiley, New York, 1989)
12. S. Bellone, F.G. Della Corte, L.F. Albanese, F. Pezzimenti, An analytical model of the forward I-V characteristics of 4H-SiC p-i-n diodes valid for a wide range of temperature and current. IEEE Trans. Power Electron. **26**(10), 2835–2843 (2011)
13. Y. Arafat, F.M. Mohammedy, M.M.S. Hassan, Optical and other measurement techniques of carrier lifetime in semiconductors. Int. J. Optoelectron. Eng. **2**(2), 5–11 (2012)
14. Z.C. Feng, J.H. Zhao, *Silicon Carbide: Materials, Processing and Devices* (Taylor & Francis, New York, 2004)
15. J. Sun, I.G. Ivanov, R. Liljedahl, R. Yakimova, M. Syväjärvi, Considerably long carrier lifetimes in high-quality 3C-SiC (111). Appl. Phys. Lett. **100**, 252101 (2012)

Chapter 40
A Comparative Study of Single Electron Threshold Logic-Based and SET-MOS Hybrid Based Half Subtractor

Arpita Ghosh, Amit Jain, N. Basanta Singh and Subir Kumar Sarkar

Abstract In the present work, we have implemented a half subtractor using two different approaches, single electron threshold logic-based approach and SET-MOS hybrid approach. The logic operation of the designed circuits is tested using TSPICE and Monte–Carlo-based simulation tool SIMON. The stability of the threshold logic-based circuit is tested using the stability plots. Further we compared the performances to characterize the advantages and disadvantages of both the approaches. The proper functioning of both the circuits is successfully verified by observing the simulated output waveforms.

Keywords Single electron tunneling (SET) · Threshold logic · SET-MOS · SIMON · MIB model · Stability plot · Half subtractor · Macro model

40.1 Introduction

Now a days, the main concern behind any circuit design lies within its size. The conventional MOS technology facing numerous problems in the nanometer regime. So for further reduction in size [1], we have to look for different new technologies such as resonant tunnel diodes (RTD), carbon nanotubes (CNT), single electron tunneling (SET [2]), etc. Another approach that can also be used to combine the advantages of conventional MOS technology and single-electron tunneling technology known as the hybrid SET-MOS [3] technology.

A. Ghosh (✉)
RCC Institute of Information Technology, Kolkata 700015, India
e-mail: arpita161@gmail.com

A. Jain · S. Kumar Sarkar
Jadavpur University, Kolkata, India

N. Basanta Singh
Manipur Institute of Technology, Manipur University, Imphal 795004, India

© Springer India 2015

K. Maharatna et al. (eds.), *Computational Advancement in Communication Circuits and Systems*, Lecture Notes in Electrical Engineering 335,
DOI 10.1007/978-81-322-2274-3_40

A half subtractor is one of the simplest but important part of any digital system circuit. It can be designed using different approaches. Throughout the paper, we have tried to explain two approaches of half subtractor design one is single electron threshold logic [4–8] based and another one is SET-MOS hybrid approach. The complete paper is divided into four parts. First part gives the introduction to the work that we have presented here. Second and third part consist of the implementation of half subtractor using both the approaches along with their corresponding simulated input–output waveforms and also their corresponding principle of operations. The fourth section includes the details of the comparative analysis between the two approaches of half subtractor design presented in this paper.

40.2 Half Subtractor Implementation with Single-Electron Threshold Logic

The single-electron threshold logic is mainly based on the comparison between the threshold and the weighted sum. Equation (40.2) gives the value of the function *Y(X)*, where *T* is the threshold. The value of *T* is calculated from Eq. (40.1) by putting the values of $C_{\Sigma}^{p} = C_b + \sum_{k=1}^{r} C_k^P$, $C_{\Sigma}^{n} = C_0 + \sum_{l=1}^{z} C_l^n$, and V_b. Here C_b capacitor is connected to bias voltage. Figure 40.1 shows of threshold logic basic symbol where $x_1, x_2, \ldots x_n$ are the inputs and their corresponding weights are denoted with $w_1, w_2, \ldots w_n$. According to Eq. (40.3) the logic values are generated.

$$T = \frac{1}{2}\left(C_{\Sigma}^{p} + C_{\Sigma}^{n}\right)e - C_{\Sigma}^{n} C_b V_b \tag{40.1}$$

$$Y(X) = \sum_{i=1}^{n} w_i x_i - T \tag{40.2}$$

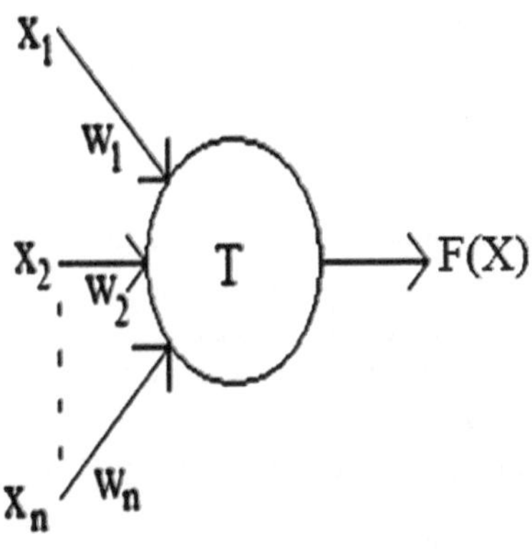

Fig. 40.1 Symbol of threshold gate

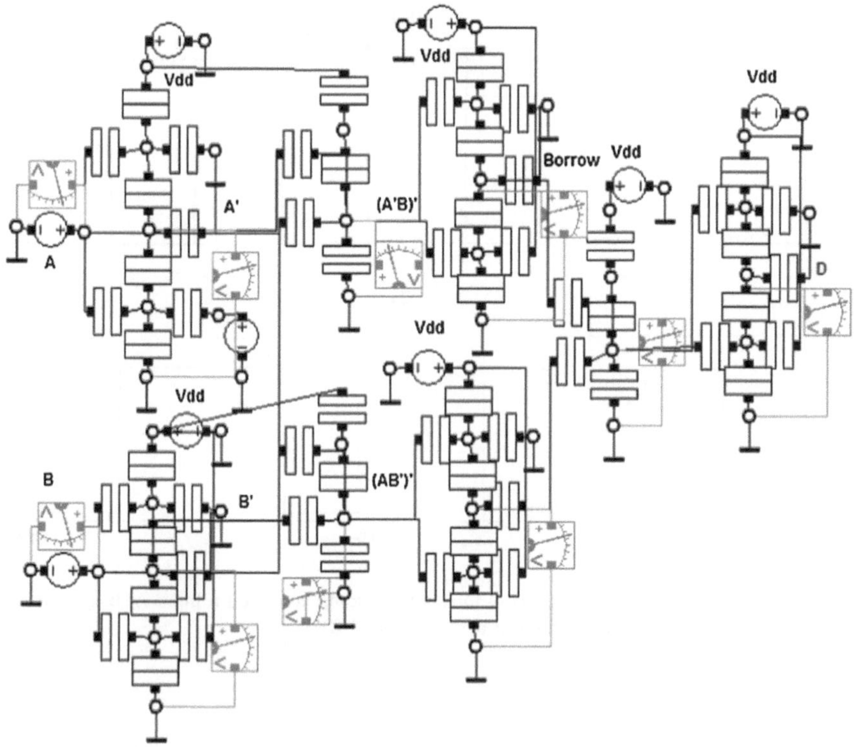

Fig. 40.2 Half subtractor circuit using single electron threshold logic

$$F(X) = \text{sgn}\{Y(\text{X})\} = \begin{cases} 0 \text{ if } Y(X) < 0 \\ 1 \text{ if } Y(X) \geq 0 \end{cases} \quad (40.3)$$

Here the half subtractor is designed using the single-electron threshold logic concept. Totally 23 tunnel junctions and 37 capacitors are used for the design with the power supply voltage V_{dd} of 0.016 V. A and B are the two inputs and D and Borrow are the two outputs. Totally three NAND gates and five inverters are required for the proposed circuit as shown in Fig. 40.2. The proposed circuit is designed and simulated in Monte–Carlo technique-based software SIMON [9, 10]. The simulated input and output waveforms are shown in Fig. 40.3a–d. The output waveforms shown in Fig. 40.3c, d ensure the logical correctness of the proposed circuit. Figure 40.3c indicates the output waveform for the difference between two input signals A and B and similarly Fig. 40.3d gives the output waveform of the borrow.

Figure 40.4 shows the stability plot [11] in which the white part indicates the combination of input value for which the output is stable. Otherwise the grey points indicate charge fluctuations. Here the range of charge fluctuation is from 0 to 4.16e−19. The points with 0 charge fluctuation are the stable points. Here V1 and V3 are, respectively, input signal A and B.

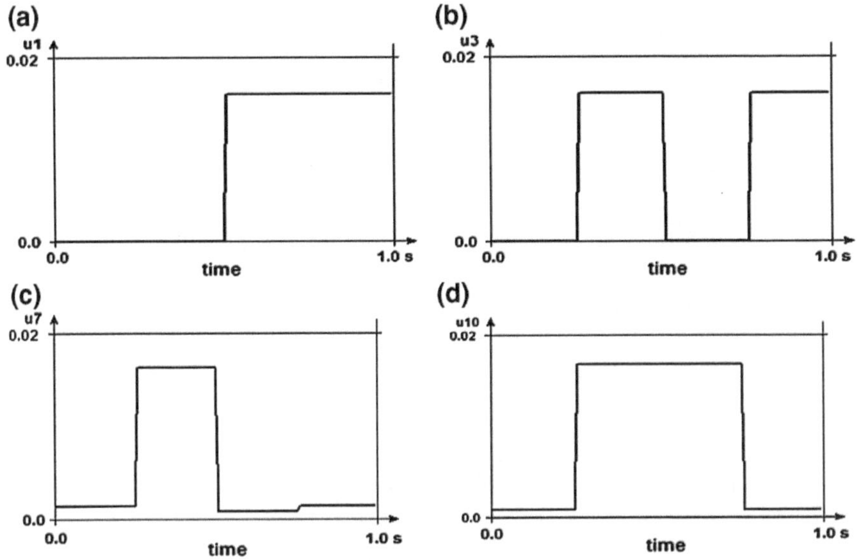

Fig. 40.3 Input and output waveforms of threshold logic-based circuit **a** Input A, **b** Input B and **c** Output borrow, **d** Output D (difference between A and B)

Fig. 40.4 Stability plot of single electron threshold logic based half subtractor

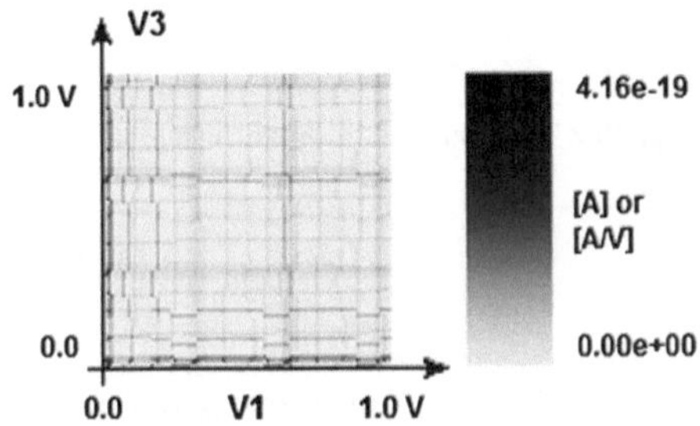

40.3 Half Subtractor Implementation with SET-MOS Hybrid

The combination of the different functionalities of SET and MOS can be achieved by combining them. Here we have shown the half subtractor circuit designed in this approach where three hybrid NAND gates and two hybrid inverters are needed. The parallel pMOS are connected to two series connected nSETs for a single NAND gate design. Totally 18 single electron tunnel junctions, 9 pMOS and 18 capacitors are required for the design along with the 0.8 V V_{dd}. For the co-simulation of the hybrid, SET-MOS different approaches are used (Fig. 40.5).

SET can be simulated using macro models, master equation, analytical models, etc. Here we have used BSIM 4.6.1 model for pMOS and MIB model for SET for

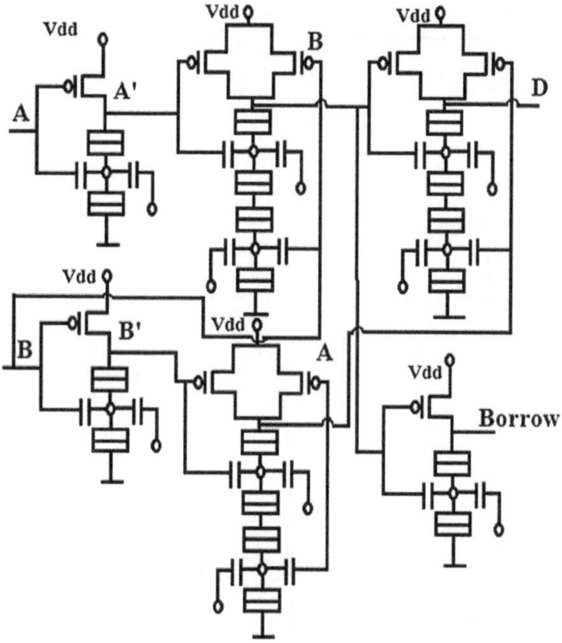

Fig. 40.5 Hybrid SET-MOS half subtractor with input A and B, and output D and borrow

simulating the complete circuit in SPICE environment. The simulated waveforms are shown in Fig. 40.6. Figure 40.6a, b shows the two input waveforms and c and d shows the difference between two inputs (D) and the Borrow respectively.

40.4 Comparative Analysis

Table 40.1 shows the comparative analysis of the half subtractor designed using two different approaches. The most important difference between the two approaches is that for a large circuit it becomes very difficult for anyone to implement it using SIMON as it uses Monte–Carlo approach for simulation and it is also time consuming. On the other hand, we can easily simulate the hybrid circuits using T-SPICE by only declaring the nodes and parameter values of SET and MOS with their corresponding model files. Moreover, the room temperature operation of SET-MOS gives much better result than the limited operation of the single-electron threshold logic only as the charge fluctuation increases with the increase in temperature. Total number of elements required, incase of hybrid SET-MOS approach design of half subtractor, is 80 whereas single-electron threshold logic based approach for half subtractor design requires 120 number of elements. Power supply requirement is also different for the two approaches, SET-MOS approach requires 0.8 V whereas the other approach needs only 0.016 V.

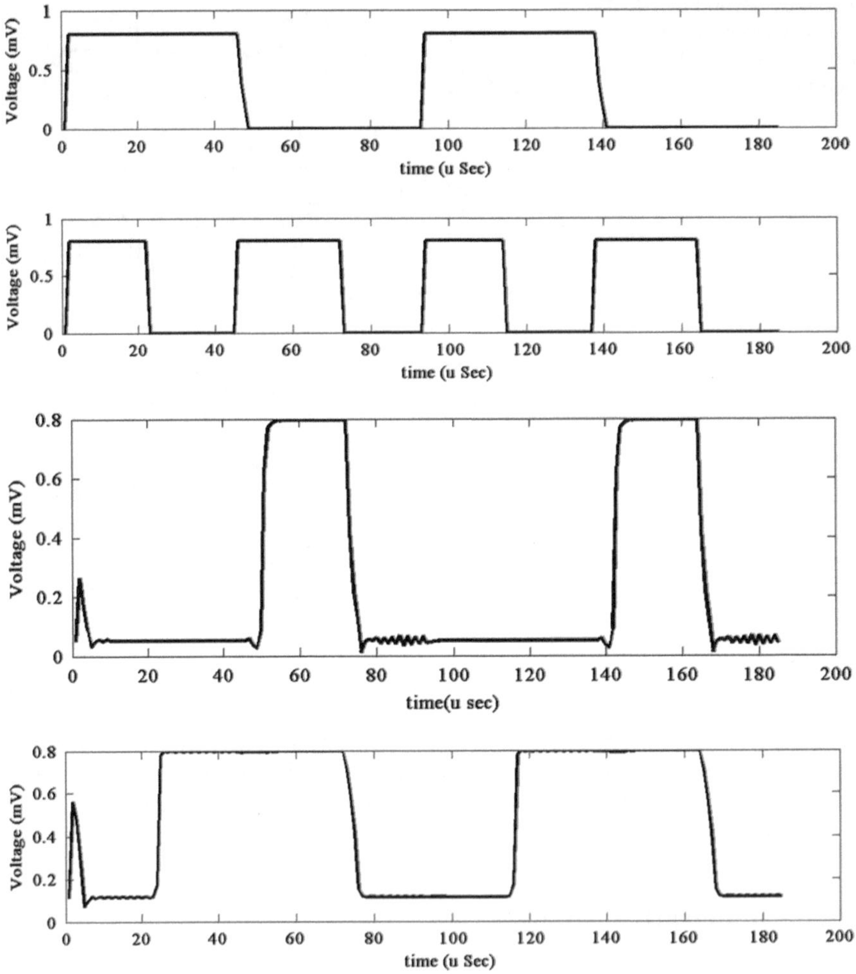

Fig. 40.6 Simulated waveforms **a** Input A, **b** Input B, **c** Output D and **d** Output borrow

Table 40.1 Comparison between half subtractor implementation with single-electron threshold logic approach and SET-MOS approach

Serial No.		Single electron threshold logic approach	SET-MOS hybrid approach
1.	Number of single electron tunnel junction	23	18
2.	Number of elements	60	45
3.	Number of capacitors	37	18
4.	Number of MOS	0	9
5.	V_{dd}	0.016 V	0.8 V
6.	Power consumption	219.073e−012 W	1.762077e−007 W

40.5 Conclusion

In this paper, we have discussed the design and analysed the performances of a half subtractor implemented using single-electron threshold logic based approach and SET-MOS hybrid-based approach. The detailed schematic diagrams for both the approaches along with verified simulation results were included. We also presented the stability plots for the input combinations, which showed that the designed half subtractor circuits are operating in a stable region for specified voltage ranges. Lastly, a comparison has been given to the performance of SET-MOS based approach and threshold logic-based design. The total number of elements, power consumption, and power supply requirement are compared and presented in the paper in a tabular form. It is observed that the power consumption of threshold logic-based circuit is three decades lower than the SET-MOS hybrid circuit.

References

1. Y. Tau, D.A. Buchanan, W. Chen, D. Frank, K. Ismail, S. Lo, G. Sai-Halasz, R. Viswanathan, H. Wann, S. Wind, H. Wong, CMOS scaling into the nanometer regime, in *Proceeding of the IEEE*, vol. 85 (1997), pp. 486–504
2. K. Likharev, Single-electron devices and their application. Proc. IEEE **87**(4), 606–632 (1999)
3. S. Mahapatra, A.M. Ionescu, *Hybrid CMOS Single—Electron-Transistor Device and Circuit Design* (Artech House, Inc., Norwood, 2006). ISBN:1596930691
4. C. Lageweg, S. Cotofana, S. Vassilladis, A linear threshold gate implementation in single electron technology, in *IEEE Computer Society Workshop on VLSI* (April 2001), pp. 93–98
5. M.J. Avedillo, J.M. Quintana, A. Rueda, E. Jiménez, A low-power CMOS threshold-gate. Electr. Lett. **31**, 2157–2159 (1995)
6. J. Fernandez Ramos, J.A. Hidalgo Lopez, M.J. Martin, J.C. Tejero, A. Gago, A Threshold logic gate based on clocked couple inverters. Int. J. Electron. **84**(4), 371–382 (1998)
7. S. Muroga, *Threshold Logic and Its Applications* (Wiley, New York, 1971)
8. C. Lageweg, S. Cotofana, S. Vassilidis, Single electron encoded latches and flip-flops. IEEE Trans. Nanotechnol. **3**(2), 237–248, (2004)
9. C. Wasshuber, *Computational Single-Electronics* (Springer, New York, 2001). ISBN:321183558X
10. C. Wasshuber, H. Kosina, S. Selberherr, SIMON–A simulator for single-electron tunnel devices and circuits. IEEE Trans. Comput. Aided Des. Integr. Circuits Syst. **16**(9), 937–944 (1997)
11. A. Jain, A. Ghosh, N.B. Singh, S.K. Sarkar, Stability and reliability analysis of hybrid CMOS-SET circuits—a new approach. J. Comput. Theor. Nanosci. **11**, 2519–2525 (American Scientific Publishers, 2014)

Chapter 41
Hybrid Single Electron Transistor-Based Low Power Consuming BCD Adder Circuit in 65 Nanometer Technology

Sudipta Mukherjee, Anindya Jana and Subir Kumar Sarkar

Abstract Traditional method of device designing is getting replaced by emerging trend of hybrid SET-CMOS logic. Mutual integration between the two has led towards ultra-dense circuitry as well as ultra low power consuming devices. Here a BCD adder circuit is realized with the help of hybrid single-electron transistor technology in 65 nm node. Power analysis and power-delay product for that adder circuit are also presented both numerically and graphically in comparison to the conventional CMOS logic, respectively.

Keywords MIB · BSIM4.6.1 · BCD adder · Hybrid SET-CMOS

41.1 Introduction

Single-electron transistor [1] is such a nano-dimensional device where electron is transferred one by one with the help of 'resonant tunneling' [2] and 'coulomb blockade' [3] effect. Proper bias voltages are to be provided to all the terminals to pass electrons from its source to island and then island to drain. This unique device when integrated with CMOS logic can produce ultra low power consuming designs. While designing this kind of device, sound attention is to be kept towards the thermal fluctuations, quantum confinement of electrons and all the concerned capacitance values attached to single electron transistor within the design.

S. Mukherjee (✉) · A. Jana · S.K. Sarkar
Department of Electronics and Telecommunication Engineering,
Jadavpur University, Kolkata 700032, India
e-mail: sudipta.conference@gmail.com

A. Jana
e-mail: anindya.jana@rediffmail.com

S.K. Sarkar
e-mail: su_sircir@yahoo.co.in

© Springer India 2015

K. Maharatna et al. (eds.), *Computational Advancement in Communication Circuits and Systems*, Lecture Notes in Electrical Engineering 335,
DOI 10.1007/978-81-322-2274-3_41

41.2 Single-Electron Transistor and Its Hybridization with CMOS

Single-electron transistor is made up of a nano-dimensional grain where charging energy is altered depending upon number of electrons residing over it as well as available thermal energy. Here, the mesoscopic grain is situated between two high-resistance tunnel junctions capacitively coupled to the gate electrode. To observe fruitful tunneling, energy to charge the island is to overcome thermal fluctuations [4]. That is, Ec >> KT (*K* = Boltzmann's constant and *T* is temperature in Kelvin) and resistance of the tunnel junction is made more valued than the quantum resistance, $h/e^2 \approx 26\ k\Omega$ (where *h* is Planck's constant). Here, $Ec = e^2/2C_\Sigma$. Where C_Σ comprises of all related capacitances values attached here with the island. Now SET being advantageous in consuming ultra low power, forming ultra dense circuitry as well as unique 'Coulomb Blockade' phenomenon, could replace MOS devices at all phase. But as it has lower current gain and not having any established room temperature operating technology, CMOS logic is needed to be integrated with this having relatively higher current driving capability. This emerging trend of hybridization [5–7] between SET and CMOS can now support all the latest inventions of science.

41.3 Concept of BCD Adder

BCD adder is such a circuit that performs addition of two distinct BCD digits and generates the sum in BCD as well. Here, if the 4-bit sum produced is equal or less than 9 then it is a perfect BCD and correction is not required. But if sum is more than 9, a corrective number is to be added to the sum. It is seen that, for decimal numbers 10–19, a corrective 6 (0110) is added. Likewise for numbers between 20 and 29, 12 (1100) is to be added; for 30–39, 18 (10010) must be added to make perfect BCD's and so on. Cause behind this can be elucidated as follows. BCD numbers are indeed a special kind of encoding that makes each digit in decimal to be represented in its respective binary sequence. That is why from decimal 0–9, all are same for both in BCD and binary. After that problematic situation arises as binary pattern continues to grow. When 10 comes, in binary it is 1010 but for BCD it is to be treated like two consecutive sequences of binary digits 1 and 0 that is, 0001 and 0000. So by ignoring the leftmost three zeroes it becomes 10,000 which is in fact 16. That is why 6 is needed to be added as corrective factor to this span of numbers. For all other phases mentioned before, corrective factors can easily be worked out also. Table 41.1 shows certain uncorrected sum output, made corrected BCD sum for each decimal numbers up to 19. This table can be easily extended to any value further.

Table 41.1 Decimal numbers with their corresponding uncorrected and corrected BCD sum

Decimal	Uncorrected sum					Corrected BCD sum				
	C4	s3	s2	s1	s0	Cout	S3	S2	S1	S0
0		0	0	0	0		0	0	0	0
1		0	0	0	1		0	0	0	1
2		0	0	1	0		0	0	1	0
3		0	0	1	1		0	0	1	1
4		0	1	0	0		0	1	0	0
5		0	1	0	1		0	1	0	1
6		0	1	1	0		0	1	1	0
7		0	1	1	1		0	1	1	1
8		1	0	0	0		1	0	0	0
9		1	0	0	1		1	0	0	1
10		1	0	1	0	1	0	0	0	0
11		1	0	1	1	1	0	0	0	1
12		1	1	0	0	1	0	0	1	0
13		1	1	0	1	1	0	0	1	1
14		1	1	1	0	1	0	1	0	0
15		1	1	1	1	1	0	1	0	1
16	1	0	0	0	0	1	0	1	1	0
17	1	0	0	0	1	1	0	1	1	1
18	1	0	0	1	0	1	1	0	0	0
19	1	0	0	1	1	1	1	0	0	1

From the above table, we are able to make K-map simplification for the condition of the 'correction needed' case and get a relationship as

$$\text{Cout} = \text{s3}^{*}\text{s2} + \text{s3}^{*}\text{s1} + \text{C4} \tag{41.1}$$

41.4 Hybrid SET-CMOS Based BCD Adder

Figure 41.1 reflects the circuit for hybrid single-electron transistor-based BCD adder. According to the desired connectivity, 6 individual full adders are replaced with their hybridized form. Moreover, 4 basic gates to produce signal Cout and S3 are also thereby designed with SET-CMOS logic. Set devices, having necessary bias voltage provided at their 2nd gate terminal, are placed instead of driver transistors. Load PMOSs are kept intact for each case.

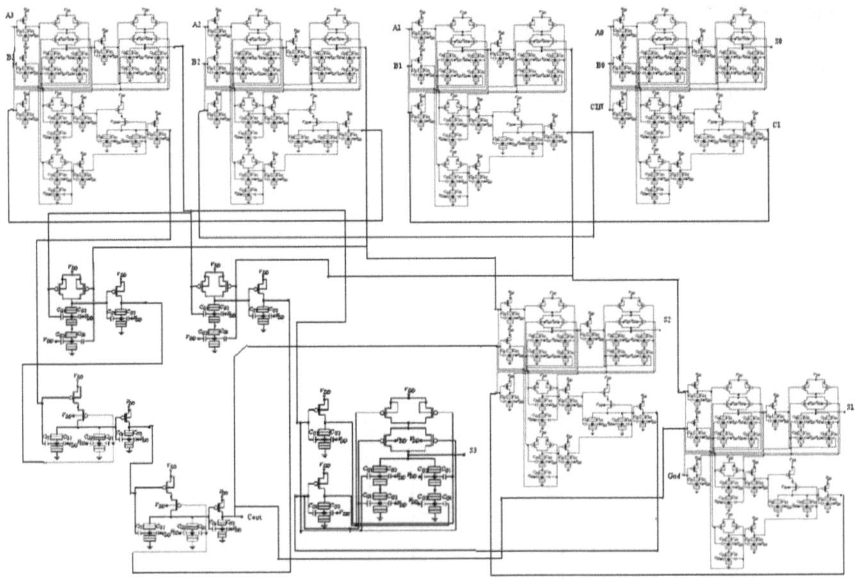

Fig. 41.1 Hybrid SET-CMOS based BCD adder circuit

41.5 Results Found at Simulation

Our design was simulated using MIB model and BSIM 4.6.1 model keeping VDD to be 0.7 V. Necessary parameters [4, 5] are tabulated below in Table 41.2.

Figure 41.2 shows input voltages from A0 to A3; B0 to B3 and Fig. 41.3 reflects the output waveforms we got at terminals S0–S3 along with Cout.

41.6 Observation

Typical power reduction is shown on Table 41.3 from conventional CMOS based circuit to the hybrid one.

Comparison has been plotted graphically below in Fig. 41.4.

Table 41.2 Parameters taken for simulation for SET and PMOS

Device	Parameters
Set	$C1$ = 0.27 aF, $C2$ = 0.125 aF, $C_D = C_S$ = 0.1 aF, $R_{TD} = R_{TS}$ = 1 MΩ
PMOS	L = 65 nm, W = 100 nm and for other parameters standard values taken from model BSIM 4.6.1

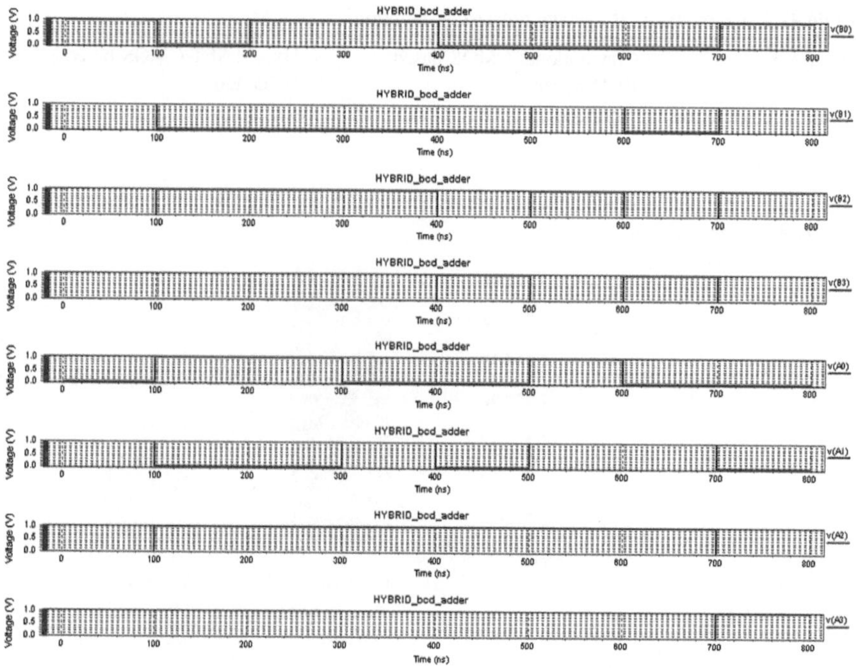

Fig. 41.2 Input waveforms

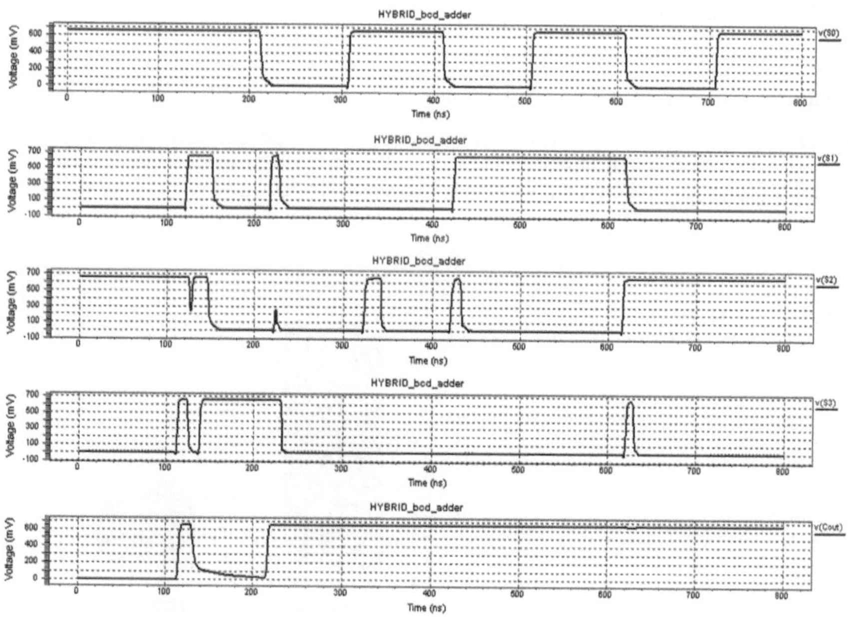

Fig. 41.3 Output waveforms

Table 41.3 Comparison of average power consumption

Parameter	Conventional CMOS based BCD adder	Hybrid SET-CMOS based BCD adder
Power consumption	3.26E-09 W	8.58E-10 W

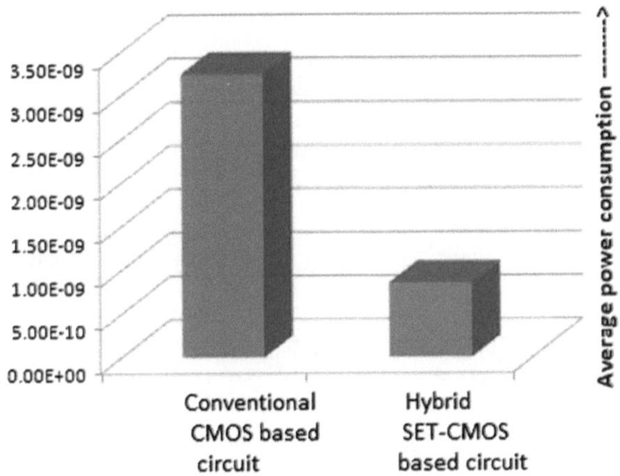

Fig. 41.4 Comparison of average power consumption

Power-delay product comparison is shown in Table 41.4 between conventional CMOS-based circuit and hybrid BCD adder circuit.

Graphical comparison has been depicted below in Fig. 41.5.

Table 41.4 Comparison of power-delay product

Parameter	Conventional CMOS based BCD adder	Hybrid SET-CMOS based BCD adder
Power-delay product	1.64E-16 W-s	5.18E-17 W-s

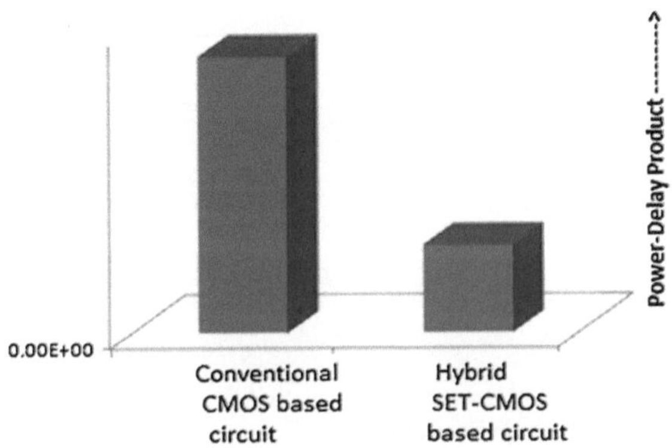

Fig. 41.5 Comparison of Power-delay Product

41.7 Conclusion

This paper indeed represents successful implementation of hybrid single-electron transistor-based BCD adder circuit in 65 nm technology node. Simulation results admit to hold the desired input–output characteristics. Reduction of the power consumption as well as PDP was observed both numerically and graphically. Reducing supply voltage VDD up to 0.7 V along with that keeping all island-related capacitance values remain in the atto-farad range [8], we actually have lowered the dynamic power consumption of our circuit, which has effectively lowered the total power consumption. Thus all these establish our proposed hybrid circuit to be used in today's ultra dense low-power designs.

Acknowledgments Subir Kumar Sarkar thankfully acknowledges the financial support for this research work obtained in form of UGC_UPE phase-II under 'Device and System' with Ref. No: R-11/43/2013.

References

1. K. Likharev, Single-electron devices and their applications. Proc. IEEE **87**, 606–632 (1999)
2. B. Jana, A. Jana, J.K. Sing, S.K. Sarkar, A comparative performance study of hybrid SET-CMOS based logic circuits for the estimation of robustness. J. Nano Electron. Phys. **5**(3), 3057 (1)–3057(6), 96–100
3. S. Mahapatra, A.M. Ionescu, *Hybrid CMOS Single-Electron-Transistor Device and Circuit Design,* (Artech House Inc., Norwood, 2006). ISBN: 1596930691
4. A.M. lonescu, M.J. Declercq, S. Mahapatra, K. Banerjee, J. Gautier, Few electron devices: towards hybrid CMOS-SET integrated circuits, in *IEEE Conference Publications* (2002), pp. 88–93
5. A. Jana, N.B. Singh, J.K. Sing, S.K. Sarkar, Design and simulation of hybrid CMOS-SET circuits. Microelectron. Reliab. (Elsevier) **53**(4), 592–599 (2013)
6. S. Mahapatra, K. Banerjee, F. Pegeon, A.M. Ionescu, A CAD framework for co-design and analysis of CMOS-SET hybrid integrated circuits, in *Proceeding of the ICCAD* (2003), pp. 497–502
7. A. Venkataratnam, A.K. Goel, Design and simulation of logic circuits with hybrid architectures of single-electron transistors and conventional MOS devices at room temperature. Microelectron. J. **39**(12), 1461–1468 (2008)
8. R. Parekh, A. Beaumont, J. Beauvais, Simulation and design methodology for hybrid SET-CMOS integrated logic at 22-nm room-temperature operation. IEEE Trans. Electron Devices **59** (4), 918–923 (2012)

Chapter 42
Design and Delay Analysis of Column Decoder Using NMOS Transistor at Nano Level for Semiconductor Memory Application

Sonali Bhowmik and Surajit Bari

Abstract Column decoder is the part of semiconductor memory used to retrieve the data from locations addressed by row decoder. In this paper 2–4 column decoder based on NMOS transistor has been designed with enhanced decoding speed at 150 nm channel length of transistor. The circuit is simulated using T-SPICE software. On voltage of the bit sources has been changed in accordance with the voltage of power supply V_{dd} to report the variation of delay for decoding column.

Keywords Column decoder · Memory · Delay · NMOS · Channel length · T-SPICE

42.1 Introduction

Data reading–writing speed and density of data are the important parameters to design semiconductor memory [1–4]. In this work we have focused the designing of 2–4 column decoder for data reading purpose from memory locations. Delay for a circuit depends on various parameters of MOS transistors like on voltage of bit source, threshold voltage, channel length, oxide capacitance, etc. In this paper we have tried to investigate the dependencies of delay on the magnitude of on voltage of bit sources used to justify the functionality of the column decoder. The circuit is designed at 150 nm channel length of MOS transistors using T-SPICE software [5].

S. Bhowmik · S. Bari (✉)
Electronics and Communication Engineering Department, Narula Institute of Technology, Kolkata, India
e-mail: surajit9bari@yahoo.co.in

S. Bhowmik
e-mail: sonali.nit.ece@gmail.com

© Springer India 2015

K. Maharatna et al. (eds.), *Computational Advancement in Communication Circuits and Systems*, Lecture Notes in Electrical Engineering 335,
DOI 10.1007/978-81-322-2274-3_42

42.2 Designing of 2–4 Column Decoder to Retrieve Data from Semiconductor Memory Locations

Column decoder is used to retrieve data from particular memory locations [6–8]. The schematic diagram of 2–4 column decoder is shown in Fig. 42.1. Two lines B1, B2 and their complements are used to select columns C1, C2, C3 and C4 according to Table 42.1. Lines B1 and B2 are driven by bit source. Also to test the column decoder each column line, C1, C2, C3 and C4, is connected with individual bit source. At a time one column line is selected. For example, when B1 = 0, B2 = 0 as shown in Fig. 42.1 transistors M1, M5 and M2 are in on state whereas transistors M3, M4 and M6 are in off state. So only column C1 is selected for B1 = 0 and B2 = 0. That means transistors M1 and M5 provide continuous path for data to propagate from column C1 only to the output terminals. Similarly, for other bit patterns of B1 and B2 columns C2, C3 and C4 are selected as shown in Table 42.1.

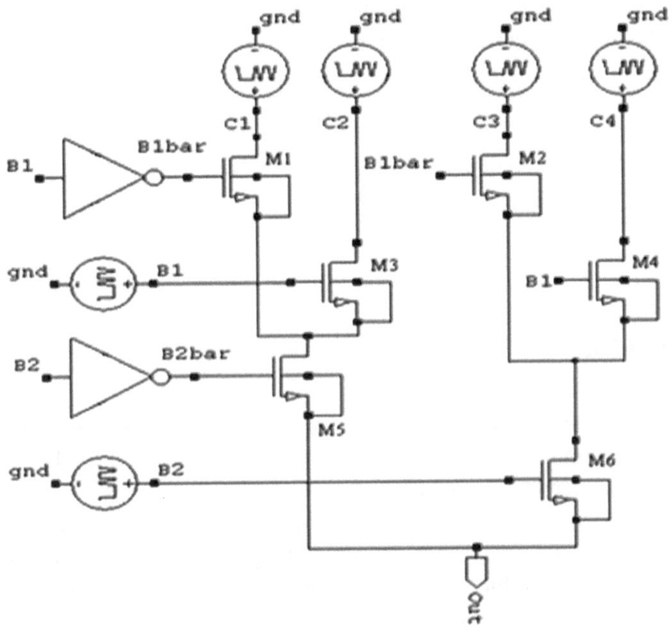

Fig. 42.1 Schematic diagram of 2–4 column decoder

Table 42.1 Selection logic of column decoder

B2	B1	C1	C2	C3	C4
0	0	1	0	0	0
0	1	0	1	0	0
1	0	0	0	1	0
1	1	0	0	0	1

In Table 42.1, logic ‘1’ against particular column means that column has been selected. In this paper we are not using memory array rather bit sources attached with each column lines have been used as data source for testing the column decoder.

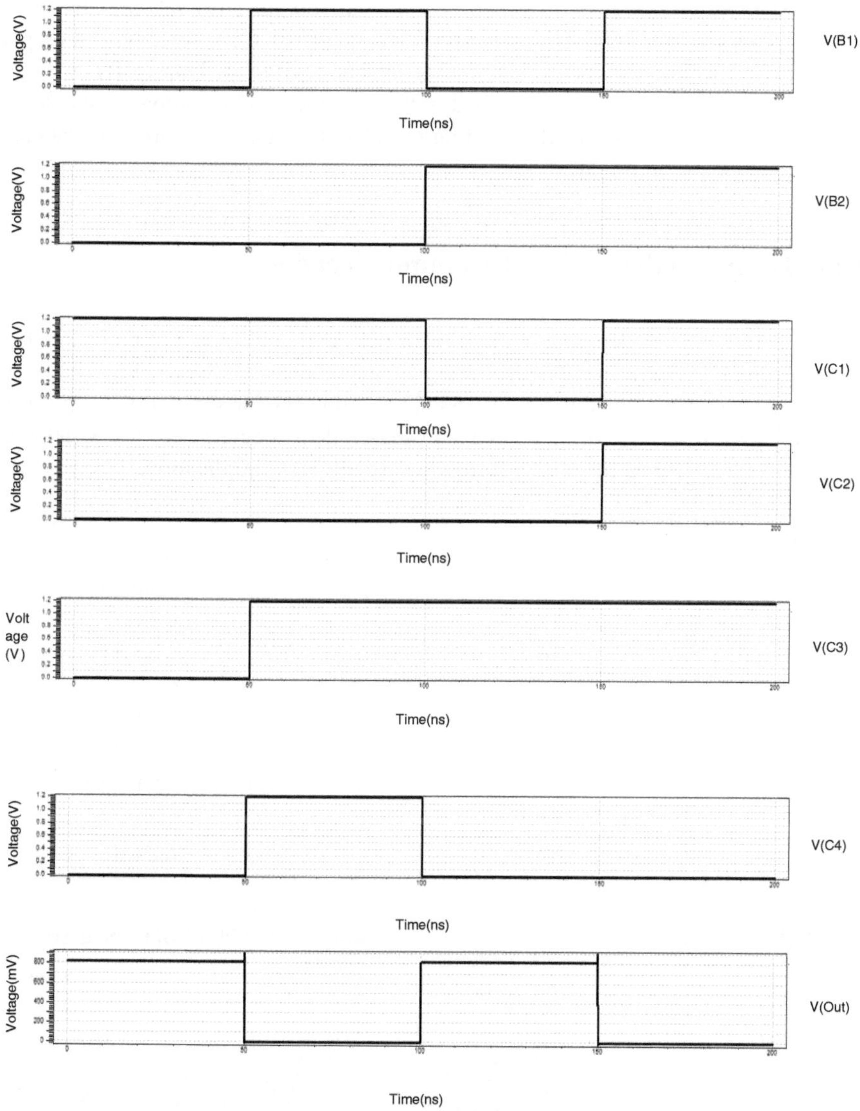

Fig. 42.2 Waveform to check functionality of 2–4 column decoder

42.3 Simulation Result

For verification of functionality the circuit has been simulated with the help of T-SPICE software. Transient analysis is performed for the circuit up to 200 ns. Bit patterns for B1, B2, C1, C2, C3, C4 and for out (output terminal) are shown in Fig. 42.2. As shown in Fig. 42.2 when B1 = 0 and B2 = 0, C1 is selected and output terminal data follow the data of bit source attached with C1. During B1 = 0 and B2 = 0, value of the bit source attached with C1 is logic '1' so output is also logic '1'. Other columns are selected according to Table 42.1 and corresponding data are retrieved at output terminal.

42.4 Delay Analysis of 2–4 Column Decoder

Variation of delay for decoding column with respect to on voltage of bit source has been reported in Table 42.2 and plotting of the same is shown in Fig. 42.3. To report delay we have considered column C2 and line B1 for simulation purpose. It is seen that as on voltage of bit source is increased delay decreases. For 1 V delay is 9.27×10^{-12} s. The result is satisfactory for small delay application.

Table 42.2 Variation of delay versus on voltage of bit source

On voltage of bit source (V)	Delay (s)
0.4	1.24E-10
0.5	6.39E-11
0.6	4.45E-11
0.7	3.12E-11
0.8	2.30E-11
0.9	1.44E-11
1	9.27E-12
1.1	5.57E-12
1.2	2.41E-12

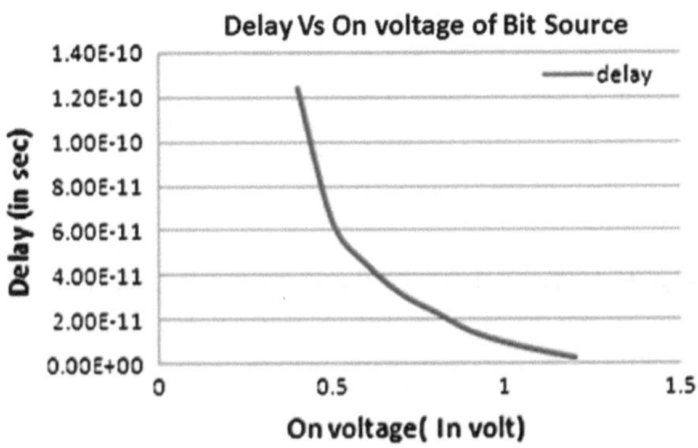

Fig. 42.3 Variation of decoding delay with respect to on voltage of bit source

42.5 Net List

The following net list is used to clarify the functionality and delay analysis of the column decoder.

```
*************** Subcircuits *****************
.subckt INV A Out Gnd Vdd
*-------- Devices: SPICE.ORDER > 0 --------
MN1 Out A Gnd 0 NH W=0.45u L=0.15u AS=0.3375p PS=2.4u
AD=0.3375p PD=2.4u
MP1 Out A Vdd Vdd PH W=0.90u L=0.15u AS=0.675p PS=3.3u
AD=0.675p PD=3.3u
.ends
********* Simulation Settings - Parameters and SPICE Op-
tions *********
*-------- Devices: SPICE.ORDER == 0 --------
XINV_1 B1 B1bar Gnd Vdd INV
XINV_2 B2 B2bar Gnd Vdd INV
*-------- Devices: SPICE.ORDER > 0 --------
MNMOS_1 C1 B1bar N_9 N_9 NH W=0.45u L=0.15u AS=0.3375p
PS=2.4u AD=0.3375p PD=2.4u
MNMOS_2 C3 B1bar N_11 N_11 NH W=0.45u L=0.15u AS=0.3375p
PS=2.4u AD=0.3375p PD=2.4u
MNMOS_3 C2 B1 N_9 N_9 NH W=0.45u L=0.15u AS=0.3375p
PS=2.4u AD=0.3375p PD=2.4u
MNMOS_4 C4 B1 N_11 N_11 NH W=0.45u L=0.15u AS=0.3375p
PS=2.4u AD=0.3375p PD=2.4u
MNMOS_5 N_9 B2bar Out Out NH W=0.45u L=0.15u AS=0.3375p
PS=2.4u AD=0.3375p PD=2.4u
MNMOS_6 N_11 B2 Out Out NH W=0.45u L=0.15u AS=0.3375p
PS=2.4u AD=0.3375p PD=2.4u
VVoltageSource_5 B1 gnd  BIT({0101}pw=50n on=1.2 off=0
rt=.1n ft=.1n delay=0n ht=50n lt=50n )
VVoltageSource_6 B2 gnd  BIT({0011}pw=50n on=1.2 off=0
rt=.1n ft=.1n delay=0n ht=50n lt=50n )
VVoltageSource_3 C3 gnd  BIT({0111}pw=50n on=1.2 off=0
rt=.1n ft=.1n delay=0n ht=50n lt=50n )
VVoltageSource_4 C4 gnd  BIT({0100}pw=50n on=1.2 off=0
rt=.1n ft=.1n delay=0n ht=50n lt=50n )
VVoltageSource_1 C1 gnd  BIT({1101} pw=50n on=1.2 off=0
rt=.1n ft=.1n delay=0n ht=50n lt=50n)
VVoltageSource_2 C2 gnd  BIT({0001}pw=50n on=1.2 off=0
rt=.1n ft=.1n delay=0n ht=50n lt=50n )
vdd vdd gnd 1.2
.include "C:\Documents and Set-
tings\Administrator\Desktop\dual.md"
.tran .01n 200n
.measure tran delay trig v(b1) rise=1 val=.6   targ
v(output1) fall=1  val=.6
.print B1 B2 C1 C2 C3 C4 out
.end
```

42.6 Conclusion

In this work 2–4 column decoder has been designed successfully. Functionality of the decoder is verified. It is seen that delay is 9.27×10^{-12} s. for on voltage 1 V. As depicted delay for decoding the column has been reduced with increasing on voltage of bit source. However, increasing the on voltage of bit source contradicts with low power design. So in addition to on voltage other parameters of MOS transistors context to delay and power have to be optimized for enhanced performance. For high density memory application our design can be extended.

References

1. S.M. Kang, Y. Leblebici, *CMOS Digital Integrated Circuits Analysis and Design* (Tata McGraw-Hill, New Delhi, 2003)
2. R.J. Baker, W.H. Li, D. Boyce, *CMOS Circuit Design, Layout, and Simulation* (PHI, New Delhi 2003)
3. J. Rabaey, A. Chandrakasan, B. Nikolic, *Digital Integrated Circuits A Design Perspective* (Pearson Education, Upper Saddle River, 2005)
4. J. Uyemura, *Introduction to VLSI Circuits and Systems* (Willey, India, 2007)
5. M.H. Rashid, *Introduction to PSpice Using OrCad for Circuits and Electronics* (PHI, New Delhi 2005)
6. T. Kobayashi, K. Arimoto, Y. Ikeda, M. Hatanaka, K. Mashiko, M. Yamada, Nakano, Takao A high-speed 64 K × 4 CMOS DRAM using on-chip self-timing techniques. IEEE J. Solid-State Circuits **SC-21**, 655–661 (1986)
7. K. Sasaki, S. Hanamura, K. Ueda, T. Oono, O. Minato, Y. Sakai, S. Meguro, M. Tsunematsu, T. Masuhara, M. Kubotera, H. Toyoshima, A 15-ns 1-Mbit CMOS SRAM. IEEE J. Solid-State Circuits **23**(5), 1067–1072 (1988)
8. F. Baba, H. Mochizuki, T. Yabu, K. Shirai, K. Miyasaka, A 64 K DRAM with 35 ns static column operation. IEEE J. Solid-State Circuits **SC-18**, 447–451 (1983)

Chapter 43
Design of Row Decoder Circuit for Semiconductor Memory at Low Power and Small Delay Using MOS Transistor at Nano Dimension Channel Length

Sonali Bhowmik and Surajit Bari

Abstract Row decoder is used to decode the address locations of memory. In this paper designing and power–delay analysis of 2–4 row decoder based on MOS transistor at channel length of 150 nm have been presented. Value of power supply voltage (V_{dd}) has been varied from 0.4 to 1.2 V to plot the variation of power consumption and delay for decoding the rows. According to simulation delay and power consumption are of the order of 10^{-11} s and 10^{-4} W, respectively. Tanner Spice (T-SPICE) software has been used to simulate the design.

Keywords Row decoder · Memory · Power · Delay · MOS · T-SPICE

43.1 Introduction

For decoding address locations of memory row decoders are used. Power and delay are the very important design metrics to design and fabricate VLSI circuits in nano dimension [1–4]. In our work designing of 2–4 memory row decoder has been focused using MOS transistor at channel length of 150 nm using T-SPICE software [5]. Power consumption and delay depend on several parameters of MOSFET like power supply voltage, channel length, threshold voltage, oxide thickness, etc. In this paper we have tried to investigate the nature of variation of power consumption and delay with respect to power supply voltage V_{dd} only.

S. Bhowmik · S. Bari (✉)
Electronics and Communication Engineering Department, Narula Institute of Technology, Kolkata, India
e-mail: surajit9bari@yahoo.co.in

S. Bhowmik
e-mail: sonali.nit.ece@gmail.com

© Springer India 2015

K. Maharatna et al. (eds.), *Computational Advancement in Communication Circuits and Systems*, Lecture Notes in Electrical Engineering 335,
DOI 10.1007/978-81-322-2274-3_43

43.2 Designing of 2–4 Row Decoder

Figure 43.1 shows the schematic diagram of 2–4 decoder using NMOS and PMOS transistors [6–8]. Two address lines A1 and A2 are used to decode four row lines R1, R2, R3 and R4. Circuit of Fig. 43.1 has been configured according to decoding logic mentioned in Table 43.1. When A1 = 0 and A2 = 0 then only R1 is at high logic and R2, R3, R4 are at logic low, i.e. only R1 is selected. Similarly, other rows are selected according to the Table 43.1. As shown in Fig. 43.1, each row line is connected with drain of PMOS transistor and PMOS transistors are always on by connecting the gates to ground. So design is made in such a way that at initial state all the row voltages are high. NMOS transistors are connected in such a fashion when particular row is selected by the bit pattern of A1 and A2 as in Table 43.1, voltage of other rows should go to logic low.

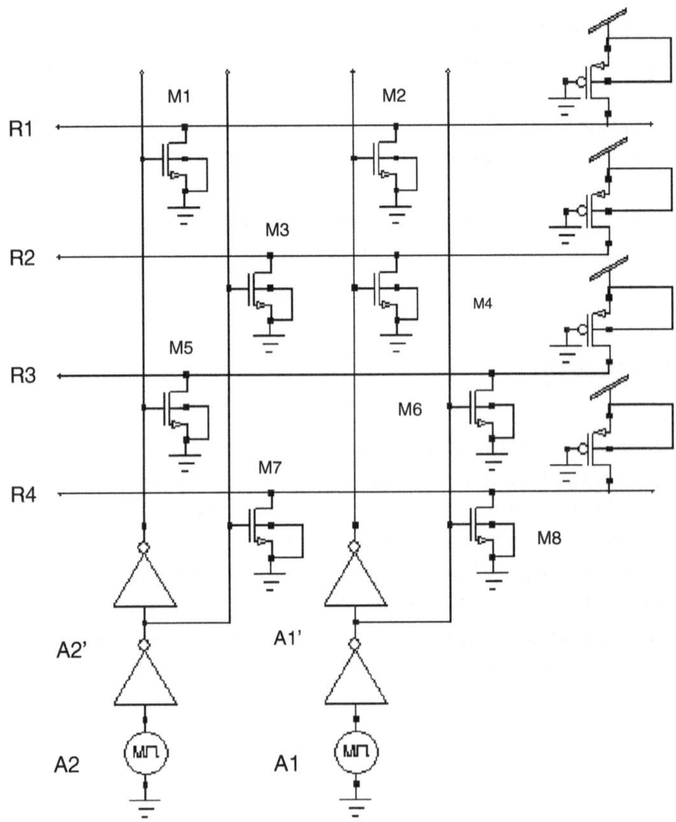

Fig. 43.1 Schematic diagram of 2–4 row decoder

Table 43.1 Decoding logic of 2–4 row decoder

A1	A2	R1	R2	R3	R4
0	0	1	0	0	0
0	1	0	1	0	0
1	0	0	0	1	0
1	1	0	0	0	1

For example, as shown in Fig. 43.1 when A1 = 0 and A2 = 0, A1 line voltage is low so M1 and M2 transistors are in off state therefore R1 is at logic high. Whereas R2, R3 and R4 rows are at logic low as transistors M3, M6, M7 and M8 are on. Similarly other rows are selected accordingly.

43.3 Simulation Results

Circuit is simulated using T-SPICE software. To display waveform for A1, A2, R1, R2, R3 and R4 transient analysis is performed up to 200 ns. Corresponding waveforms are shown in Fig. 43.2. Waveforms in Fig. 43.2 indicate the functional verification of the row decoder according to Table 43.1.

43.4 Delay and Power Analysis of 2–4 Row Decoder

Power supply voltage V_{dd} has been varied from 0.4 to 1.2 V to plot variation of power consumption and delay for row decoding. To measure delay for decoding, A1 has been considered as triggering node and R3 is considered as target node. Simulation results are documented in Table 43.2 to plot the variation of power consumption and delay for decoding rows with respect to V_{dd} as shown in Figs. 43.3 and 43.4, respectively.

It is seen that when V_{dd} increases delay decreases whereas power consumption increases.

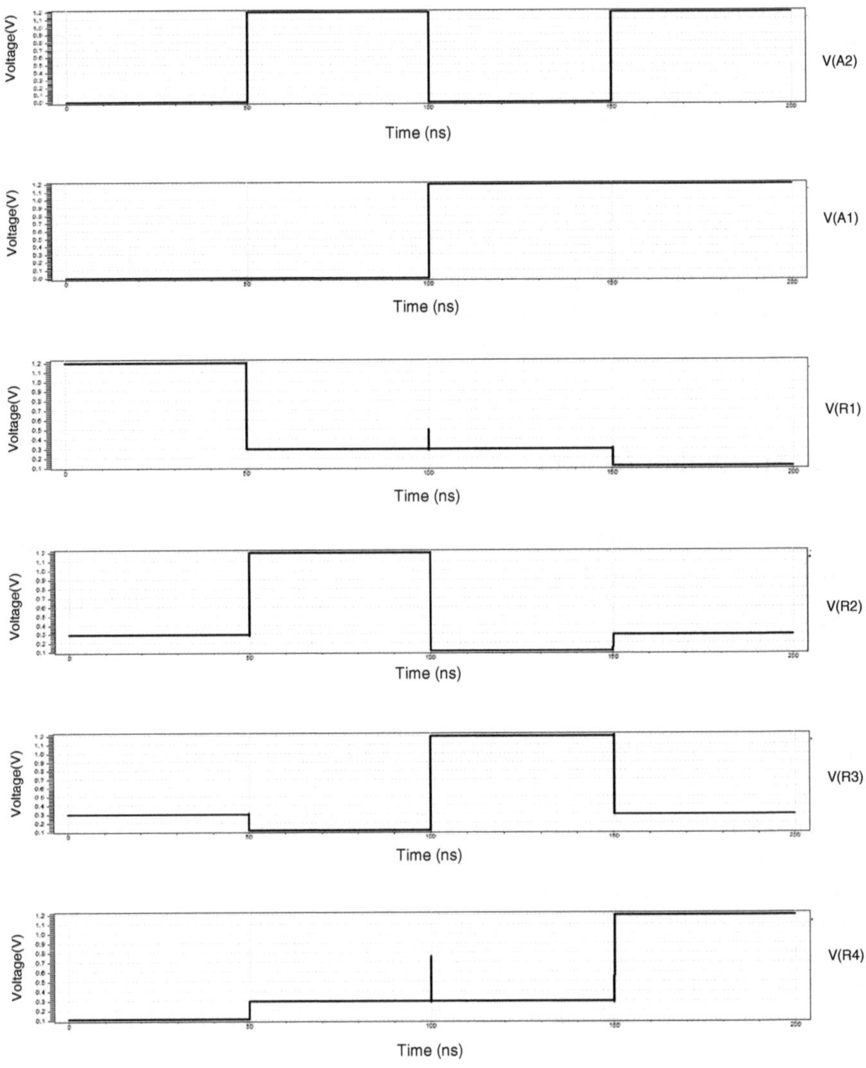

Fig. 43.2 T-Spice generated wave form for A1, A2, R1, R2, R3 and R4

Table 43.2 Variation of average power dissipation and delay for row decoding

V_{dd} (V)	Average power dissipation across row decoder (W)	Delay to decode row R3 (s)
0.4	4.79E-06	6.05E-10
0.5	2.90E-05	1.57E-10
0.6	8.59E-05	7.65E-11
0.7	1.71E-04	5.21E-11
0.8	2.78E-04	4.14E-11
0.9	4.07E-04	3.55E-11
1	5.57E-04	3.19E-11
1.1	7.27E-04	2.94E-11
1.2	9.18E-04	2.77E-11

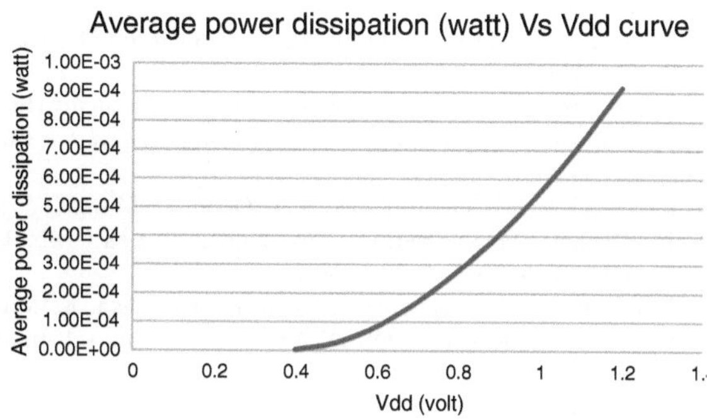

Fig. 43.3 Average power dissipation versus V_{dd}

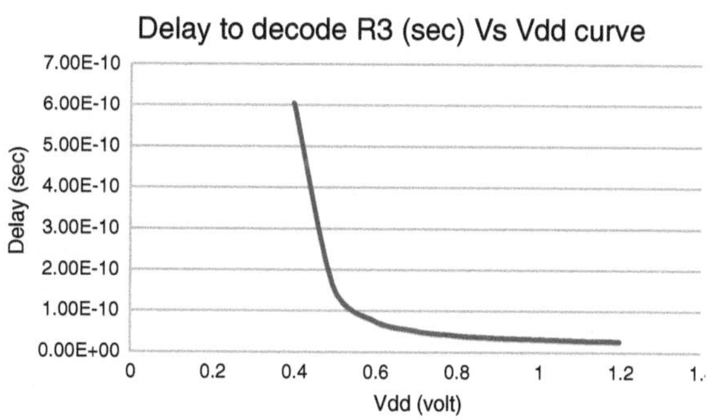

Fig. 43.4 Delay for decoding row R3 versus V_{dd}

43.5 Conclusion

In this work 2–4 row decoder has been designed and simulated at 150 nm channel length of MOS transistor to check the functionality. To analyse the variation of delay and power consumption with respect to power supply voltage, V_{dd} has been varied from 0.4 to 1.2 V. Value of power consumption and delay at 1 V is

5.57×10^{-4} W and 3.19×10^{-11} s, respectively. From the analysis it is seen that power consumption can be reduced by decreasing V_{dd}. Where as delay can be reduced by increasing V_{dd}. For small delay consideration higher value of V_{dd} contradicts with low power design. So optimization of power-delay product with respect to V_{dd} will enhance the performance of the design. Our design can be extended for very high density memory design.

References

1. S.M. Kang, Y. Leblebici, *CMOS Digital Integrated Circuits Analysis and Design* (Tata McGraw-Hill, New Delhi, 2003)
2. J. Rabaey, A. Chandrakasan, B. Nikolic, *Digital Integrated Circuits A Design ersptive* (Pearson Education, 2005)
3. R. Jacob Baker, H.W. Li, D. Boyce, *CMOS Circuit Design, Layout, and Simulation* (PHI, New Delhi, 2003)
4. J. Uyemura, *Introduction to VLSI Circuits and Systems* (India, 2007)
5. M.H. Rashid, *Introduction to PSpice using OrCad for Circuits and Electronics* (PHI, New Delhi, 2005)
6. K. Sasaki, S. Hanamura, K. Ueda, T. Oono, O. Minato, Y. Sakai, S. Meguro, M. Tsunematsu, T. Masuhara, M. Kubotera, H. Toyoshima, A 15-ns 1-Mbit CMOS SRAM. IEEE J. Solid-State Circuits **23**(5), 1067–1072 (1988)
7. T. Kobayashi, K. Arimoto, Y. Ikeda, M. Hatanaka, K. Mashiko, M. Yamada, T. Nakano, A high-speed 64 K × 4 CMOS DRAM using on-chip self-timing techniques. IEEE J. Solid-State Circuits **21**(5), 655–661 (1986)
8. G. Samson, N. Ananthapadmanabhan, S.A. Badrudduza, L.T. Clark, Low-power dynamic memory word line decoding for static random access memories. IEEE J. Solid-State Circuits **43**, 2524–2532 (2008)

Chapter 44
Constrained Optimization of CMOS Analog Circuits via All-Inversion Region MOS Model

Magnanil Goswami and Sudakshina Kundu

Abstract In this paper we furnish a novel approach of design automation and optimization of CMOS analog integrated circuits. This method offers an effective amalgamation between the principles of orthogonal-convex optimization and the all-inversion region MOS transistor model. We have observed that the constituting equations of the all-inversion region MOS model share great resemblance with the standard representation of orthogonal-convex functions. Therefore, these design equations, emerging from various device and circuit specifications can be modelled as the constraints of an orthogonal-convex optimization problem and can be evaluated automatically to ensure a globally optimal solution over a range of design scenarios. Additionally, there is a provision for design feasibility analysis with this semi-empirical approach.

Keywords All-inversion region MOS model · Analog integrated circuit · Design constraint · Orthogonal-convex optimization

44.1 Introduction

New trends in scaling technology have created an incessant demand for mixed-mode and analog integrated circuits making their computation an increasingly complex issue [1, 2]. Automatic analog circuit synthesis as an aid to improve the designer's productivity has been the theme of active research for more than two decades resulting in numerous optimization methodologies [3, 4] such as—Classical Optimization, Knowledge-Based Optimization, Global Optimization, etc.

M. Goswami (✉) · S. Kundu
Department of CSE and IT, West Bengal University of Technology, Kolkata, India
e-mail: magnanil.goswami.in@ieee.org

S. Kundu
e-mail: skundu@wbut.ac.in

© Springer India 2015
K. Maharatna et al. (eds.), *Computational Advancement in Communication Circuits and Systems*, Lecture Notes in Electrical Engineering 335,
DOI 10.1007/978-81-322-2274-3_44

Depending on how circuit performance and constraints are evaluated, an optimization approach can be broadly classified into simulation-based and equation-based optimization [5, 6].

Our discussion converges on the equation-based optimization technique, where the analog circuit design problem is formulated in terms of a set of device and circuit equations. These equations can either be an exact formulation or an approximation so as to obtain a compatible representation [5]. These equations are in turn used as the objective and constraint functions for the orthogonal-convex optimization problem. The outcome is overwhelming, as this semi-empirical optimization technique is capable of fetching prompt and accurate globally optimal design solutions [5, 7].

It is observed from [1, 2, 6] that orthogonal-convex optimization technique is extensively used for CMOS analog circuit optimization with a stringent presumption that all MOS transistors are operating in the strong inversion region. This assumption however truncates the design space to provide ease of modelling but precision gets compromised. Our approach countervails this impediment by dint of an all-inversion region transistor model to attain a truly global solution, which is practically more viable from the perspective of analog circuit modelling.

This paper is organized as follows: in Sect. 44.2, we briefly elaborate the method of orthogonal-convex optimization. Section 44.3 highlights the concept of all-inversion region MOS transistor model in context with the optimization strategy adopted here. The above bases are attributed to optimize the power consumption of a common-source amplifier [8, 9] in Sect. 44.4. Obtained results are interpreted and tallied with SPICE simulation data and the results are acquired from strong inversion MOS model-based optimization [1, 10] in Sect. 44.5. Our discussion is inferred in Sect. 44.6.

44.2 Orthogonal-Convex Optimization

Orthogonal-convex optimization is a special form of objective and constraint function-based mathematical optimization [7], also referred to as geometric programming [5]. The fundamental idea of modelling any practical problem using this method is to express the design objective and constraint functions in a commensurate manner prior to taking them for optimization. The modelling may take place either by exact formulation or by power law-based approximation of the functions. Although it is not certain whether every design aspect can be formulated as a compatible function, any practical problem, when duly formulated, can be solved efficiently and reliably using this method. Objectives and constraints are either represented by a monomial and/or a posynomial and/or a positive fractional power or pointwise maximum of posynomials [5, 7]. An orthogonal-convex optimization problem is of the form:

$$\begin{array}{lll} \text{Optimize} & f_0(x) & \\ \text{subject to} & f_i(x) \leq 1, & i = (1, \ldots, m) \\ & g_i(x) = 1, & i = (1, \ldots, p) \\ & x_i > 0, & i = (1, \ldots, n) \end{array} \quad (44.1)$$

where f_1, …, f_m are posynomial and g_1, …, g_p are monomial constraints of the vector x comprising n real positive variables x_1, …, x_n. The objective function f_0 is either a posynomial to minimize (only) or a monomial to maximize or minimize (scenario specific).

A function $g(x)$: $\mathbf{R^n} \rightarrow \mathbf{R}$ is said to be a monomial [5, 7] if its domain is the set of vectors with positive components and its values take a form given by the following power law expression:

$$g(x) = cx_1^{a_1} \ldots x_n^{a_n} \quad (44.2)$$

where $c > 0$ is the coefficient and vector $a = a_1, \ldots, a_n$ is the exponent of the monomial.

A function $f(x)$: $\mathbf{R^n} \rightarrow \mathbf{R}$ is said to be a posynomial [5, 7] if its domain is the set of vectors with positive components and its values take a form of non-negative sum of monomials:

$$f(x) = \sum_{k=1}^{K} c_k g_k(x) \quad (44.3)$$

where $g_k(x)$ are monomials and $c_k \geq 0$ for $k = 1, \ldots, K$.

Due to their non-convex property, monomials and posynomials are not used in optimization straightaway. They are converted into convex and affine functions, respectively, by logarithmic transformation of variables. Then the highly efficient primal-dual interior point method [7] is used to find the global solution for the problem.

After transformation of variables, objective and constraint functions of the problem definition is rewritten as a convex optimization problem:

$$\begin{array}{ll} \text{optimize} & f_0(y) = \log\left(\sum_{k=1}^{K_0} e^{a_{0k}^T y + b_{0k}}\right) \\ \text{subject to} & f_i(y) = \log\left(\sum_{k=1}^{K_0} e^{a_{ik}^T y + b_{ik}}\right) \leq 0, \quad i = (1, \ldots, m) \\ & g_i(y) = a_i^T y + b_i = 0, \quad i = (1, \ldots, p) \end{array} \quad (44.4)$$

where the new variables $y_i = \log x_i$, such that $x_i = e^{yi}$.

Orthogonal-convex optimization process, in conjunction with the analog CMOS circuit optimization, is summarized as a flowchart in Fig. 44.1.

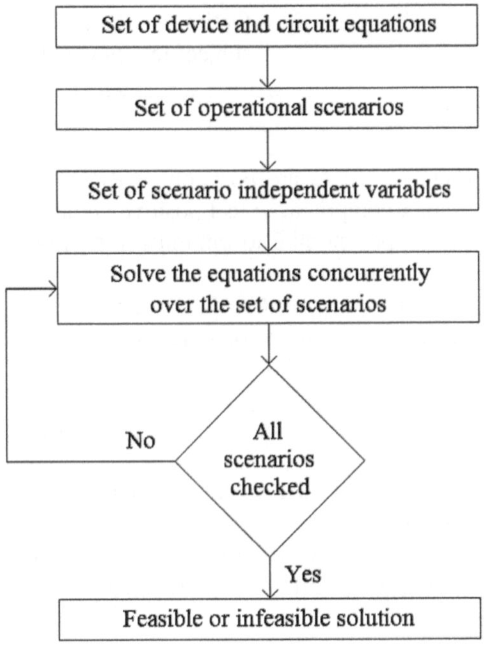

Fig. 44.1 Electronic circuit optimization process flowchart

44.3 All Inversion Region MOS Model

The advent of modern fabrication processes have caused zodiac shrinking of MOS transistor dimensions, leading to a 'design gap' in modern analog CMOS circuit technology due to higher electric field, decreased supply voltage headroom, reduced dynamic range, lower gain and similar detrimental attributes. Consequently, the square law-based MOS equations have failed to produce expected results from precise realization of these aspects [11]. This debilitation is superseded by a more coherent approach based on MOS Inversion Coefficient (IC) [11, 12]. It can arbitrate among the countervailing requirements of modern deep submicron CMOS technology by means of a complete characterization of MOS transistor process in lesser iterative steps.

MOS inversion coefficient (IC) is defined as follows:

$$\mathrm{IC} = \frac{I_D}{I_0 S} \tag{44.5}$$

where I_D is the drain–source current, S is the transistor aspect ratio and I_0 is the technology current, given as

$$I_0 = 2n\mu C_{\mathrm{OX}} U_T^2 \tag{44.6}$$

where n is slope factor, μ is low-field carrier mobility, C_{OX} is oxide capacitance per unit area and U_T is thermal voltage, approximately 26 mV at 300 K.

Generally, in forward saturation, IC < 0.1 corresponds to weak inversion and IC > 10 corresponds to strong inversion. Moderate inversion is represented by 0.1 < IC < 10. Transconductance efficiency (g_m/I_D) is a fundamental way of interpreting the MOS inversion coefficient. Thus (g_m/I_D) is useful in obtaining an insight into the MOS operation and all the key circuital elements of an analog designer's interest [11, 13, 14]. Despite its semi-empirical nature, (g_m/I_D) based approach enables designers to conduct an accurate analysis on the MOS behaviour [12, 14]. Simultaneously, the idea of the device operating regime can be procured from this method.

Using the current-based MOS model in [15], (g_m/I_D) in terms of inversion coefficient can be formulated as follows:

$$\frac{g_m}{I_D} = \frac{1}{n\phi_T} \frac{2}{1 + \sqrt{\mathrm{IC} + 1}} \tag{44.7}$$

The (g_m/I_D) expression in (44.7) is almost a process-technology-independent relationship for MOSFETs [15].

44.4 Design Formulation: Common Source Amplifier

The common-source amplifier we consider is illustrated in Fig. 44.2. Although it can be handled without the sledgehammer of orthogonal-convex optimization, its lucidity offers a clear percept of the optimization methodology. The objective is to optimize its power dissipation (P) subject to specified gain-bandwidth (GBW) and DC voltage gain (A_V) constraints for a given load capacitance (C_L).

Equation (44.8) is formulated in accordance with the standard representation of orthogonal-convex function and posynomial inequality is used in lieu of the equality constraint, to provide a more generalized design space.

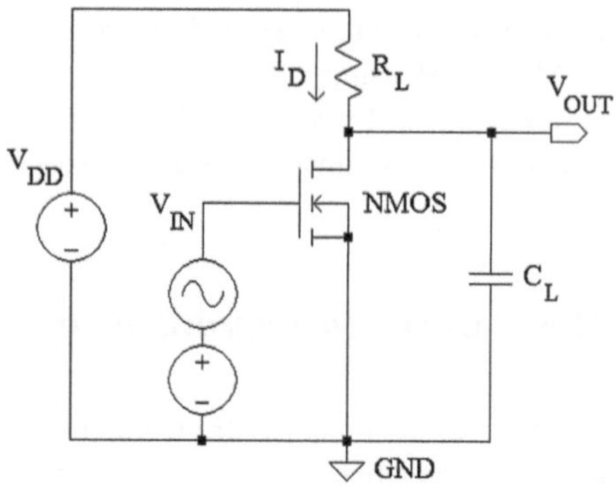

Fig. 44.2 Schematic of common-source amplifier

$$\left(\frac{n\phi_T}{2}\right)^2\left(\frac{g_m}{I_D}\right)^2(\text{IC}) + (n\phi_T)\left(\frac{g_m}{I_D}\right) \leq 1 \tag{44.8}$$

The DC voltage gain (A_V) is expressed as follows:

$$A_V = \left(\frac{g_m}{I_D}\right)V_A \tag{44.9}$$

where V_A is Early voltage. The expression in (44.9) is a monomial and it is constrained by a minimum gain requirement ($A_{V\min}$). Further accuracy can be introduced to (44.9) by considering the effect of the resistance (R_L). In that case, A_V may be treated as an inverse posynomial constraint.

The gain-bandwidth (GBW) is formulated as a monomial given by (44.10):

$$\text{GBW} = \frac{\left(\frac{g_m}{I_D}\right)I_D}{2\pi C_L} \tag{44.10}$$

where the value of C_L is circuit specific. The GBW expression in (44.10) is also constrained by a minimum gain-bandwidth requirement ($\text{GBW}_{\min}$).

Upper and lower bounds are imposed on device dimensions; i.e., $L_{\min} \leq L \leq L_{\max}$ and $W_{\min} \leq W \leq W_{\max}$. The final design problem formulation comprises five optimization variables, two equalities and six inequalities in monomial, and one inequality in posynomial and another inequality in inverse posynomial.

44.5 Result Interpretation

The device parameters used here are augmented from an LTspice level-12 model for 0.35 μm CMOS process at 300 K. Specifications and the constraints for the design example are listed in Table 44.1. The problem is modelled and implemented on MATLAB using GGPLAB toolbox [5] and the envisaged optimum values are summarized in Table 44.2. We have opted for a two-tier verification strategy to assess our approximated design approach:

44.5.1 Observed Improvements

Optimal values obtained from our method are compared with the approach presented by Hershenson et al. [10]. The fundamental difference between the two methods is that while Hershenson et al. method is valid for strong inversion only, our approach holds good for all three regions of inversion.

Table 44.1 Design constraints and specifications

Constraint	Specification
Device length	0.35 μm ≤ L ≤ 3.5 μm
Device width	1.0 μm ≤ W ≤ 10.0 μm
Supply voltage	3.3 V
Load capacitance	2 pF
Oxide thickness	7.575 nm
DC voltage gain	≥20 dB
Gain-bandwidth	≥10 MHz
Power	Minimize

Table 44.2 Optimal solution for design example

Parameter	Optimal value
Device length	0.35 μm
Device width	10 μm
Load resistance	208.01 kΩ
Drain current	6.1009 μA
Transconductance efficiency	20.5976 s/A
Inversion coefficient	2.1682
DC voltage gain	27.7199 dB
Gain-bandwidth	10 MHz
Power	20.133 μW

Using the fundamentals of Sect. 44.3, it can be inferred from the value of inversion coefficient in Table 44.2 that the NMOS device used in our circuit operates in moderate inversion. Comparison between the optimal values obtained from the two design approaches is summarized in Table 44.3. It can be clearly distinguished from Fig. 44.3 that our design approach significantly reduces the power consumption of the amplifier circuit under consideration.

Table 44.3 Comparison of design approaches

Parameter	Our method	Hershenson's method
Device length (μm)	0.35	1.0995
Device width (μm)	10	3.1853
Resistance (kΩ)	208.01	144.44
Current (μA)	6.1009	12.566
Gain (dB)	27.7199	24.2959
Gain-bandwidth (MHz)	10	10
Power (μW)	20.133	41.469

Fig. 44.3 Power consumption versus input voltage

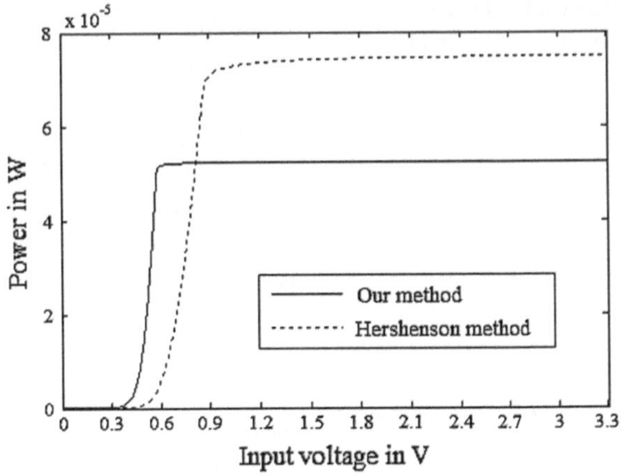

44.5.2 Verification Through Higher Fidelity Approach

We consider the estimated device dimension and resistance from the all-inversion region model-based orthogonal-convex optimization method to construct a common-source amplifier using a more sophisticated LTspice level-12 model for 0.35 μm process. To verify the coherence between the design approaches, we set an input-bias such that the drain current obtained from SPICE operating point analysis complies with the optimal drain current value in Table 44.2. An abridged operating point analysis, as illustrated in Fig. 44.4, is presented in this regard. In this context, the transconductance efficiency (g_m/I_D) versus input voltage (V_{IN}) graph is plotted in Fig. 44.5 and the DC voltage gain (A_V) versus frequency graph is plotted in Fig. 44.6.

Table 44.4 compares the key performance measures achieved by the proposed method with those acquired from the SPICE simulation data. Observed deviations between the two results are due to the approximations made in our design by ignoring some of the parasitic effects to ensure modelling compatibility. Yet, the performance measures predicted by our method are in proximal agreement with that of the higher fidelity SPICE simulations.

Fig. 44.4 Operating point analysis

```
V(vout):     2.03093          voltage
V(vin):      0.513585         voltage
V(vdd):      3.3              voltage
Id(M1):      6.10099e-006     device_current
Ig(M1):      0                device_current
Ib(M1):      -4.15795e-012    device_current
Is(M1):      -6.10098e-006    device_current
I(C1):       4.06187e-024     device_current
I(R1):       6.10099e-006     device_current
I(Vdd):      -6.10099e-006    device_current
I(Vg):       0                device_current
```

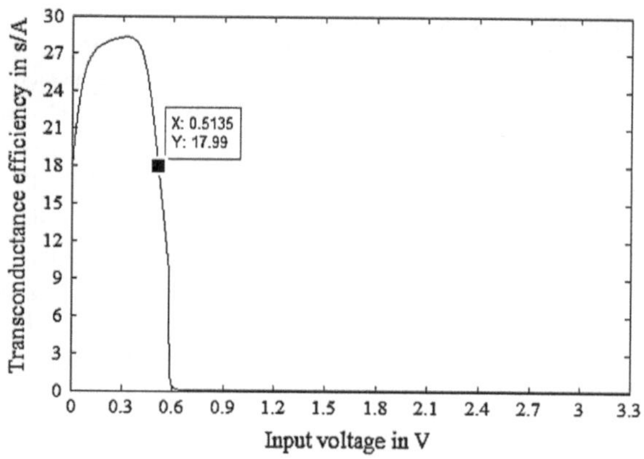

Fig. 44.5 Transconductance efficiency versus input voltage

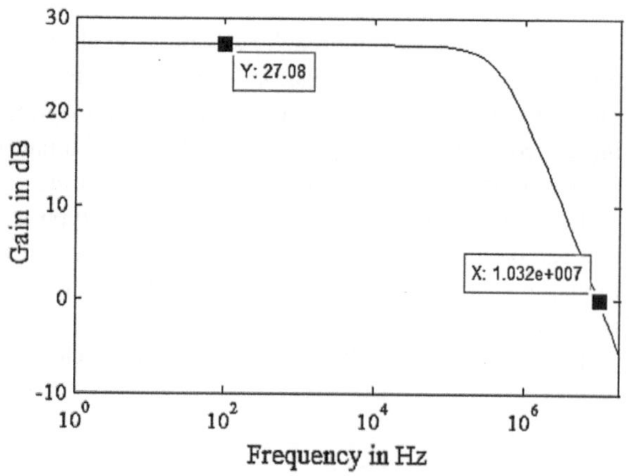

Fig. 44.6 Gain versus frequency

Table 44.4 Comparison of correctness

Performance specification	Our data	SPICE data
Transconductance efficiency (s/A)	20.5976	17.99
Gain (dB)	27.7199	27.0874
Gain-bandwidth (MHz)	10	10.327

Finally, Table 44.5 collates the percentage error arising in the key performance metrics of our approach alongside the method proposed by Hershenson et al., when tallied with LTspice level-12 and level-8 model based simulations respectively. It is evident from the results that our approach makes reasonably better predictions of the circuit behaviour compared to that of Hershenson's method.

Table 44.5 Percentage error estimation

Metric	Value	% Error
Gain	Our method: 27.7199 dB	2.34
	Level-12 simulation: 27.0874 dB	
	Hershenson's method: 24.2959 dB	14.6
	Level-8 simulation: 21.98 dB	
Gain-bandwidth	Our method: 10 MHz	3.17
	Level-12 simulation: 10.327 MHz	
	Hershenson's method: 10 MHz	40.13
	Level-8 simulation: 7.136 MHz	

44.6 Inference

The semi-empirical approach we propose here exploits the effectiveness of orthogonal-convex optimization and the flexibility of all-inversion region MOS model to pertain efficient and reliable globally optimal solutions to CMOS analog integrated circuit design problems with little human oversight. This amalgamated optimization strategy can be applied to demanding optimization problems as well.

Perhaps the only source of limitation is the type of constraints and specifications that can be handled. Hence, there is always room for incorporating higher precision modelling strategies. Since our method is valid for all three regions of MOS inversion, it offers greater flexibility in terms of the device operating regime and is certainly a higher fidelity approach over the existing alternatives.

References

1. M. Hershenson, S. Boyd, T. Lee, Optimal design of a CMOS op-amp via geometric programming. IEEE Trans. Comput. Aided Des. Integr. Circuits Syst. **20**, 1–21 (2001). http://www.stanford.edu/~boyd/papers/opamp.html
2. M. Hershenson, CMOS analog circuit design via geometric programming, in *Proceedings of the American Control Conference*, Boston, Massachusetts, (2004), pp. 3266–3271
3. B.A.A. Antao, Trends in CAD of analog ICs. IEEE Circuits Devices Mag. **12**, 31–41 (1996)
4. L.R. Carley, R.A. Rutenbar, How to automate analog IC designs. IEEE Spectr. **25**, 26–30 (1988)
5. S. Boyd, S.J. Kim, L. Vandenberghe, A. Hassibi, A tutorial on geometric programming. Technical Report, Stanford University and University of California (2005). http://www.stanford.edu/~boyd/gptutorial.html
6. P. Mandal, V. Visvanathan, CMOS op-amp sizing using a geometric programming formulation. IEEE Trans. Comput. Aided Des. Integr. Circuits Syst. **20**, 22–38, (2001)
7. S. Boyd, L. Vandenberghe, *Introduction to Convex Optimization with Engineering Applications* (Stanford University, Stanford, 1997). http://www.stanford.edu/~boyd/cvxbook/

8. P.R. Gray, R.G. Meyer, MOS operational amplifier design—a tutorial overview. IEEE J. Solid State Circuits **SC-17**, 969–982 (1982)
9. P.E. Allen, D.R. Holberg, *CMOS Analog Circuit Design* (Oxford University Press, Oxford, 1987)
10. M. Hershenson, S.M. Sunderajan, Automated circuit design using active set solving process. US 8307309 B1, (2012). http://www.google.co.uk/patents/US8307309
11. C. Enz, F. Krummenacher, E. Vittoz, An analytical MOS transistor model valid in all regions of operation and dedicated to low-voltage and low-current applications. J. AICSP **8**, 83–114 (1995)
12. D. Binkley, M. Bucher, D. Foty, Design-oriented characterization of CMOS over continuum of inversion level and channel length, in *Proceedings of 7th IEEE International Conferences on Electronics, Circuits and Systems*, Kaslik, Lebanon, (2000), pp. 161–164
13. F. Silveira, D. Flandre, P.G. Jespers, A (gm/ID) based methodology for the design of CMOS analog circuits and its application to the synthesis of a silicon-on-insulator micropower OTA. IEEE J. Solid-State Circuits **31**, 1314–1319 (1996)
14. D. Foty, M. Bucher, D. Binkley, Re-interpreting the MOS transistor via the inversion coefficient and the continuum of gms/ID, in *Proceedings of International Conferences on Electronics, Circuits & Systems* (2002), pp. 1179–1182
15. A. Cunha, M. Schneider, C.G. Montoro, An MOS transistor model for analog circuit design. IEEE J. Solid-State Circuits **33**, 1510–1519 (1998)
16. M. Bucher, C. Lallement, C. Enz, F. Théodoloz, F. Krummenacher, The EPFL-EKV MOSFET model equations for simulation: Model Version 2.6. Technical Report, Swiss Fed. Inst. of Tech. (EPFL), Lausanne, Switzerland (1999)

Chapter 45
Thermal Modeling of III-V WBG-Based p-i-n Switch

Abhijit Kundu, Maitreyi Ray Kanjilal, Payel Biswas and G.C. Nandy

Abstract Breakdown voltage and power dissipation of a semiconductor device determine the range of voltage, frequency and temperature over which it can operate keeping in track its electrical and optical characteristics. The power dissipation of p-i-n diode has been studied using thermal modelling with heat sink. The electrical properties of the switch are determined by its internal resistance and depend on geometrical structure of the device. This type of switch can control the high power from k_u band through millimetre wave frequency in the verity of switch configuration such as single-pole single-throw switch (SPST), single-pole double-throw switch (SPDT) and single-pole multi-throw switch (SPMT). It offers high breakdown voltage for which it enables to operate at high temperature and becomes capable to handle the high power.

Keywords GaN · 3C-SiC · p-i-n diode · Effective diffusion length · Stored charge · Intrinsic impedance · Heat sink

A. Kundu (✉)
ECE Department, Abacus Institute of Engineering and Management,
Kolkata, India
e-mail: abhijituday@yahoo.co.in

M.R. Kanjilal · P. Biswas
ECE Department, Narula Institute of Technology, Kolkata, India
e-mail: mr.kanjilal@gmail.com

P. Biswas
e-mail: biswas.payel@gmail.com

G.C. Nandy
ECE Department, West Bengal University of Technology,
Kolkata, India
e-mail: nandy_gc@yahoo.com

© Springer India 2015
K. Maharatna et al. (eds.), *Computational Advancement in Communication Circuits and Systems*, Lecture Notes in Electrical Engineering 335,
DOI 10.1007/978-81-322-2274-3_45

45.1 Introduction

The characteristics as well as application of p-i-n device are determined by its intrinsic region width which is sandwiched between two oppositely doped regions and also depend on minority carrier diffusion length. The size of p-i-n diode can be made very small compared to wavelength at high frequency. This type of p-i-n diode is more reliable to control high power and high temperature for its inherent electric properties. It can be used as switch in a multidirectional path without any distortion of the signal [1, 2].

In a switch the power dissipation is an important factor when it is used at high voltage, high frequency and high temperature regions. Thus semiconductor devices like p-i-n diode are applicable at high voltage and high frequency. The precaution should be taken to control the increase of temperature within the device especially at the junction. The p-i-n device is used as switch or inferred detector in the high frequency region therefore high power dissipation may occur for which temperature is increased and as a result the performance of diode can be damaged. Hence temperature analysis is essential for high power operation from which one can understand the operation temperature limit. It has been observed that the p-i-n diode is capable to handle high power in the presence of good heat sink and this has been developed in the present work (Fig. 45.1).

In the present work wide band gap III-V material GaN is considered as the base material of the p-i-n switch. The results have been compared with reference to another WBG IV-IV material-based p-i-n diode.

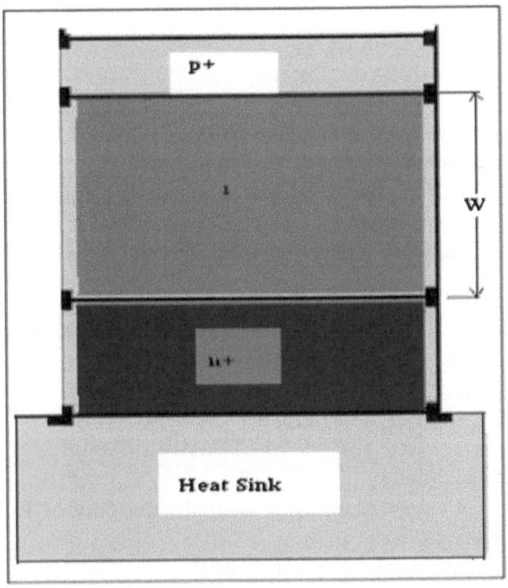

Fig. 45.1 Symmetric structure of p-i-n diode with heat sink

45.2 Theory

The p-i-n diode can operate in two modes 'On state' and 'Off state' depending on the frequency range and the device is considered as forward bias in the 'On state' and as reverse bias in the 'Off state'. With increase of forward bias current (I_0) more charge is injected in the i-region which offers the low series impedance of the diode [3, 4] and the injected charge in i-region ($Q = I_0\tau$) also depends on carrier life time (τ).

The resistance of the switch is

$$R_T = R_i + 2R_j \tag{45.1}$$

The intrinsic resistance of p-i-n diode depends on width (W) of the i-region and the carrier diffusion length (L) in this region and also on forward bias current (I_0). The intrinsic resistance (R_i) is expressed as

$$R_i = \left({}^{kT}/_{qI_0}\right)\left({}^{W}/_{L}\right)\tanh\left({}^{W}/_{2L}\right) \tag{45.2}$$

The width of the i-region can be extracted approximately by the expression of depletion region width of p-n junction [3]. The carrier diffusion length is an important parameter which controls the electrical properties. The diffusion length can be expressed as [2]

$$L = \sqrt{D_{\text{eff}}\tau} \tag{45.3}$$

The junction resistance (R_j) varies with frequency and it can be written as

$$R_j(f) = \left({}^{kT}/_{qI_0}\right)\beta\tanh\left({}^{W}/_{2L}\right)\cos(\phi - \theta/2) \tag{45.4}$$

The internal resistance plays a crucial role for power dissipation within the device. In p-i-n switch the power dissipation is mainly controlled by the junction temperature [3, 5, 6]. The analysis on power dissipation and operation range can be done for both the continuous power (CW) and pulse power (PW) systems. At ambient temperature (T_a) for a given power dissipation (P_d), the junction temperature (T_j) can be expressed as

$$T_j = T_a + \theta_j P_d \tag{45.5}$$

Power dissipation as a function of available power P_{av}, series resistance R_s and a characteristic impedance Z_0 of p-i-n diode is given by

$$P_d = \frac{4R_T Z_0}{(Z_0 + 2R_T)^2} \tag{45.6}$$

It is required to study the thermal impedance for p-i-n switch for its useful application. The thermal impedance of heat flow can be determined by the conductivity of the medium (K), the cross-sectional area of heat sink (A) and the length of heat flow path (x). It can be written as

$$\theta_j = \int_0^x \frac{\mathrm{d}x}{KA} \tag{45.7}$$

The heat capacity of the switch also depends on specific heat (C_p) and the heat density (ρ) of WBG semiconductor. Considering the volume of the switch as V it can be expressed as

$$H_c = C_p \rho V \tag{45.8}$$

The p-i-n switch requires high breakdown voltage to handle high power at high frequency [7]. The WBG semiconductor offers high breakdown voltage which can be written as

$$V_b = \int_0^W E(z)\mathrm{d}z \tag{45.9}$$

where electric field $E(z)$ is constant and z represents the width of i-region. So it can be written as

$$V_b = E \cdot W \tag{45.10}$$

Using Gauss law

$$\nabla \cdot E = {}^{\rho}\!/_{\varepsilon} \tag{45.11}$$

where all the notations are same. The electric field in the i-region of p-i-n switch can be determined by the charge which is distributed by ionized donors (N_i) in the i-region then [8, 9]

$$E(z) = {}^{zN_i q}\!/_{\varepsilon} \tag{45.12}$$

45.3 Result and Analysis

The resistance of the switch has been estimated at 1 mA forward bias current. The analysis of resistance at higher frequency of two different compound semiconductors has been shown in Fig. 45.2. GaN is wider bandgap material compared to 3C-SiC

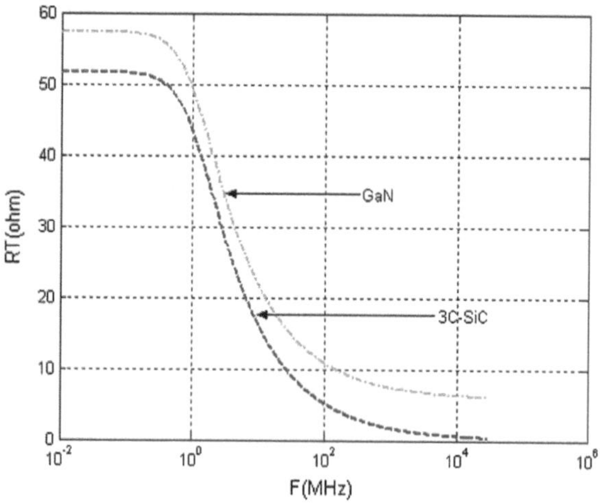

Fig. 45.2 Total resistance of p-i-n diode at 1 mA forward current

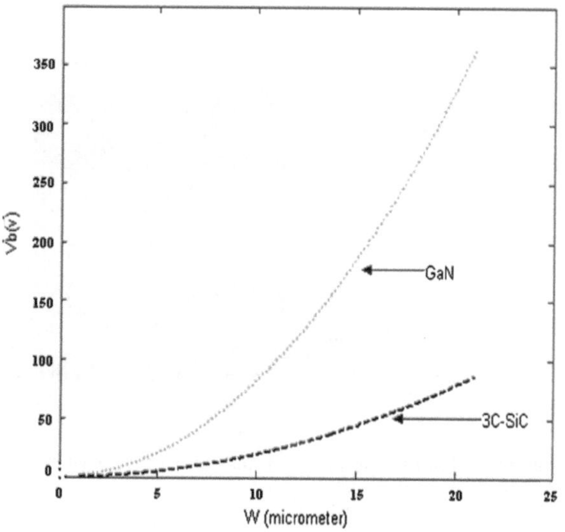

Fig. 45.3 Breakdown voltage for GaN and 3C-SiC p-i-n diode

and offers better application at higher frequency. Again it has been observed that GaN has high breakdown voltage than 3C-SiC at same width of the i-region as shown in Fig. 45.3.

Heat is generated by the dissipation of power at the junction of the p-i-n diode. It absorbs the small amount of power and the junction temperature increases. Analysis of power under PW and CW system is performed by thermal imaging of p-i-n switch with thermal analysis software. In each case GaN shows less thermal power dissipation than 3C-SiC (Fig. 45.4), GaN is preferred for p-i-n switch in the high voltage and high frequency range. Figures 45.5 and 45.6 show the steady state thermal image of p-i-n switch under high power condition. The flow of heat occurs

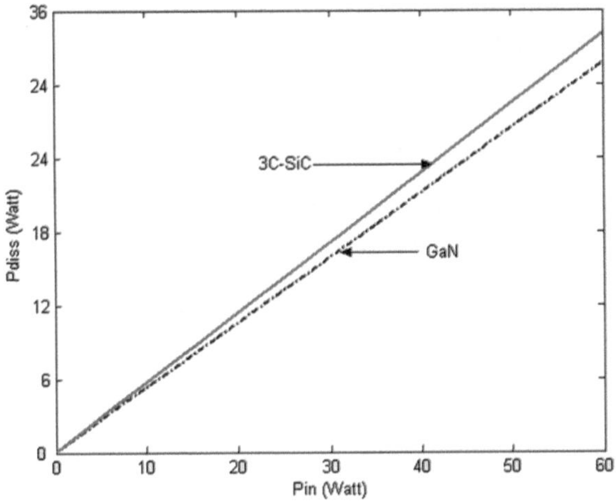

Fig. 45.4 Power dissipation for GaN and 3C-SiC p-i-n diode

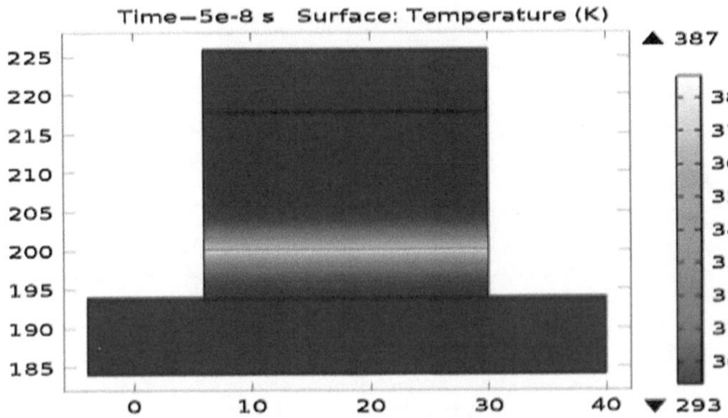

Fig. 45.5 The thermal image of GaN-based p-i-n switch operating

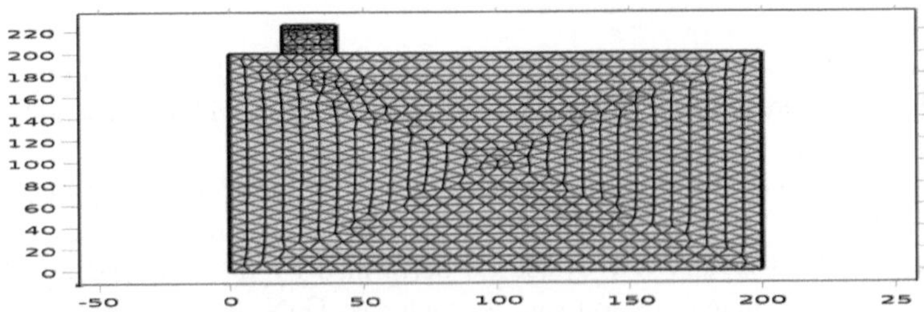

Fig. 45.6 Mesh formation for heat diffusion in the device

within the device presenting a mesh structure and where the p-i-n switch is mounted on Cu substrate. It can be observed that mesh should be in such a way that n number of nodes and elements are present through the thickness of each portion of switch geometry.

References

1. A. Iturri-Hinojosa, L.M. Resendize, T.V. Torchynska, Numerical analysis of the performance of p-i-n diode microwave switches based on different semiconductor materials. Int. J. Pure Appl. Sci. Technol. **1**(2), 93–99 (2010)
2. R. Caverly, G. Hiller, Microwave resistance of gallium arsenide and silicon p-i-n diode. IEEE MTT-S Digest **2**, 591–594 (1987)
3. A. Kundu, M.R. Kanjilal, M. Mukherjee, Insertion loss and isolation of p-i-n switch based on SiC family. J. Electron. Devices **18**, 1568–1574 (2013)
4. D. Leenov, The silicon p-i-n diode as a microwave radar protector at megawatt levels. IEEE Trans. Electron Devices **11**(2), 53–61 (1964)
5. E. Gated et al., An improved physics- based formulation of the microwave p-i-n diode impedance. IEEE Microwave Wirel. Compon. Lett. **17**(3), 211–213 (2007)
6. A. Kundu, M.R. Kanjilal, A. Das, J. Kundu, M. Mukherjee, Cubic structure sic p-i-n diode as RF switch, in *International Conference IET* (2013), pp. 119–121
7. A. Ashegi, K. Kurabyashi, R. Kasnavi, K.E. Goodson, Thermal conduction in doped single crystal silicone films. J. Appl. Sci. **91**(8), 5097–5088 (2002)
8. D.A. Neamen, *Semi Conductor Physics and Devices* (McGraw-Hill, New York, 1967)
9. A.S. Grov, *Physics and Technology of Semi Conductor Device* (Wiley, New York, 1967)

Chapter 46
Electrical Characteristics of MESFET Using GaAs, InP and GaN as Substrates

Puspak Pain, Dipayan Purakait, Nilanjan Chatterjee and Maitreyi Ray Kanjilal

Abstract The theoretical analysis of electrical properties of MESFETs has been carried out with different materials of narrow as well as wide energy bandgap III–V binary compound semiconductors. The gate-to-source capacitance (C_{gs}) and transconductance (g_m) have been studied as a function of gate-to-source voltage (V_{gs}). It is revealed that wide bandgap semiconductor is best suited at high frequency because it has lower capacitance. The transconductance also depends on temperature and decreases with temperature. This is because of the increasing scattering phenomenon within the channel. The wide bandgap semiconductors offer higher transconductance which enables them to be used as high-speed switch. The aim of this work is to improve the understanding of MESFET device performances based on different III–V binary compound materials and also to find out the most suitable material combination for MESFET.

Keywords MESFET · Transfer characteristics · Gate-to-source capacitance · Transconductance · Wide bandgap semiconductor

P. Pain (✉) · M.R. Kanjilal
ECE Department, Narula Institute of Technology, Kolkata 700109, India
e-mail: puspak1985@gmail.com

M.R. Kanjilal
e-mail: mr.kanjilal@gmail.com

D. Purakait
India Government Mint, Kolkata, India
e-mail: dipayan.purkait@yahoo.in

N. Chatterjee
Tech Mahindra Ltd, Pune, India
e-mail: nilanjan.chatterjee71@gmail.com

© Springer India 2015

K. Maharatna et al. (eds.), *Computational Advancement in Communication Circuits and Systems*, Lecture Notes in Electrical Engineering 335,
DOI 10.1007/978-81-322-2274-3_46

46.1 Introduction

Metal Semiconductor Field Effect Transistor (MESFET) becomes an emerging device applicable both at low- and high-frequency regime. It attains high value of carrier mobility with high transconductance and transit frequency [1]. The fundamental approach is to evaluate a comparative study on the different electrical parameters such as transconductance (g_m) and capacitance–voltage characteristics exhibited by III–V binary compound semiconductors GaAs, InP and also wide bandgap GaN MESFETs. Device dimensions and material properties both affect the output resistance [1, 2]. Again the magnitude of the device output conductance is directly proportional to device gate width and inversely with gate length, hence short gate length typically results in lower output resistance. The output resistance can also be reduced by increasing channel doping concentrations (N_D), or the device epi-thickness (a) [2]. The transconductance (g_m), transit time delay τ, and output resistance R_{ds} can be calculated from the controlled drain–source current, I_{ds}. The transconductance, one of the most important indicators of device quality for microwave applications is defined in terms of the slope of the $I_{\text{ds}} \sim V_{\text{gs}}$ characteristics with constant drain–source voltage [3, 4].

On the other hand, the circuit model can be divided into two parts: extrinsic and intrinsic parameters. The intrinsic parameters characterize the active region under the gate and are functions of biasing conditions, whereas at least to first-order approximation the extrinsic parameters depend on technological parameters. C_{ds} is the drain–source capacitance dominated by geometric capacitance and is often treated as a constant. C_{gs} and C_{gd} are the channel capacitances, which in general are nonlinear [2] elements for their strong dependence on V_{gs} and V_{ds}.

46.2 Theory

A schematic cross-sectional view of the equivalent circuit of MESFET [3] models showing physical origin of each element is illustrated in Fig. 46.1.

As charge is redistributed with changing gate–source and gate–drain voltages, various electrical parameters can be calculated from capacitance–voltage characteristics.

46.2.1 Drain Characteristics

The drain current for different values of gate-to-source voltage can be given by the following equations, for the linear region it can be expressed as [2]:

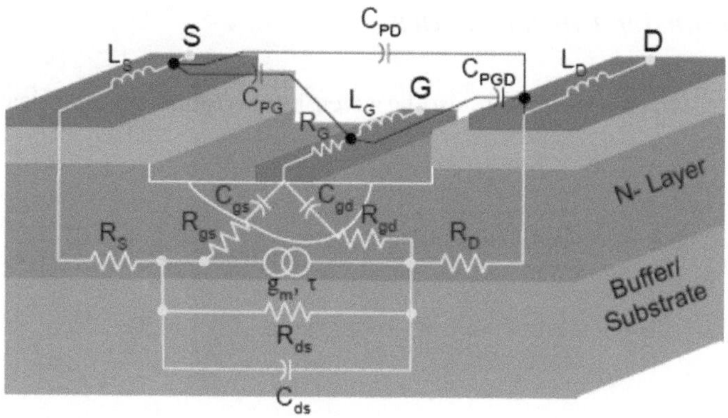

Fig. 46.1 Equivalent circuit of MESFET

$$I_d(V_{\rm ds}, V_{\rm gs}) = K_n\left[\frac{3V_{\rm ds}}{V_{\rm bi} - V_p} - 2\left\{\frac{(V_{\rm ds} - V_{\rm gs} + V_{\rm bt})^{3/2}}{(V_{\rm bi} - V_p)^{3/2}} - \frac{(V_{\rm bi} - V_{\rm gs})^{3/2}}{(V_{\rm bi} - V_p)^{3/2}}\right\}\right] \tag{46.1}$$

The n-channel MESFET with a homogeneously doped channel region is in the linear regime for $V_{\rm ds} \le V_{\rm gs} - V_p$, and it is in the saturated regime, for $V_{\rm ds} > V_{\rm gs} - V_p$, and the drain current I_d for the saturation region is expressed by [2]:

$$I_d(V_{\rm ds}, V_{\rm gs}) = K_n\left[1 - 3\left\{\frac{(V_{\rm bi} - V_{\rm gs})}{(V_{\rm bi} - V_p)} + 2\frac{(V_{\rm bi} - V_{\rm gs})^{3/2}}{(V_{\rm bi} - V_p)^{3/2}}\right\}\right] \tag{46.2}$$

where,

$$K_n = \frac{q^2 N_D^2 \mu_n a^3 W_G}{6\varepsilon_s L_G}$$

a thickness of the channel
$V_{\rm bi}$ built-in potential
V_p pinch-off voltage

N_D the doping concentration is assumed to be constant throughout the channel and the other parameters are as usual.

46.2.2 Transfer Characteristics

The transfer characteristics can also be given by the following empirical relationship [2]:

$$I_d = N_D v_{\mathrm{sat},n} W_G a \left(1 - \sqrt{\frac{V_{\mathrm{bi}} - V_{\mathrm{gs}}}{V_{\mathrm{bi}} - V_p}}\right) \tag{46.3}$$

where $v_{\mathrm{sat},n}$ = drift saturation velocity of electrons (Fig. 46.2).

46.2.3 Capacitance–Voltage Characteristics

The extent and dimensions of the depletion region beneath the gate of the MESFET are determined by the bias applied to the device terminals, because in a MESFET only one region of depletion charge is shared between gate–source and gate–drain capacitances. The gate–source capacitance is defined as

$$C_{\mathrm{gs}} = \frac{\mathrm{d}Q_{\mathrm{g}}}{\mathrm{d}V_{\mathrm{gs}}} \Big| V_{\mathrm{gd=constant}} \tag{46.4}$$

$$C_{\mathrm{gd}} = \frac{\mathrm{d}Q_{\mathrm{g}}}{\mathrm{d}V_{\mathrm{gd}}} \Big| V_{\mathrm{gs=constant}} \tag{46.5}$$

The definitions expressed by Eqs. (46.4, 46.5) are not the only possible definitions applied to the MESFET capacitance. Under typical operating conditions, the MESFET source terminal is grounded with negative bias to the gate and positive bias to the drain. Therefore, the gate–source and drain–source DC bias voltages also control the capacitance and the capacitance is increasing monotonically with

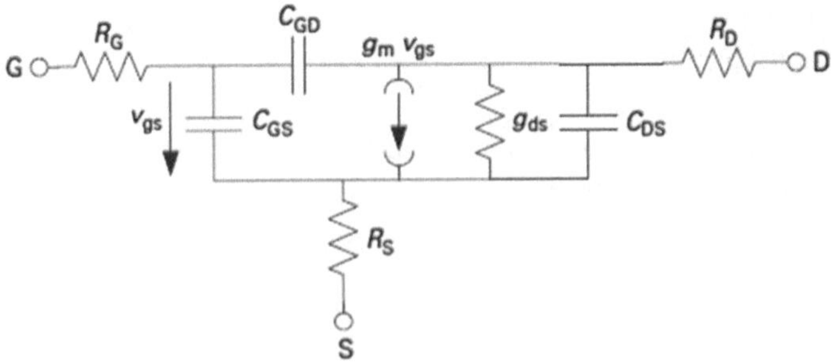

Fig. 46.2 Simple small-signal equivalent circuit of MESFET

gate–source voltage from pinch-off towards zero bias. For this reason, the gate–source capacitance is often defined as

$$C_{gs'} = \frac{dQ_g}{dV_{gs}} | V_{ds=\text{constant}} \tag{46.6}$$

Equations (46.4) and (46.6) are not equivalent, but slightly different quantities with a minor distinction. It can be significant if calculations are based on a physically based model in which the depletion charge is defined by a mathematical expression [1, 2].

The capacitance for different gate-to-source voltage is given by the following equation [2]:

$$C_{gs} = \frac{qN_D W_G L_G a}{2\sqrt{(V_{bi} - V_p) \times (V_{bi} - V_{gs})}} \tag{46.7}$$

46.2.4 Transconductance

Transconductance (also known as mutual conductance, g_m) is defined as the ratio of the current change at the output port to the voltage change at the input port.

$$g_m = \frac{\Delta I_{out}}{\Delta V_{in}} \tag{46.8}$$

The transconductance of MESFET for different gate-to-source voltages is given by the following equation [2]:

$$g_m = \frac{qN_D v_{sat,n} W_G a}{2\sqrt{(V_{bi} - V_p) \times (V_{bi} - V_{gs})}} \tag{46.9}$$

At a particular bias point transconductance depends on the slope of the tangent line of the transfer characteristic curve of the FET, at that point, while all other characteristics are equal, a device with high transconductance will provide greater gains and become superior in high-frequency performance.

In practice, device technology advances that have allowed the production of shorter gate length devices have also required higher doping densities in the channel for optimum scaling. While shorter gate lengths produce smaller C_{gs} values, higher doping densities result in higher capacitance. These two trends have tended to cancel each other, resulting in a gate–source capacitance to gate width ratio, (C_{gs}/w) for microwave devices that has remained nearly constant at a value near to 1 pF/mm [2, 5–7].

46.3 Results and Analyses

The results based on a generalized simulation analysis are discussed here to evaluate the electronic property such as transconductance (g_m), variation of gate-to-source capacitance (C_{gs}) with gate-to-source voltage (V_{gs}).

A comparative study of electrical characteristics of a n-channel MESFET with a homogeneously doped channel region using GaAs, InP and GaN as potential substrate has been illustrated in terms of V_{gs} and temperature (T) with the investigation of drain, transfer, capacitance–voltage characteristics. Various parameters used for the simulations are mentioned in Table 46.1 [7–9].

Figures 46.3, 46.4 and 46.5 show the different characteristics of MESFET using different substrate materials of Gr. III–V. The results of the variation of

Table 46.1 Parameter values of different materials

Material properties with (symbol) (unit)	Comparison of different Semiconductors		
	GaAs	InP	GaN
Electron mobility (μ_n) (m^2/V-s)	0.85	0.54	0.125
Hole mobility (μ_P) (m^2/V-s)	0.04	0.02	0.085
Bandgap energy (E_g) (eV)	1.42	1.35	3.44
Saturated carrier drift velocity (V_s) ($\parallel$ to c-axis) (10^5 ms^{-1})	1.2	2.5	2.2
Critical electric breakdown field (E_C) (10^7 V/m)	4.0	5.0	20.0
Thermal conductivity (K) (Wm^{-1} K^{-1})	46.0	69.0	225.0
Relative dielectric constant (ϵ_r)	13.1	12.5	8.9

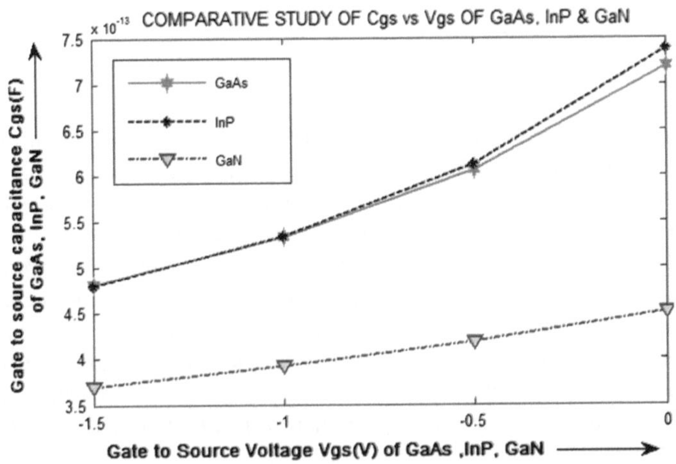

Fig. 46.3 Capacitance–voltage characteristics for MESFETs

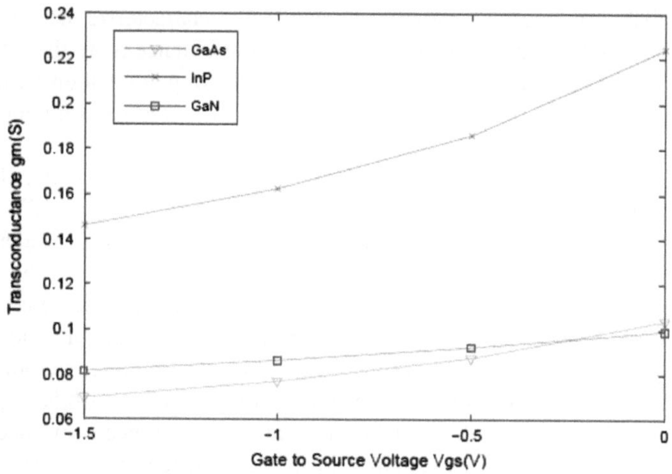

Fig. 46.4 Transconductance (g_m) versus gate-to-source voltage (V_{gs}) characteristics for MESFETs

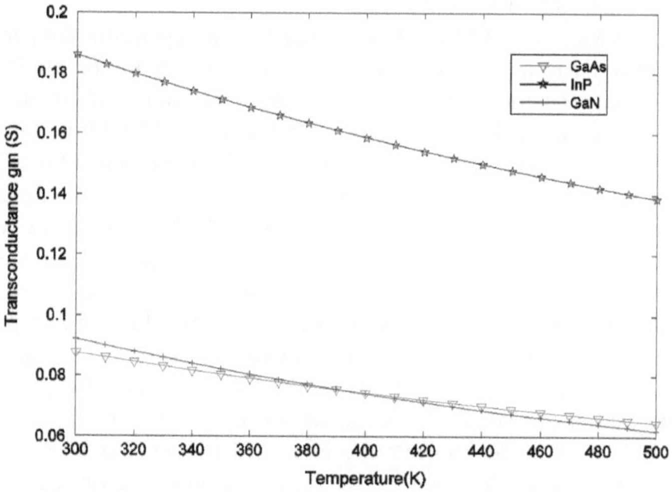

Fig. 46.5 Variation of transconductance (g_m) with temperature (T)

transconductance (g_m) over a range of gate-to-source voltage, and temperature are shown in Figs. 46.4 and 46.5, respectively, with $N_D = 10^{17}/\text{cm}^3$. Here gate-to-source voltage V_{gs} is varied from −1.5 to 0 V, for a temperature range of 300–500 K, with a constant drain to source voltage V_{ds}.

All the characteristics show the similar nature as expected. The high current carrying capability of the wide bandgap materials has been established in the analysis. It is interesting to observe that among the three substrate materials of Gr. III–V, InP having the highest gate-to-source capacitance (C_{gs}) in comparison with

GaAs and GaN. The intrinsic parameters include C_{gs}, characterize the active region under the gate and are functions of biasing conditions. Hence more the reverse bias of gate-to-source voltage (V_{gs}) lesser will be the value of gate-to-source capacitance (C_{gs}). Here different drift saturation velocities are considered for different substrate materials as mentioned in Table 46.1.

The comparative analysis of capacitance–voltage characteristics on MESFET is shown in Fig. 46.3 using different substrates from Eqs. (46.6, 46.7) and it is revealed that the gate-to-source capacitance (C_{gs}) is increasing stridently with higher saturation drift velocity, mobility and permittivity (ε_s) of the substrate material used. It is evident that the influence of higher terminal gate-to-source capacitance (C_{gs}) leads to higher drain current over a range of gate-to-source voltage (V_{gs}). The variation of transconductance with V_{gs} and temperature for the materials GaAs, InP and GaN have been shown in Figs. 46.4 and 46.5, respectively. Among the different materials used in this analysis, it has been revealed that the wider bandgap material GaN exemplify lower dependence on g_m with increasing V_{gs}, whereas the slope of the curve using InP as substrate is maximum. For potential substrate GaN, transconductance (g_m) increases more than that of GaAs. It is clear from the representation that for InP, the transconductance of MESFET increases more rapidly than that of GaN and GaAs.

The performance of MESFET is influenced by its operating temperature significantly. In order to achieve the most favourable application of MESFETs with different substrate material, it is necessary to get temperature effects on it.

Figure 46.5 realizes the performance simulation of MESFETs under varied temperatures using different materials as substrate. The higher rate of degradation of transconductance for InP MESFET than in GaN and GaAs is observed with the increase in temperature. In GaAs mobility is high (0.85 m^2/V-s) due to the less phonon scattering which in turn produces higher drain current compared to InP, GaN structure counterparts. Moreover, maximum transconductance for MESFET devices with variation in V_{gs} and temperature occurs when Indium Phosphide (InP) is used as a substrate. The value of C_{gs} is proportional to both gate length and gate width. Furthermore, the extent and dimensions of the region of depletion charge beneath the gate of the MESFET are determined by the bias applied to the device terminals. It is used to demonstrate the impact of transconductance and gate-to-source capacitance with the variation of V_{gs} and temperature on the small-signal voltage gain.

46.4 Conclusion

The core factor differentiating the results for different semiconductor-based MESFETs is the material as well as gate-to-source capacitance having its chief concerning aspect in the performance of MESFETs at high frequency. The utmost steepness in variation of transconductance in InP among the III–V semiconductor materials corresponds to the most suitable potential substrate related to the voltage gain. To increase the high-frequency characteristics of MESFETs, the reduction of gate–source capacitance is desirable to produce optimal gain. The GaN followed by the GaAs can offer reduced gate–source capacitance comparable to the InP with moderate gain and efficiency. Comparison between the different semiconductor materials used as substrates with the simulated data at different bias voltages, temperatures and their influences on gain, efficiency and overall device performance are presented to optimize the scaling, design and fabrication technique of MESFETs.

Acknowledgments The authors would like to acknowledge their affiliating institutes for valuable support and cooperation.

References

1. A. Sayed, Ultra Wideband 5 W Hybrid Power Amplifier Design Using Silicon Carbide MESFETs (2005), pp. 11–29
2. S. Prasad, H. Schumacher, A. Gopinath, *High-Speed Electronics and Optoelectronics: Devices and Circuits* (Cambridge university Press, Cambridge, New York), pp. 46–66
3. Y.A. Khalaf, *Systematic Optimization Technique for MESFET Modeling* (Blacksburg, Virginia, 2000), pp. 12–37
4. J.M. Golio, *Microwave MESFETs and HEMTs* (Artech House, MA, 1991)
5. M. Altay, Comparison and Evaluation of Various Mesfet Models-a thesis for the degree of Master of Science in the Department of Electrical and Electronics Engineering of Graduate School of Natural and Applied Sciences, of Middle East Technical University, March 2005
6. E.N. Ganesh, Simulation of GaAs MESFET and HEMT devices for RF applications. Int. J. Emerg. Trends Technol. Comput. Sci. **2**(1), 123–128 (2013). ISSN 2278-6856
7. P. Sarkar, A. Das, K.M. Ray, Electrical characteristics of AlGaN/AlNandAlGaN/GaN HEMTs-international. J. Sci. Eng. Res. **4**(5), (2013). ISSN 2229-5518
8. S.M. Sze, *Semiconductor Devices Physics and Technology* (Wiley, New York, 1985)
9. M. Mukherjee (ed.), Wide Band Gap Semiconductor Based High Power ATT diodes In The MM-wave and THz Regime: Device Reliability, Experimental Feasibility and Photo-sensitivity, Advanced Microwave and Millimeter Wave Technologies Semiconductor Devices Circuits and Systems. ISBN:978-953-307-031-5 (2010)

Chapter 47
Design and Simulation of Two-Stage Low-Power CMOS Op-amp in Nanometre Range

Soumen Pal and Pinky Ghosh

Abstract This paper presents a novel design procedure of two-stage CMOS op-amp with robust bias circuit for low-power application. The aim of the design methodology in this paper is to propose straightforward yet accurate equations for the design of CMOS op-amp in nanometre dimension. The method handles a wide variety of specifications and constraints. The SPICE simulation results of designed op-amp provide expected characteristics for given specification.

Keywords Two-stage op-amp · CMOS op-amp · Low-power op-amp · Op-amp in nm scale

47.1 Introduction

An operational amplifier (op-amp) is a direct-coupled high-gain electronic voltage amplifier with a differential input and, usually, a single-ended output. An op-amp produces an output potential that is typically hundreds or thousands times larger than the potential difference between its input terminals. Operational amplifiers had their origins in analog computers, where they were used to do mathematical operations in many linear, non-linear and frequency-dependent circuits. Characteristics of a circuit using an op-amp are set by external components with little dependence on temperature changes or manufacturing variations in the op-amp itself, which makes op-amps popular building blocks for circuit design. Op-amps are the most widely used electronic device today, being used in a vast array of consumer, industrial and scientific devices.

S. Pal (✉) · P. Ghosh
Narula Institute of Technology, Kolkata 700109, India
e-mail: soumen.vlsi@gmail.com

P. Ghosh
e-mail: Pinkey.ghosh321@gmail.com

© Springer India 2015

K. Maharatna et al. (eds.), *Computational Advancement in Communication Circuits and Systems*, Lecture Notes in Electrical Engineering 335,
DOI 10.1007/978-81-322-2274-3_47

Operational amplifiers (Op-amps) are basic building blocks of a wide range of analog and mixed signal systems. Op-amp is a versatile device that can be used to amplify DC as well as AC input signals and was originally designed for performing mathematical operations. It was found that the application of negative feedback around a high-gain DC amplifier would produce a circuit with a precise gain characteristic that depended only on the feedback used. By proper selection of the feedback components, operational amplifier circuits could be used to add, subtract, integrate and differentiate. As new generations of CMOS technology tend to have shorter transistor channel length and scaled-down supply voltage, the design of op-amps stays a challenge for designers. The designing of op-amps puts new challenges in low-power applications with reduced channel length devices. In some applications, the gain and or the output swings provided by cascade op-amps are not adequate. In such cases, we resort to "two stage" op-amps, with the first stage providing a high gain and also better noise and offset performance; the second stage provides the conversion of differential to single-ended output and large output swing. In contrast to cascade op-amps, a two-stage configuration isolates the gain and swing requirements as well to provide more gain and large output swing. The two-stage op-amp is also one of the most popular configurations because of its simple structure and robustness. The design of an op-amp involves two distinct activities. The first part includes creating basic structure of the op-amp, i.e. the diagram that describes the interconnections of all transistors used to construct required basic blocks of op-amp such as differential transconductance amplifier, bias circuit, compensation circuit and buffer circuit. The second activity is the selection of proper DC currents and sizing of transistors. Finally the design of compensation circuit may be modified in order to achieve close loop stability as op-amp uses negative feedback in many applications [1, 2].

In designing an op-amp, numerous electrical characteristics, such as supply voltages, DC gain, load capacitance, unity gain frequency, phase margin, slew rate, input common mode range and output swing, all have to be taken into consideration. Op-amps are designed to be operated with negative feedback connection; so frequency compensation is necessary for closed loop stability. So the designing of an op-amp that satisfies all the electrical specifications requires a good compensation strategy and design technique [3].

47.2 Design Steps of Two-Stage Op-amp

A design step for two-stage op-amp (Fig. 47.1) can be constructed as follows [3].

Step 1: $C_c = 0.22C_L$
Step 2: $I_{D5} = C_c\mathrm{SR}$
Step 3: $I_{D7} = \mathrm{SR}(C_c + C_L)$
Step 4: $(W/L)_6 = \frac{2\mathrm{SR}(C_c + C_L)}{\mu_p C_{ox}\left(V_{DD} - V_{out(max)}\right)^2}$

Step 5: $(W/L)_{1,2} = \frac{\omega_u^2 C_c}{\mu_n C_{ox} \mathrm{SR}}$

Step 6: $(W/L)_{5,8} = \frac{2\mathrm{SR}C_c}{\mu_n C_{ox}\left(V_{in(max)} - V_{ss} - V_{tn} - \mathrm{SR}/\omega_u\right)^2}$

Step 7: $(W/L)_7 = \left(\frac{C_c + C_L}{C_c}\right)(W/L)_{5,8}$

Step 8: $(W/L)_{3,4} = \frac{(W/L)_6}{2(W/L)_7}(W/L)_{5,8}$

Step 9: $(W/L)_{12} = \frac{(W/L)_6}{(W/L)_7}(W/L)_8$

Step 10: $(W/L)_{14} = 4(W/L)_8$

Step 11: $(W/L)_9 = \left(\frac{C_c}{C_c + C_L}\right)(W/L)_6$

Step 12: $R_B = \frac{1}{\sqrt{2\mu_n C_{ox}(W/L)_8 \mathrm{SR} C_c}}$

Step 13: $(W/L)_{12} = (W/L)_{10} = (W/L)_{11} = (W/L)_{13}$

where,

C_c	Compensation capacitor.
ω_u	Unity gain frequency of the op-amp.
SR	Slew rate of the op-amp.
C_L	Load capacitance of the op-amp.
I_{D5}	Drain current of 5th transistor in Fig. 47.1.
I_{D7}	Drain current of 7th transistor Fig. 47.1.
μ_p	Mobility of PMOS.
μ_n	Mobility of NMOS.
C_{ox}	Oxide capacitance per unit area.
V_{ss}	Negative supply voltage.
V_{DD}	Positive supply voltage.

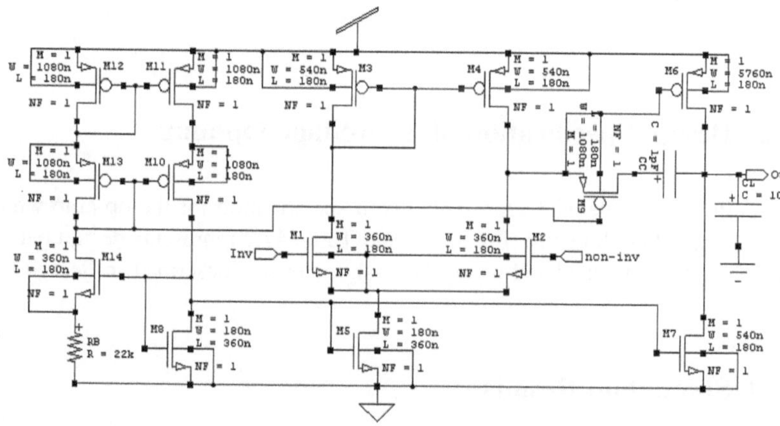

Fig. 47.1 Circuit diagram of two-stage CMOS op-amp designed in tanner tools

Table 47.1 Specification of CMOS op-amp

Electrical parameters	Expected values
Supply voltage	±1.8 V
Load capacitance	10 pF
Unity gain frequency	5 MHz
Slew rate	±5 V/μs
Input common mode range	±0.7 V
Output swing	±1.5 V

Table 47.2 Design parameters of op-amp as shown in Fig. 47.1

Parameters	Value	Unit
C_c	1	pF
R_B	22	kΩ
(W/L)1,2	360/180	nm/nm
(W/L)3,4	540/180	nm/nm
(W/L)5,8	180/360	nm/nm
(W/L)6	5760/180	nm/nm
(W/L)7	540/180	nm/nm
(W/L)9	1080/180	nm/nm
(W/L)10–13	1080/180	nm/nm
(W/L)14	360/180	nm/nm

$V_{\text{out(max)}}$ Maximum output swing.
$V_{\text{in(max)}}$ Maximum common mode range.
R_B Bias resistance.
$(W/L)_i$ Aspect ratio of ith transistor.

47.3 Design Specification of Two-Stage Op-amp

The following Table 47.1 represents given specification for the op-amp which has been designed as shown in (Fig. 47.1) and Table 47.2 represents design parameters which have been calculated using the design steps as described in Sect. 47.2.

47.4 Simulation Results

For the given specifications of op-amp as shown in Table 47.1 and using the design steps as mentioned in Sect. 47.2, the design parameters of op-amp as shown in Table 47.2 has been calculated and a two-stage CMOS op-amp (Fig. 47.1) with a

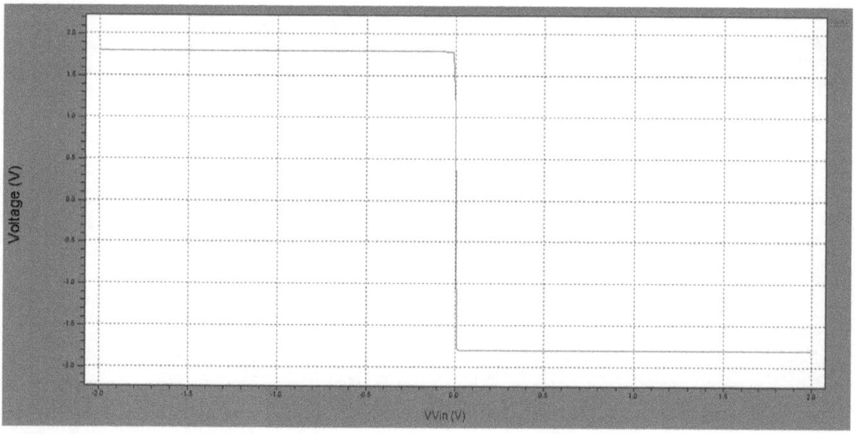

Fig. 47.2 Voltage transfer curve of designed op-amp in inverting mode

robust bias circuit has been designed and simulated using TANNER TOOL version-13. The T-SPICE simulation results are discussed as follows.

We can get (Fig. 47.2), when op-amp is operated in inverting mode and non-inverting terminal is grounded, i.e. 0 V and variable DC voltage is applied in inverting terminal.

We can get (Fig. 47.3), when op-amp is operated in non-inverting mode, a variable DC voltage is applied in non-inverting terminal.

For open loop configuration of op-amp it will work as comparator and voltage gain of op-amp will be very high. A sinusoidal voltage is applied in non-inverting input of the op-amp. Then, Fig. 47.4 shows the transient analysis.

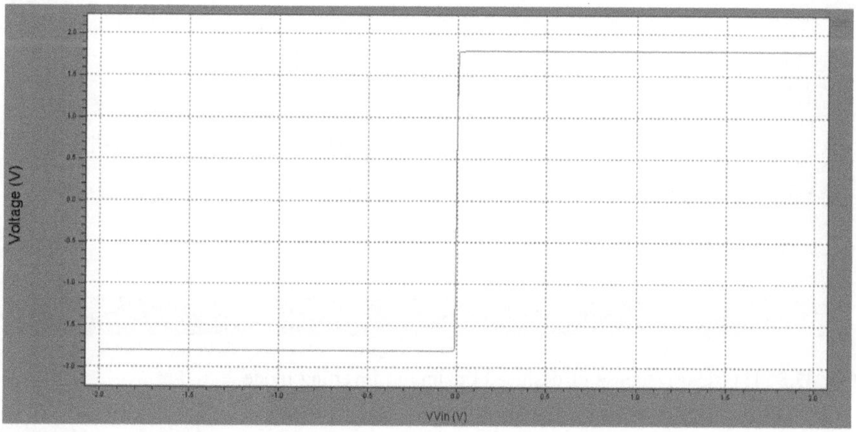

Fig. 47.3 Voltage transfer curve of designed op-amp in non-inverting mode

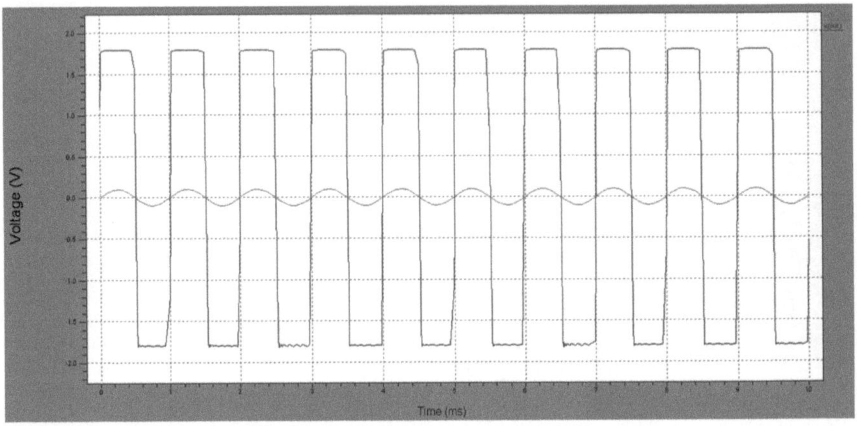

Fig. 47.4 Transient response of designed op-amp in non-inverting mode

Figure 47.5 show the transient analysis when a sinusoidal voltage is applied in inverting input of the op-amp.

Figure 47.6 shows the frequency response of designed op-amp. From this graph, −3 dB bandwidth can be measured which is 6.6 MHz.

Figure 47.7 shows the input common mode range of designed op-amp. From this graph that can be measured which is −1.6 to +1.7.

Figure 47.8 shows the output swing of designed op-amp. From this graph that can be measured which is −1.58 to 1.78.

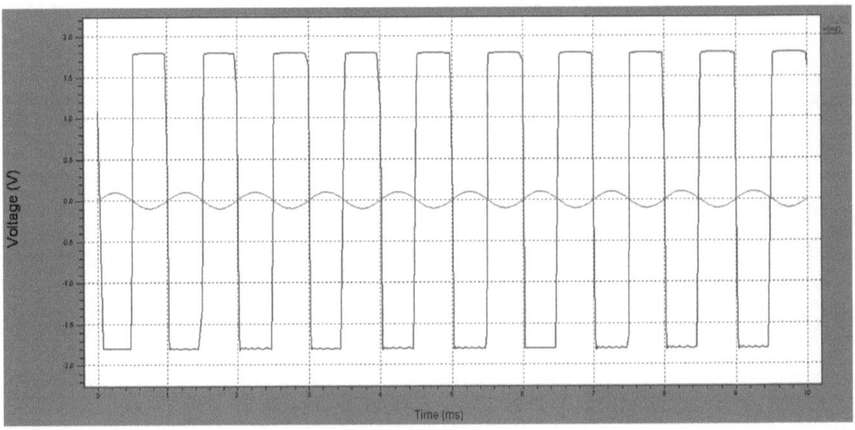

Fig. 47.5 Transient response of designed op-amp in inverting mode

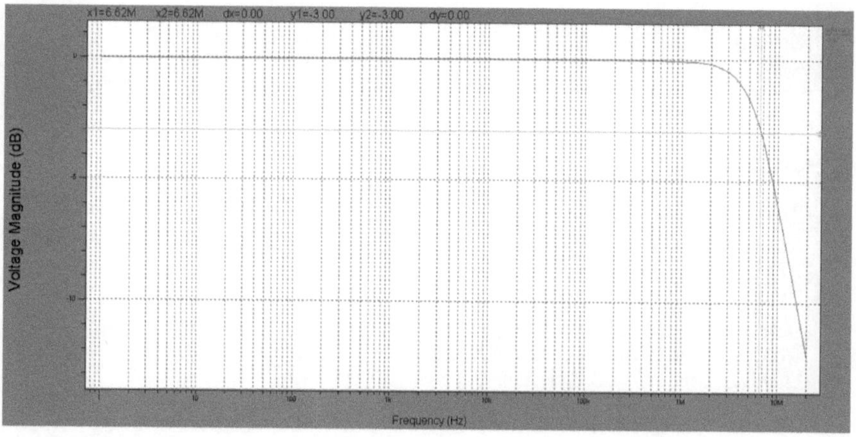

Fig. 47.6 Frequency response of designed op-amp

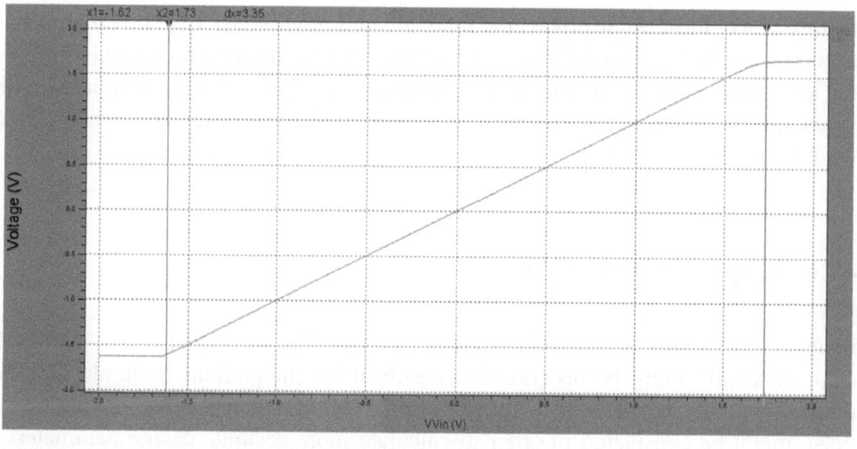

Fig. 47.7 Input common mode range of designed op-amp

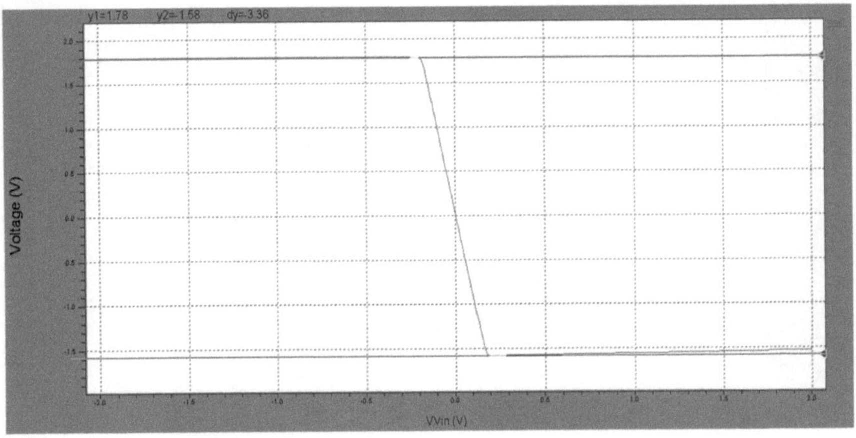

Fig. 47.8 Output swing of designed op-amp

47.5 Conclusion

In this work, the design procedure for two-stage low-power CMOS op-amp has been presented. The simulation result confirms that the design procedure is suitable for low-power CMOS op-amp design in nm range. The results are compared with given specifications with the help of table and graph. The simulation results agree with theoretical prediction.

47.6 Scope for Future Work

In this work, 180 nm technology has been used. Further reduction in MOSFET channel length might be necessary to ensure more integration. It should also be pointed out that power supply rejection ratio, offset voltage and contribution of noise might be considered in order to calculate more accurate design parameters.

Acknowledgments The authors would like to thank Mr. Avijit Dey, Narula Institute of Technology, Agarpara, Kolkata, for his inspiration and special help throughout the work.

References

1. P.E. Allen, D.R. Holberg, *CMOS Analog Circuit Design* (Oxford University Press, Oxford, 2002)
2. B. Razavi, *Design of Analog CMOS Integrated Circuits* (McGraw-Hill, New Yok, 2002)
3. J. Mahattanakul, J. Chutichatuporn, Design procedure for two stage CMOS op-amp with flexible noise power balancing scheme. IEEE Trans. Circuits Syst. **52**(8) (2005)

Chapter 48
TOAD-Based All-Optical Reversible New Multiplexer

Ashis Kumar Mandal, Supriti Samanta and Goutam Kumar Maity

Abstract Recently the reversible logic has formed as an unconventional form of computing. It is relatively new in the area of extensive applications in low power CMOS, quantum computing, DNA computing, digital signal processing (DSP), nanotechnology, communication, computer graphics, etc. Here we present a new reversible multiplexer circuit with the help of all-optical Toffoli gate in all-optical domain and also in this paper we have explained their principle of operations and used a theoretical model to fulfil this task, finally supporting through numerical simulation. In the field of ultrafast all-optical signal processing Terahertz Optical Asymmetric Demultiplexer (TOAD) has an important function. The different logical (realization of Boolean function) functions can be executed by this multiplexer in the domain of reversible logic-based information processing.

Keywords Reversible logic gates · Terahertz optical asymmetric demultiplexer (TOAD) · Toffoli gate · Multiplexer

A.K. Mandal (✉)
Department of Physics, Chakur Haris Seminary High School,
Howrah, West Bengal, India
e-mail: ashiskumarmandal7@gmail.com

S. Samanta
Department of Physics, Chaipat Girls' High School, Daspur,
West Bengal, India
e-mail: samantasupriti@yahoo.com

G.K. Maity
Electronics and Communication Engineering,
MCKV Institute of Engineering, Howrah, West Bengal, India
e-mail: goutam123_2005@yahoo.co.in

© Springer India 2015

K. Maharatna et al. (eds.), *Computational Advancement in Communication Circuits and Systems*, Lecture Notes in Electrical Engineering 335,
DOI 10.1007/978-81-322-2274-3_48

48.1 Introduction

All-optical switching—the switching of one beam of light by another—is an essential operation for transparent fibre optic networks and for all forms of optical information processing [1–4]. To overcome the electronic bottlenecks and fully exploit the advantages of optical fibre communication, it is necessary to move towards networks where the transmitted data would remain exclusively in the optical domain without optical-electrical-optical (OEO) conversions [5]. In conventional computers, majority of the computation operations are irreversible, i.e. once a logic block generates the output bits based on certain input combinations, the later bits are lost. A gate is reversible if the gate's inputs and outputs have a one-to-one correspondence, i.e. there is a distinct output assignment for each distinct input combination. Therefore, a reversible gate's inputs can be uniquely determined from its outputs. Reversible logic gates must have an equal number of inputs and outputs. Then the output rows of the truth table of a reversible gate can be obtained by permutation of the input rows. Reversible logic circuits have been emerged as a promising technology in the field of information processing. Irreversible computation results in energy dissipation due to data loss [6]. On the other hand, the reversible logic circuits offer an alternative form that allows computation with arbitrary small energy dissipation [7].

48.2 Operational Principle of TOAD-Based Optical Switch

TOAD-based gate has taken an important role in optical communication and information processing [8–10]. Sokoloff et al. [9] demonstrated a TOAD capable of demultiplexing data at 50 Gb/s. The TOAD consists of a loop mirror with an additional intraloop 2 × 2 (ideally 50:50) coupler. The loop contains a control pulse (CP) and a non-linear element (NLE) that is offset from the loop's midpoint by a distance Δx as shown in Fig. 48.1a. A signal with field $E_{\text{in}}(t)$ at angular frequency ω is split in coupler. It travels in clockwise (cw) and counterclockwise (ccw) direction through the loop. The electrical field at port-1 and port-2 can be expressed as follows:

$$\underline{E}_{\text{out},1}(t) = \underline{E}_{\text{in}}(t - t_d) \cdot e^{-j\omega t_d} \cdot \left[d^2 \cdot \underline{g}_{\text{cw}}(t - t_d) - k^2 \cdot \underline{g}_{\text{ccw}}(t - t_d)\right] \tag{48.1}$$

$$\underline{E}_{\text{out},2}(t) = jdk\underline{E}_{\text{in}}(t - t_d) \cdot e^{-j\omega t_d} \cdot \left[\underline{g}_{\text{cw}}(t - t_d) + \underline{g}_{\text{ccw}}(t - t_d)\right] \tag{48.2}$$

where t_d is pulse round trip time within the loop as shown in the Fig. 48.1. Coupling ratios k and d indicate the cross and through coupling, respectively.

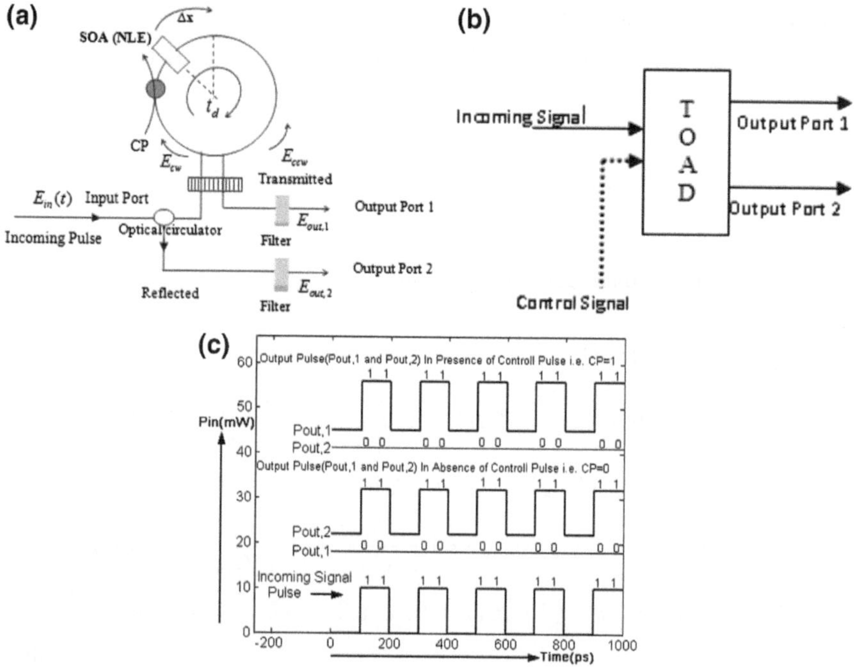

Fig. 48.1 **a** TOAD-based optical switch. **b** Schematic diagram of TOAD-based optical switch. **c** Simulation result: (*a*) output pulse (P_{out}, 1 and P_{out}, 2) in presence of control pulse, i.e. CP = 1 (*b*) output pulse (P_{out}, 1 and P_{out}, 2) in absence of control pulse, i.e. CP = 0 (*c*) incoming signal pulse

The cw signal be amplified by the complex field gain. $\underline{g}_{\text{cw}}(t)$, while ccw by $\underline{g}_{\text{ccw}}(t)$. The output power at port-1 can be expressed as,

$$\begin{aligned} P_{\text{out},1}(t) &= \frac{P_{\text{in}}(t-t_d)}{4} \cdot \left\{ G_{\text{cw}}(t) + G_{\text{ccw}}(t) - 2\sqrt{G_{\text{cw}}(t) \cdot G_{\text{ccw}}(t)} \cdot \cos(\Delta\varphi) \right\} \\ &= \frac{P_{\text{in}}(t-t_d)}{4} \cdot SW(t) \end{aligned} \tag{48.3}$$

where, $SW(t)$ is the transfer function. The phase difference between cw and ccw pulse is defined by $\Delta\varphi = (\varphi_{\text{cw}} - \varphi_{\text{ccw}})$. The symbols $G_{\text{cw}}(t), G_{\text{ccw}}(t)$ indicate the respective power gains. Power gain is related with the field gain as $G = g^2$ and $\Delta\varphi = -\frac{\alpha}{2} \cdot \ln\left(\frac{G_{\text{cw}}}{G_{\text{ccw}}}\right)$

Table 48.1 Truth table of TOAD

Incoming signal	Control signal	Output port-1	Output port-2
0	0	0	0
0	1	0	0
1	0	0	1
1	1	1	0

Now we will calculate the power at port-2

$$\begin{aligned}P_{\text{out},2}(t) &= \frac{1}{2}\underline{E}_{\text{out},2}(t)\cdot\underline{E}^{*}_{\text{out},2}(t)\\ &= d^2k^2\cdot P_{\text{in}}(t-t_d)\cdot g^2_{\text{cw}}(t-t_d)\cdot\left\{1+\frac{g^2_{\text{ccw}}(t-t_d)}{g^2_{\text{cw}}(t-t_d)}+2\cdot\frac{g_{\text{ccw}}(t-t_d)}{g_{\text{cw}}(t-t_d)}\right.\\ &\quad\left.\times\cos[\varphi_{\text{cw}}(t-t_d)-\varphi_{\text{ccw}}(t-t_d)]\right\}\\ &= d^2k^2\cdot P_{\text{in}}(t-t_d)\cdot G_{\text{cw}}\cdot\left\{1+\frac{G_{\text{ccw}}}{G_{\text{cw}}}+2\cdot\sqrt{\frac{G_{\text{ccw}}}{G_{\text{cw}}}}\cdot\cos[\Delta\varphi]\right\}\\ &= d^2k^2\cdot P_{\text{in}}(t-t_d)\cdot\left\{G_{\text{cw}}+G_{\text{ccw}}+2\cdot\sqrt{G_{\text{ccw}}\cdot G_{\text{cw}}}\cdot\cos[\Delta\varphi]\right\}\end{aligned} \tag{48.4}$$

For ideal 50:50 coupler, $d^2 = k^2 = 1/2$. In the absence of a control signal, data signal (incoming signal) enters the fibre loop, passes through the SOA at different times as they counter-propagate around the loop, experiences the same unsaturated amplifier gain G_0 and recombines at the input coupler, i.e. $G_{\text{ccw}} = G_{\text{cw}}$. This leads to $\Delta\varphi = 0$. So expression for $P_{\text{out},1}(t) = 0$ and $P_{\text{out},2}(t) = G_0 \cdot P_{\text{in}}$. It shows that data are reflected back towards the source. When a control pulse is injected into the loop, it saturates the SOA and changes its index of refraction. As a result, the two counter-propagated data signals will experience a differential gain saturation profiles, i.e. $G_{\text{ccw}} \neq G_{\text{cw}}$. Therefore, when they recombine at the input coupler, the data will exit from the output port-1. For this case, the mathematical forms of two output powers can be expressed as, $P_{\text{out},1}(t) = \frac{P_{\text{in}}(t-t_d)}{4}\cdot SW(t)$ and $P_{\text{out},2}(t) \approx 0$. Result of numerical simulation with Matlab7.0 has been shown in Fig. 48.1c. In this simulation linewidth enhancement factor of SOA (α) was taken 9.5 and the ratio $G_{\text{ccw}}/G_{\text{cw}}$ was taken 0.52 (Table 48.1).

48.3 Toad-Based Toffoli Gate (TG)

Figure 48.2b (Schematic diagram) and a (TOAD-based circuit) show the Toffoli gate which is a 3 * 3 gate. It has the mapping inputs (*A*, *B*, *C*) to outputs P = A, Q = B and R = AB ⊕ C. A Toffoli gate can be implemented using three

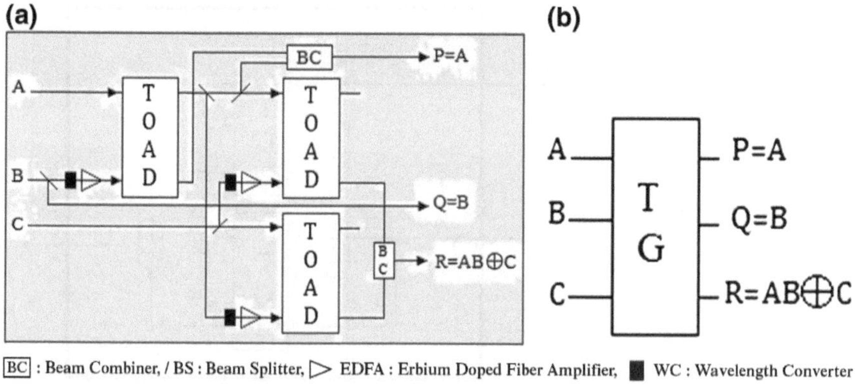

Fig. 48.2 **a** TOAD-based TG. **b** Schematic diagram

Table 48.2 Truth table of Toffoli gate

A	B	C	P	Q	R
0	0	0	0	0	0
0	0	1	0	0	1
0	1	0	0	1	0
0	1	1	0	1	1
1	0	0	1	0	0
1	0	1	1	0	1
1	1	0	1	1	1
1	1	1	1	1	0

TOAD-based all-optical switches, two Beam Combiners and four Beam Splitters in all-optical reversible computing. As the Toffoli gate can be implemented using three TOAD-based optical switches the optical cost of this gate is considered as 3 and the delay of the optical Toffoli gate is considered as 2Δ as two out of three TOAD switches work in parallel. Table 48.2 truth table of TG.

48.4 Reversible Logic Gate (RLG)-Based Multiplexer (MUX)

Using reversible only two TGs we can make all-optical 2:1 multiplexer (Fig. 48.3). Here '*X*' is the control input, by which we can select the input '*Y*' or '*Z*' to the output '*O*'. The basic operation of 2:1 multiplexer can be expressed as: *O* = {*Y* if *X* = 0 and *Z* if *X* = 1}, where *Y*, *Z* {0,1}. The truth table of 2:1 multiplexer is shown in Table 48.2. The logical equation of the output is $\boldsymbol{0} = (\boldsymbol{Y}.\overline{\boldsymbol{X}} + \boldsymbol{Z}.\boldsymbol{X})$. Case I: When *X* = '0', i.e. no light is given to this input port '*X*' then $\boldsymbol{0} = (\boldsymbol{Y}.\overline{\boldsymbol{0}} + \boldsymbol{Z}.\boldsymbol{0}) = Y$.

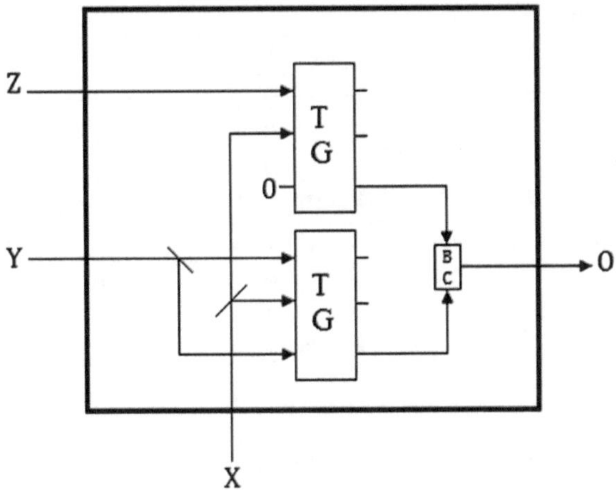

Fig. 48.3 TOAD-based new MUX circuit

Table 48.3 Truth table of MUX

Inputs			Output	
X	*Y*	*Z*	O	Remarks
0	0	0	0	***O* = *Y***
0	0	1	0	
0	1	0	1	
0	1	1	1	
1	0	0	0	***O* = *Z***
1	0	1	1	
1	1	0	0	
1	1	1	1	

Hence the circuit copies the input '*Y*' to the output '*O*'. Case II: When $X = 1$, i.e. constant pulsed light is given to this input port '*X*' then $\boldsymbol{0} = (\boldsymbol{Y}.\overline{\mathbf{1}} + \boldsymbol{Z}.\mathbf{1}) = Z$. Hence the circuit copies the input '*Z*' to the output '*O*' (Table 48.3).

48.5 Application of MUX as Universal Gate

Table 48.4 Universal application of 2-to-1 MUX.

48.6 Crossover Switch

Figure 48.4a shows the block diagram of a 2 × 2 crossover switch.

Table 48.4 Universal application

Inputs			Output	
Z	*Y*	*X*	*O*	Boolean function
1	*B*	*A*	$A + B$	OR
B	0	*A*	AB	AND
0	1	*A*	$\overline{A}$	NOT
0	$\overline{B}$	*A*	$\overline{(A + B)}$	NOR
$\overline{B}$	1	*A*	$\overline{AB}$	NAND
$\overline{B}$	*B*	*A*	$A \oplus B$	EX-OR
B	$\overline{B}$	*A*	$A \odot B$	EX-NOR

When the control input C is '0' then $Y_1 = X_1$ and $Y_2 = X_2$ and for C = 1, $Y_1 = X_2$ and $Y_2 = X_1$.

Therefore, we may write,

$$Y_1 = X_1\overline{C} + X_2C \quad \text{and}$$

$$Y_2 = X_2\overline{C} + X_1C$$

These two equations can be realized using two 2-to-1 MUX as shown Fig. 48.4b.

48.7 Simulation and Result

The simulation is done by Matlab-7.

[vertical axis in dBm(power), horizontal axis in ps(time scale)] (Figs. 48.5 and 48.6).

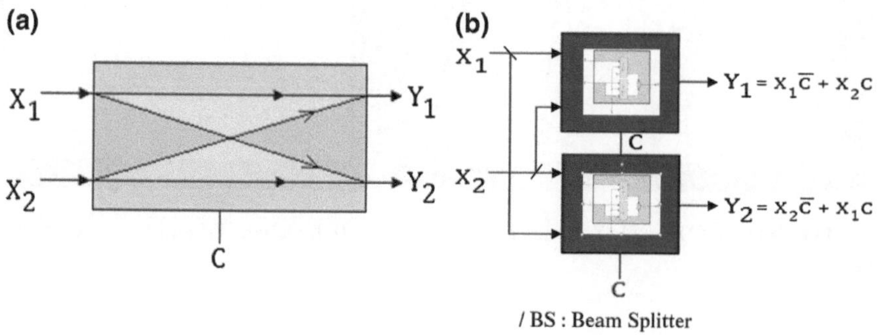

Fig. 48.4 **a** Block diagram of a 2 × 2 crossover switch. **(b)** MUX-based crossover switch

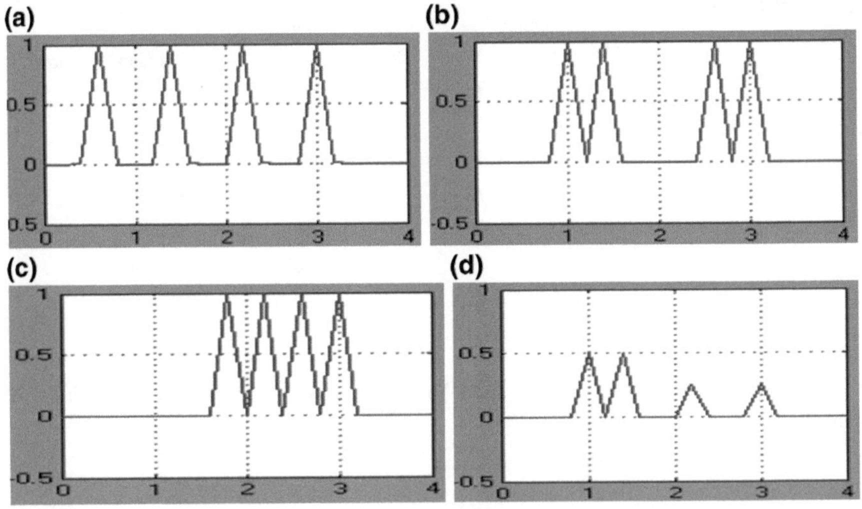

Fig. 48.5 **a** Input '*Z*'. **b** Input '*Y*'. **c** Input '*X*'. **d** Output '*O*'

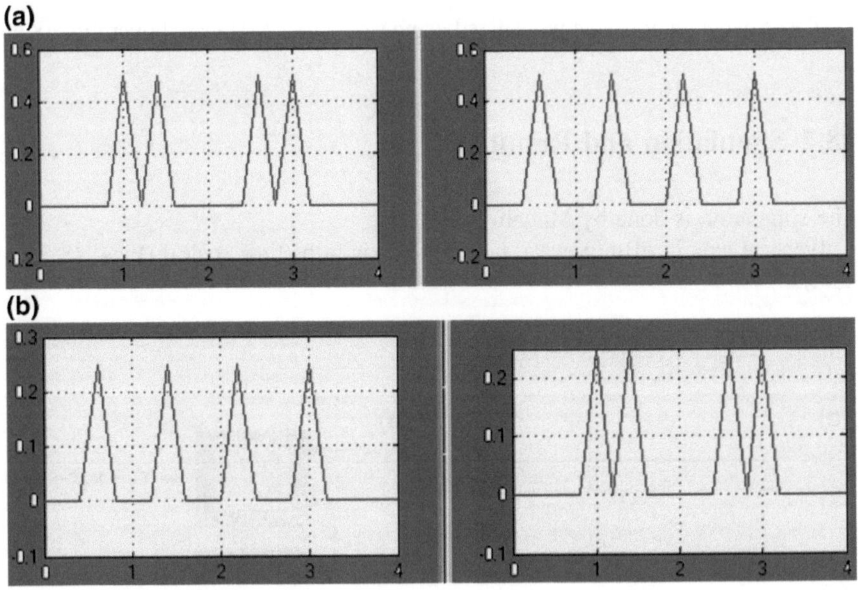

Fig. 48.6 **a** Output 'Y_1' and 'Y_2' when control input, $C = 0$. **b** Output 'Y_1' and 'Y_2' when control input, $C = 1$

48.8 Conclusion and Discussion

In this paper, the all-optical scheme of reversible multiplexer is proposed and explained. Simulation result verifies the functionality of those designed gates with verified reversibility. This is important that the above explanations are based on simple model. The theoretical models developed and the results obtained numerically are useful for future all-optical reversible logic computing system. Different logic operations in reversible system can easily be performed with this reversible multiplexer. The major constraints in reversible logic are to minimize the number of reversible gates used and garbage outputs produced. The output, which is not used for further computations, is known as garbage output. Future work would concentrate on realization of various Boolean expressions and arithmetic operations using MUX.

References

1. J. Hardy, J. Shamir, Optics inspired logic architecture. Opt. Express **15**(1), 150–165 (2007)
2. F.T.S. Yu, S. Jutamulia, S. Yin, *Introduction to Information Optics* (Academic Press, San Diego, 2001)
3. G.P. Agrwal, *Applications of Nonlinear Fibre Optics* (Academic press, India [an imprint of Elsevier, San Diego, USA], 2001)
4. M.A. Karim, A.A.S. Awal, *Optical Computing: An Introduction* (Wiley, New York, 1992)
5. K.E. Zoiros, T. Houbavlis, M. Kalyvas, Ultra-high speed all-optical shift registers and their applications in OTDM networks. Opt. Quant. Electron. **36**, 1005–1053 (2004)
6. R. Landauer, Irreversibility and heat generation in the computational process. IBM J. Res. Dev. **5**, 183–191 (1961)
7. C.H. Bennett, Logical reversibility of computation. IBM J. Res. Dev. **17**, 525–532 (1973)
8. D.P. Vasudavan, P.K. Lala, J. Parkerson, J.P. Di, Reversible-logic design with online testability. IEEE Trans. Instrum. Meas. **55**(2), 406–414 (2006)
9. J.P. Sokoloff, P.R. Prucnal, I. Glesk, M. Kane, A terahertz optical asymmetric demultiplexer (TOAD). IEEE Photonics Technol. Lett. **5**(7), 787–789 (1993)
10. J.N. Roy, G.K. Maity, D. Gayen, T. Chattopadhyay, Terahertz optical asymmetric demultiplexer based tree-net architecture for all-optical conversion scheme from binary to its other 2^n radix based form. Chin. Opt. Lett. **6**(7), 536–540 (2008)

Chapter 49
Implementation of High Performance Vedic Multiplier and Design of DSP Operations Using Vedic Sutra

Supriyo Srimani, Diptendu Kumar Kundu, Saradindu Panda and B. Maji

Abstract Digital signal processing (DSP) operations are very important part of engineering as well as medical discipline. Designing of DSP operations have many approaches. For the designing of DSP operations, multiplication plays a important role to perform signal processing operations such as convolution and correlation. The aim of this paper is to design a multiplier circuit based on Vedic sutras and method for DSP operations based on ancient Vedic mathematics is contemplated. In this paper, we have given the design up to multipliers based on Vedic multiplication sutra 'Urdhva-Tiryakbhyam' the design of 4 × 4 has been sketched in DSCH2 and all the outputs have been given. The layout of those circuits has also been generated by Microwind. The internal circuit diagram of all the blocks has been explained. The noise power have been calculated by T-Spice-13 in 45 nm Technology. This algorithm is implemented in MATLAB and also compared with the inbuilt functions in MATLAB.

Keywords DSP · Vedic mathematics · Vedic multiplier

S. Srimani (✉)
Department of Radio Physics and Electronics, Rajabazar Science College, Kolkata, India
e-mail: srimani.supriyo@gmail.com

D.K. Kundu
Electronics and Telecommunication Department, Narula Institute of Technology, Kolkata, India
e-mail: diptendukumarkundu@gmail.com

S. Panda
ECE Department, Narula Institute of Technology, Kolkata, India
e-mail: saradindupanda@gmail.com

B. Maji
ECE Department, National Institute of Technology, Durgapur, India
e-mail: bmajiecenit@yahoo.com

© Springer India 2015

K. Maharatna et al. (eds.), *Computational Advancement in Communication Circuits and Systems*, Lecture Notes in Electrical Engineering 335,
DOI 10.1007/978-81-322-2274-3_49

49.1 Introduction

A multiplier is one of the key hardware blocks in most of the digital signal processor systems. Multiplication is an operation of scaling one number by another multiplication operation such as convolution, Discreet Fourier Transform, Fast Fourier transform, etc. Design of all digital signal processing (DSP) operations with the help of high-speed Vedic multiplier which increases the efficiency of system and reduces the processing time [1]. The Sanskrit word 'Veda' means 'knowledge'. The Vedas consist of a huge number of documents called 'Ganita sutras' (the name 'Ganita' means mathematics), were devoted to mathematical knowledge. Sri Bharati Krishna Tirtha Maharaja introduce Vedic Mathematics is based on 16 sutras dealing with mathematics related to arithmetic, algebra and geometry [2].

In this paper, we have designed a multiplier based on Vedic Sutra. The first section describes the basic circuits of designing a multiplier in minimal transistor i.e. AND gate, Half Adder, Full Adder. In the next section, we have discussed about the Vedic sutra both in decimal and binary. On the basis of that we have implemented the circuit of 4 × 4 Vedic Multiplier. All the waveforms corresponding each circuit are given separately. Noise, Power, Delay and Area have also been calculated and given in the result analysis section. Then this algorithm is implemented in MATLAB and comparison is done with the conventional functions present in the MATLAB.

49.2 Circuit Techniques

Urdhva-Tiryakbhyam sutra is a general multiplication formula applicable to all cases of multiplication [3]. It literally means 'vertically and crosswise'. The main advantage of utilizing this algorithm in comparison with the existing multiplication techniques is the fact that it utilized only logical AND operation, half address, full address to complete the operation. Also the partial products required for multiplication are generated in parallel and a priori to the actual addition, thus saving as lot of processing time.

The Urdhva Tiryakbhyam sutra is applicable for decimal system as well as for binary system. For implementing this algorithm, we expressed the numbers in a different way. Such as $4 \times 2 = 8 = 0^8$ and $9 \times 3 = 27 = 2^7$. Let us consider two examples to describe the algorithm. In the first example Fig. 49.1a, No carry are generated from the previous stage, i.e. 0 carry has been generated from the previous stage. But for the second example, for each step there is a previous carry and the carry is added to the next stage and so on. Finally we get the result [4].

In case of binary Fig. 49.1b, the representation is slightly deferent such as $1 \times 1 = 1 = 0^1$ and $(1 \times 1) + (1 \times 1) + (1 \times 1) + (1 \times 1) = 1 + 1 + 1 + 1 = 100 = 10^0$ let us take as example to describe the binary multiplication.

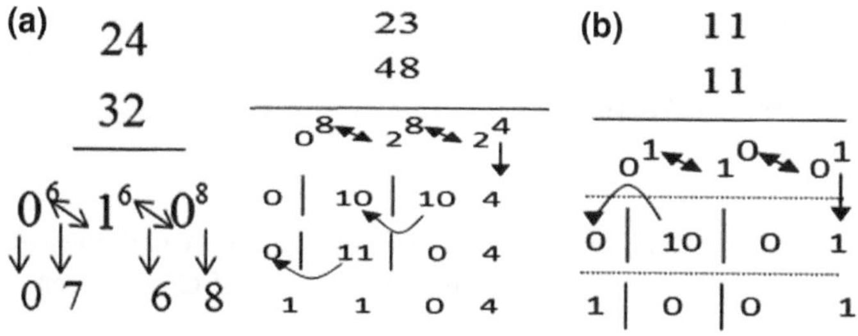

Fig. 49.1 **a** Sutra for decimal numbers. **b** Sutra for binary numbers

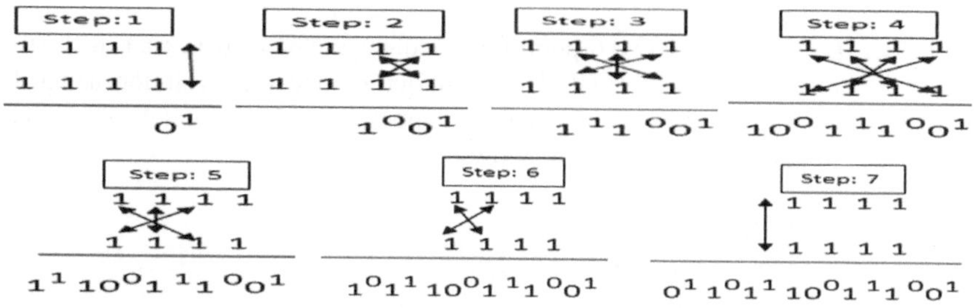

Fig. 49.2 Steps regarding 4 × 4 multiplication

Here the binary addition is done by half address and full address and we know any adding operator results a sum and a carry. From the representation of the number, we can easily found the relation that the representation in nothing but $\text{Carry}^{\text{Sum.}}$

Now we have designed 4 × 4 Vedic multiplier using the Vedic Sutra. Let us see the algorithm first for that we have considered two numbers as usual A (1111) and B (1111) Fig. 49.2.

Now the result comes like $0^1 1^0 1^1 10^0 1^1 1^0 0^1$. For the final result, here we used two arrows: One-sided arrow shows the final result and both-sided arrow shows the intermediate stage and we have numbered each step using Alphabets. And for marking addition, we have used the +sign; Fig. 49.3 shows the complete steps for addition and result.

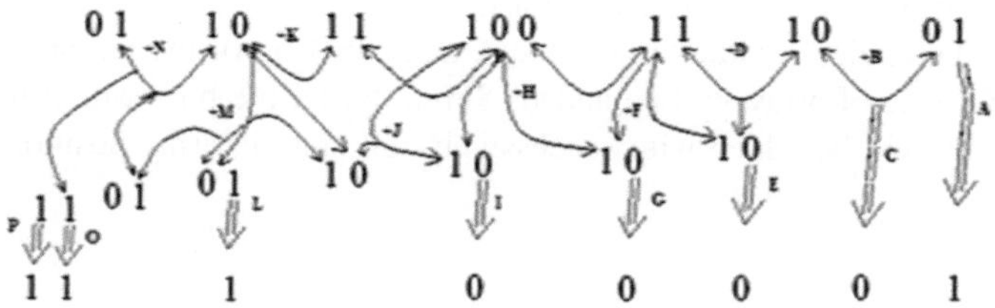

Fig. 49.3 Final steps regarding 4 × 4 Vedic multiplication

49.3 DSP Operations

49.3.1 Linear Convolution

Let us consider two input sequence $x(n) = [x(0), x(1), \ldots, x(L - 1)]$ and $h(n) = [h(0), h(1), \ldots, h(M + 1)]$. The convolution of the length-L input X with the order-M filter h will output the sequence $Y(n)$. Then the Direct form of Convolution

$$y(n) = \sum_{m=\max(0,n-L+1)}^{\min(M,n)} h(m)x(n-m)$$

$n = 0, 1,\ldots,L + M - 1$. For two 6-point input sequence $x(n) = [x(0), x(1), x(2), x(3)]$ and $h(n) = [h(0), h(1), h(2), h(3)]$. The technique of linear convolution using the Urdhava-Triyagbhyam sutras of Vedic mathematics. The convolved outputs sequences are given by the equation show below.

$$\begin{aligned}
y(0) &= x(0)h(0) \\
y(1) &= x(0)h(1) + x(1)h(0) \\
y(2) &= x(0)h(2) + x(1)h(1) + x(2)h(0) \\
y(3) &= x(0)h(3) + x(1)h(2) + x(2)h(1) + x(3)h(0) \\
y(4) &= x(1)h(3) + x(3)h(1) + x(2)h(2) \\
y(5) &= x(3)h(2) + x(2)h(3) \\
y(6) &= x(3)h(3)
\end{aligned}$$

49.3.2 Circular Convolution

Circular Convolution perform the following steps are done.

1. The middle term of the output of convolution is first marked. The output term $y(3)$ is circled.
2. Before the middle term output sequence $y(0)$, $y(1)$, $y(2)$ consist array which is at the left side of array and after the middle term output sequence $y(4)$, $y(5)$, $y(6)$ consist array which is right side of array.
3. Put up the circled middle term is fixed, and MSB bit of the left-side array and the right-side of array will be added. Similarly, all the bit position in the right side array will be added with successively left side bit array position.

The convolved outputs sequences are given by the equation show below.

$$v(0) = y(0) + y(4)$$
$$v(1) = y(1) + y(5)$$
$$v(2) = y(2) + y(6)$$
$$v(3) = y(3)$$

49.4 Results

The circuit that does the multiplication function is given in Fig. 49.4. The process has been implemented in circuit and the output has been soon via glowing of LED also the layout has been drawn and the output of layout is also given in Fig. 49.5.

Multiplier name	Vedic multiplier
Transistor count	139
Noise	30 μV
Power	9.2×10^{-5} W
Delay	0.060 ns
Area	5.01 mm^2

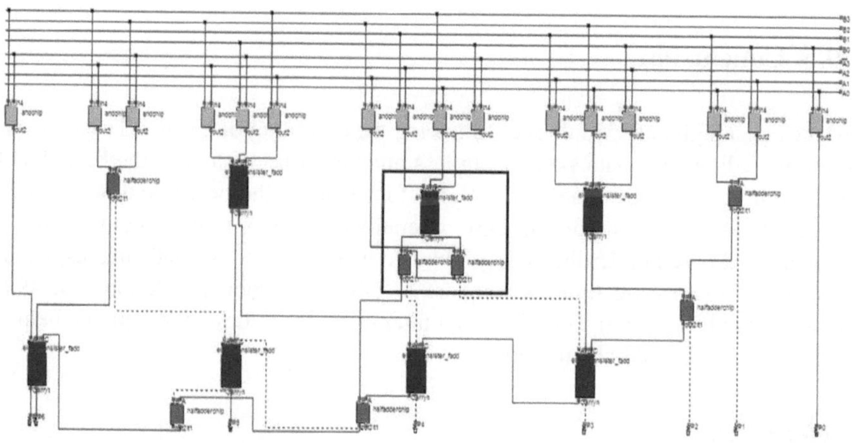

Fig. 49.4 Circuit implementation of 4 × 4 Vedic multiplier

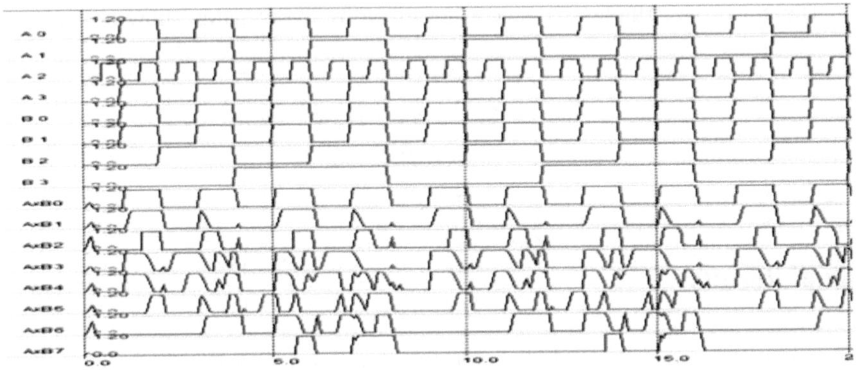

Fig. 49.5 Output of the layout of 4 × 4 Vedic multiplier

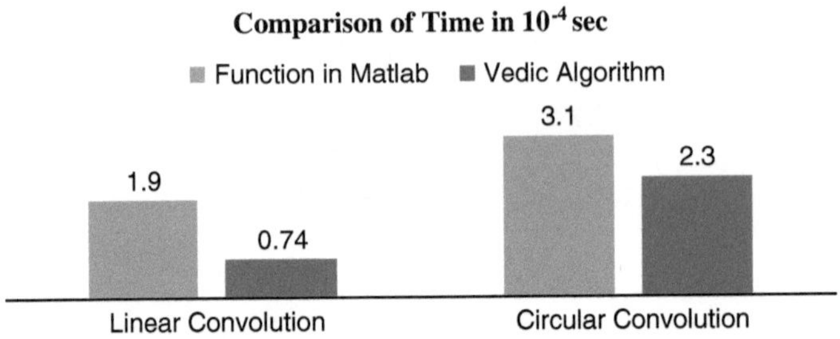

Fig. 49.6 Comparison of DSP application

49.5 Conclusion

We design a high-speed multiplier following the algorithm of Urdhva Tiryakbhyam sutra of Vedic mathematics and using less number of transistors of the basic block of the multiplier we designed the circuits, for reducing the power and increase the speed. A fast computation of DSP operations of two finite length sequence implemented based on Urdhava-Triyagbhyam method of Vedic mathematics, which reduces the processing time as compare to inbuilt function of Matlab Fig. 49.6. In future, the fast Fourier Transform and filter operation is designed with the help of Vedic Urdhva Tiryakbhyam method.

References

1. R. Saligram, T.R. Rakshith, Optimized reversible Vedic multipliers for high speed low power operations, in *Proceedings of IEEE Conference on Information and Communication Technologies* (*ICT 2013*) (2013)
2. H.D. Tiwari, G. Gankhuyag, C.M. Kim, Y.B. Cho, Multiplier design based on ancient Indian Vedic mathematics, in *IEEE International SoC Design Conference*, 978-1-4244-2599-0/08/ $25.00 ©2008

3. A. Kanhe, S.K. Das, A.K. Singh, Design and implementation of floating point multiplier based on Vedic multiplication technique, in *International Conference on Communication, Information & Computing Technology* (*ICCICT*), Mumbai, India, 19–20 Oct 2012
4. P. Mehta, D. Gawali, Conventional versus Vedic mathematical method for hardware implementation of a multiplier, in *International Conference on Advances in Computing, Control, and Telecommunication Technologies*

Part VII
Advances in Signal Processing

Chapter 50
Does Music Affect HRV Impulse? A Time Domain Study

Anilesh Dey, Anwesha Banerjee, D.K. Bhattacharya and D.N. Tibarewala

Abstract The two-dimensional (2D) Poincaré plot of HRV signal reflects the effect of different external stimuli on ANS. The classification is generally done by fitting an 'ellipse' on the dense region of the constructed Poincaré plot of HRV signal. However, 2D Poincaré plot sometimes fails to describe the proper behaviour of the system. One such example is 2D Poincaré plot of HRV signal in pre-music and on-music condition. In fact, 2D Poincaré plots in pre and on-music condition look almost similar for few subjects. So a proper classification tool is sought for. In this article, an improved technique called '3D Poincaré plot with proper delay' has been applied to properly distinguish the pre-music and on-music state of some normal healthy subjects.

Keywords Average mutual information · Ellipse fit · Ellipsoid fit · HRV signal · 2D Poincaré plot · 3D Poincaré plot

---

A. Dey (✉)
Electronics and Communication Engineering Department, Narula Institute of Technology, Kolkata 700109, India
e-mail: anilesh.dey@gmail.com

A. Banerjee · D.N. Tibarewala
School of BioScience and Engineering, Jadavpur University, Kolkata 700032, India
e-mail: anwesha.banerjee@ymail.com

D.N. Tibarewala
e-mail: biomed.ju@gmail.com

D.K. Bhattacharya
Instrumental Music Department, Rabindra Bharati University, Kolkata 700050, India
e-mail: dkb_math@gmail.com

© Springer India 2015

K. Maharatna et al. (eds.), *Computational Advancement in Communication Circuits and Systems*, Lecture Notes in Electrical Engineering 335,
DOI 10.1007/978-81-322-2274-3_50

50.1 Introduction

Nowadays, stress and anxiety have a large negative impact on our society and most of our diseases originate from psychosomatic disorder. Therefore reduction of stress is necessary for the well-being of our society. Since music has a direct connection with human feeling and mood, it can be used for the reduction of stress of human being. An essential question that may arise at this point is how does music affect human physiological condition and what are the important parameters needed to classify the pattern? During the last two decades, the effect of different types of music to promote relaxation has been studied throughout the globe [1, 2]. The primary focus of those researches was to relate to the subjective responses of the subject, such as pleasure and changes in mood [3] rather than objective physiological responses. However, the results regarding the effect of music were found to be very much conflicting and confusing in relating to the exact circumstances and variables that affect the body's response to music, such as the type of music [4] and the subject's involvement in the music [3]. In some literature, it was found that music decreases the sympathetic nervous system (SNS) and increases the parasympathetic nervous system (PNS) activity as measured by heart rate (HR), blood pressure (BP) and heart rate variability (HRV), indicating physiological relaxation [3], while in some other literature it was found that music increases SNS activity and it also increases the HR in subjects who listened to some preferred music after exercise [1]. In fact, it was also found that two other factors—respiratory rhythm [1, 5] and gender [1]—affect the human physiological response to music.

HRV [6–9] is a popular non-invasive tool to assess different heart conditions. Due to its non-invasive character, HRV has become an attractive tool for use in the study of human physiological response to different stimuli [10, 11]. HRV is the variation of time between two consecutive heartbeats. It is a useful tool to know the overall cardiac health and the status of the autonomic nervous system (ANS). There are two branches of the ANS—the sympathetic and the parasympathetic. The sympathetic branch increases HR and the parasympathetic branch decreases it. Thus, at any instant, the observed HRV is an indicator of the dynamic interaction and balance between these two nervous systems. In the resting condition, both the sympathetic and parasympathetic systems are active with parasympathetic dominance. The balance between them is constantly changing to optimize the effect of all internal and external stimuli [12].

Although the effects of music on mind are mostly realized in brain through Central Nervous system (CNS), music also affects the conditions of heart through the dominance of parasympathetic nerves of ANS. So it is no less important to study the effect of music through analysis of HRV data extracted from the corresponding ECG signals of the heart in the time domain, when we listen to music. In this article, for the purpose of analysing the effect of music on HRV signals under pre-music and on-music states, a very simple method has been proposed which clearly analyses the effect of music on HRV signals. In fact, traditional 2D Poincaré plot with proper delay is applied for the analysis of the effect of music on HRV signals. But no such distinction between the pre-music and on-music states can be made visually or even by fitting an ellipse in the cloud region of the Poincaré plot of HRV signals [11]. Hence, 3D Poincaré plot with proper delay is applied which gives satisfactory result in this context.

50.2 Method

50.2.1 Acquisition of HRV Data

At first ECG data are collected from different subjects (age between 20 and 30 years male and female). All subjects are basically students and academicians. The digitized form of ECG data are collected from those subjects and recorded by 'HRV data logger' machine (made by School of BioScience and Engg., Jadavpur University). All signals are taken at School of BioScience and Engg., Jadavpur University under normal room temperature and least noisy environment. Signals have been collected in two stages. In the first stage, ECG are taken at normal condition from healthy subjects. Then in the second stage, ECG signals are taken when subjects are listening to music. All signals are taken in 10 min duration. Then recorded signals are processed by MATLABR2010a software using moving window integration of a digital filter and converted into HRV signals.

50.2.2 2D Poincaré Plot, Average Mutual Information and 3D Poincaré Plot

To explore the HRV dynamics on 'beat-to-beat' basis, the original idea of 2D Poincaré plot included a delay/lag of one beat only [13]. Later on, 2D Poincaré plot with non-unit lag was developed [14]. But there was no specific basis of choosing this non-unit lag, the choice was purely arbitrary. Possibly, the reason was to get comparatively better form of 2D Poincaré plot. If, however, proper quantification of the 2D Poincaré plot is required for the purpose of interpretation of the behaviour of the data, then the best form of Poincaré plot is to be obtained in any way. For example, when quantification of 2D Poincaré plot is made [15, 16] by the process of 'ellipse fit' then for this ellipse, independent coordinates are to be sought for from the data itself. Naturally, for quantification by this method, the Poincaré plot itself should have some form, which should not be of much irregular shape. Hence, there is a necessity for selecting proper lag for constructing best 2D Poincaré plot. In this case, the standard methods available for obtaining the proper lag are the method of minimum autocorrelation, a linear measure and the method of average mutual information [14, 17].

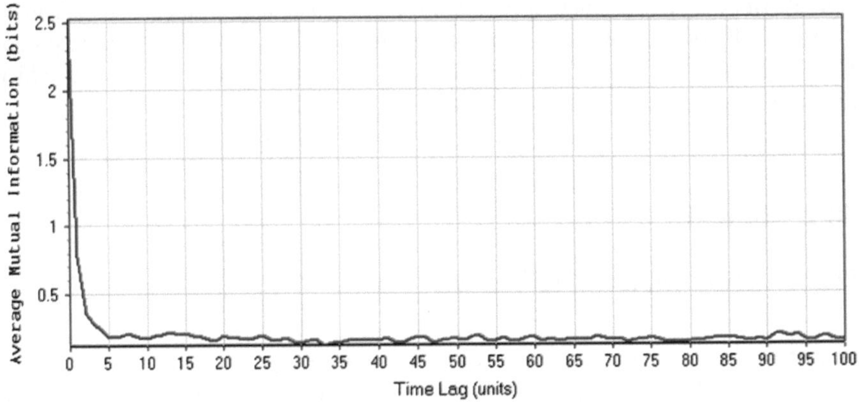

Fig. 50.1 $I(\tau)$ versus τ plot

Mutual information function can be used to determine the 'optimal' value of the time delay for the state space reconstruction.

Suppose $\{x(t)\}_{t=1}^{N}$ is given time series. A good choice for the delay τ is one that, given the state of the system $x(t)$, provides maximum new information with the measurement at $x(t+\tau)$.

$$I(\tau) = \sum_{t=1}^{N-\tau} P[x(t), x(t+\tau)] \log\left(\frac{P[x(t), x(t+\tau)]}{P[x(t)]P[x(t+\tau)]}\right)$$

It was suggested in [17] that the value of delay, where $I(\tau)$ reaches its first minimum is used for the 2D Poincaré reconstruction (Fig. 50.1).

2D Poincaré plot is constructed with the independent coordinates $(x(t), x(t+\tau))$ and 3D Poincaré plot is plotted with the independent coordinates $(x(t), x(t+\tau), x(t+2\tau))$.

50.2.3 Quantification of 2D and 3D Poincaré Plots

50.2.3.1 Quantification of 2D Poincaré Plots

Let $\{X(j)\}_{j=1}^{N}$ be a discrete signal. Let the 2D Poincaré plot be constructed by subdividing this signal into two groups as x^+, x^- with same time delay τ, where

$$x^+ = \{|X(j)|\}_{j=1}^{N-\tau}, \quad x^- = \{|X(j)|\}_{k=1+\tau}^{N}, \quad \tau = 1, 2, \ldots, (N-1).$$

Now this coordinate system is transformed by a two-dimensional rotation with same angle $\frac{\pi}{4}$ with respect to X axis. The transform is given by

$$\begin{pmatrix} x_m \\ x_n \end{pmatrix} = \begin{pmatrix} \cos\frac{\pi}{4} & -\sin\frac{\pi}{4} \\ \sin\frac{\pi}{4} & \cos\frac{\pi}{4} \end{pmatrix} \begin{pmatrix} x^+ \\ x^- \end{pmatrix}$$
$$= \frac{1}{\sqrt{2}} \begin{pmatrix} 1 & -1 \\ 1 & 1 \end{pmatrix} \begin{pmatrix} x^+ \\ x^- \end{pmatrix}$$

Thus, a new coordinate system (x_m, x_n) is formed. Let $\overline{x_m} = \text{Mean}(x_m), \overline{x_n} = \text{Mean}(x_n)$ and $SD_1\sqrt{\text{Var}(x_m)}, SD_2\sqrt{\text{Var}(x_n)}$. Lastly, an ellipse centred at $(\overline{x_m}, \overline{x_n})$ with two axes of length SD_1, SD_2 is taken for quantification of the existing 2D Poincaré plot.

50.2.3.2 Quantification of 3D Poincaré Plots

Quantification of 3D Poincaré plot is generally done by the method of 'ellipsoid fit'. Axes of the ellipsoid stand as a strong indicator of the changing energy dynamics of HRV. Figure 50.3 shows how an ellipsoid is fitted to the dense region of the phase space.

Let $\{X(j)\}_{j=1}^{N}$ be a discrete HRV signal. Let the three-dimensional frequency delay plot be constructed by subdividing this signal into three groups as x^+, x^-, x^{--} with same time delay τ, where

$$x^+ = \{|X(j)|\}_{j=1}^{N-2\tau}, \quad x^- = \{|X(j)|\}_{k=1+\tau}^{N-\tau},$$
$$x^{--} = \{|X(j)|\}_{j=1+2\tau}^{N}, \ \tau = 1, 2, \ldots, (N-1).$$

Now this coordinate system is transformed by a three-dimensional rotation with same angle $\frac{\pi}{4}$ with respect to X, Y and Z axes. The transform is given by

$$\begin{pmatrix} x_m \\ x_n \\ x_p \end{pmatrix} = \begin{pmatrix} \cos\frac{\pi}{4}\cos\frac{\pi}{4} & \cos\frac{\pi}{4}\sin\frac{\pi}{4}\sin\frac{\pi}{4} - \cos\frac{\pi}{4}\sin\frac{\pi}{4} & \cos\frac{\pi}{4}\cos\frac{\pi}{4}\sin\frac{\pi}{4} + \sin\frac{\pi}{4}\sin\frac{\pi}{4} \\ \cos\frac{\pi}{4}\sin\frac{\pi}{4} & \cos\frac{\pi}{4}\cos\frac{\pi}{4} + \sin\frac{\pi}{4}\sin\frac{\pi}{4}\sin\frac{\pi}{4} & -\cos\frac{\pi}{4}\sin\frac{\pi}{4} + \cos\frac{\pi}{4}\sin\frac{\pi}{4}\sin\frac{\pi}{4} \\ -\sin\frac{\pi}{4} & \cos\frac{\pi}{4}\sin\frac{\pi}{4} & \cos\frac{\pi}{4}\cos\frac{\pi}{4} \end{pmatrix} \begin{pmatrix} x^+ \\ x^- \\ x^{--} \end{pmatrix}$$
$$= \frac{1}{2\sqrt{2}} \begin{pmatrix} 2\sqrt{2} & -(\sqrt{2}-1) & (\sqrt{2}+1) \\ 2\sqrt{2} & (\sqrt{2}+1) & -(\sqrt{2}-1) \\ -2 & \sqrt{2} & \sqrt{2} \end{pmatrix} \begin{pmatrix} x^+ \\ x^- \\ x^{--} \end{pmatrix}$$

Hence,

$$x_m = \frac{1}{2}\cdot x^+ + \left(\frac{1}{2\sqrt{2}} - \frac{1}{2}\right)\cdot x^- + \left(\frac{1}{2\sqrt{2}} + \frac{1}{2}\right)\cdot x^{--} = \frac{2\sqrt{2}\cdot x^+ - (\sqrt{2}-1)\cdot x^- + (\sqrt{2}+1)\cdot x^{--}}{2\sqrt{2}};$$
$$x_n = \frac{1}{2}\cdot x^+ + \left(\frac{1}{2\sqrt{2}} + \frac{1}{2}\right)\cdot x^- + \left(\frac{1}{2\sqrt{2}} - \frac{1}{2}\right)\cdot x^{--} = \frac{2\sqrt{2}\cdot x^+ + (\sqrt{2}+1)\cdot x^- - (\sqrt{2}-1)\cdot x^{--}}{2\sqrt{2}};$$
$$x_p = \left(-\frac{1}{\sqrt{2}}\right)\cdot x^+ + \frac{1}{2}\cdot x^- + \frac{1}{2}\cdot x^{--} = \frac{-2\cdot x^+ + \sqrt{2}\cdot x^- + \sqrt{2}\cdot x^{--}}{2\sqrt{2}}.$$

Thus, a new coordinate system (x_m, x_n, x_p) is formed. Let and $\overline{x_m} = \text{Mean}(x_m)$, $\overline{x_n} = \text{Mean}(x_n)$, $\overline{x_p} = \text{Mean}(x_p)$ and $\text{SD}_1 = \sqrt{\text{Var}(x_m)}, \text{SD}_2 = \sqrt{\text{Var}(x_n)}$, $\text{SD}_3 = \sqrt{\text{Var}(x_p)}$. Lastly, an ellipsoid centred at $(\overline{x_m}, \overline{x_n}, \overline{x_p})$ with three axes of length SD_1, SD_2 and SD_3 are taken for quantification of the existing 3D Poincaré plot.

50.3 Result and Discussion

50.3.1 2D Poincaré Plot with Proper Delay of HRV Signal in Pre-music and On-Music State

Figure 50.2 shows the 2D Poincaré plot with proper delay of HRV signals in pre-music and on-music state.

It is observed that both the Poincaré plots are dense in some area with some outliers. Actually, there is no known canonical way to eliminate these outliers of the plots except by manual supervision and visual inspection. However, these things are not at all important here. Rather, it is necessary to focus on the main cluster because the important, relevant and necessary information in this context is hidden within the orientation of the main cluster. We have tried to quantify these plots by fitting an ellipse to their main cluster. So, we have computed the lengths of the major axis and minor axis in each case and finally taken the ratio of two axes as a quantifying parameter, which are given in Tables 50.1 and 50.2.

It is observed from Tables 50.1 and 50.2 that the ratio of the axis length SD2/SD1 decreases most of the subjects except a6 and a2 in on-music state. So it is not able to quantify all the cases in on-music state. So 2D Poincaré plot with proper delay is not a proper tool for distinguishing the pre-music and on-music states.

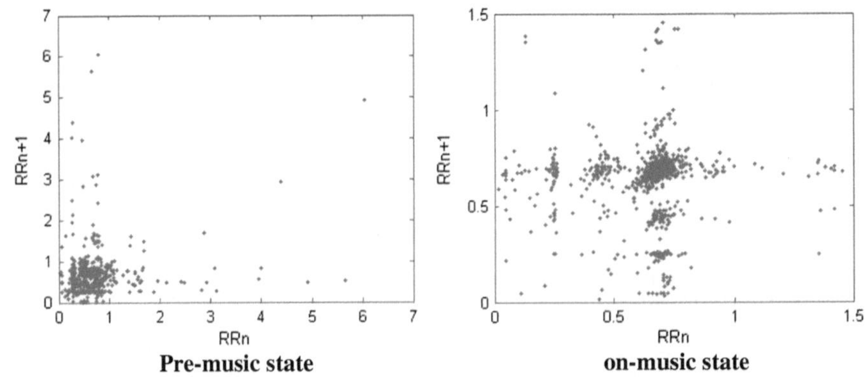

Fig. 50.2 2D Poincaré plot with proper delay of HRV signals in pre-music and on-music state

Table 50.1 Quantification table of 2D Poincaré plot with proper delay of HRV signals in pre-music state

Subject	SD_1	SD_2	SD_2/SD_{11}
a1	0.445502	0.518924	1.164807
a2	0.14233	0.1527	1.072858
a3	0.180022	0.177443	0.985675
a4	0.563637	0.713092	1.265162
a5	0.23555	0.244572	1.038301
a6	0.181542	0.186061	1.024894
a7	0.274641	0.287152	1.045553
a8	0.154729	0.157943	1.020775
a9	0.207647	0.251096	1.209247

Table 50.2 Quantification table of 2D Poincaré plot with proper delay of HRV signals and on-music state

Subject	SD_1	SD_2	SD_2/SD_1
a1m	0.182648	0.18935	1.036693
a2m	0.11955	0.129195	1.080674
a3m	0.24773	0.23288	0.940056
a4m	0.134011	0.168468	1.257123
a5m	0.218786	0.209284	0.95657
a6m	0.211619	0.270061	1.276162
a7m	0.241921	0.234208	0.968116
a8m	0.256898	0.252189	0.981669
a9m	0.263063	0.258643	0.9832

So we switch from 2D Poincaré plot with proper delay to 3D Poincaré plot with proper delay. Before this, 3D Poincaré plot with unit delay is attempted but it is not able to quantify all the cases in on-music state.

50.3.2 Poincaré Plot with Proper Delay of HRV Signal in Pre-music and On-Music State

From the above figure (Fig. 50.3), it is evident that both the plots are well formed and dense compared to the previously obtained 2D Poincaré plots in pre-music and on-music states. So we have tried to quantify these plots by fitting an ellipsoid to their main cluster. For this purpose, we have computed the lengths of three axes (SD_1, SD_2, SD_3) and then taken the ratio of SD_2/SD_1 and SD_2/SD_3. Finally, the quantifying parameter is taken as the average of the two aforesaid ratios. This is given in Tables 50.3 and 50.4.

It is observed from Tables 50.3 and 50.4 that the average value of SD_2/SD_1 and SD_2/SD_3 decreases in all of the cases in on-music state. So 3D Poincaré plot with proper delay is a proper tool for the purpose of distinction of the pre-music and on-music sates.

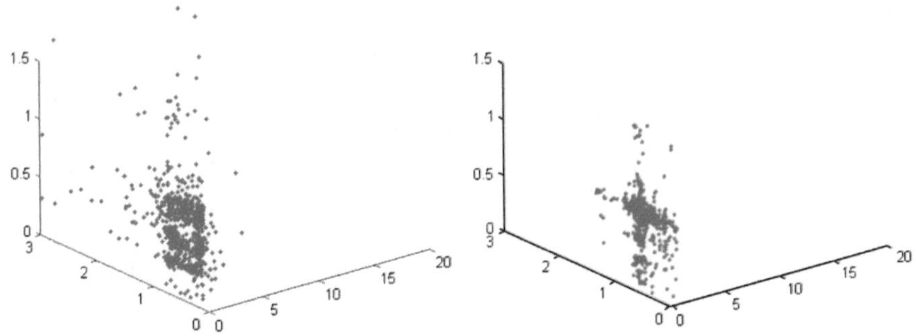

Fig. 50.3 3D Poincaré plot with proper delay of HRV signals in pre-music and on-music state

Table 50.3 Quantification table of 3D Poincaré plot with proper delay of HRV signals in pre-music state

Sub	SD_1	SD_2	SD_3	SD_2/SD_1 (a1)	SD_2/SD_3 (b1)	(a1 + b1)/2
a1	0.6398	0.6773	0.4685	1.05861	1.445667	1.2521367
a2	0.1704	0.17	0.1198	0.99791	1.419758	1.2088355
a3	0.2413	0.2346	0.1755	0.97246	1.336872	1.1546641
a4	0.92	0.9141	0.5688	0.99357	1.607158	1.3003659
a5	0.3102	0.3247	0.2422	1.04693	1.340795	1.1938643
a6	0.3406	0.3456	0.2175	1.01488	1.588942	1.3019119
a7	0.3605	0.38	0.2854	1.05405	1.331293	1.1926706
a8	0.2135	0.2083	0.153	0.97556	1.361346	1.1684511
a9	0.3245	0.3228	0.2086	0.99478	1.547641	1.2712123

Table 50.4 Quantification table of 3D Poincaré plot with proper delay of HRV signals in on-music state

Sub	SD_1	SD_2	SD_3	SD_2/SD_1 (a2)	SD_2/SD_3 (b2)	(a2 + b2)/2
a1 m	0.2449	0.2489	0.1844	1.0166	1.349781	1.183189
a2 m	0.1986	0.1969	0.1385	0.99164	1.422129	1.206886
a3 m	0.3458	0.3053	0.2276	0.8828	1.341541	1.112171
a4 m	0.2173	0.2162	0.1352	0.99491	1.598874	1.296893
a5 m	0.2872	0.2778	0.2136	0.96741	1.300821	1.134115
a6 m	0.2413	0.2465	0.1845	1.02136	1.335565	1.17846
a7 m	0.3155	0.3112	0.2393	0.98645	1.30068	1.143565
a8 m	0.3343	0.3349	0.2533	1.00182	1.321874	1.161848
a9 m	0.3343	0.3349	0.2533	1.00182	1.321874	1.161848

50.4 Conclusion

It is to be noted that the average value of SD_2/SD_1 and SD_2/SD_3 of the fitted ellipsoid reduces in case of on-music state as compared to the corresponding value of the subjects in the normal condition. The possible reason is the reduction of stress in the on music states. Thus, it may be inferred that reduction in the average value of SD_2/SD_1 and SD_2/SD_3 of the fitted ellipsoid is a sufficient criteria for minimization of stress. However, this study is limited to a smaller sample size of data and so further investigation is required on a larger sample size of data to substantiate the present work.

References

1. T. Kaplan, Signal Processing. Warwick- exercises tex. 5(2004)
2. L. Glass, M.C. Mackey, *From Clocks to Chaos: The Rhythms of Life* (Princeton University Press, Princeton, 1988)
3. J.F. Christini, M. Bennett, K.R. Lutchen, H.M. Ahmed, J.M. Hausdorff, N. Oriol, Application of linear and nonlinear time series modeling to heart rate dynamics analysis. IEEE Trans. Biomed. Eng. **42**, 411 (1995)
4. J.P. Eckmann, D. Ruelle, Ergodic theory of chaos and strange attractors. Rev. Mod. Phys. **57**, 617 (1985)
5. U.R. Acharya, K.P. Joseph, N. Kannathal, L.C. Min, J.S. Suri, *Advances in Cardiac Signal Processing: Heart Rate Variability* (Springer, New York, 2007), pp. 121–165
6. J.E. Sanderson, Heart rate variability in heart failure. Heart Fail. Rev. **2**, 235–244 (1998)
7. J.P. Saul, Beat-to-beat variations of heart rate reflects modulation of cardiac autonomic outflow. News Physiol. Sci. **5**, 32–37 (1990)
8. L.C. Andrews, B.K. Shivamoggi, *Integral Transforms for Engineers* (Prentice-Hall of India, New Delhi, 2005)
9. M. Weeks, *Digital Signal Processing* (Infinity Science Press, Massachusetts, 2007)
10. P.J. Schwartz, S.G. Priori, Sympathetic nervous system and cardiac arrhythmias, in *Cardiac Electrophysiology, From Cell to Bedside*, ed. by D.P. Zipes, J. Jalife (Saunders, W.B, Philadelphia, 1990), pp. 330–343
11. T. Kaplan, L. Glass, *Understanding Nonlinear Dynamics* (Springer, New York, 1995)
12. S. Mukherjee, S.K. Palit, A new scientific study towards distinction of ECG signals of a normal healthy person and of a congestive heart failure patient. J. Int. Acad. Phys. Sci. **15**(4), 413–433 (2011)
13. V. Anishchenko, A. Neiman, T. Vadivasova, V. Astakhov, G. Schimansky, *Dynamics of Chaotic and Stochastic Systems*, 2nd edn. (Springer, Berlin, 2007)
14. G.L. Baker, J.P. Gollub, *Chaotic Dynamics: An Introduction* (Cambridge University Press, Cambridge, 1998). [62]
15. M.P. Tulppo, T.H. Makikallio, T.E.S. Takala, T. Seppanen, H.V. Huikuri, Quantitative beat-to-beat analysis of heart rate dynamics during exercise. Am. J. Physiol. **271**, H244–H252 (1996)
16. M. Marchal, Determinism, Random, Chaos, Freedom. Henri Poincare and the Revolution of Scientific ideas in the Twentieth Century. Regul. Chaotic Dyn. **10**, 227–236 (2005)
17. H.D.I. Abarbanel, *Analysis of Observed Chaotic Data* (Springer, New York, 1997)

Chapter 51
A Novel Design Approach of Subband Coder and Decoder of Speech Signal Using Log Normal Probability Distribution

Sangita Roy and Sheli Sinha Chaudhuri

Abstract In modern speech communication less bandwidth and low data rate is very essential from portable systems with limited storage capacity. Researchers are concerned with the tradeoff between bandwidth and SNR (Signal-to-Noise), BER (Bit Error Rate). Speech signal can be compressed below 64 Kbps taking care of SNR above 30 dB, and BER below 10^{-5}. Here the authors proposed Log Normal Distribution in the design of Subband Coder and Decoder of Speech Signal taking care of the SNR and BER criterion with data rate 9.3316 Kbps.

Keywords Subband · BER · Probability distribution · SNR

51.1 Introduction

Speech is the basic form of human communication. Speech communication is of immense importance as the speech signal is different from any other sounds. In the last few decades speech communication has been of great concern due to the fast growing technologies. The communication channel is probabilistic in nature. Therefore, measuring parameters are essential to ensure the quality of the speech signal through the channel. The parameters for good speech signal transmission are (i) low bit rate, (ii) more than 30 dB SNR and (iii) BER less than 10^{-5} [1–3]. Speech signals follows some probability density functions (PDF), i.e. Gaussian [4], Rayleigh [5], Log Normal [6], etc. The authors identified Log Normal distribution for characterization of speech signal and compared it with tested speech signal distributions, i.e. standard power spectral density of speech signal, Gaussian

S. Roy (✉)
ECE Department, Narula Institute of Technology, WBUT, Kolkata, India
e-mail: roysangita@gmail.com

S.S. Chaudhuri
ETCE Department, Jadavpur University, Kolkata, India
e-mail: shelism@rediffmail.com

© Springer India 2015

K. Maharatna et al. (eds.), *Computational Advancement in Communication Circuits and Systems*, Lecture Notes in Electrical Engineering 335,
DOI 10.1007/978-81-322-2274-3_51

Distribution [7], Rayleigh Distribution [8]. By employing these PDFs (Probability Density Functions) communication channels as well as PSD (power spectral density) of speech signal can be used efficiently. Subband coding (SBC) is a kind of transform coding. A signal is divided into a number of different frequency bands and encodes each one independently. It enables data reduction by discarding information about frequencies which are masked. The result differs from the original signal, but if the discarded information is chosen carefully, the difference will not be noticeable, or more importantly, objectionable [9–12].

51.2 Literature Survey

The paper—"A low-complexity audio data compression technique using subband coding (SBC) and a recursively indexed quantizer (RIQ)" compared SBC and RIQ with conventional coding techniques. The system shows SNR 2–5 dB higher than that of SNRs of other coders of similar computational complexity of wideband audio signals [7]. The basic concept of "Frequency Domain Coding of Speech" methods is to divide the speech into frequency components by a filter bank (subband coding), or by a suitable transform (transform coding), and then encode them using adaptive PCM (Pulse Code Modulation). Three basic factors of the design of coders are: (1) the type of the filter bank or transform, (2) the choice of bit allocation and noise shaping properties and (3) the control of the step-size of the encoders. Short-time analysis/synthesis, practical realizations of subband and transform coding are interpreted within this framework. Spectral estimation, models of speech production, perception and the "side information" can be most efficiently represented and utilized in the design of the coder (particularly the adaptive transform coder) to control the dynamic bit allocation and quantizer step-sizes. Recent developments and examples of the "Vocoder-driven" adaptive transform coder for low bit-rate applications is also discussed [8]. In digital telecommunication systems different signals are processed with different sampling rates, leading to significant errors. In "Subband Coding of Speech Signals Using Decimation and Interpolation"—a structure of a two-channel quadrature mirror filter with low pass filter, high pass filter, decimators and interpolators, is proposed to perform subband coding of speech signals in the digital domain. The performance of the proposed structure is compared with the performance of delta-modulation encoding systems. The results show that the proposed structure significantly reduces error and achieves considerable performance improvement compared to delta-modulation encoding systems [13]. Gaussian Distribution is well suited for describing the Power Spectral Density of Speech Signal. In statistical voice activity detection (VAD) Rayleigh Distribution has been used as the distribution has longer asymmetric tail than Gaussian distribution. MMSEEs (Minimum Mean Square Estimators) for speech enhancement have employed various PDFs, such as Gaussian Distribution, Log Normal Distributions, etc.

51.3 Basic Principles of the Proposed System Model [8]

51.3.1 Design Procedure for Subband Coding for Speech Signal

The Power Spectral Density (PSD) of a voice signal has been considered to be restricted to 3.5 kHz only, Power Spectral Density to be in watt/Hz or dB (Fig. 51.1).

In this figure frequency axis is divided into a number of subbands (say 0–*f*1, *f*1–*f*2, *f*2–*f*3, *f*3–*f*4, etc.). The frequency band (0–*f*1) is baseband signal, whereas (*f*1–*f*2), (*f*2–*f*3), (*f*3–*f*4), etc. are bandpass signals. Each band will be translated into baseband by multiplying with the lowest frequency component of the said subband. Here seven subbands have been considered (Fig. 51.2).

Fig. 51.1 PSD of speech signal

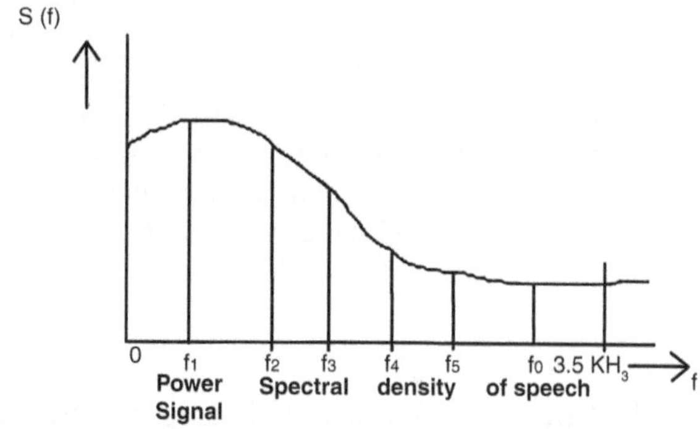

Fig. 51.2 Block diagram of subband coder

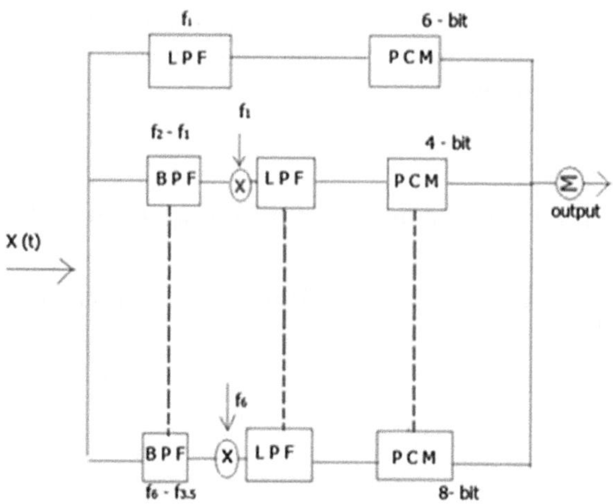

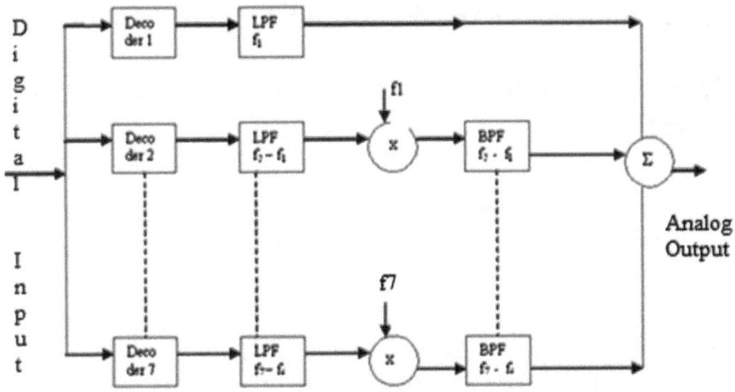

Fig. 51.3 Block diagram of subband receiver

The transmitter consists of one LPF and six BPFs. All BPFs outputs are multiplied by the lowest frequency component of those bands at the multiplier block. Then outputs are PCM and then added by summer. Finally the summed output is put into channel (Fig. 51.3).

At the receiver signals are decoded by seven decoders. Then each signal is passed through LPF of cut-off frequency $f1$, $f2$–$f1$, $f3$–$f2$, etc. From the second to the seventh signal outputs are multiplied by their respective lowest frequency components and then passed through BPFs of $f2$–$f1$, $f3$–$f2$, etc. Then the outputs are summed up to get a replica of the original signal.

51.4 Proposed Method with Log Normal Distribution PSD

Speech Coding follows different probability distributions. Authors have already worked with Gaussian and Raleigh Distributions [7, 8]. Here, they have chosen Log Normal Distribution and followed the same procedure as earlier. The results are shown below.

51.4.1 Mathematical Validation Using MATLAB Simulation

(See Figs. 51.4, 51.5 and 51.6; Tables 51.1 and 51.2)

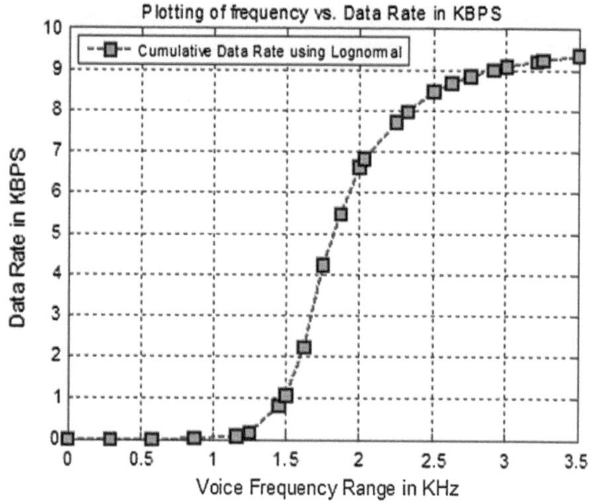

Fig. 51.4 Cumulative data rate versus frequency

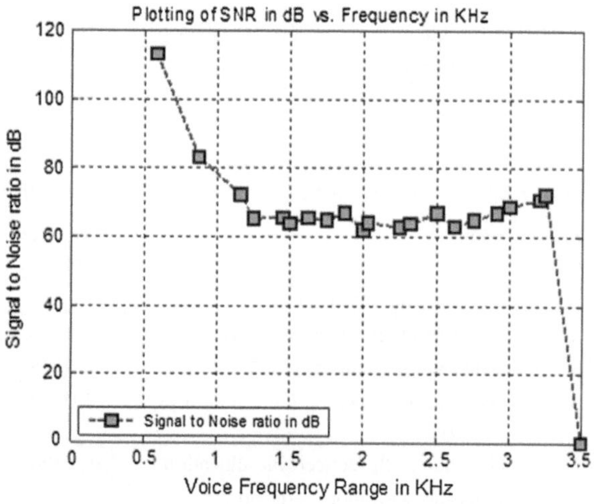

Fig. 51.5 SNR versus frequency

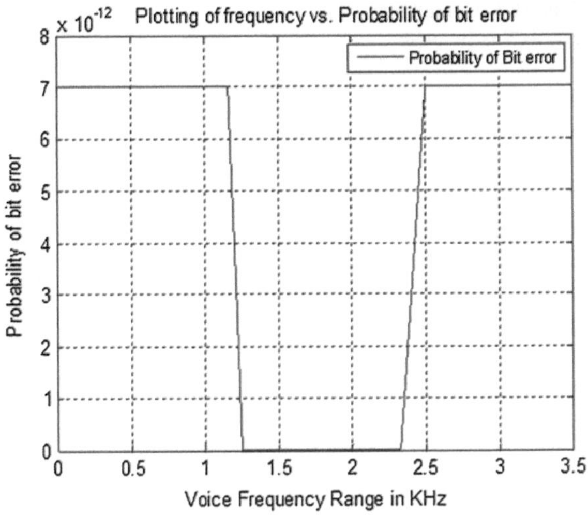

Fig. 51.6 Probability of bit error

Table 51.1 Data Rate, SNR_{min}, BER of Log Normal Distribution

Nature of speech	Bit rate (Kbps)	SNR_{min} (dB)	Probability of bit error
Log normal distribution	9.3316	61.9169	10^{-5}

Table 51.2 Comparative list of different distribution with their data rates

Sl no.	Name	Data rate (Kbps)	SNR_{min} (dB)	Probability of bit error
1	Existing data rate [7, 8]	64	49	Below 10^{-5}
2	Staircase approximation [7, 8]	19.5	49	Below 10^{-5}
3	Gaussian distribution [7, 8]	12.0128	56.5828	Below 10^{-5}
4	Gaussian with perception criterion [8, 13–15]	11.0959	48.9566	Below 10^{-5}
5	Gaussian with reduced bit allocation	10.9494	48.9566	Below 10^{-5}
6	Gaussian with more reduced bit allocation	10.7175	48.9566	Below 10^{-5}
7	Gaussian with more reduced bit allocation	10.4867	48.9566	Below 10^{-5}
8	The BELL system technical journal, october 1976 [14]	9.6	–	–
9	Rayleigh distribution [8]	9.4875	68.99	Below 10^{-5}
10	Log normal distribution	9.3316	61.9169	Below 10^{-5}

51.5 Conclusion

It is evident from the above discussion that both subband coding and existing 64 Kbps line have almost negligible probability of bit error but subband offers lowest data rate bandwidth ever possible. Authors have used different probability distribution for speech coding for validation. Log Normal Distribution shows the least. Therefore it can be deduced that subbanding generates all the possible significant footsteps towards data rate as well as bandwidth savings without losing any significant information and probability of bit error is also least or may be said negligible. PCM requires high bandwidth as well as data rate. But PCM and DM have almost the same SNR up to 30 dB. After 30 dB PCM shows performance-wise better results than DM. It has been shown by Matlab program. If more subbands are used, data rate can be reduced more and more accurate approximation of the original voice signal can be reconstructed. Therefore, authors can conclude that communication engineering will be immensely benefited by using this scheme. There are a lot more distribution support speech signals. These distributions can be simulated and results can be found out.

References

1. L.W. Couch II, *Modern Communication Systems Principles and Applications* (Prentice–Hall of India Private Limited, Delhi, 1995)
2. G.J. Proakis, *Digital Communications*, 4th edn. (Mcgraw-Hill International Edition, New York)
3. H. Taub, D.L. Schilling, *Principles of Communication Systems*, 2nd edn. (Tata Mcgraw-Hill publishing company limited, Noida)
4. I. Tashev, A. Aecero, *Speech Technology Group* (Microsoft Research, Redmond)
5. Y. LI, J. Chen, H. Tan, Voice activity detection under Rayleigh distribution. J. Electron. (Springer), **26**(4), 552–556 (2009)
6. M. Suman, T.V. Bhargava, G.P. Teja, K.B.N.P. Kumar, Speech enhancement and recognition compressed speech signal in noisy reverberant conditions, IRJSP **2**(2) (2011). ISSN 2249-6505
7. S. Roy, D.B. Gupta, P.K. Banerjee, Studies and implementation of subband coder and decoder of speech signal, in *Proceedings of NCECS* (2012), pp. 8–16
8. S. Roy, D.B. Gupta, S.S Chaudhuri, Studies and implementation of subband coder and decoder of speech signal using raleigh distribution, LNEE298, Springer, doi:10.1007/978-81-322-1817-3_2
9. Z. Peric, J. Nikolic, An adaptive waveform coding and its application in speech coding. Digit. Signal Process. (Elsevier) **22**, 199–209 (2012)
10. Y.-J. Chen, R.C. Maher, *Sub-band Coding of Audio using recursively indexed quantization* (Department of Electrical Engineering and Center for Communication and Information Science, University of Nebraska, Lincoln, 1995)
11. M.J. Tribolet, R.E. Crochiere, Frequency domain coding of speech. IEEE Trans. Acoust. Speech Signal Process. **Assp-27**(5), 550–558 (1979)
12. A.M. Aziz, Subband Coding of Speech Signals Using Decimation and Interpolation, in *13th International Conference ASAT–13*, Military Technical College, KobryElkobbah, Cairo, 26–28 May 2009

13. B. Rivet, L. Girin, C. Jutten, Log—Rayleigh Distribution: a simple and efficient statistical representation of log-spectral coefficients, in *IEEE Transactions on Audio, Speech, and Language Processing*, vol. 15 March 2007
14. R.E. Crochiere, S.A. Webber, N. Flanagan, Digital coding of speech in subbands. BELL Syst. Tech. J. (1976)
15. E. Zwicker, U. Zwicker, *Psychoacoustics, Facts and Models* (Springer, Berlin, 1990)
16. P.G. Knutson, K. Ramaswamy, J.W. Richardson, Subband adpcm voice encoding and decoding, PCT/US2000/034410, July 2001
17. C.F. Szczutkowski, Subband encoding method and apparatus. EP **0178608**, A2 (1986)

Chapter 52
Effect of Audio Cue on Electrooculogram-Based Eye Movement Analysis of Visual Memory Recall

Anwesha Banerjee, Anilesh Dey, Shreyasi Datta and D.N. Tibarewala

Abstract Context aware ubiquitous computing systems are capable of assisting people by sensing human cognitive context. In this work, visual memory recall of human beings is identified by analysing their eye movements. Electrooculogram signals, potential difference produced in the surrounding region of eye socket for eye ball movement, are recorded to collect eye movement data. Electrooculogram signals while viewing 'repeated' and 'non-repeated' visual stimuli were classified for 'with' and 'without' audio cue sections. Adaptive autoregressive parameters, power spectral density, Hjorth parameters and wavelet coefficients are extracted from these signals as features. A combined feature space is formed comprising all four signal features. A maximum accuracy of 88.70 % is obtained on an average over five participating subjects using SVM-RBF classifier for 'without audio' visual memory recall. From this study, it is evident that this auditory effect leaves an impact on EOG signal patterns so that to make reduction in the recognition performance.

Keywords Electrooculogram · Eye movement analysis · Context aware ubiquitous computing systems · Visual memory recall

A. Banerjee (✉) · D.N. Tibarewala
School of Bioscience and Engineering, Jadavpur University, Kolkata 700032, Bengal, India
e-mail: anwesha.banerjee@gmail.com

D.N. Tibarewala
e-mail: biomed.ju@gmail.com

A. Dey
Department of Electronics and Communication Engineering, Narula Institute of Technology, Kolkata 700109, West Bengal, India
e-mail: anilesh.dey@gmail.com

S. Datta
Department of Electronics and Telecommunication Engineering, Jadavpur University, Kolkata 700032, West Bengal, India
e-mail: shreyasidatta@gmail.com

© Springer India 2015

K. Maharatna et al. (eds.), *Computational Advancement in Communication Circuits and Systems*, Lecture Notes in Electrical Engineering 335,
DOI 10.1007/978-81-322-2274-3_52

In eye-based context aware systems visual memory [1] recall finds an interesting application. Recognition of cognitive context, that is related to perception, memory and learning forms the base of context aware ubiquitous computing systems [2]. Cognitive context recognition and analysis has been a challenging issue. In decoding cognitive activities, brain–computer Interfacing (BCI) [3, 4] techniques such as electroencephalogram (EEG) [5] magnetic resonance imaging (MRI) and functional magnetic resonance imaging (fMRI) [6] have been used. For online application, due to having large computational complexities, such systems are not much suitable to be used.

Attention, relational memory or learning, the main aspects of cognitive context recognition, is highly related to a person's visual behaviour [7, 8]. From literatures, it is evident that a high degree of information regarding a person's activities and context [9–11] is conveyed by eye movements. Hence eye movement analysis can be used as a tool to recognizing the cognitive context of a person. Visual Memory recall detecting systems can be used in eye-based cognitive context aware systems for assisting people with memory stupor in various situations such as recognizing faces or locations.

Infrared video system (IRVS), infrared oculography (IROG), optical-type eye tracking system, Purkinje dual-Purkinje-image (DPI) and electrooculography (EOG) [12] are different techniques to measure eye movements. EOG comprises of surface electrodes to be placed on skin surrounding the eye socket. Being cost-effective, easy to acquire, simple to process and work in real time, noninvasive and also portable, EOG [13] has proved to be a very efficient tool for a eye movement measurement. Eye movements and the corresponding EOG amplitude maintain a linear relationship up to a certain degree. Detection and assessment of many diseases [14, 15] as well as eye movement controlled human computer interfaces and neuro-prosthetic aids [16, 17] are significant applications of EOG signal analysis include.

The process of visual memory recall has been detected in this present work. It has been done by classifying EOG signals acquired from a series of experiments by presenting visual stimulus to discriminate between new visuals and previously seen visuals. At first, a set of pictures are shown to the subjects and after a period of rest another set of pictures are shown that contain some new pictures and others repeated from the previous set. In the second stage, the discrimination was done in repeated and non-repeated pictures by providing audio cue. EOG signals are acquired using a two-channel system developed in the laboratory. Four standard signal features, namely adaptive autoregressive parameters, wavelet coefficients, power spectral density and Hjorth parameters and support vector machine have been used for classification. A maximum classification accuracy of 88.70 % has been obtained using SVM-RBF classifier for a combination of all the features.

The rest of the paper is structured as follows: Section 52.2 explains principles and the methodology followed in the course of the work. Section 52.3 covers the experiments and results. Finally in Sect. 52.4, the conclusions are drawn and future scopes of work are stated. Figure 52.1 depicts the experimental procedure.

Fig. 52.1 Experimental procedure

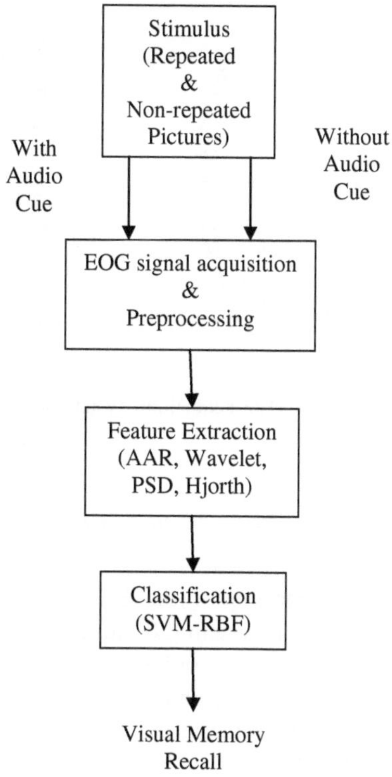

52.1 Methodology

This section describes the process of electrooculogram signal acquisition and processing to for visual memory recall detection

52.1.1 Electrooculogram

Electrooculogram [12, 13], [16–21] is the standing potential which is detected by the surface electrodes placed around the eye socket. From the characteristics of EOG signal, it is known that the frequency range of the signal is 0.1–20 Hz and the amplitude lies between 100 and 3,500 μv [5]. The data acquisition system is shown in Fig. 52.2b.

EOG signal is acquired at a sampling frequency of 256 Hz using 5 Ag-AgCl disposable electrodes, two for the horizontal channel, two for the vertical channel and one as reference. Electrooculogram data has been acquired in the LabView 2012 platform for processing in the computer using National Instruments 12-bit ADC.

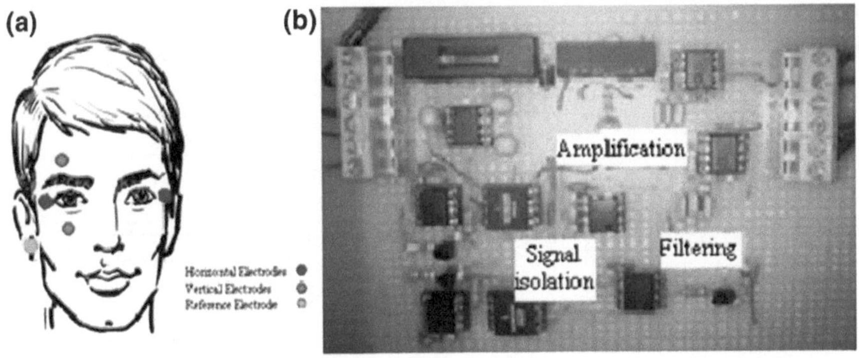

Fig. 52.2 Data acquisition system showing **a** placement of electrodes and **b** acquisition circuit snapshot

52.1.2 Data Acquisition and Preprocessing

The recording of the EOG signal has been done through a two-channel data acquisition system [20] developed in the laboratory, using five Ag/AgCl disposable electrodes at a sampling frequency of 256 Hz. Two electrodes are used for acquiring horizontal EOG and two for vertical EOG, while one electrode placed on earlobe acts as the reference as shown in Fig. 52.2 [7].

To eliminate undesirable noise and obtain EOG in the frequency range of 0.1–15 Hz, the range where maximum information is contained, bandpass filtering has been implemented using a six-order Elliptical band-pass filter in the band 0.1–15 Hz.

52.1.3 Feature Extraction and Classification

AAR parameters, power spectral density, Hjorth parameters and wavelet coefficients have been used as signal features. A combined normalized feature space comprising of all four signal features is constructed [22].

The feature spaces need to be classified to obtain the distinction between EOGs corresponding to new pictures and that of old. Hence two-stage binary classification process is done as there are two classes to be considered: EOG corresponding to new pictures and that for old pictures, in order to evaluate visual memory recall and then in which case it was better, i.e. 'with' or 'without' audio cue. Support vector machine classifier [23, 24] has been used with radial basis function (RBF) kernel using the width of the Gaussian as 1.

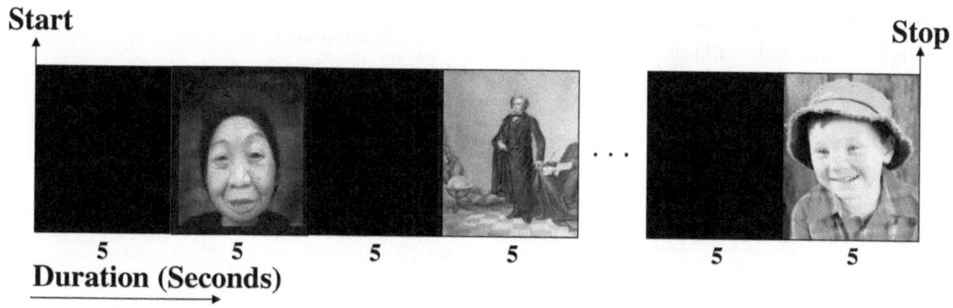

Fig. 52.3 Sample of visual stimulus

52.2 Experimental Setup

EOG data is collected from five healthy subjects, three male and two female in the age group of 25 ± 5 years with their consent. In the first phase, a visual stimulus in the form of 10 pictures is presented. After 10 min of relaxation, another set of 10 pictures are presented where 5 are completely new and the rest of 5 are selected randomly from the set of pictures used in the first phase. During this phase, EOG data is acquired for analysis. The known and the unknown pictures are produced in a completely random order. During each phase of experiment, the stimulus begins with 5 s of black screen for relaxation. Each picture is kept for 5 s and another 5 s of black screen (for relaxation) is kept before the appearance of the next picture. Such a sample visual cue is shown in Fig. 52.3. After that, in the second stage the discrimination was done in repeated and non-repeated pictures by providing audio cue. The pictures used in the experiments include various portraits and are displayed using a screen with a projector in a dimly lit room. Data is taken from each subject for 10 days.

From the acquired EOG data for each subject, over 10 days feature spaces are constructed and classified to recognize old and new pictures. In each case after feature extraction, the obtained feature space is normalized with respect to the respective maximum value. For classification, training and testing instances are obtained by cross-validation on the obtained feature space consisting of EOG data of 10 days for each subject.

52.3 Analysis of Audio Effect on Visual Memory Recall

The results of classification in terms of classification accuracy or CA average over 10 subjects for combined feature space for 'with' and 'without' audio cue is shown in Table 52.1. It is observed that the use of audio cue increases the performance. Each of the feature spaces are normalized with respect to its maximum value and concatenated to form the combined feature spaces. A significant decrease in the

Table 52.1 Classification results of 'with' and 'without' audio cue stages

Subject id	Classification accuracy (%)	
	With audio cue	Without audio cue
1	86.70	89.50
2	83.60	88.66
3	85.22	87.90
4	86.41	88.20
5	84.70	89.24
Average	85.33	88.70

performance is observed by providing audio cue as shown in Table 52.1. The highest classification accuracy of 88.70 % is observed without audio cue. From this study, it is evident that this auditory effect leaves an impact on EOG signal patterns so that to make reduction in recognition performance.

52.4 Conclusion

The present work is concerned with the recognition of the process of visual memory recall on the basis of eye movement analysis using electrooculogram signals. EOG signals are acquired while producing visual stimuli consisting of new pictures and pictures that were previously shown to them and the data is classified to determine the distinction between the EOG signals, while visualizing these two classes of pictures with or without audio cue. AAR parameters, power spectral density, Hjorth parameters and wavelet coefficients have been used to form a combined signal feature and classification of the EOG signals is done SVM-RBF with a maximum accuracy of 88.70 %. Auditory effect significantly reduces the process of visual recognition.

This work is important with respect to the research in cognitive context-awareness, applicable in ubiquitous computing, using eye movement study. EOG is simple to acquire and process in real time and very cost-effective thereby justifying its use. Future works in this direction include the implementation of this scheme in real-time environment, the use of other features that can provide better results and also using other bio-signals individually or together with EOG for the same purpose.

Acknowledgment This work is supported by Jadavpur University and Council of Scientific and Industrial Research (CSIR), India.

References

1. S. Magnussen, Cognition and neurosciences: implicit visual working memory. **50**, 532–542 (2009)
2. N. Davies, D.P. Siewiorek, R. Sukthankar (eds.), special issue on activity based computing, IEEE Pervasive Comput. **7**(2), 20–21 (2008)

3. G. Dornhege (ed.), *Toward Brain-Computer Interfacing* (MIT press, Cambridge, 2007)
4. J.-D. Haynes, G. Rees, Decoding mental states from brain activity in humans. Nat. Rev. Neurosci. **7**(7), 523–534 (2006)
5. W. Klimesch, EEG alpha and theta oscillations reflect cognitive and memory performance: a review and analysis. Brain Res. Rev. **29**(2), 169–195 (1999)
6. R.A. Poldrack, The role of fMRI in cognitive neuroscience: where do we stand? Curr. Opin. Neurobiol. **18**(2), 223–227 (2008)
7. A. Bulling, D. Roggen, Recognition of visual memory recall processes using eye movement analysis, in *UbiComp* (2011) pp. 455–464
8. J.M. Henderson, A. Hollingworth, Eye movements and visual memory: detecting changes to saccade targets in scenes. Percept. Psychophys. **65**(1), 58–71 (2003)
9. A. Bulling, J.A. Ward, H. Gellersen, G. Tröster, Eye movement analysis for activity recognition using electrooculography. IEEE Trans. Pattern Anal. Mach. Intell. **33**(4), 741–753 (2011)
10. A. Bulling, H. Gellersen, Toward mobile eye-based human-computer interaction. IEEE Pervasive Comput. **9**(4), 8–12 (2010)
11. A. Bulling, J.A. Ward, H. Gellersen, G. Tröster, *Robust Recognition of Reading Activity in Transit Using Wearable Electrooculography* (Springer, Berlin, 2008) pp. 19–37
12. L.Y. Deng, C.L. Hsu, T.C. Lin, J.S. Tuan, S.M. Chang, EOG-based Human-Computer Interface system development. Expert Syst. Appl. **37**(4), 3337–3343 (2010)
13. G.B. Arden, P.A. Constable, The electro-oculogram. Prog. Retinal Eye Res. **25**(2), 207–248 (2006)
14. P. Stavrou, P.A. Good, E.J. Broadhurst, S. Bundey, A.R. Fielder, S.J. Crews, ERG and EOG abnormalities in carriers of X-linked retinitis pigmentosa. Eye **10**(5), 581–589 (1996)
15. O. Rascol, M. Clanet, J.L. Montastruc, M. Simonetta, M.J. Soulier-Esteve, B. Doyon, A. Rascol, Abnormal ocular movements in parkinson's disease evidence for involvement of dopaminergic systems. Brain **112**(5), 1193–1214 (1989)
16. J. Keegan, E. Burke, J. Condron E. Coyle, Improving Electrooculogram-Based Computer Mouse Systems: The Accelerometer Trigger. Bioengineering in Ireland (Galway, 2011)
17. R. Barea, L. Boquete, M. Mazo, E. López, L.M. Bergasa, EOG guidance of a wheelchair using neural networks. IEEE Int. Conf. Pattern Recognit. **4**, 668–671 (2000)
18. A.W. North, Accuracy and precision of electrooculographic recording. Invest. Ophthalmol. **4**, 343–348 (1965)
19. A. Banerjee, S. Chakraborty, P. Das, S. Datta, A. Konar, D.N. Tibarewala, R. Janarthanan, Single channel electrooculogram (EOG) based interface for mobility aid, in *4th IEEE International Conference on Intelligent Human Computer Interaction* (*IHCI*) (2012) pp. 1–6
20. A. Banerjee, S. Datta, M. Pal, A. Konar, D.N. Tibarewala, R. Janarthanan, Classifying electrooculogram to detect directional eye movements. Procedia Technol. **10**, 67–75 (2013)
21. S. Roy Choudhury, S. Venkataramanan, H.B. Nemade, J.S. Sahambi, Design and development of a novel EOG biopotential amplifier, in *IJBEM* **7**(1) (2005)
22. A. Banerjee, S. Datta, A. Konar, D.N. Tibarewala, R. Janarthanan, Electrooculogram based detection of visual memory recall, in *International Conference Communication Signal Processing* (2014), pp. 711–715
23. C.J. Burges, A tutorial on support vector machines for pattern recognition. Data Min. Knowl. Disc. **2**(2), 121–167 (1998)
24. M. Tom, *Machine Learning* (McGraw Hill, 1997)

Part VIII
Advances in Image Processing

Chapter 53
Segmentation Approach for Iris Recognition in Less Constrained Environment

Navjot Kaur and Mamta Juneja

Abstract As demand for security is increasing day by day, the methods for security are also improving to meet the challenging needs for security. Iris recognition is one of the emerging technologies to be used for security. The most challenging step in the process of iris recognition is iris localization as accuracy in iris localization significantly affects further processing of feature extraction and template matching stages. Traditional algorithms accurately locate iris as iris images were taken under ideal conditions. But their accuracy is affected when eye images are taken in unconstrained environment. The proposed algorithm starts with the extraction of iris even in the presence of specular highlights, eyelids, eyelashes and pupil. The accuracy for the proposed work has enhanced due to the use of intuitionistic fuzzy-based clustering for iris segmentation on UBIRIS v2 images.

Keywords Intuitionistic fuzzy · Fuzzy C-mean clustering · Iris recognition and biometric

53.1 Introduction

In this technological era, the demand for security in every field is in danger and it is very challenging to deal with this security-related problem. This security-related problem can be tackled with the help of emerging biometric technologies. The biometric technologies include finger print recognition, palm vein recognition, voice recognition, iris recognition and many more [1, 2]. There is a large interest in improved, authentic, secure and genuine identification methods which are more useful. The demand in the area of iris recognition is increasing as iris is the most

N. Kaur (✉) · M. Juneja
CSE Department UIET, Panjab University, Chandigarh, India
e-mail: navjotkaur2611@gmail.com

M. Juneja
e-mail: mamtajuneja@pu.ac.in

© Springer India 2015

K. Maharatna et al. (eds.), *Computational Advancement in Communication Circuits and Systems*, Lecture Notes in Electrical Engineering 335,
DOI 10.1007/978-81-322-2274-3_53

unique feature of human body which remains stable throughout life even though the person ages, his iris pattern remains same. Iris has the highest degrees of freedom among all the other biometric features [1–3]. Even the iris pattern of both the eyes of an individual is unique. The probability of getting same iris is one in 10^{72} which is very low. Also iris has the highest degrees of freedom having value of 266, which is greater among all the facial features. Some desirable properties like uniqueness, stability etc. make iris recognition suitable for highly reliable and accurate human identification [1, 3]. Iris has a great advantage in mathematics also as its pattern variability is enormous among different persons. Iris as a recognition system was introduced in 1987 for the first time [1]. Many researchers like Daugman [2–6], Bowyer et al. [7] etc. proposed many powerful iris recognition algorithms. Some of these algorithms need user cooperation and also take care that user is properly positioned with respect to camera when image is taken. These traditional algorithms work efficiently with eye images taken under ideal conditions only but they failed to segment the iris area properly when eye images were taken under non-ideal conditions. For using this technology in real-life scenario, it should work effectively and efficiently under non-ideal conditions also [8–10].

The proposed algorithm works well for the eye images taken under non-ideal situations. The main inspiration behind this paper is to propose an algorithm to handle eye images which are noisy and taken under non-ideal situations. In this, intuitionistic fuzzy C-mean algorithm is used for segmentation of eye image to extract iris area accurately. Circular Hough transform is used to detect the inner and the outer boundaries of the iris. Iris and pupil are assumed to have circular shapes. Eyelids and eyelashes are also removed efficiently; hence, the accuracy of the algorithm has greatly enhanced.

53.2 Proposed Scheme

In the proposed algorithm, the most important is the iris area segmentation. Various steps involved in the proposed scheme are shown in Fig. 53.1 and explained below.

53.2.1 Image Segmentation

Image segmentation is done by using fuzzy C-mean clustering technique [11]. The concept of Fuzzy set was given in 1965 [12]

$$S = \{(x, \mu_s(x)) | x \in X\} \tag{53.1}$$

In Eq. (53.1), $\mu_s(x)$ is the degree of membership and non-membership degree is given by $1 - \mu_s(x)$. However for real-life scenarios, there always exist a degree of hesitation or uncertainty associated with this degree of membership which is not

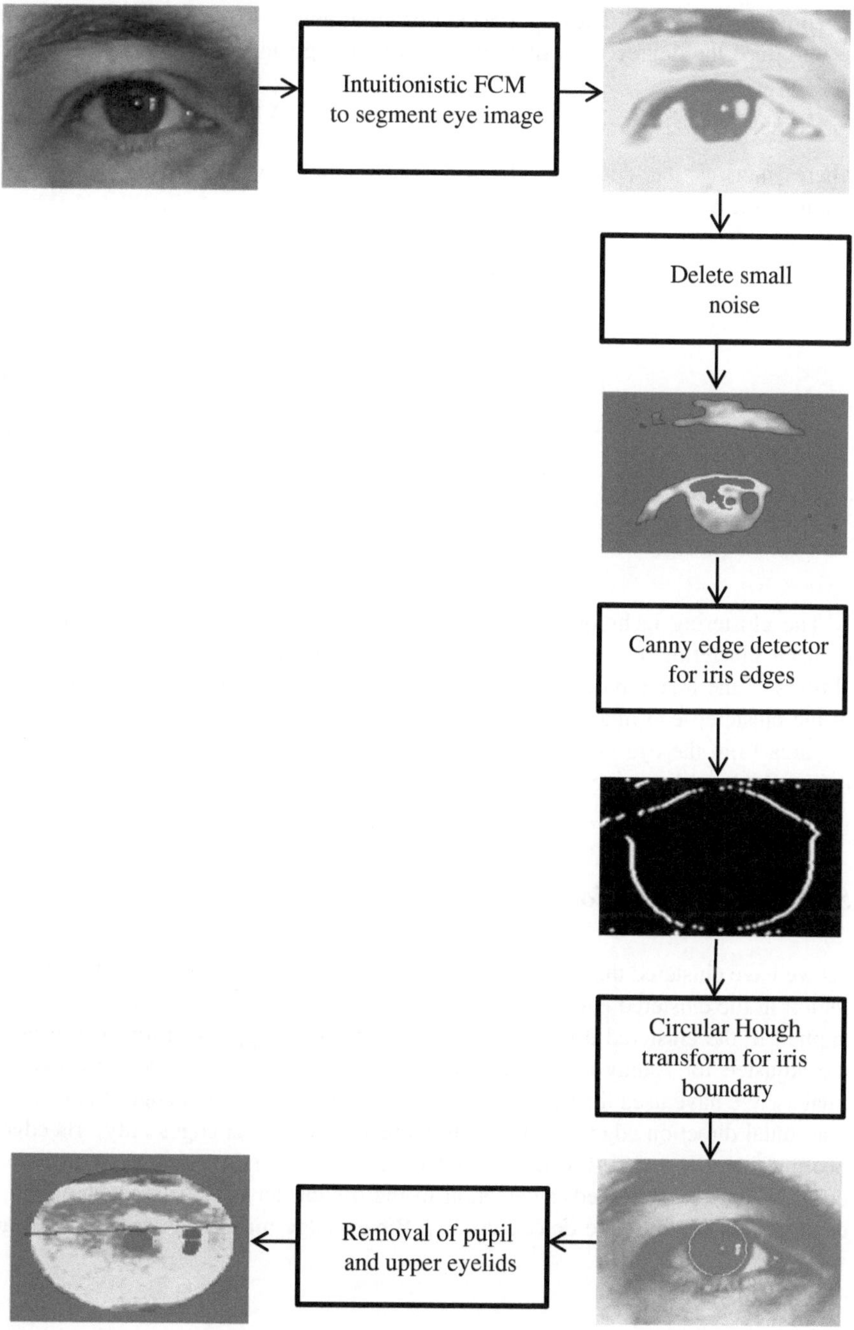

Fig. 53.1 Steps of the proposed algorithm

given in any fuzzy set. Thus the concept of intuitionistic fuzzy set (IFS) was given to introduce the degree of hesitation [13] which is given as B.

$$B = \{(x, \mu_B(x), \nu_B(x)) \,|\, x \in X\} \quad (53.2)$$

where $\mu_B(x)$ is the degree of belongingness and $\nu_B(x)$ is the degree of non-belongingness.

$$\mu_B : X \to [0, 1], x \in X \to \mu_B(x) \in [0, 1] \quad (53.3)$$

$$\nu_B : X \to [0, 1], x \in X \to \nu_B(x) \in [0, 1] \quad (53.4)$$

$$\text{for all } x \in X;\ \mu_B(x) + \nu_B(x) \leq 1 \quad (53.5)$$

$$\Pi_B(x) = 1 - \mu_B(x) - \nu_B(x) \quad (53.6)$$

where Π_B (x) is hesitation degree of x to B, IFS becomes fuzzy set if

$$\Pi_B(x) = 0 \quad (7)$$

The clustering technique used in our proposed scheme is not a simple fuzzy C-mean clustering but it is based on intuitionistic logic. Thus we create three clusters of the image based on the intensity values. These three clusters are shown in the clustered eye image. After the clustering of the image, segmentation of the iris area from the eye image becomes easier as it is in one separate cluster. Small noise is deleted by using morphological operations.

53.2.2 *Edge Detection*

As we have clustered the image, the iris edge detection becomes easier. The darkest region in the clustered eye image represents the iris region. Canny edge detector is applied to the clustered image after removal of noise. Upper and lower thresholds are adjusted for Canny edge detector so as to make them suitable for noisy eye images. We have used the higher values for the vertical direction edges than that for horizontal direction edges as our main interest is in vertical edges only, iris edges. Scaling factor is reduced to half to reduce image size and consequently to reduce error. We have applied edge detection to the iris cluster only. Image edge map is created after Canny edge detection [14]. Binary edge maps of various images are shown in Fig. 53.2.

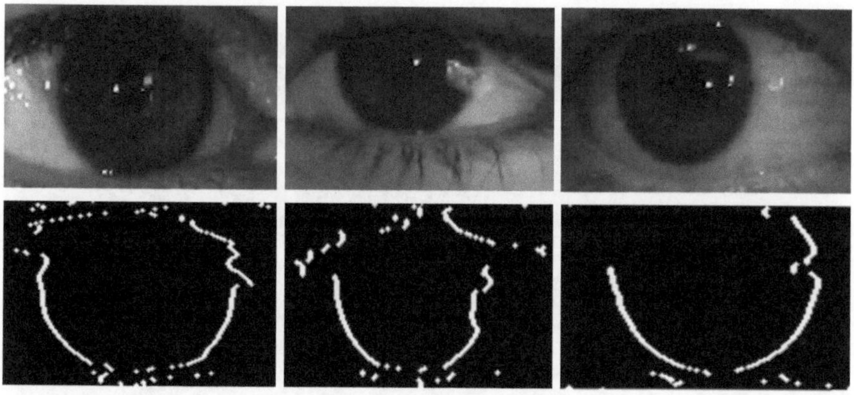

Fig. 53.2 Binary edge map for iris image

53.2.3 Circular Hough Transform (CHT)

Circular Hough Transform (CHT) is applied after the Canny edge detector to get circular iris boundaries. As eye images of UBIRIS v2 dataset [15] are taken in non-ideal conditions, so many types of noise like specular refection, eyelashes, eyelids, lightening effects etc. are present in the image. This noise needs to be removed to avoid iris localization errors and further processing and iris matching stages. Circular Hough transform is very expensive in time so we have reduced its execution time by following steps:

- Reduce the scaling factor to half to reduce the search time for Hough Transform. Scaling factor reduces the image size to half and thus reduces edges in the image.
- Clustering the image reduced the search area for Hough Transform and reduced edges for Canny edge detector.
- Morphological operators are used to remove small noise blocks from the image and make it easier for CHT to create circular edges.

Figure 53.3 shows the localization of iris region after the application of circular Hough transform on the eye images of UBIRIS v2 dataset.

53.2.4 Eyelid, Eyelash and Pupil Removal

For the eye images taken in non-ideal conditions, noise like specular reflections, pupil, upper and lower eyelids are present. These noisy factors must be removed so that they do not affect the iris segmentation stage and the whole iris recognition system.

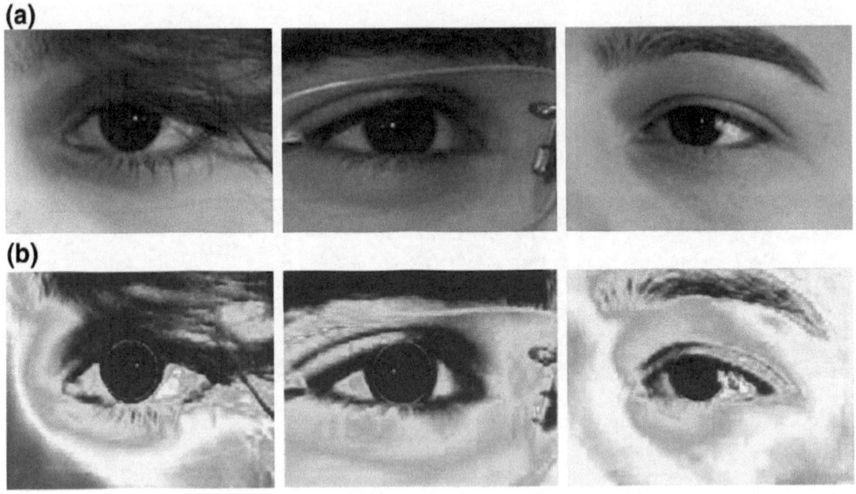

Fig. 53.3 Illustration of accurate iris localization **a** original eye image **b** iris localization after application of clustering and circular Hough transform

We proposed a new method for upper eyelids detection, in this we have used the intensity contrast between the sclera and eyelashes. We mark the eyelashes in the sclera region as the intensity contrast is high between these two regions. Mark two points on the sclera region where upper eyelid and eyelash fall on the two ends of the iris region. Then apply linear Hough transform and draw a horizontal line taking these two as reference points. Thus in this way, we can easily locate and delete upper eyelid and eyelash from the iris region.

Localization of lower eyelid is much easier than that of the upper eyelid, as they obstruct the least part of iris. To localize the lower eyelid, we first apply Canny edge detector to the lower part of the iris region. Then find a best line fitting the region by using line Hough transform. If there is no such line in the lowest region, it means no lower eyelid occlude the iris region.

Pupil removal is done at the last step, because intensity contrast between pupil and iris is very low for noisy eye images. So to manage this high contrast, image is first enhanced to make pupil appear properly. Then apply median filtre to reduce noise from the pupil area and keep edges. Canny edge detector is used to get the edges and circular Hough transform is used to locate pupil region. Pupil radius is approximately 1/10 of the iris radius.

53.2.5 Template Matching

We are able to extract the iris region exactly from the eye image. We have successfully deleted all the noise present in the eye image and the iris region we get is free of all such noise like pupil, eyelids and eyelashes. Then this image is normalized and encoded as a template and stored in the database which is later on used for iris recognition for individuals. Template matching is done by using Hamming distance measure (HD), a measure of dissimilarity.

By calculating the HD between two bit patterns, a decision about the exact match or no match between the templates can be made.

53.2.6 Experimental Results

The implementation was done on the eye images taken from UBIRIS v2 eye image dataset [15]. The dataset contained images taken under unconstrained conditions. The accuracy and the average segmentation time for the proposed algorithm and various previous algorithms are given in Table 53.1.

We assume that pupil and iris have circular boundaries and every circle is best described by the values of its radius and centre. After application of the proposed algorithm on UBIRIS v2 images, we get segmented eye images. After segmentation of iris, we can calculate the accuracy of the segmentation algorithm for the proposed scheme. Table 53.1 gives the accuracy of the proposed segmentation algorithm in comparison with some previous algorithms. Time given in this table is the time taken by the algorithm to accurately extract the iris region after the removal of all small noise blocks from the image.

As the results show, the accuracy of the proposed algorithm is better than the previous algorithms. At the same time, the execution time has also decreased. The decreased execution time is mainly due to the use of clustering technique as it reduced the searching time for Hough transform. First, the segmented iris image is normalized and encoded as a template and stored. To calculate the match and non-match distributions for the dataset, each iris image is matched for comparison with all the other iris images of the database. For comparison, Hamming Distance (HD) is used as a measure of dissimilarity between two iris codes: code A and code B

$$\mathrm{HD} = \frac{\|(\text{code A XOR code B}) \cap \text{mask A } \cap \text{mask B}\|}{\|\text{mask A} \cap \text{mask B}\|} \tag{53.8}$$

Here mask A and mask B are the two corresponding masks for code A and code B, respectively. Mask is composed of whether iris area is occluded by any type of

Table 53.1 Comparison of proposed algorithm with previous algorithm

Method	Accuracy (%)	Time (s)
Daugman	95.22	2.73
Wildes	98.68	1.95
S.A. Sahmoud	98.76	1.49
Proposed	98.80	1.36

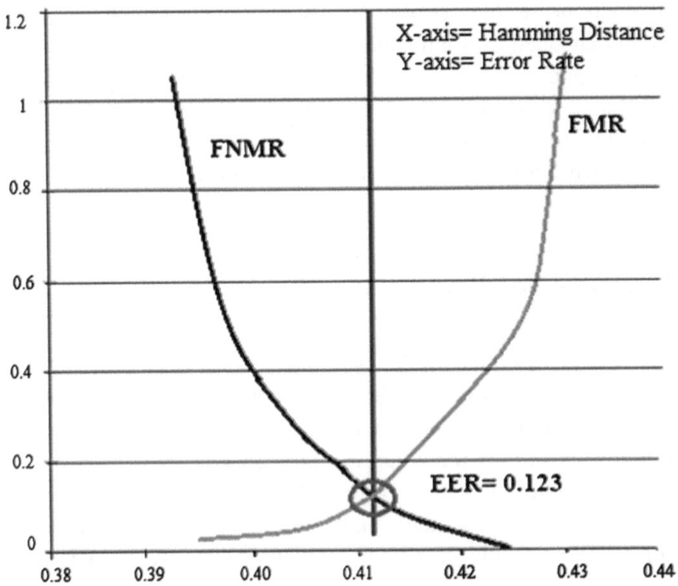

Fig. 53.4 Equal error rate value at equal FMR and FNMR

noise like eyelashes, eyelids etc. Hence, it affects the achieved results. HD is a fractional measure of the dissimilarity after removal of noise from the iris area.

Figure 53.4 shows the EER of the iris recognition system when our proposed algorithm is used. EER is the point where FMR and FNMR are calculated at a

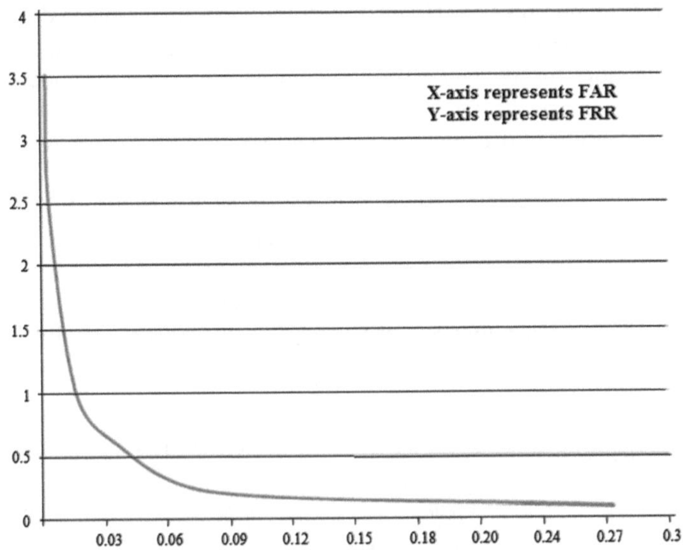

Fig. 53.5 ROC curve showing relation between FAR and FRR

single point. The value for EER is very low when proposed algorithm is applied. This shows that our proposed algorithm accurately isolates noise from the iris region. EER is low so FMR and FNMR are also low. We also calculated the FAR and FRR for our system. Figure 53.5 shows ROC showing the relation between FAR and FRR when our proposed algorithm is used. The value for FRR and FAR has significantly reduced only because our algorithm handles noise accurately. It is very good for the accuracy of iris recognition system.

53.3 Conclusion

This paper proposed a new algorithm for iris recognition for images taken in less constrained environment. These eye images have several types of artifacts like noise, specular reflections, varying lightening conditions and varying eye to camera distances. This scheme proposed a new segmentation method based on intuitionistic fuzzy C-mean algorithm to segment the eye image and extract iris region accurately. Experimental results show the accuracy of the proposed scheme in comparatively less time.

The time taken by the proposed scheme had reduced greatly as clustering is followed by circular Hough transform and thus search time for Hough transform has reduced and so far the segmentation.

Acknowledgments I would like to thank God first and all the people who are involved in the fulfilment of this research work.

References

1. L. Flom, A. Safir, Iris recognition system, U.S. Patent 4641394 (1987)
2. J. Daugman, Biometric personal identification system based on iris analysis, U.S. Patent 5291560 (1994)
3. J. Daugman, High confidence visual recognition of persons by a test of statistical independence. IEEE Trans. Pattern Anal. Mach. Intell. **15**(11), 1148–1161 (1993)
4. J. Daugman, New methods in iris recognition. IEEE Trans. Syst. Man Cybern. Part B Cybern. **37**(5), 1167–1175 (2007)
5. J. Daugman, Statistical richness of visual phase information: update on recognizing persons by their iris patterns. Int. J. Comput. Vision **45**(1), 25–38 (2001)
6. J. Daugman, Demodulation by complex-valued wavelets for stochastic pattern recognition. Int. J. Wavelets Multiresolut. Inf. Process. **1**(1), 1–17 (2003)
7. K.W. Bowyer, K. Hollingsworth, P.J. Flynn, Pupil dilation degrades iris biometric performance. Comput. Vis. Image Underst. **113**(1), 150–157 (2009)
8. N. Kaur, M. Juneja, A review on iris recognition, in *Proceedings of IEEE Conference on Recent Advances in Engineering and Computational Sciences (RAECS)* (2014), pp. 1–5
9. S.A. Sahmoud, I.S. Abuhaiba, Efficient iris segmentation method in unconstrained environment. Pattern Recogn. **46**(12), 3174–3185 (2013)

10. R. Wildes, Iris recognition: an emerging biometric technology. Proc. IEEE **85**(9), 1348–1363 (1997)
11. N.R. Pal, J.C. Bezdek, On cluster validity for the fuzzy C-means model. IEEE Trans. Fuzzy Syst. **3**(3), 370–379 (1995)
12. L.A. Zadeh, Fuzzy sets. Inf. Control **8**, 338–353 (1965)
13. G. Deschrijver, E.E. Kerre, On the representation of intuitionistic fuzzy T-norms and T-conorms. IEEE Trans. Fuzzy Syst. **12**(1), 45–61 (2004)
14. J. Canny, A computational approach to edge detection. IEEE Trans. Pattern Anal. Mach. Intell. **8**(6), 679–678 (1986)
15. H. Proença, S. Filipe, R. Santos, J. Oliveira, L.A. Alexandre, The UBIRIS.v2: a database of visible wavelength iris images captured on-the-move and at-a distance. IEEE Trans. Pattern Anal. Mach. Intell. (2009). doi:10.1016/j.imavis.2009.03.003

Chapter 54
NIR Spectrometry-Based Milk Fat Content Classification Using Bagging Ensembles

Dwaipayan Chakraborty, Sankhadip Saha and Sayari Ghoshal

Abstract The short-wave near-infrared spectroscopy at 540–910 nm region is investigated for non-destructive multivariate analysis of fat content for packaged milk in four categories: Double toned, full cream, standard and toned. Visible near-infrared spectrometry is used in the discrimination (classification) of milk fat content while red, green, blue component spectra are recorded for each sample under each aforesaid category. Features are extracted considering the highest 30 peaks of each spectra—red, green, blue component. Ensembles of classifier based on bagging strategy is employed here for the classification of samples. Two types of base classifier used here namely, support vector machine and multi-layer perceptron network. Result shows that support vector machine supersede multi-layer perceptron as individual learner in terms of classification accuracy. Single classifier performance is also compared with their native bagging-based ensemble. It is found that the bagging-based ensemble of classifier exhibits promising result in improving the prediction accuracy.

Keywords Fat content · Spectrometry · Bagging · SVM · MLP

D. Chakraborty (✉)
Department of Electronics and Instrumentation, Netaji Subhash Engineering College, Kolkata, India
e-mail: dwaipayanchakraborty@hotmail.com

S. Saha · S. Ghoshal
Department of Electrical Engineering, Netaji Subhash Engineering College, Kolkata, India
e-mail: imsankhadip@gmail.com

S. Ghoshal
e-mail: sayari.ghoshal@gmail.com

© Springer India 2015

K. Maharatna et al. (eds.), *Computational Advancement in Communication Circuits and Systems*, Lecture Notes in Electrical Engineering 335,
DOI 10.1007/978-81-322-2274-3_54

54.1 Introduction

The composition of milk may vary throughout the year due to regional and seasonal factors and breed of cow. Most of the dairy manufacturers standardize the levels of fat and protein in the milk to meet consumers' conscious expectations of a product which bears consistent composition and taste throughout the year. One way of doing this is to add or remove fat. The food standards code allows manufacturers to add or withdraw milk components to or from milk as long as the total fat level remains at least 3.2 % and the protein at least 3 %.

In [1], an alternative approach is proposed for total protein classification for powdered milk using near-infrared reflectance spectrometry (NIRS) and variable selection methods. Principal component regression (PCR), partial least squares (PLS) multivariate calibrations and soft independent modelling of class analogy (SIMCA) are used to predict the total protein. In [2], Adaboost [3] based ensemble of classifier is employed successfully in classifying different cigarette brands. Results indicate that in comparison with principal component analysis [3] when kernel principal component analysis (KPCA) [3] is a applied for feature extraction from spectrometer data, adaboost classifier gives better accuracy. Short-wave NIR spectrometry-based classification model using multi-criteria quadratic programming (MCQP) and support vector machine (SVM) with PCA-based feature extraction technique is used in [4] for classification of different Thai varieties of orange. The vital element is the instability of the prediction method. Hence bagging [5] based predictors incorporate multiple learner by making bootstrap replicates of the learning set and using these as new learning sets. Majority voting amongst the base learners outcomes the aggregated result. If perturbing the learning set can cause considerable changes in the predictor constructed, then bagging can improve classification accuracy [6].

In our work, we focused on total fat content classification of packaged milk using NIR spectrometer. Simple technique is used for feature extraction where ensemble of classifier strategy is adapted for improving the prediction accuracy of our model.

54.2 Theory and Experimental Setup

54.2.1 *Chemical Composition and Types of Milk*

According to chemical analogy, milk is an emulsion of butterfat globules within a water-based fluid consisting of dissolved carbohydrates and protein aggregates with minerals. Hence the solid component of milk can be separated into two major classes: Fat and solid non-fats (SNF). The solid non-fats consist of proteins, carbohydrates, vitamins, minerals and salts. The different types of milk available in the Indian market for direct ingestion are almost identical in solid non-fat content, but can be distinguished on the basis of fat content. Milk varieties that have been considered for this investigation, with corresponding fat contents, are provided in Table 54.1.

Table 54.1 Varieties of milk used

Milk type	Fat content (%)	Corresponding class in the dataset
Double toned	1.5	1
Full cream	6.0	2
Standard	3.5	3
Toned	3.0	4

54.2.2 Near-Infrared Spectrometry

Near-infrared spectrometry [1, 2] is a prominent non-destructive analytical method which utilizes the absorption characteristics of different materials in the near-infrared region of the electromagnetic spectrum. In our work, we have used an inexpensive and portable USB webcamera-based spectrometer. A broad-spectrum light (usually from a halogen or incandescent lamp) is shone through the sample to be analysed. Depending on the molecular composition of the sample, some colours are absorbed more than others. This light then passes through a diffraction grating which splits it into different colours, which are then measured separately. This is done by using a webcam which measures each colour and graphs their intensities on a laptop computer where it is processed further, in order to obtain better results during classification depicted in Fig. 54.1.

54.2.3 Bagging (Bootstrap Aggregation)

Bagging or bootstrap aggregating is an ensemble-based meta-algorithm used in machine learning [3]. Proposed by Breiman [5], it improves stability and accuracy of machine learning algorithms. From the given set of samples, some samples are chosen randomly with replacement. Then a learning model is trained with the chosen samples, which is then saved. This method is repeated for a number of

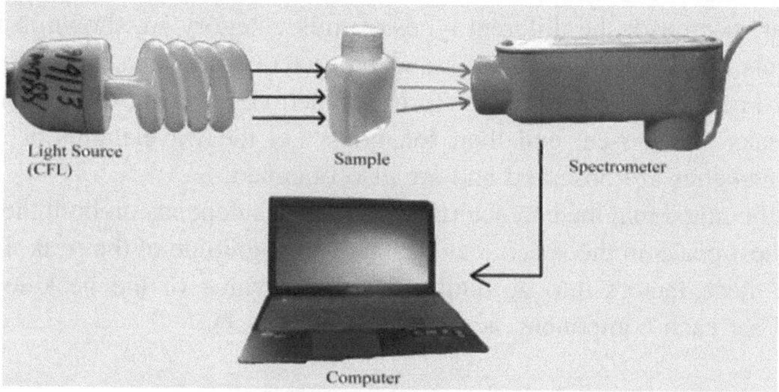

Fig. 54.1 Experimental set-up for data collection

Table 54.2 Bootstrap aggregation algorithm

Input:
1. Set of m samples $S = \{(x_1, y_1), (x_2, y_2), \ldots ,(x_m,y_m)\}$.
2. These samples have the class label $y_i \in Y = \{1, \ldots, k\}$, where i is the index of the sample, k is the number of the class.
3. Let T be the total number of iterations.

Training:
4. Do for t = 1, 2,..., T
5. Draw m bootstrap samples from S.
6. Train a new model with the drawn samples.

Output:
7. Provide test samples to the trained models saved in the ensemble.
8. Obtain final hypothesis by simple majority voting.

iterations, as specified by the user. When the test samples are provided to the base learners, each of them provide a hypothesis, which are then combined and simple majority voting is used to obtain the final hypothesis for the ensemble. In this work, we use support vector machines (SVM) [3, 6] as well as scaled gradient conjugate back-propagation trained multi-layer perceptron (MLP) [3, 7] as the base learning model during bootstrap aggregation. The algorithm for bagging is given in Table 54.2.

54.3 Data Processing

The acquired waveform is a plot of corresponding colour intensities at particular wavelengths. From the experimental observations, we have seen that most of the peaks and troughs in the waveforms lie in the wavelength range of 540–910 nm. The rest of the data is either insignificant or noise from the ambient light. So after procuring the raw data, the waveforms are chopped to only have values in the mentioned range. The chopped waveforms containing the red, green and blue component spectra for the different types of milk category are shown in Fig. 54.2.

After this, the red (R), green (G) and blue (B) components of the spectra are separated. From the respective waveform of each component, a given number of highest peaks are chosen, and their locations, i.e. the wavelengths at which the greatest intensities are observed and are also obtained.

It must be noted that the characteristic of a sample depends on both the locations of the highest peaks in the spectra, as well as the magnitude of the peak. In order to take both these factors into account, a modified value of the peak locations is calculated for each component, according to Eq. (54.1).

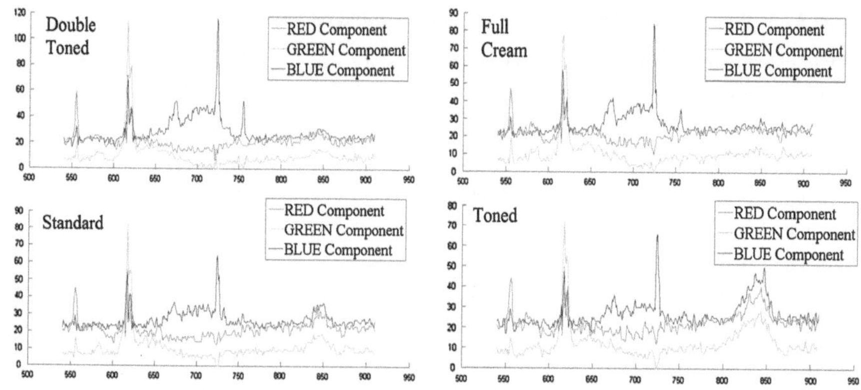

Fig. 54.2 Spectral waveform for different types of milk

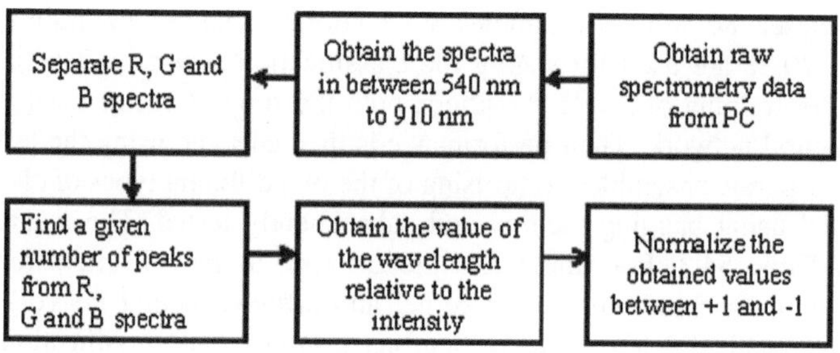

Fig. 54.3 Block diagram of data processing

$$\text{Modified value} = (\text{Peak location}) \times (\text{Peak Magnitude}), \qquad (54.1)$$

where peak magnitude is the magnitude of the peak being considered, the peak location is the wavelength at which it occurs and the modified value is the value dependent on both the peak location and peak magnitude.

We consider the 30 peaks from each of the red, green and blue component spectra, which give us a total of 90 modified values for each sample. These 90 values form the feature vector for each sample. The steps for data processing are shown in Fig. 54.3.

54.4 Experimental Results

After data acquisition and processing, a comprehensive dataset is formed containing total of 60 samples, taking 15 samples from each class of milk. Each of these samples are represented by a vector containing 90 features and a class label. From

Table 54.3 Performance of single classifier and ensemble on test samples

Single classifier/ensemble	SVM accuracy (%)	MLP accuracy (%)
Single classifier	58.33	22.0833
Bagging, $T = 5$	70	65.4167
Bagging, $T = 10$	71.25	66.25
Bagging, $T = 15$	69.1667	67.5
Bagging, $T = 20$	72.0833	70
Bagging, $T = 25$	70.8333	71.6667
Bagging, $T = 30$	73.333	73.333
Bagging, $T = 35$	72.9167	70.833
Bagging, $T = 40$	71.6667	74.5833
Bagging, $T = 45$	75.8333	72.9167
Bagging, $T = 50$	71.6667	73.333

the full dataset, 80 % of the samples are randomly selected for training and the remaining 20 % are used for testing the classification accuracy. The samples are first used for training of single classifiers, both the support vector machine and the artificial neural network. Their performance is then checked using the test samples. Similarly, separate ensembles comprising of the two different types of classifiers are then trained using bagging method and subsequently tested. The results are presented in Table 54.3. For single classifier as well as base learners for bagging, support vector machine with linear kernel and artificial neural network with one hidden layer and 100 hidden layer neurons is used. The classification accuracy obtained is the mean accuracy of 20 random runs, in all the cases.

The comparative performance of bagging using both types of base classifiers with respect to the bagging iterations, i.e. the total number of classifiers in the ensemble, is presented in Fig. 54.4. We observe that the best performance is

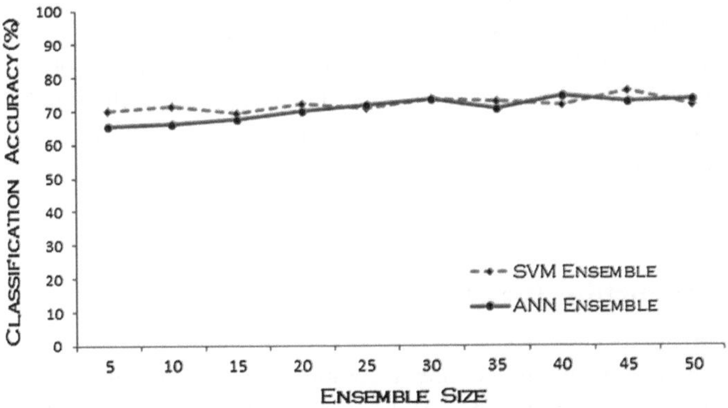

Fig. 54.4 Variation of performance with ensemble size

achieved by an ensemble consisting of 45 classifiers in case of support vector machine and 40 classifiers for artificial neural networks. The classification accuracy has a tendency to increase with a greater ensemble size.

54.5 Conclusion

To the best of our knowledge, this study is the first that combines NIR spectrometer data and bagging-based machine learning technique for milk fat content classification. From the comparative results, it is also noted that the support vector machine-based bagging ensemble exhibit superior classification performance over multi-layer perceptron-based bagging classifier. In our case, 15 samples for each class are taken for experimentation. Obviously, classification accuracy can be increased when more samples taken. Our work will be extended to the discrimination of raw and fresh milk from the packaged milk.

References

1. M.R.C. Inácio, M.F.V. Moura, K.M.G. Lima, Classification and determination of total protein in milk powder using near infrared reflectance spectrometry and the successive projections algorithms for variable selection. Vib. Spectrosc. **57**(2), 342–345 (2011)
2. C. Tan, M. Li, X. Qin, Study of the feasibility of distinguishing cigarettes of different brands using an adaboost algorithm and near-infrared spectroscopy. Anal. Bioanal. Chem. **389**(2), 667–674 (2007)
3. E. Alpaydin, *Introduction to Machine Learning*, 2nd edn. (PHI, 2012)
4. W. Suphamitmongkol, G. Nie, R. Liu, S. Kasemsumran, Y. Shi, An alternative approach for the classification of orange varieties based on near infrared spectroscopy. Comput. Electron. Agric. **91**, 87–93 (2013)
5. L. Breiman, Bagging predictors. Mach. Learn. **24**(2), 123–140 (1996)
6. H. Kim, S. Pang, H. Je, D. Kim, S. Bang, Support vector machine ensembles with bagging. Pattern Recogn. Support. Vector Mach. Lect. Notes Comput. Sci. **2388**, 397–408 (2002)
7. G. Fumera, F. Roli, A. Serrau, A theoretical analysis of bagging as a linear combination of classifiers. IEEE Trans. Pattern Anal. Mach. Intell. **30**(7), 1293–1299 (2008)

Chapter 55
Performance Improvement of Reversible Watermarking Using Convolution Coding and Lifting

Amit Phadikar, Poulami Jana and Goutam K. Maity

Abstract Reversible data embedding has drawn lots of interest recently especially in the field of protection for cultural heritage and medical image. Being reversible, the original digital content can be completely restored. This paper proposes Tian's algorithm of difference expansion as a reversible watermarking scheme for the cultural heritage image. Lifting together with convolution coding is used to increase the detection performance of watermark bits. Experimental results and performance comparison with other reversible data hiding schemes are presented to demonstrate the validity of the proposed algorithm.

Keywords Reversible watermarking · Lifting · Convolution coding · Viterbi decoding

55.1 Introduction

For some critical applications such as the law enforcement, medical and military imaging system, it is crucial to restore the original image without any distortions. The watermarking techniques satisfying those requirements are referred to as 'reversible watermarking'. Reversible watermarking is designed so that it can be removed completely to restore the original image [1–6]. Considering the age of reversible watermarking, which is just a decade to count, it has fetched enormous

---

A. Phadikar (✉)
Department of IT, MCKV Institute of Engineering, Liluah, Howrah, India
e-mail: amitphadikar@rediffmail.com

P. Jana
Department of ECE, Bengal Institute of Technology, Kolkata, India
e-mail: poulami.mtech@gmail.com

G.K. Maity
Department of ECE, MCKV Institute of Engineering, Liluah, Howrah, India
e-mail: goutam123_2005@yahoo.co.in

© Springer India 2015

K. Maharatna et al. (eds.), *Computational Advancement in Communication Circuits and Systems*, Lecture Notes in Electrical Engineering 335,
DOI 10.1007/978-81-322-2274-3_55

attention to researchers. In reversible watermarking, we embed a watermark in a digital image **I**, and obtain the watermarked image I′. Before sending it to the content authenticator, the image I′ might or might not have been tampered by some intentional or unintentional attack. If the authenticator finds that no tampering is happened in I′ (i.e. I′ is authentic), the authenticator can remove the watermark from I′ to restore the original image, which results a new image I″. By the definition of reversible watermark, the restored image I″ would be exactly the same as the original image I, pixel by pixel or bit by bit.

This paper proposes Tian's algorithm [7] of difference expansion as a reversible watermarking scheme for the cultural heritage image. It is to be noted that Tian's algorithm is applied on a spatial domain. So the scheme is fragile in nature. Whereas, in this work, lifting together with convolution coding is used to increase the detection performance of watermark bits after common image processing operations.

The rest of the paper is organized as follows: Sect. 55.2 describes the proposed scheme. Sections 55.3 and 55.4 present some experimental results and conclusions along with the scope of the future works, respectively.

55.2 Proposed Scheme

The proposed watermarking scheme, like other watermarking method, consists of two parts namely watermark encoding and decoding. Block diagram of image encoding and image decoding is shown in Fig. 55.1.

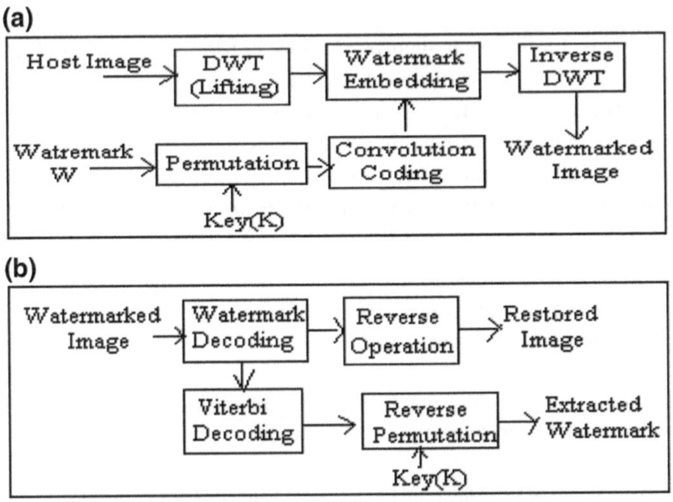

Fig. 55.1 Block diagram of the proposed scheme: **a** image encoding; **b** image decoding

A. Encoding Process. The encoding process consists of the following steps.

Step 1: *Watermark Permutation.* A binary image that is chosen as a watermark is processed using a secret key (*K*) because the attacker can easily forge a watermark if he/she has the knowledge of it.
Step 2: *Channel Coding of Watermark.* The permuted binary watermark image is then convolution coded before embedding to increase the data transmission reliability.
Step 3: *Image Transformation.* 2-level lifting base discrete wavelet transform (DWT) is performed on the original image. The present scheme used lifting as it provided integer coefficients after transformation that ultimately results a complete elimination of watermark bits at decoder as opposed to traditional DWT.
Step 4: *Watermark Bit Embedding.* The present scheme uses Tian [7] method for data embedding due to its simplicity and ease of implementation over other reversible schemes. In this technique, low–high (LH) and high–low (HL) sub-bands are used for data embedding to compromise between robustness and data imperceptibility. If x and y be the coefficients values of lifting, then integer average (l) and differences (h) are defined as:

$$l = \lfloor (x + y)/2 \rfloor \tag{55.1}$$

$$h = x - y \tag{55.2}$$

By appending an information bit b to the LSB of the difference h, a new LSB can be created. The watermarked difference is

$$h' = 2h + b \tag{55.3}$$

Finally, we compute the new values, based on the new difference value h' and the original integer average value.

$$x' = l + \lfloor (h' + 1)/2 \rfloor \tag{55.4}$$

$$y' = l - (h'/2) \tag{55.5}$$

After watermark embedding, inverse lifting is calculated and the watermarked image is obtained.

B. Decoding Process. The decoding process consists of the following steps.

Step 1: *Watermark Bit Extraction and Image Restoration.* From the embedded pair (x', y'), we can extract the embedded bit and restore the original pair (x, y). To achieve that goal, again we compute the integer average and difference

$$h = \lfloor h'/2 \rfloor \tag{55.6}$$

$$b = lsb(h')$$

$$\text{or,} \quad b = 2h - h' \tag{55.7}$$

$$x = l + \lfloor (h + 1)/2 \rfloor \tag{55.8}$$

$$y = l - (h/2) \tag{55.9}$$

After restoration of wavelet coefficients, inverse lifting is calculated and restored image is obtained.

Step 2: *Viterbi Decoding and Reverse Permutation.* Extracted watermark bits are Viterbi decoded and reversely permuted using the same secret key (K).

Step 3: *Decoding Reliability for the Extracted Watermark.* We calculate the normalized cross-correlation (NCC) between the original watermark image (b) and the decoded watermark image ($\hat{b}$) to quantify the visual quality of the extracted watermark. The NCC is defined by Eq. (55.10) below.

$$NCC = \frac{\sum_i \sum_j b_{ij}\hat{b}_{ij}}{\sum_i \sum_j (b_{ij})^2} \tag{55.10}$$

55.3 Experimental Results

The performance of the proposed scheme is evaluated over four test images namely Lena, Pepper, Baboon and Boat having varied image characteristics. Figure 55.2 shows various test images. All experimentations are conducted in Pentium IV, 2.80 GHz processor, with 512 MB RAM using MATLAB 7. The present study uses

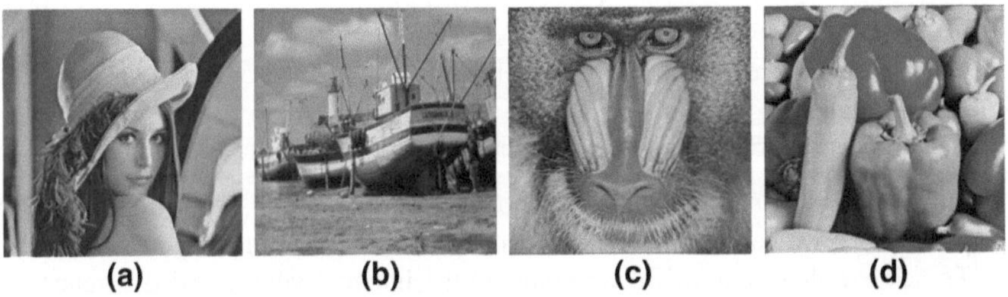

Fig. 55.2 Test host images **a** Lena, **b** Boat, **c** Baboon, **d** Pepper

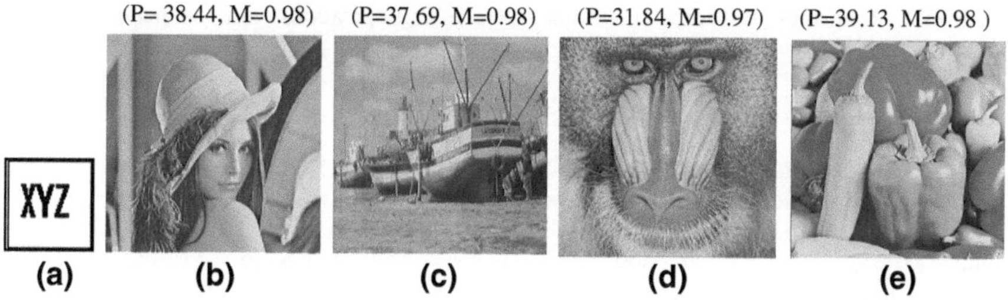

Fig. 55.3 **a** Watermark image, **(b–e)**: watermarked images, **b** Lena, **c** Boat, **d** Baboon, **e** Pepper. (*P* PSNR (in dB), *M* MSSIM)

Fig. 55.4 **a** Extracted watermark from Fig. 55.3b–e, **(b–e)** images after reverse operation, **b** Lena, **c** Boat, **d** Baboon, **e** Pepper. (*P* PSNR (in dB), *M* MSSIM)

peak signal-to-noise-ratio (PSNR) and mean-structure-similarity-index-measure (MSSIM) [8] as distortion measure for the watermarked image.

Figure 55.3a shows the watermark image, while Fig. 55.3b–e show the watermarked images. Figure 55.4a shows the extracted watermark image from all watermarked image, while Fig. 55.4b–e show the restored images. Table 55.1 show the PSNR (in dB) and MSSIM values of the restored images. Form Figs. 55.3b–e, 55.4b–e and Table 55.1, it is cleared that images are resorted completely, after reverse operations. Table 55.2 shows the robustness performance for different image

Table 55.1 Experimental results in term of PSNR (dB) and MSSIM, without image processing and channel coding. *Inf* Infinite

Image	Watermarked image		Recovered image	
	PSNR (dB)	MSSIM	PSNR (dB)	MSSIM
Lena	38.4492	0.9827	Inf	1
Boat	37.6965	0.9824	Inf	1
Baboon	31.8429	0.9730	Inf	1
Pepper	39.1302	0.9850	Inf	1

Table 55.2 Robustness test for different image processing operations with channel coding

Image	NCC	
	Without convolution coding	With convolution coding
JPEG 60	0.5666	0.6322
Wiener	0.8844	0.9780
Medfilter (3 × 3)	0.5786	0.6356
Imfilter (3 × 3)	0.8866	0.9736
High pass filter	0.8560	1
Histogram equalization	0.7288	0.7578
Image resize	0.9234	1
Speckle	0.5822	0.6622
Salt and pepper	0.9251	0.9511
Gaussian noise	0.5330	0.6711
Dynamic range change	0.9258	1
JPEG 70	0.5141	0.5956
JPEG 80	0.5727	0.6267

Table 55.3 *A* JPEG 60, *B* Wiener, *C* Medfilter (3 × 3), *D* Imfilter (3 × 3), *E* High pass filter, *F* Histogram equalization, *G* Image resize, *H* Speckle, *I* Salt and pepper, *J* Gaussian noise, *K* Dynamic range change, *L* JPEG 70, *M* JPEG 80

Attacks	A	B	C	D	E	F	G	H	I	J	K	L	M
Tian's [7]	0.56	0.88	0.57	0.88	0.85	0.72	0.92	0.58	0.92	0.53	0.92	0.51	0.57
Proposed	0.63	0.97	0.63	0.97	1	0.75	1	0.66	0.95	0.67	1	0.59	0.62

processing operations with the channel coding. Table 55.3 lists the robustness comparison of the proposed scheme with Tian's [7] technique. It is seen that robustness is improved with the help of lifting and channel coding.

55.4 Conclusions

In this paper, we have presented a reversible date-embedding technique with the integration of lifting and channel coding. It is seen that the use of lifting with the integration of channel coding increases the robustness performance of the proposed scheme. Future work can be done for further performance improvement of the proposed scheme using source channel coding.

References

1. D.M. Thodi, J.J. Rodriguez, Expansion embedding techniques for reversible watermarking. IEEE Trans. Image Process. **3**, 776–730 (2007)
2. A.M. Alattar, Reversible watermark using the difference expansion of a generalized integer transform. IEEE Trans. Image Process. **13**, 1147–1156 (2004)
3. Z. Ni, Y.Q. Shi, N. Ansari, W. Su, Reversible data hiding. IEEE Trans. Circuits Syst. Video Technol. **16**, 354–362 (2006)
4. C. Vleeschouwer, J.F. Delaigle, B. Macq, Circular interpretation on histogram for reversible watermarking, in *IEEE International Multimedia Signal Processing Workshop* (2001), pp. 345–350, France
5. S. Weng, Y. Zhao, J. Pan, R. Ni, Reversible watermarking based on invariability and adjustment on pixel pairs. IEEE Signal Process. Lett. **15**, 721–724 (2008)
6. S. Lee, C.D. Yoo, T. Kalker, Reversible image watermarking based on integer-to-integer wavelet transform. IEEE Trans. Inf. Forensics Secur. **2**, 321–331 (2007)
7. J. Tian, Reversible data embedding using a difference expansion. IEEE Trans. Circuits Syst. Video Technol. **3**, 890–896 (2003)
8. Z. Wang, A.C. Bovik, H.R. Sheikh, E.P. Simoncelli, Image quality assessment: from error measurement to structural similarity. IEEE Trans. Image Process. **13**, 1–14 (2004)

Chapter 56
Image Noise Removal Using Principle of Suprathreshold Stochastic Resonance

Anil K. Pandey, ParamDev Sharma, S.K. Sharma, Kaushik Sarkar, Akshima Sharma, Rakesh Kumar and C.S. Bal

Abstract In this paper, we have developed an algorithm for noise cleaning based on the principle of suprathreshold stochastic resonance. Stochastic resonance is the phenomenon in which the addition of right type and right amount of noise improves the detection of the signal in the system performance. In suprathreshold stochastic there are a number of array of detectors, all are subjected to the input signal and the same noise intensity distribution, then the noise-added signals are threshold, and a threshold noise added signal from each detectors are averaged to get the output image. We get the enhanced output when the noise-added input signal are threshold with respect to mean of the noise-added input signal. This algorithm was implemented on FreeMat open source software. PET scan of the Jaszack deluxe phantom image is performed and used as a input noisy image for the validation of the developed algorithm. The algorithm is successful in removing the noise from the image.

Keywords Suprathreshold stochastic resonance · Medical imaging

56.1 Introduction

The nuclear medicine images are noisy because of the acquisition of a limited number of photons (Fig. 56.2b). Noise removal is essential to derive accurate information about the scanned object. The positron emission tomography (PET)

A.K. Pandey (✉) · S.K. Sharma · A. Sharma · R. Kumar · C.S. Bal
Department of Nuclear Medicine, A.I.I.M.S., New Delhi, India
e-mail: pandeyanilkumar@rediffmail.com

P. Sharma
Department of CS, SGTB Khalsa College, University of Delhi, Delhi, India
e-mail: pdsharma@sgtbkhalsa.du.ac.in

K. Sarkar
Department of ECE, Narula Institute of Technology, Kolkata, India
e-mail: Kaushik.srkr@gmail.com

© Springer India 2015

K. Maharatna et al. (eds.), *Computational Advancement in Communication Circuits and Systems*, Lecture Notes in Electrical Engineering 335,
DOI 10.1007/978-81-322-2274-3_56

image quality and standardized uptake value (SUV) calculations are affected by many factors such as technical errors of equipment, biological factors of the patients and physical factors of scanning protocols; technical errors like relative calibration of PET scanner and dose calibrator, intravenous administration of F-18 FDG; biological factors like blood glucose level, uptake period, patient motion or breathing. Physical factors like scan acquisition parameters, image reconstruction parameters, ROI, normalization factors for SUV, etc. [1].

PET scans reconstructed using iterative ordered subset and expectation maximization (OSEM) algorithm. The choice of reconstruction parameters represents a compromise between resolution and noise and has a major impact on the performance of clinically important tasks such as lesion detection [2]. The improvement in one parameter is frequently achieved only at the expense of degrading other [3]. There always exists a grey area while reporting the scan where the noise cleaning in the image or algorithm which can help in differentiating the presence or absence of the lesion is required.

There is a need to remove the noise from the reconstructed transverse slices. In nuclear medicine, noise removal from the image by adding noise to the input noisy image has not been investigated sufficiently. Stochastic resonance is a phenomenon in a detection process in which the addition of just the right amount and type of noise improves the performance. Too little or too much noise results in degraded performance. This concept was first put forward by Benzi et al. to address the periodicity of ice ages [4]. Since then, the idea of utilizing noise to enhance the performance of a nonlinear system has been widely investigated and applied to various scientific fields such as physics, chemistry, biomedical sciences and financial market system [5–8].

Suprathreshold stochastic resonance is the phenomenon in which random noise is added to the noisy image and then the images are threshold with respect to mean pixel intensity value. In this way a number of frames are generated, and then all the frames are averaged to get the final output image. The output signal has an enhanced peak where the resonance have occurred.

In this paper, we have applied the concept of suprathreshold stochastic resonance to remove the noise which is available in the image after image reconstruction using OSEM in deluxe Jaszack phantom study.

56.2 Materials and Methods

56.2.1 Suprathreshold Stochastic Resonance

Suprathreshold stochastic resonance was first demonstrated in arrays of identical threshold devices in the year 2000 [9]. A sufficient condition for suprathreshold stochastic resonance is that the array has more than one threshold device, and all threshold levels are set to the same value. Stochastic resonance occurs regardless of whether the input signal is entirely subthreshold or not, and indeed the

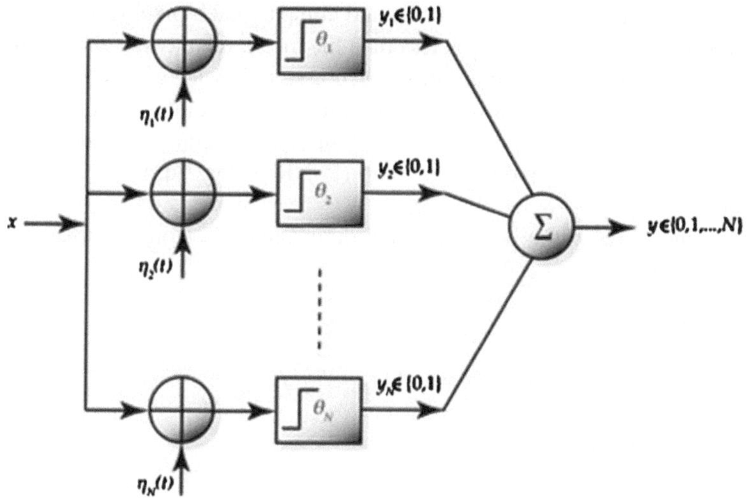

Fig. 56.1 Typical example of a network in which suprathreshold stochastic resonance can be observed

suprathreshold stochastic resonance peak in performance is maximized when threshold levels are set equal to the signal mean, and diminishes otherwise.

The most studied model is the network shown in Fig. 56.1. It consists of N identical threshold devices (i.e. single bit quantizers, either zero or one), each operating on a common signal x subject to independent additive noise. The overall output y is the sum of the N noisy binary outputs and is an integer between zero and N. The fidelity with which y can represent x is dependent on the noise distribution, the signal distribution, the input signal-to-noise-ratio, N, and the threshold level. Performance depends on the conditional distribution of y given x.

56.2.2 Algorithm

1. Read the input grey images and convert the data into double, say $I = f(x, y)$
2. Generate the white Gaussian noise with zero mean and a particular variance and put it into the matrix size equal to the input image.

$$\text{noise} = n(x, y).$$

3. Add the input image and noise matrix

$$G(x, y) = f(x, y) + n(x, y) \text{ and store } G(x, y)$$

4. Calculate the mean of image $G(x, y)$ and compare each pixel of $G(x, y)$ images with this mean, if pixel value is less than the mean, then assign zero to the pixel value, and if it is greater than mean, then assign 255 to the pixel value
5. Repeat the steps 2–4, N number of times, this will generate N number of frames
6. Final output image is equal to sum of N numbers of frames divided by N
7. Calculate the peak signal-to-noise ratio (PSNR) using Eq. (56.1) in the output image.
8. At the end, display the both input and output images, store the images for the further analysis.

Then repeat the steps 2–7 for the next noise variance.

$$\mathrm{PSNR} = 10\log_{10}\left[255^2/\mathrm{MSE}\right] \tag{56.1}$$

MSE is the mean square error.

56.2.3 Experiment

56.2.3.1 Image Acquisition and Reconstruction

A cylindrical Lucite Deluxe Jaszczak phantom was used for this study. The diameter of this phantom was 21.6 cm containing two types of inserts: Cold rod inserts having diameters 4.8, 6.4, 7.9, 9.5, 11.1 and 12.7 mm and cold spheres with diameters 9.5, 12.7, 15.9, 19.1, 25.4, and 31.8 mm. The phantom was filled with water and 2 mci (74 MBq) of F-18 FDG was added in it. Radioactivity was properly mixed in water by shaking the phantom carefully. Scan was done on BiographmCT (Siemens Healthcare) PET/CT scanner with LSO crystal and 64 slices CT. Scan was acquired in routine whole-body PET/CT protocol. The default reconstruction parameters were iterations 2, subsets 21, matrix size 200 × 200, zoom 1 and FWHM 2 mm.

56.2.3.2 Algorithm Implementation

The algorithm for noise removal using suprathreshold stochastic resonance was implemented using FreeMat open source software. FreeMat is an interpreter, matrix-oriented development environment for engineering and scientific applications, similar to the commercial package MATLAB. FreeMat provides visualization, image manipulation and plotting as well as parallel programming [10]. One transverse slice corresponding to the cold sphere of the jaszack phantom (Fig. 56.2a) was taken as input image for processing and exhaustive experimentation. We generated noise matrix with mean zero and noise variance in the steps of 32, 64, 128, 255, 510, 1020, 1275, 3187.5, 3825, 5100, 6375, 12750, 38250, 63750, 127500 and 1275000, respectively. For each noise variance, 5, 10, 20, 50 and 100 frames were generated. We reviewed the output image each time having a

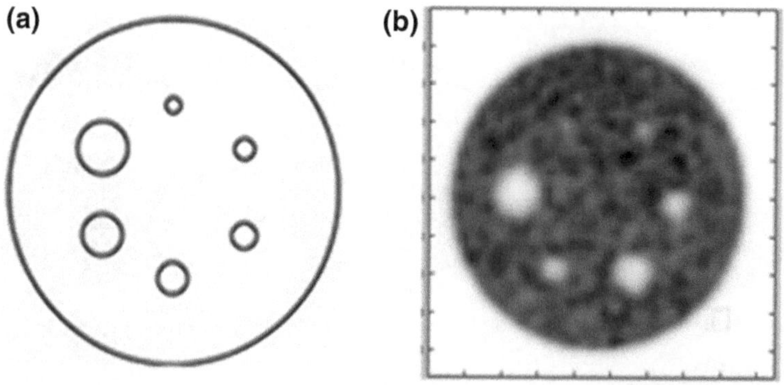

Fig. 56.2 Transverse section of deluxe Jasczak phantom: **a** showing six solid spheres in increasing diameters (9.5, 12.7, 15.9, 19.1 and 31.8 mm) from *top* in clockwise direction. **b** Corresponding transverse slice reconstructed using the OSEM reconstruction algorithm

priori knowledge of the object in mind. The plot of PSNR with respect to the variance of Gaussian white noise was created.

56.3 Result

56.3.1 Visual Analysis

The phantom has six solid sphere of increasing diameter (Fig. 56.2a). When the activity is filled in the phantom and given sufficient time for uniform mixing, there will be no activity inside the solid sphere and will have uniform activity around the solid sphere. Therefore, no photons will be registered from spheres and in the image corresponding sphere will appear black and remaining areas will be white. We have taken input noisy image as inverse image of the phantom (Fig. 56.2b).

It can be seen from the Fig. 56.3c that this algorithm successfully cleans the noise and all the six spheres are visible in the image with much better contrast than that of the original image. A typical feature of the stochastic resonance was observed in which at right amount of noise variance, the output image, resembles the input image (Fig. 56.3a, e). Then with further increase in noise intensity images deteriorates significantly (Fig. 56.3f).

56.3.2 Noise Intensity Versus PSNR

Figure 56.4 demonstrates the typical nature of stochastic resonance phenomenon that is the signal which improves with the addition of noise intensity in increasing

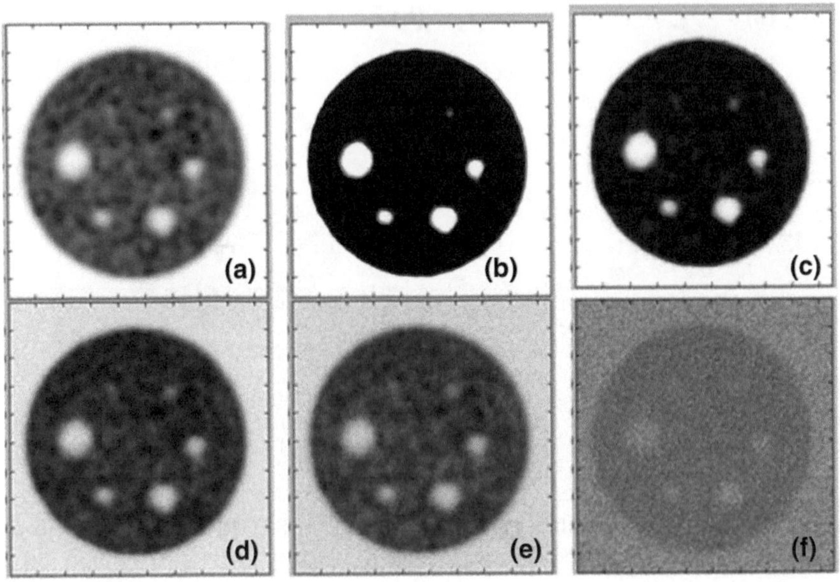

Fig. 56.3 Showing the effect of noise variance on visual quality. **a** Input image, **b** five spheres are visible with much better contrast, all noise removed only object is visible, noise variance = 255. **c** All six spheres are visible but with some noise, noise variance = 1,275. **d** Noise variance = 6,375, started resemblance with input image. **e** Resemblance with input image less noise than input image. Addition of right amount of noise, the output image resembles with the input image (**a**, **e**). **f** Noise variance = 1,275,000, significant deterioration

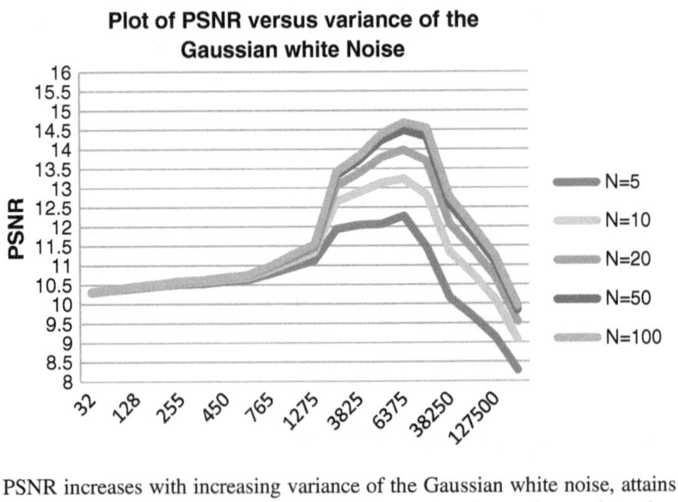

Fig. 56.4 PSNR increases with increasing variance of the Gaussian white noise, attains peak at a noise variance equals to 6,375 and then decreases. For each noise variance, the value of PSNR attain maximum as the number of frames increases (N = 100)

order and attains a peak and thereafter decreases. We obtained maximum value of PSNR at noise variance equal to 6,375.

56.4 Discussion

In this study, we have added noise to the input noisy image to clean the noise using the principle of suprathreshold stochastic resonance. The results demonstrate that it is possible to clean the noise using the algorithm used in this study. This is a very basic or preliminary work in which we have optimized the value of right amount of noise interactively viewing the output image having a priori knowledge about the object being scanned.

56.4.1 Explanation of the Result

Suprathreshold stochastic resonance occurs when there is a tradeoff between degradation of input image due to random noise and degradation due to deformation by nonlinearity of the scanner on which the images were acquired.

Each pixel is a signal. Each pixel value is a result of the photons detected by the PET scanner; therefore, the pixel value is due to the nonlinearities of the scanner corresponding to sensor location and random noise. In the absence of random noise, all degradations in the image is due to nonlinearities.

If at each pixel nonlinearity is identical, the overall system's performance is the same as single pixel. For intermediate noise levels, the random noise alters the properties of the nonlinearities, so that each is effectively no longer identical. Since the multiple devices (i.e. multiple frames generated after adding noise to input noisy image) operate on a common signal (same input image), they allow essentially a nonlinear averaging effect to occur. Hence, the performance improves for every additional device (as the number of frames increases the PSNR value also increases). The end result is that performance increases with the increasing noise, up to some optimal noise level. For large noise, while performance still increases with number of devices, degradation of the signal due to the randomness of the noise outweighs the effects of the noise of the nonlinearity, and performance only decreases with increasing noise.

56.4.2 Comparison with Literature

Pandey et al., added Poisson noise and have shown appreciable visual changes in the output images in the real patients PET images and was validated by the nuclear medicine physicians. Their algorithm and type of noise added were completely different from this study [11]. Therefore, comparison cannot be made with their study. Jha et al. [12] has worked with the image denoising algorithm using the principle of stochastic resonance, their input was the synthetic images. Our result is similar to them; however we have worked with real image data.

56.4.3 Usefulness

Nuclear medicine physician makes diagnosis based on the presence or absence of photons (counts) and also grade the tumour based on the greylevel information in the image. Our validated algorithm cleans the noise and improves the contrast in the image; however, it makes the output image binary [0, 1]. Therefore, this method might help the nuclear medicine physician in identifying the presence or absence of disease. Our future task will be to find the effectiveness of this algorithm in identifying the presence or absence of disease in real patient data.

56.5 Conclusions

The algorithm is successful in removing noise from the PET phantom image.

References

1. R. Boellaard, Standards for PET image acquisition and quantitative data analysis. J. Nucl. Med. **50**, 115–205 (2009)
2. D.J. Kadramas, P.E. Christian, Comparative evaluation of lesion detectability for 6 PET imaging Platforms using highly reproducible whole body phantom with Na-22 lesions and localization ROC analysis. J. Nucl. Med. **43**, 1545–1554 (2002)
3. M.A. Lodge, A. Rahmim, R.L. Wahl, Simultaneous measurement of noise and spatial resolution in PET phantom images. Phys. Med. Biol. **55**, 1069–1081 (2010)
4. R. Benzi, A. Sutera, A. Vulpani, J. Phys. A: Math. Gen. **14**, L453–L457 (1981)
5. L. Gammaitoni, P. Hanggi, P. Jung, F. Marchesoni, Rev. Mod. Phys. **70**, 223 (1998)
6. J.D. Harry, J.B. Niemi, A.A. Priplata, J.J. Collins, IEEE Spectr. **42**, 36 (2005)
7. R.P. Morse, E.F. Evans, Nat. Med. **2**, 928 (1996)
8. X.M. Mao, K. Sun, Q. Ouyang, Chin. Phys. **11**, 1106 (2002)
9. N.G. Stocks, Suprathreshold stochastic resonance in multilevel threshold systems. Phys. Rev. Lett. **84**, 2310–2313 (2000)
10. C. Nguyen, D. Kyriakis, E. Ingerman, J. Pack, S. Basu, T.J. Cyders, S. Tom, http://sourceforge.net/p/freemat/wiki/Home/FreeMat. Accessed 9 July 2014
11. A.K. Pandey, S.K. Sharma, P. Sharma, H. Singh, C. Patel, K. Sarkar, R. Kumar, C.S. Bal, Enhancement of positron emission tomography image quality using the principle of stochastic resonance, under publication Indian journal of nuclear medicine (2014)
12. R.K. Jha, P.K. Biswas, B.N. Chatterji, Image denoising using stochastic resonance, in *Proceedings of the International Conference on Cognition and Recognition*. http://www.researchgate.net/publication/240719142_Image_Denoising_using_Stochastic_Resonance. Accessed 09 July 2014

Chapter 57
Study the Effect of Parameters Used in Stochastic Resonance to Enhance an Image

Mrityunjoy Roy, Partha Sarkar and Kaushik Sarkar

Abstract Enhancement is one of the challenging factors in image processing. The objective of enhancement is to improve the structural appearance of an image without any degradation in the input image. The enhancement techniques make the identification of key features easier by removing noise. One of the methods to enhance an image is the stochastic resonance (SR). In SR, noise is added randomly to the image to get back the information of the processed image which has been distorted. But the parameter choice of SR is too difficult to realize. In our paper we study the effect of those parameters over an image.

Keywords Stochastic resonance · Image enhancement · Noise

57.1 Introduction

Noise, in general, is unwanted in any domain. Image Enhancement on the other hand is essential for any image processing task [1–4]. Stochastic resonance (SR) is one of the techniques where noise is used in positive sense to enhance an image [5, 6]. Surprisingly, this is somewhat unconventional way to think. Without filtering the noise, in SR, we have added noise signals to enhance the image [7–9]. Actually, we want to detect or recover all the information present in an image. We have added noise in random or stochastic manner. At a certain point the value of the pixels gets enhanced relatively and the objects get prominent. So a resonance is occurring at that point. This method is quite relevant to the conception of Dynamic Stochastic

M. Roy (✉) · P. Sarkar · K. Sarkar
Narula Institute of Technology, Agarpara, Kolkata 700109, India
e-mail: mrtnjroy@gmail.com

P. Sarkar
e-mail: parthasarkar.info@gmail.com

K. Sarkar
e-mail: kaushik.srkr@gmail.com

© Springer India 2015

K. Maharatna et al. (eds.), *Computational Advancement in Communication Circuits and Systems*, Lecture Notes in Electrical Engineering 335,
DOI 10.1007/978-81-322-2274-3_57

Resonance [10–12]. In this procedure, the performance of an input signal can be improved by addition of external noise. But SR would be fruitful if the parameters tuning is known prior to the processing. Otherwise, it may lead to very inappropriate results. The problem is to tune the parameters. We present our paper in the following way: In Sect. 57.2, we introduce the basics of SR. Our observations are discussed in Sect. 57.3, followed by the conclusions in Sect. 57.4.

57.2 Dynamic Stochastic Resonance

It was traditionally believed that the presence of noise can only make a system worse [13]. However, recent studies have convincingly shown that in non-linear systems, noise can induce more ordered regimes that cause amplification of weak signals and increase the signal-to-noise ratio [5].

In order to exhibit SR, a system should possess three basic properties: a non-linearity in terms of threshold, sub-threshold signals like signals with small amplitude and a source of additive noise [5]. This phenomenon occurs frequently in bistable systems or in systems with threshold—like behaviour. The general behaviour of SR mechanism shows that at lower noise intensities the weak signal is unable to cross the threshold, thus, giving a very low SNR. For large noise intensities the output is dominated by the noise, also leading to a low SNR. But, for moderate noise intensities, the noise allows the signal to cross the threshold giving maximum SNR at some optimum additive noise level.

57.2.1 Mathematical Formulation of Dynamic Stochastic Resonance for Contrast Enhancement

A classic one-dimensional nonlinear dynamic system that exhibits SR is modelled with the help of lingering equation of motion [5] in the form of equation

$$\frac{\mathrm{d}x(t)}{\mathrm{d}t} = \frac{\mathrm{d}U(x)}{\mathrm{d}x} + \sqrt{D}\,\xi(t) \tag{57.1}$$

where $U(x)$ is a bistable potential, D is the noise variance and $\xi(t)$ is the noise.

$$U(x) = -a\frac{x^2}{2} + b\frac{x^2}{4} \tag{57.2}$$

Here, a and b are positive bistable double-well parameters. The double—well system is stable at $x_m = \pm\sqrt{\frac{a}{b}}$ separated by a barrier of height $\Delta U = \frac{a^2}{4b}$ when the $\xi(t)$ is zero. Addition of a periodic input signal $[B\ \sin(\omega t)]$ to the bistable system makes it time dependent whose dynamic are governed by Eq. (57.3).

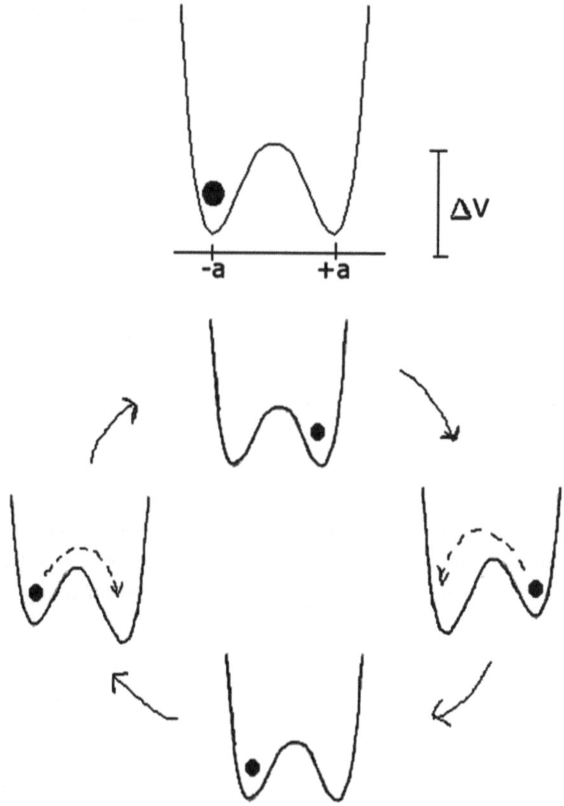

Fig. 57.1 Double well concept of stochastic resonance

$$\frac{\mathrm{d}x(t)}{\mathrm{d}t} = -\frac{\mathrm{d}U(x)}{\mathrm{d}x} + B\,\sin(\omega t) \tag{57.3}$$

where B and ω are the amplitude and frequency of the periodic signal, respectively.

It is assumed that the signal amplitude is small enough so that in the absence of noise it is sufficient to force a particle of unit mass as shown in the Fig. 57.1 to move from one well to another. It, therefore, fluctuates around its local stable states.

$$\frac{\mathrm{d}x(t)}{\mathrm{d}t} = -\frac{\mathrm{d}U(x)}{\mathrm{d}x} + \sqrt{D}\xi(t) + B\,\sin(\omega t) \tag{57.4}$$

When a weak periodic force and noise are applied to the unit mass particle in the bistable potential well, noise driven switching between the potential wells take place only at some 'resonant' value of noise. This noise induced hopping is synchronized with the average waiting time, between two noise driven inter-well transitions that satisfies the time—scale matching between signal frequency and the residence times of the particle in each well. Maximum SNR is achieved when $a = 2\sigma_0^2$. Thus, SNR has maximum value at an intrinsic parameter, a, of the dynamic double well system.

The other parameter b can be obtained using parameter a. For weak input signal, condition $b = 4a^3/27$ is required to ensure sub-threshold condition. In the following Fig. 57.1 the double well conception is schematically explained.

Solving the stochastic differential equation in (57.4) after substituting $U(x)$ from Eq. (57.2) using the stochastic version of Euler-Maruyama's iterative discredited method [5] as follows

$$x(n+1) = x(n) + \Delta t\left[ax(n) - bx^3(n) + \mathrm{Input}(n)\right] \tag{57.5}$$

where $\mathrm{Input}(n) = \sqrt{D}\,\xi(t) + B\,\sin(\omega t)$ denotes the sequence of input signal and noise, with initial condition being $x(0) = 0$. Here Δt is the sampling time.

Considering the image enhancement scenario, one can experiment by proper selection of parameters a, b gives the strong signal state or enhanced images. The objective is to add stochastic fluctuation or noise to the pixel value of the weak signal state so that the pixel particle is activated, jumps over the detector threshold and transits to strong signal state or enhanced state.

So, it is evident from the above discussion, that the parameter tuning of 'a', 'b', and the sampling time 'Δt' are the key of success to deploy SR. In the next section, we study the effect of these parameters on an image.

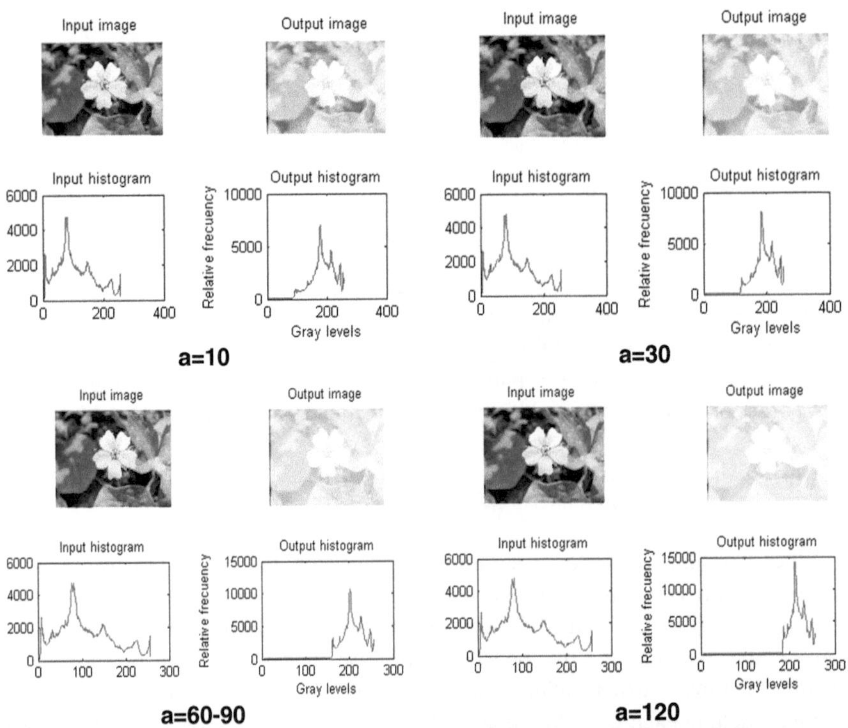

Fig. 57.2 Different image quality for the variation of parameter 'a' with their histogram

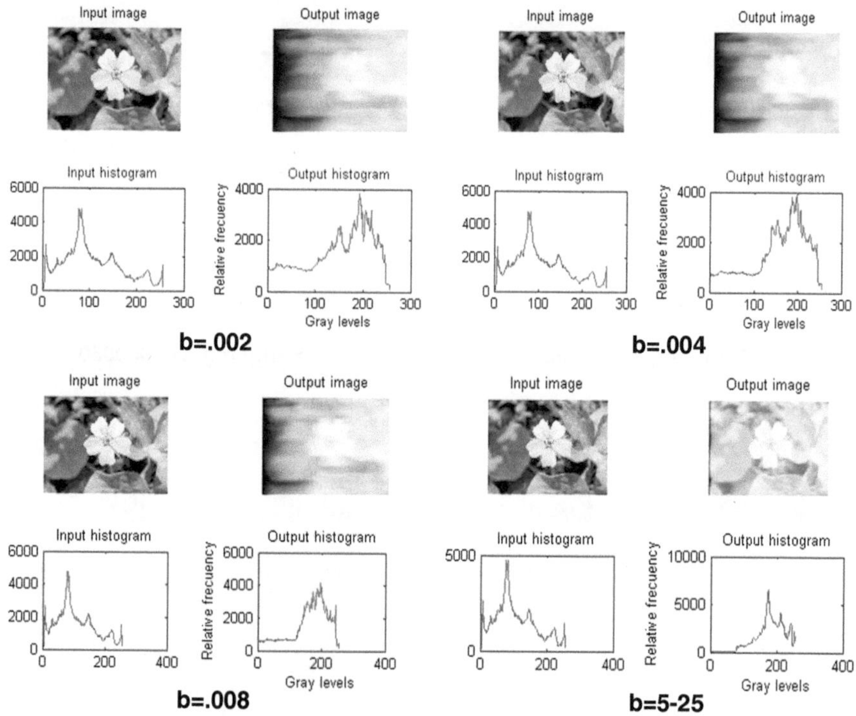

Fig. 57.3 Different image quality for the variation of parameter '*b*' with their histogram

57.3 Study the Effect of the Parameters of SR

To study the effect of the parameters, we take a grey scale image and observe the effect of image using histogram. In Fig. 57.2, we vary the parameter '*a*' from 10 to 120, keeping the value '*b*' = 50 and 'Δt' = 0.0015. It is seen that the image quality is degraded in terms of contrast as the value is increased. The corresponding histogram is shown in Fig. 57.2.

Next, we take '*a*' = 0.3 and 'Δt' = 0.0015 as constant and vary the parameters '*b*' ranging from a very low value of 0.002–25. It is quite interesting to see that at very low value image quality is very poor compared to very high value, e.g., 5–25. So, there is a dynamic shift of the parameter.

Again, we vary the sampling rate 'Δt' from 0.000025 to 0.001 keeping '*a*' = 0.5 and '*b*' = 75. Here, the image is enhanced as we increase the sampling rate (Figs. 57.3 and 57.4).

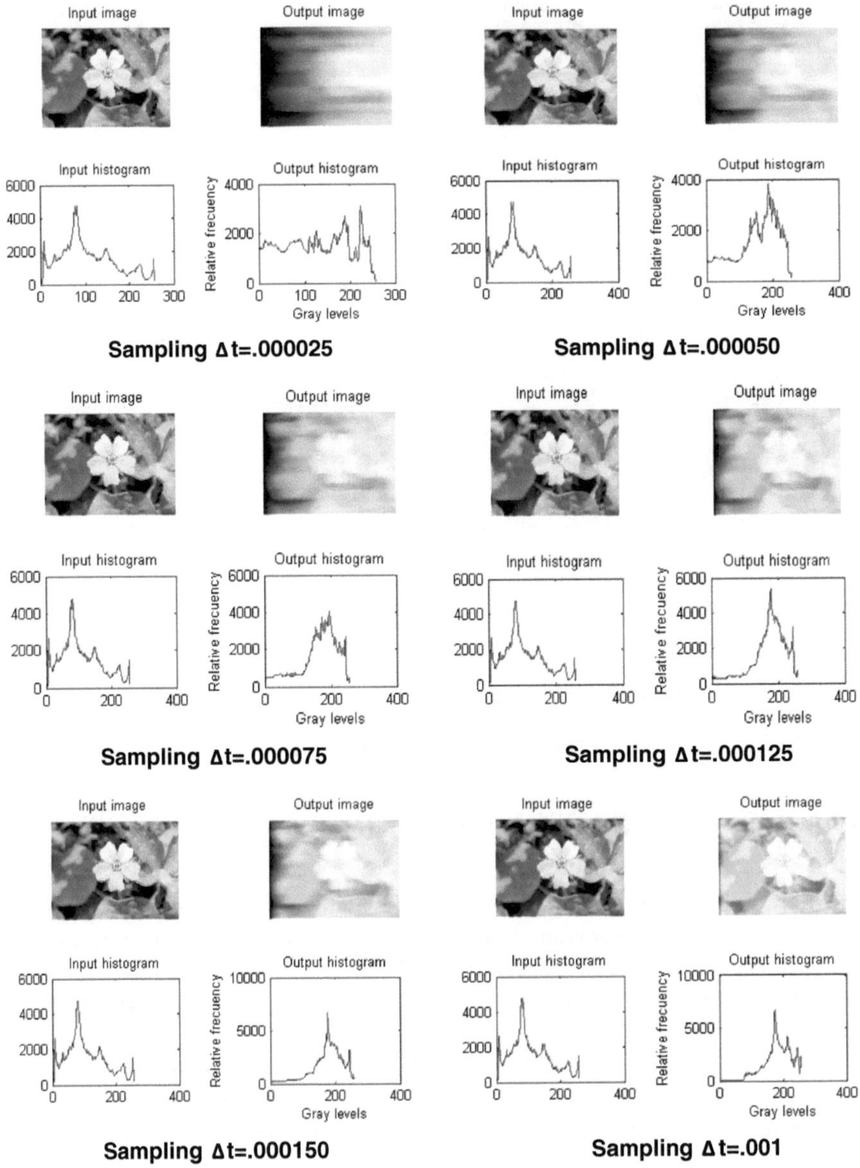

Fig. 57.4 Different image quality for the variation of parameter 'Δt' with their histogram

57.4 Conclusions

We observe that the image quality is being varied with the variation of all the parameters. In case of the parameter 'a' as we increase the value the image is enhanced relatively but after a certain value it is degraded again. In case of parameter 'b', the image gets better and better along with the increased value of it. Similar thing is also happening with the change in the sampling time 'Δt'. It also enhances the image with its ascending value. In all the cases, a drastic change has been observed in the histogram.Therefore, it is a challenging task to estimate the parameters of SR prior to the processing of an image. It is still an open area for the researchers.

References

1. OgeMarques, *Practical Video and Image Processing Using MATLAB* (Wiley, New Jersy, 2011)
2. G.A. Baxes, *Digital Image Processing: Principles and Applications* (Wiley, New York, 1994)
3. R.C. Gonzales, E. Woods, *Digital Image Processing* (Prentice Hall, Addison-Wesley, MA, 1992)
4. P. Chakraborty, Histogram equalization by cumulative frequency distribution. Int. J. Sci. Res. Publ. **2**(7) (2012)
5. R.K. Jha, R. Chouhan, Noise-induced contrast enhancement using stochastic resonance on singular values. Signal Image Video Process. **8**(2), 339–347 (Springer, 2012). doi: 10.1007/s11760-012-0296-2 [online]
6. P. Hanggi, P. Jung, F. Marchesoni, Stochastic resonance. Rev. Mod. Phys. **70**, 223–270 (1998)
7. R. Peng, H. Chen, P.K. Varshney, Stochastic resonance: an approach for enhanced medical image processing. IEEE-NIH Life Sci. Syst. Appl. Workshop **1**, 243–246 (2007)
8. V.P.S. Rallabandi, P.K. Roy, Magnetic resonance image enhancement using stochastic resonance in fourier domain. Magn. Reson. Imaging **28**, 1361–1373 (2010)
9. R.K. Jha, R. Chouhan, P.K. Biswas, K. Aizawa, Internal noise-induced contrast enhancement of dark images, in *Proceedings International Conference on Image Processing* (2012)
10. E. Simonotto, M. Riani, S. Charles, M. Roberts, J. Twitty, F. Moss, Visual perception of stochastic resonance. Phys. Rev. Lett. **78**, 1186–1189 (1997)
11. M. Piana, M. Canfora, M. Riani, Role of noise in image processing by the human perceptive system. Phys. Rev. E **62**, 1104–1109 (2000)
12. Q. Ye, H. Huang, X. He, C. Zhang, Image enhancement using stochastic resonance, in *Proceedings International Conference on Image Processing (ICIP 2004)*, vol. 1 (2004) pp. 263–266
13. C. Ryu, S.G. Konga, H. Kimb, Enhancement of feature extraction for low quality fingerprint images using stochastic resonance. Pattern Recogn. Lett. **32**, 316–320 (2008)

Index

© Springer India 2015
K. Maharatna et al. (eds.), *Computational Advancement in Communication Circuits and Systems*, Lecture Notes in Electrical Engineering 335,
DOI 10.1007/978-81-322-2274-3